LONDON MATHEMATICAL SOCIETY LECTURE NOTE SERIES

Managing Editor: Professor N. J. Hitchin, Mathematical Institute,
University of Oxford, 24-29 St Giles, Oxford OX1 3LB, United Kingdom

The titles below are available from booksellers, or from Cambridge University Press at www.cambridge.org/mathematics

London Mathematical Society Lecture Note Series: 329

Spaces of Kleinian Groups

Edited by

YAIR N. MINSKY
Yale University

MAKOTO SAKUMA
Osaka University

CAROLINE SERIES
University of Warwick

CAMBRIDGE
UNIVERSITY PRESS

CAMBRIDGE
UNIVERSITY PRESS

Shaftesbury Road, Cambridge CB2 8EA, United Kingdom

One Liberty Plaza, 20th Floor, New York, NY 10006, USA

477 Williamstown Road, Port Melbourne, VIC 3207, Australia

314–321, 3rd Floor, Plot 3, Splendor Forum, Jasola District Centre, New Delhi – 110025, India

103 Penang Road, #05–06/07, Visioncrest Commercial, Singapore 238467

Cambridge University Press is part of Cambridge University Press & Assessment, a department of the University of Cambridge.

We share the University's mission to contribute to society through the pursuit of education, learning and research at the highest international levels of excellence.

www.cambridge.org
Information on this title: www.cambridge.org/9780521617970

First published 2006

A catalogue record for this publication is available from the British Library

ISBN   978-0-521-61797-0   Paperback

Contents

Preface

This volume is the proceedings of the programme *Spaces of Kleinian Groups and Hyperbolic 3-Manifolds* held at the Isaac Newton Institute in Cambridge, 21 July–15 August 2003. It is a companion volume to *Kleinian Groups and Hyperbolic 3-Manifolds*, London Mathematical Society Lecture Notes **299**, the proceedings of a conference with the same title held at the Mathematics Institute, University of Warwick, 11–15 September 2001.

The period surrounding these two conferences has seen a series of remarkable advances in our understanding of hyperbolic structures on 3-manifolds. Many of the outstanding issues immediately preceding the Newton Institute meeting related to difficulties in extending results from manifolds with incompressible boundary to the general case. Proofs of Thurston's ending lamination conjecture and the Bers–Sullivan–Thurston density conjecture for general tame groups were announced at the meeting, and the picture was completed not long after the Newton programme, with two independent proofs of Marden's tameness conjecture. As a result, we now have a very clear understanding of the internal geometry of hyperbolic 3-manifolds, combined with an increasingly detailed, but quite intricate, picture of the topology and geometry of the associated deformation spaces of discrete groups.

The Newton Institute meeting turned out to be the international gathering at which many of these new results were disseminated. Almost all the primary contributors took part. Quite how rapid progress has been only became apparent to many of us during the meeting, which will be remembered as a milestone at which all of the new ideas were brought together.

This volume contains articles and expositions which it is hoped will give some impression of the breadth and scope involved. Contributions have been arranged bringing similar themes together, starting with topology and geometry of 3-manifolds, moving through curve complexes and classical Ahlfors–Bers theory, to computer explorations and projective structures.

The editors, who were also the organisers of the SKG programme, would like to extend thanks on behalf of all the participants to the Newton Institute for hosting us in such pleasant surroundings so conducive to mathematical interaction. We enjoyed generous funding not only from the EPSRC but also the EU, the NSF, the Leverhulme Trust, the London and Edinburgh Mathematical Societies and, through various individual grants, the JSPS. We acknowledge with thanks the support of all these bodies. Finally, we are extremely grateful to David Sanders for his skilled editorial assistance, without which this volume would not have been produced.

Yair Minsky, Makoto Sakuma & Caroline Series
April 2005

Spaces of Kleinian Groups
Lond. Math. Soc. Lec. Notes **329**, 1–27

Cambridge University Press
Y. Minsky, M. Sakuma & C. Series (Eds.)

Drilling short geodesics in hyperbolic 3-manifolds

K. Bromberg[1]

Abstract

We give an expository account of the deformation theory of geometrically
finite, 3-dimensional hyperbolic cone-manifolds and its application to three classical conjectures about Kleinian groups.

1. Introduction

In a series of papers ([HK98, HK02, HK]), Hodgson and Kerckhoff developed a deformation theory for 3-dimensional hyperbolic cone-manifolds which they used to prove
various important results about closed and finite volume hyperbolic 3-manifolds. This
deformation theory was extended to infinite volume, geometrically finite hyperbolic
cone-manifolds in [Bro04b, Bro04a]. In this setting the deformation theory has had a
number of applications to classical conjectures about Kleinian groups.

Here is an example of a basic problem that can be addressed via the deformation
theory. Let (M, g) be a geometrically finite hyperbolic 3-manifold that contains a
simple closed geodesic γ. Let $\hat{M} = M \backslash \gamma$ be the complement of γ. There will be then
be a unique, geometrically finite, complete hyperbolic metric $\hat{g}$ on $\hat{M}$ such that the
conformal boundaries of (M, g) and $(\hat{M}, \hat{g})$ agree. We have the following theorem

Theorem 1.1 ([BB04]). *For each $K > 1$ there exists an $\ell > 0$ such that if the length
of γ in (M, g) is less then ℓ then there exists a K-bi-Lipschitz map*

$$\phi : (M \backslash \mathbb{T}, g) \longrightarrow (\hat{M} \backslash \hat{\mathbb{T}}, \hat{g})$$

where $\mathbb{T}$ and $\hat{\mathbb{T}}$ are Margulis tubes about γ and the rank two cusp, respectively.

We call such a theorem a "drilling theorem" for we have drilled the geodesic γ out
of the hyperbolic manifold (M, g).

The way we obtain geometric control of the metric $\hat{g}$ is to interpolate between
g and $\hat{g}$ using *hyperbolic cone-metrics*. The Hodgson-Kerckhoff deformation theory
gives means to bound the change in geometry as this one-parameter family of metrics

---

[1] Supported by a grant from the NSF.

varies. The first part of this paper will be an exposition of this deformation theory emphasizing the most geometric parts. For an expository account of Hodgson and Kerckhoff's work see [HK03]. To keep this paper somewhat self-contained there is some necessary overlap between the two papers.

In the second part of the paper we will apply the deformation theory to a collection of classical conjectures in Kleinian groups: the density conjecture, density of cusps on the boundary of quasiconformal deformation spaces and the ending lamination conjecture. Rather than discussing these conjectures in their full generality we will restrict to the special case of a Bers' slice. This will allow us to demonstrate how the deformation theory plays a role in approaching the conjectures in a simpler setting.

Acknowledgments. This paper is an expanded version of a talk given at the workshop on Spaces of Kleinian Groups and Hyperbolic 3-Manifolds held at the Newton Institute in August 2003. The author would like to thank Caroline Series, Yair Minsky and Makoto Sakuma for organizing the workshop and their solicitation of this article.

The author would also like to thank his collaborator, Jeff Brock, with whom he did much of the work described in this paper.

2. Deformations of hyperbolic metrics

We will begin by examing the various different ways one can study a family of hyperbolic metrics: as Riemannian metrics, as (G, X)-structures and as representations of the fundamental group in the space of hyperbolic isometries. We will see the advantages of each viewpoint and the connections between the different viewpoints. A reference for this material is §1 and §2 of [HK98].

In the final subsection we will discuss complex projective structures on surfaces. These arise naturally as the boundary of hyperbolic 3-manifolds and will play an important role in the extension of the Hodgson-Kerckhoff deformation theory to infinite volume and geometrically finite hyperbolic cone-manifolds.

2.1. One-parameter families of metrics

We start with a family of metrics, $g_t : V \times V \longrightarrow R$, on a finite dimensional vector space V. For each t there is a unique $\eta_t \in \hom(V, V)$ such that

$$\frac{dg_t(v, w)}{dt} = 2g_t(v, \eta_t(w)). \qquad (2.1)$$

Since g_t is symmetric, η_t is self-adjoint, i.e.

$$g_t(\eta_t(v), w) = g_t(v, \eta_t(w)).$$

We measure the size of η_t using the metric g_t. Let $\{e_1, \ldots, e_n\}$ be an orthonormal basis for V in the g_t metric. Then define the norm of η_t by the formula

$$\|\eta_t\|^2 = \sum g_t(\eta_t(e_i), \eta_t(e_i)). \tag{2.2}$$

For any $v \in V$ we then have

$$g_t(v, \eta_t(v)) \leq \|\eta_t\| g_t(v, v).$$

By integrating (2.1) we see that if $\|\eta_t\| \leq K$ for all $t \in [0, T]$ then

$$e^{-2KT} g_0(v, v) \leq g_T(v, v) \leq e^{2KT} g_0(v, v).$$

In particular the identity map on V is a KT-bi-Lipschitz map from the g_0-metric to g_T-metric.

The trace of η_t is the *divergence* and it is the derivative of the volume. The traceless part of η_t is the *strain* and it measures the change in the conformal structure.

2.2. Metrics on a manifold

Now we apply the above work to a family of metrics, g_t, on a differentiable manifold M. In this setting η_t is a one-parameter family in $\hom(TM, TM)$. Let $\|\eta_t(p)\|$ be the pointwise norm of η_t. Let $\phi_t : (M, g_0) \longrightarrow (M, g_t)$ be the identity map on M. If $\|\eta_t(p)\| \leq K$ for all $p \in M$ and all $t \in [0, T]$ then ϕ_t is a KT-bi-Lipschitz diffeomorphism.

The identity map on M may not have the smallest bi-Lipschitz constant of all maps from (M, g_0) to (M, g_t). In particular for an arbitrary family of metrics there is no reason to hope that we can control the norm of η_t. The driving idea behind the Hodgson-Kerckhoff deformation theory is to find one-parameter families of hyperbolic metrics g_t where the derivative η_t is a *harmonic strain field*. As we will see below, this extra structure will allow us to control the norm of η_t.

2.3. Hyperbolic metrics on a manifold

Let $\mathcal{H}(M)$ be the space of all hyperbolic metrics on M. Two metrics g and h in $\mathcal{H}(M)$ are equivalent if there is a diffeomorphism $\psi : M \longrightarrow M$ isotopic to the identity such that $h = \psi^* g$. Given two equivalence classes of metrics we want to find an efficient path between them. That is we want to find a path g_t that minimizes the derivative η_t. The last statement can be interpreted in a number of ways. For example, we could try to minimize the pointwise or L^2-norm of η_t. However, if M is not compact then both of these norms can and will be infinite. Our efficient paths will have two properties. First, they will be divergence free so that η_t is a strain field. Second they will be harmonic. We will not formally define harmonic. Informally, one can think of a harmonic strain field as locally minimizing the L^2-norm (see Appendix B of [McM96]).

A harmonic strain field satisfies the following important equation:

Theorem 2.1. *Let (M,g) be a compact hyperbolic manifold with boundary and let η be a harmonic strain field. Then*

$$\int_M \|\eta\|^2 + \|\nabla\eta\|^2 = \int_{\partial M} *\nabla\eta \wedge \eta. \qquad (2.3)$$

This formula is very important because it allows us to compute the L^2-norm of a strain field by only knowing information on the boundary. We also note that η is harmonic if it satisfies (2.3) for all compact submanifolds.

Another feature of harmonic strain fields is that they satisfy a mean value inequality:

Theorem 2.2. *Let (M,g) be a hyperbolic manifold and η a harmonic strain field. If B is a ball in M of radius $R > \frac{\pi}{\sqrt{2}}$ centered at p then*

$$\|\eta(p)\| \le \frac{3\sqrt{2(B)}}{4\pi f(R)} \sqrt{\int_B \|\eta\|^2 dV}$$

where $f(R) = \cosh(R)\sin(\sqrt{2}R) - \sqrt{2}\sinh(R)\cos(\sqrt{2}r)$.

Together, Theorems 2.1 and 2.2 will allow us to get pointwise bounds on the the norm of η, at least for points in the thick part of (M,g).

2.4. Developing maps

Another way to think of a hyperbolic structure is as a (G,X)-structure, where X is hyperbolic space and G the group of hyperbolic isometries. A (G,X) structure is an atlas of charts to X with transition maps which are restrictions of elements of G. A (G,X)-structure determines a developing map and a holonomy representation.

Here's how it works for a hyperbolic 3-manifold: A *developing map* is a local diffeomorphism,

$$D : \tilde{M} \longrightarrow \mathbb{H}^3,$$

and the *holonomy representation* is a representation of the fundamental group,

$$\rho : \pi_1(M) \longrightarrow PSL_2\mathbb{C} = \mathrm{Isom}^+(\mathbb{H}^3).$$

The developing map commutes with the action of the fundamental group where the fundamental groups acts on $\tilde{M}$ as deck transformations and on $\mathbb{H}^3$ via the holonomy representation. That is

$$D(\gamma(x)) = \rho(\gamma)D(x) \qquad (2.4)$$

for all $\gamma \in \pi_1(M)$. Let $\tilde{g}$ be the pull back of the hyperbolic metric. Then (2.4) implies that $\tilde{g}$ is equivariant and descends to a hyperbolic metric g on M.

Conversely, a hyperbolic manifold, (M,g), determines a developing map and holonomy representation. The developing map is unique up to post-composition with hyperbolic isometries. If we post-compose the developing with an isometry $\alpha \in PSL_2\mathbb{C}$ then we conjugate the holonomy by α.

Given a smooth family of hyperbolic metrics (M,g_t), there is a smooth family of developing maps D_t, and holonomy representations ρ_t. The derivative of the developing maps determines a family of vector fields v_t on $\tilde{M}$ in the following way. For a point $x \in \tilde{M}$, $D_t(x)$ is smooth path in $\mathbb{H}^3$. Let $v_t(x)$ be the pull-back, via D_t, of the tangent vector of this path at time t. These vector fields are not equivariant. However, they do satisfy the following *automorphic* property. For all $\gamma \in \pi_1(M)$ the difference, $\gamma_* v_t - v_t$, is an infinitesimal isometry in the $\tilde{g}_t$-metric. That is, the flow of the vector field $\gamma_* v_t - v_t$ is an isometry. This follows directly from differentiating (2.4).

The automorphic vector fields v_t, lead to the connection between the developing maps and the derivative, η_t, of the metrics g_t. The covariant derivative, $\nabla_t v_t$, is an element of $\hom(T\tilde{M}, T\tilde{M})$. Let $\operatorname{sym}\nabla_t v_t$ be its symmetric part. The covariant derivative of an infinitesimal isometry is skew. Therefore, the automorphic property of v_t implies that $\operatorname{sym}\nabla_t v_t$ is equivariant and descends to an element of $\hom(TM, TM)$. By noting that the derivative $\frac{dg_t(v,w)}{dt}$ is the Lie derivative $\mathcal{L}_{v_t} g_t(v,w)$ we see that $\operatorname{sym}\nabla_t v_t = \eta_t$.

2.5. Holonomy representations

Let $\mathcal{R}(M)$ be the space of representations of $\pi_1(M)$ in $PSL_2\mathbb{C}$. We are only interested in representations up to conjugacy so we would like to study the quotient of $\mathcal{R}(M)$ under the action of $PSL_2\mathbb{C}$ by conjugacy. Unfortunately, this quotient may not be a nice object. For instance it may not even by Hausdorff. Instead one takes the *Mumford quotient* of $\mathcal{R}(M)$ which we denote $R(M)$. The Mumford quotient is an algebraic variety and its Zariski tangent space at a representation ρ is the cohomology group $H^1(\pi_1(M); \operatorname{Ad}\rho)$. It will turn out, that at all points were are interested in, $R(M)$ is simply the topological quotient of $\mathcal{R}(M)$ by conjugacy. Furthermore, at these points $R(M)$ will be a differentiable manifold and the the Zariski tangent space will be naturally identified with the differentiable tangent space. For this reason we will ignore the distinction between the Mumford quotient and the topological quotient.

By differentiating a smooth family of representations ρ_t we can see how the differentiable tangent space at each ρ_t is identified with $H^1(\pi_1(M); \mathrm{Ad}\rho_t)$. Let γ be an element of $\pi_1(M)$. Then $\rho_t(\gamma)$ is a smooth path in $PSL_2\mathbb{C}$. Each tangent space of $PSL_2\mathbb{C}$ is canonically identified with the Lie algebra $sl_2\mathbb{C}$. Therefore the derivative $\dot{\rho}_t$ can be thought of as a map

$$\dot{\rho}_t : \pi_1(M) \longrightarrow sl_2\mathbb{C}$$

for each t. This map satisfies the cocyle condition

$$\dot{\rho}_t(\gamma\beta) = \dot{\rho}_t(\gamma) + \mathrm{Ad}\rho_t(\gamma)\dot{\rho}_t(\beta)$$

for all γ and β in $\pi_1(M)$ and therefore determines a cohomology class in $H^1(\pi_1(M); \mathrm{Ad}\rho_t)$.

We also remark that $\dot{\rho}_t(\gamma)$ corresponds to the vector field $\gamma_* v_t - v_t$. The latter vector field is identified with an element of $sl_2\mathbb{C}$ by pushing forward $\gamma_* v_t - v_t$ via D_t. This push foward is an infinitesimal isometry on $\mathbb{H}^3$ and the space of infinitesimal isometries of $\mathbb{H}^3$ is canonically identified with $sl_2\mathbb{C}$.

2.6. Complex projective structures

A *complex projective structure* on a surface S is an atlas of charts to the Riemann sphere, $\widehat{\mathbb{C}}$, where the transition maps are restrictions of elements of $PSL_2\mathbb{C}$. A projective structure is another example of (G,X)-structure where $G = PSL_2\mathbb{C}$ and $X = \widehat{\mathbb{C}}$. Let $P(S)$ be the space of projective structures on S. Since the action of $PSL_2\mathbb{C}$ is conformal, a projective structure also determines a conformal structure on S so there is a map

$$P(S) \longrightarrow T(S)$$

where $T(S)$ is the *Teichmüller space* of marked conformal structures on S. One is often interested in the space of projective structures with a fixed conformal structure X. We denote the space of such structures $P(X)$.

Elements of $PSL_2\mathbb{C}$ take round circles in $\widehat{\mathbb{C}}$ to round circles. Therefore, there is a well defined notion of a round circle on a projective structure. A conformal map f between two projective structures Σ and Σ' will distort these round circles. The *Schwarzian derivative*, Sf, measures this distortion. We will not give an exact definition of Sf although we will describe an infinitesimal version below. We will however state the key properties of the Schwarzian derivative that we will use. First, Sf is a holomoprhic quadratic differential on X. The quotient of the absolute value of a holomorphic quadratic differential and a metric is a function. Using the unique hyperbolic metric on X we can take the sup-norm of this function to a define the sup-norm, $\|Sf\|_\infty$, of the Schwarzian. This determines a metric on $P(X)$ by setting $d(\Sigma, \Sigma') = \|Sf\|_\infty$. Furthermore, given any holomorphic quadratic differential Φ on X there is a projective structure Σ' such that for the conformal map $f : \Sigma \longrightarrow \Sigma'$, $Sf = \Phi$. Therefore $P(X)$

is isomorphic to the vector space $Q(X)$ of holomorphic quadratic differentials on X.

A projective structure is *Fuchsian* if it is the quotient of a round disk in $\widehat{\mathbb{C}}$. There is a unique Fuchsian projective structure, Σ_F, in each $P(X)$. We will often be interested in the distance between an arbitrary projective structure $\Sigma \in P(X)$ and this unique Fuchsian projective structure. We therefore let $\|\Sigma\|_F = d(\Sigma, \Sigma_F)$.

As with any (G,X)-structure, a projective structure Σ on S determines a developing map

$$D : \tilde{S} \longrightarrow \widehat{\mathbb{C}}$$

and a holonomy representation

$$\rho : \pi_1(S) \longrightarrow PSL_2\mathbb{C}$$

satisfying (2.4). Now let Σ_t be a smooth path of projective structures in $P(X)$. Then there is a smooth path of developing maps D_t which determine vector fields v_t on $\tilde{S}$. The developing maps, D_t, can be chosen to be conformal maps from $\tilde{X}$ to $\widehat{\mathbb{C}}$ which will make the vectors fields v_t conformal on $\tilde{X}$.

Let $v(z)$ be a conformal vector field on a domain in $\widehat{\mathbb{C}}$. Then $v(z) = f(z)\frac{\partial}{\partial z}$ where f is a holomorphic function. A conformal vector field is *projective* if its flow consists of elements of $PSL_2\mathbb{C}$. The space of projective fields is the Lie algebra $sl_2\mathbb{C}$ and $v(z)$ will be projective if and only if $f(z)$ is a quadratic polynomial. At each point z in the domain let $s(z)$ be the unique projective vector field that best approximates v at z. Note that $s(z)$ is obtained by taking the first three terms of the Taylor series of f at z. Differentiating $s(z)$ we obtain an $sl_2\mathbb{C}$-valued 1-form which can be canonically associated with a holomorphic quadratic differential. This quadratic differential is the Schwarzian derivative, Sv, of the vector field v.

We now return to our path of projective structures Σ_t in $P(X)$. The Schwarzian derivative of the conformal vector fields v_t will be equivariant and therefore Sv_t will be a holomorphic quadratic differential on X. The norm $\|Sv_t\|_\infty$ is the infinitesimal version of the metric on $P(X)$ and if we can bound it for all t we bound the distance between Σ_0 and Σ_1.

We need one final fact about projective structures. The holonomy representation defines a map from $P(S)$ to the space $R(S)$ of representations of $\pi_1(S)$ in $PSL_2\mathbb{C}$ modulo conjugacy. We the have the following theorem.

Theorem 2.3 ([Hej75, Ear81, Hub81]). *The holonomy map*

$$\text{hol} : P(S) \longrightarrow R(S)$$

is a holomorphic, local homeomorphism.

3. Hyperbolic cone-manifolds

3.1. Geometrically finite hyperbolic cone-manifolds

Let N be a compact manifold with boundary, C a collection of simple closed curves in the interior of N and M the interior of $N\backslash C$. Let g be a complete metric on the interior of N that is a smooth Riemannian metric on M. We say that g is a hyperbolic *cone-metric* if the following holds: First g is a hyperbolic metric on M. Second, for points on C the metric has the form

$$dr^2 + \sinh^2 rd\theta^2 + \cosh^2 rdz^2$$

where θ is measured modulo some *cone-angle* α. Note that the cone-angle must be locally constant on C. Therefore there is a cone-angle associated to each component of C.

Since the metric g is complete the boundary ∂N consists of tori and higher genus surfaces. Let $\partial_0 N$ denote the higher genus components of the boundary. To develop a good deformation theory we need to assume that there metric g has certain asymptotic behavior as we approach $\partial_0 N$. We say that a hyperbolic, cone-metric g is *geometrically finite* if the hyperbolic structure extends to a *projective structure* on $\partial_0 N$. More explicitly g is geometrically finite if for each $p \in \partial_0 N$ there exists an open neighborhood of p in N and a map $\psi : V \longrightarrow \mathbb{H}^3 \cup \widehat{\mathbb{C}}$ that is a homeomorphism onto its image and is an isometry on $V \cap \mathrm{int}\,M$. The restriction of ψ to $V \cap \partial_0 N$ will determine an atlas of charts to $\widehat{\mathbb{C}}$. Since hyperbolic isometries of $\mathbb{H}^3$ extend to projective transformations of $\widehat{\mathbb{C}}$ this atlas will determine a projective structure on $\partial_0 N$.

Let $GF(N, C)$ be equivalence classes of geometrically finite hyperbolic cone-manifolds on the pair (N, C). If g is a hyperbolic cone-metric on (N, C) we refer to the induced projective structure on $\partial_0 N$ as the *projective boundary*. The projective structure induces a conformal structure on $\partial_0 N$. This is the *conformal boundary*.

Note that the round circles in the projective boundary are the boundary at infinity of hyperbolic planes in the hyperbolic manifold. As the 3-dimensional hyperbolic metric deforms these planes will not stay totally geodesic. This will be detected by the change in the projective boundary.

3.2. Deformations of hyperbolic cone-manifolds

A *meridian* for the pair (N, C) is a simple closed curve $\gamma \subset \mathrm{int}\,N$ that bounds a disk in N which intersects C in a single point. Each component of C has a unique meridian up to homotopy in $M = \mathrm{int}\,N\backslash C$. Furthermore if ρ is the holonomy of a cone-manifold structure on (N, C) then $\rho(\gamma)$ will be elliptic (or the identity if the cone angle is a multiple of 2π) for all meridians γ.

On the other hand there certainly will be representations where not all meridians are elliptic. For this reason we let $R_e(M)$ be the subset of $R(M)$ where the meridians are elliptic or the identity. We then have the following theorem which is essentialy due to Thurston ([Thu80]).

Theorem 3.1. *The holonomy map*

$$\mathrm{hol} : GF(N, C) \longrightarrow R_e(M)$$

is a local homeomorphism.

With this theorem our next goal is to give a local parameterization of $R(M)$. To do this we first need to define parameters. This local parameterization will be of a neighborhood in $R(M)$, not just a neighborhood in $R_e(M)$. These more general representations also have geometric signifigance. They correspond to Thurston's *generalized Dehn surgery singularities*. We will not explain the geometry of these singularities here.

Let

$$\mathcal{L}_M : R(M) \longrightarrow \mathbb{C}^k$$

be the holomorphic map which assigns to each representation the k-tuple of complex lengths of the k-meridians of (N, C). This is our first set of parameters.

The second set of parameters comes from the conformal boundary. Given a component S of $\partial_0 N$ we can define a map from $GF(N, C)$ to the Teichmüller space $T(S)$. This map assigns to each geometrically finite cone-manifold the conformal boundary structure on S. If $\rho \in R(M)$ is the holonomy of a cone-manifold in $GF(N, C)$ then by pre-composing this map with hol^{-1}, we obtain a map ∂_S from a neighborhood of ρ in $R_e(M)$ to $T(S)$. Here we choose the unique branch of hol^{-1} that takes ρ to the given geometrically finite cone-manifold. There is then a unique holomorphic extension of ∂_S to a neighborhood of ρ in $R(M)$.

Repeating the construction for each component of $\partial_0 N$ and combining the maps we have a single map

$$\partial : R(M) \longrightarrow T(\partial_0 N).$$

Strictly speaking ∂ is only defined for a neighborhood of ρ in $R(M)$. We also note that there are examples of distinct geometrically finite hyperbolic cone-manifolds with the same holonomy representation. When this happens each manifold will define a different boundary map ∂.

Now we combine our two parameters. Define

$$\Phi : R(M) \longrightarrow \mathbb{C}^k \times T(\partial_0 N)$$

by $\Phi(\rho) = (\mathcal{L}_{\mathcal{M}}(\rho), \partial(\rho))$.

Theorem 3.2 ([HK98, HK, Bro04b]). *Let ρ be the holonomy of a geometrically finite cone-manifold. If the cone-angle is $\leq 2\pi$ or the tube radius of the singular locus is $\geq \sinh^{-1} 1/\sqrt{2}$ then the map Φ is a holomorphic local homeomorphism.*

Sketch of proof of theorem 3.2. By a theorem of Thurston

$$\dim_{\mathbb{C}} R(M) \geq k + \dim_{\mathbb{C}} T(\partial_0 N).$$

Since the map Φ is holomorphic if we can show that the derivative, Φ_*, is injective at ρ then Φ will be a local homeomorphism at ρ.

The first step in proving this injectivity is a Hodge theorem: Any tangent vector of $R(M)$ at ρ that is in the kernel of ∂_* is represented by a harmonic strain field η on (M, g_α). Note there are some subtle issues to proving this Hodge theorem since our manifold is not compact and the metric is not complete. In particular, the harmonic strain field η is only unique after making some choice of boundary conditions for the solution.

Next we would like to calculate the L^2-norm of η on M. Theorem 2.1 tells how to calculate the L^2-norm of a harmonic strain field on a compact manifold with boundary. We can obtain a similar formula for harmonic strain fields on a geometrically finite manifold if the strain field fixes the conformal boundary. Analytically this is equivalent to $\partial_* \eta = 0$ where ∂_* is the tangent map of the boundary map ∂ from $R(M)$ to $T(\partial_0 N)$. The pointwise norm of such conformal deformations will decay exponentially and the boundary term in (2.3) will limit to zero for surfaces exiting the geometrically finite end. This allows us to calculate the L^2-norm of η even on the non-compact geometrically finite ends. In particular, we have

$$\int_{M \backslash U} \|\eta\|^2 + \|\nabla \eta\|^2 = \int_{\partial U} *\nabla \eta \wedge \eta$$

where U is tubular neighborhood of the singular locus, even though $M \backslash U$ is not compact. Note that in general the L^2-norm will be infinite on all of M.

The final step is to calculate the boundary term. This is done in the following way. In a tubular neighborhood of the singular locus we can decompose η as the sum of two strain fields, $\eta = \eta_0 + \eta_c$. The first term, η_0, is an explicit model deformation completely determined by the derivatives of the complex lengths of the components of the singular locus and the meridians. The second term, η_c, is a correction term. It does not affect the complex length of the singular locus or the meridians. In particular, there is a vector field v on a tubular neighborhood of the singular locus such that $\eta_c = \text{sym} \nabla v$.

The advantage of this decomposition is that we can now decompose the boundary term:

$$\int_{\partial U} *\nabla \eta \wedge \eta = \int_{\partial U} *\nabla \eta_0 \wedge \eta_0 + \int_{\partial U} *\nabla \eta_c \wedge \eta_c. \tag{3.1}$$

The first term on the right can be calculated explicitly and will be non-positive if $(\mathcal{L}_{\mathcal{M}})_* \eta = 0$. The hard work is to show that the second term will always be non-positive. Together this implies if $\Phi_* \eta = 0$ then $\eta \equiv 0$ and therefore Φ_* is injective. $\square$

The following is a simple corollary of Theorem 3.1 and 3.2.

Corollary 3.3. *Let M_α be a geometrically finite cone-manifold with cone angle α whose singular locus has a tubular neighborhood of radius $\geq \sinh^{-1} \sqrt{2}$. Then, for t near α, there exists a one-parameter family of cone-manifolds M_t with cone-angle t and conformal boundary fixed.*

We now set some notation that will be used throughout the rest of the paper. For any essential simple closed curve γ in M, $L_\gamma(t)$ is the the length of γ in M_t and $\mathcal{L}_\gamma(t)$ is the complex length of γ. The imaginary part of $\mathcal{L}_\gamma(t)$ is denoted $\Theta_\gamma(t)$. For the special case of the singular locus, $L_C(t)$ is the the sum of the lengths of all the components of the singular locus. Let $U_t(R)$ be the union of the R-tubular neighborhoods of the components of the singular locus. The n components of the conformal boundary are denoted $X^1, \ldots, X^n$. The corresponding components of the projective boundary of M_t are denoted $\Sigma_t^1, \ldots, \Sigma_t^n$.

The next theorem is key in controlling the geometry of the one-parameter family of cone-manifolds M_t.

Theorem 3.4. *The one parameter family of cone-manifolds M_t can be realized by metrics g_t with derivatives η_t such that η_t is a harmonic strain field outside of a radius 1 tube of the singular locus and*

$$\int_{M_t \setminus U_t(R)} \|\eta_t\|^2 + \|\nabla_t \eta_t\|^2 \leq \frac{2L_C(t)}{t^2 \sinh^2 R}$$

for all $R \geq 1$.

Sketch of proof of theorem 3.4. The proof has two parts: The construction of the metrics g_t and the estimate on the L^2-norm of the strain field η_t. We will skip the first part and focus on the second.

The bound on the L^2-norm of η_t follows the same pattern as the completion of the proof of Theorem 3.2. For each t we decompose η_t in U_t as $\eta_t = \eta_0 + \eta_c$ where η_0 is a model deformation and η_c is a correction term. We have the same decomposition of the boundary term as in (3.1) and once again the correction term makes a non-positive contribution. The one difference we have is that the cone angle is now decreasing and so the term coming from the model deformation will be positive. However, we can make an explicit calculation to bound this positve number and see that

$$\int_{\partial U_t(R)} *\nabla_t \eta_0 \wedge \eta_0 \le \frac{2L_C(t)}{t^2 \sinh^2 R}$$

which gives the theorem.   □

We remark that the only significance of the tube radius 1 in the above theorem is that $1 > \sinh^{-1} 1/\sqrt{2}$.

4. The drilling theorems

We call the process of decreasing the cone angle "drilling". In the three drilling theorems that follow we control various geometric quantites as we drill. Note that these drilling theorems only apply where the one-parameter family of cone-manifolds M_t is defined. To be useful we need to know that we can drill the cone-angle a definite amount, say from 4π to 2π or 2π to 0. As we will see one consequence of the drilling theorems is that under certain conditions we can drill this definite amount.

In the first drilling theorem we estimate how the lengths of geodesics change as we drill.

Theorem 4.1 ([Bro04a]). *For each $L > 0$ there exists an $\varepsilon > 0$ and an $A > 0$ such that if γ is a simple closed curve in M with $L_\gamma(\alpha) \le L$ and $L_C(\alpha) \le \varepsilon$ then*

$$e^{-AL_C(\alpha)} L_\gamma(\alpha) \le L_\gamma(t) \le e^{AL_C(\alpha)} L_\gamma(\alpha)$$

and

$$(1 - AL_C(\alpha))\Theta_\gamma(\alpha) \le \Theta_\gamma(t) \le (1 + AL_C(\alpha))\Theta_\gamma(\alpha)$$

for all $t \le \alpha$.

Sketch of proof of theorem 4.1. To prove the first statement we need to bound the derivative $L'_\gamma(t)$. There are two cases. The first case is when the length of γ is bounded but not very short. In this case γ will be in the thick part of M_t. We then use the L^2-bounds given by Theorem 3.4 along with Theorem 2.2 to find a pointwise bound on η_t for all points on γ. This, in turn, bounds the derivative.

The second case is when γ is very short. By a version of the Margulis Lemma, γ will have a large tubular neighborhood U. We decompose η_t on U into a model term, η_0, and a correction term, η_c, as before. A bound on the L^2-norm of the model will bound the derivative $L'_\gamma(t)$. The model term, η_0, is like a deformation of a component of the singular locus that does not change the cone angle. As in the proof of Theorem 3.2 this determines the sign of the boundary term. However, in this case the sign will be positive since the torus ∂U has the opposite orientation of the boundary torus in Theorem 3.2. This is because we are calculating the L^2-norm on U rather than its complement. The sign of the boundary term for η_c will also be positive for the same reason. This last fact, together with Theorem 3.4 gives the desired bound on the L^2-norm of η_0 on U.

The second statement of the theorem is proved by a similar method.   □

In the next drilling theorem we bound the change in the projective boundary of M_t as we drill. This should be thought of as controlling the geometry of the geometrically finite ends.

Theorem 4.2 ([Bro04a]). *There exists a C depending only on* α, *the injectivity radius of the unique hyperbolic metric on* X^i *and* $\|\Sigma^i_\alpha\|_F$ *such that*

$$d(\Sigma^i_\alpha, \Sigma^i_t) \leq CL_C(\alpha)$$

for all $t \leq \alpha$.

Sketch of proof of theorem 4.2. The derivative of the path Σ^i_t in $P(X)$ is a path of quadratic differentials Φ^i_t in $Q(X^i)$. We will bound the size of Φ^i_t.

A embedded round disk D in Σ^i_t bounds an embedded half space H in M_t. The first step is to show that a bound on the L^2-norm of η_t on H implies a bound on the sup norm of Φ^i_t with respect to the hyperbolic metric on D. The proof of this fact follows our previous theme. We decompose the harmonic strain field η_t into a model term, η_0, completely determined by Φ^i_t and a correction term η_c. Once again the L^2-norm of η_t on H is the sum of the L^2-norms of η_0 and η_c so Theorem 3.4 bounds the L^2-norm of η_0. Since η_0 is explicitly determined by Φ^i_t this bounds the sup norm of Φ^i_t.

Notice that we have only bounded the sup norm with respect to the hyperbolic metric on D, not with respect to the hyperbolic metric on X^i. To finish the proof we need to compare the two metrics. In particular for every point z we can find a disk D containing z where the ratio of the two metrics is bounded by constants depending only on the injectivity radius of X^i and $\|\Sigma^i_t\|_F$.   □

Together, the previous two results give enough control to prevent any degeneration as the cone angle decreases. In particular we have the following theorem:

Theorem 4.3 ([Bro04a]). *For any $\alpha > 0$ there exists an $\ell > 0$ such that if M_α is a geometrically finite cone-manifold with $L_C(\alpha) \leq \ell$ and tube radius $> \sinh^{-1} 1/\sqrt{2}$ then the one parameter family is defined for all $t \in [0, \alpha]$.*

The final drilling theorem is also the strongest. Theorem 1.1 is a special case. It essentially implies the previous two drilling theorems although the dependence of the constants on the length of the singular locus is not so clear.

Theorem 4.4 ([BB04]). *For any $K > 1$ there exists an $\ell > 0$ depending only on K and α such that the following holds. If $L_C(\alpha) \leq \ell$ there is for each $t \in [0, \alpha]$ a standard neighborhood $\mathbb{T}_t(C)$ of the singular locus C and a K-bi-Lipschitz diffeomorphism of pairs*

$$h_t : (M_\alpha \backslash \mathbb{T}_\alpha(C), \partial \mathbb{T}_\alpha(C)) \longrightarrow (M_t, \mathbb{T}_t(C), \partial \mathbb{T}_t(C)).$$

Sketch of proof of theorem 4.4. Recall that in Theorem 3.4 we constructed a family of metrics, $M_t = (M, g_t)$, whose derivative was the harmonic strain fields η_t. For points in the thick part of M_t the combination of Theorems 3.4 and 2.2 bound the pointwise norm of η_t. Therefore, on the thick part of M_α the identity map on M is a K-bi-Lipschitz map from (M, g_α) to (M, g_t) when the singular locus is sufficiently short.

We are left to extend h_t to the thin parts of M_α which will be a collection of Margulis tubes. This is done by hand. The maps h_t are K-bi-Lipschitz on the boundary of these Margulis tubes and we build an explicit extension of this map inside the tube. The construction is somewhat tedious and we will not describe it here. $\square$

5. Geometric inflexibility

A nice application of the boundary formula of Theorem 2.1 is to show exponential decay of the L^2-norm. Essentially, the formula shows that the L^2-norm of a harmonic strain field on the 3-manifold is equal to its L^2-norm on the boundary. A function whose integral equals its boundary values will be exponential. This leads to the exponential decay of harmonic strain fields. Here is the precise theorem:

Theorem 5.1 ([BB]). *Let M be a complete hyperbolic 3-manifold with boundary and η a harmonic strain field on M that has finite L^2-norm. Let $M(t)$ be the subset of M consisting of the points that are distance t or greater from ∂M. Then*

$$\int_{M(t)} \|\eta\|^2 + \|\nabla \eta\|^2 \leq e^{-2t} \int_M \|\eta\|^2 + \|\nabla \eta\|^2.$$

Sketch of proof of theorem 5.1. The first step is to see that Theorem 2.1 applies to M and $M(t)$ to get

$$\int_{M(t)} \|\eta\|^2 + \|\nabla \eta\|^2 = \int_{\partial M(t)} *\nabla \eta \wedge \eta$$

even thought $M(t)$ is not compact. The second fact we need is the following inequality

$$\|\eta\|^2 + \|\nabla\eta\|^2 \geq 2\| * \nabla\eta \wedge \eta\|.$$

Now let

$$f(t) = \int_{M(t)} \|\eta\|^2 + \|\nabla\eta\|^2$$

which we can rewrite as

$$f(T) = \int_T^\infty \int_{\partial M(t)} (\|\eta\|^2 + \|\nabla\eta\|^2) dA dt$$

where dA is the area form on $\partial M(t)$. Differentiating we have

$$
\begin{aligned}
-f'(t) &= \int_{\partial M(t)} (\|\eta\|^2 + \|\nabla\eta\|^2) dA \\
&\geq 2\int_{\partial M(t)} *\nabla\eta \wedge \eta \\
&\geq 2f(t).
\end{aligned}
$$

Integrating both sides of the final inequality gives the theorem. $\qquad\square$

McMullen has proven a similar theorem, using entirely different methods, for harmonic strain fields arising from quasi-conformal deformations of complete hyperbolic 3-manifolds. He calls his theorem "geometric inflexibilty" which we follow.

One application of the geometric inflexibility theorem is stronger versions of the drilling theorems. For example, the bounds on the change in length of a closed geodesic given by Theorem 4.1 will decay exponentially in the distance of the geodesic from the singular locus.

To apply geometric inflexibility to Theorem 4.2 we need another definition. A geometrically finite cone-manifold will have a convex core which will be a submanifold with boundary consisting of convex surfaces. There will be one component of the boundary the convex core facing each component of the projective boundary. For γ a closed geodesic and Σ a component of the projective boundary let $d(\gamma, \Sigma)$ be the shortest distance from γ to the component of the boundary of the convex core facing Σ.

Theorem 5.2. *There exists C_1 and C_2 depending only on α, the injectivity radius of the unique hyperbolic metric X^i and $\|\Sigma_\alpha^i\|_\infty$ such that*

$$d(\Sigma_\alpha^i, \Sigma_t^i) \leq C_1 e^{-C_2 d(C, \Sigma_\alpha^i)} L_C(\alpha)$$

for all $t \leq \alpha$.

6. Applications to the Bers' slice

A *Kleinian group* is a discrete subgroup of $PSL_2\mathbb{C}$. Here we will restrict to the special class of Kleinian groups that arise as the image of holonomy representations of projective structures on a closed surface S. The advantage of restricting to this class is that we can use the topology and metric on the space of projective structures to study the family of Kleinian groups.

For a more precise definition let $U(X)$ be the set of projective structures in $P(X)$ whose developing map is injective. For every projective $\Sigma \in U(X)$ the image of the holonomy representation $\rho(\pi_1(S))$ will act properly discontinuously on the image of the developing map, $D(\tilde{S})$. Since the developing map is injective, $D(\tilde{S})$ will be an open topological disk in $\widehat{\mathbb{C}}$. A group that acts properly discontinuously on a open subset of $\widehat{\mathbb{C}}$ will be discrete and therefore $\rho(\pi_1(S))$ is a Kleinian group.

A Kleinian group is *quasifuchsian* if it acts properly discontinuously on two disjoint open disks in $\widehat{\mathbb{C}}$. Let $T(X)$ be the subset of $U(X)$ where the image of the holonomy representation is quasifuchsian. The space $T(X)$ is a *Bers' slice* of the space of all quasifuchsian groups. Let ρ be the holonomy of a projective structure in $T(X)$ and let Ω be the open disk, disjoint from $D(\tilde{S})$, on which $\rho(\pi_1(S))$ acts properly discontinuously. Then $\Omega/\rho(\pi_1(S))$ defines a projective structure and hence a conformal structure on $\bar{S}$, where $\bar{S}$ is the oriented surface S with the orientation reversed. This defines a map $T(X) \longrightarrow T(\bar{S})$ which we call the *Bers' isomorphism* for reasons that the following Theorem make apparent.

Theorem 6.1 ([Ber60]). *The map* $T(X) \longrightarrow T(\bar{S})$ *is a homeomorphism.*

The Bers' slice, $T(X)$, is the simplest example of a *quasi-conformal deformation space*. The above theorem implies that $T(X)$ is canonically identified with Teichmüller space. Since $U(X)$ is bounded in $P(X)$ the closure $\overline{T(X)}$ is a compactification of Teichmüller space.

In what follows we will continually refer to various objects determined by a projective structure Σ in $U(X)$. First there is the holonomy representation ρ. Since its image is a Kleinian group isomorphic to $\pi_1(S)$ the quotient $\mathbb{H}^3/\rho(\pi_1(S))$ is a hyperbolic 3-manifold M homotopy equivalent to S. By the previous theorem if Σ is in $T(X)$ it will also determine a conformal structure Y in $T(\bar{S})$. If we are examining sequences of projective structures we will add indices and decorations to Σ. These will be promoted to all the corresponding objects.

We now state three conjectures about these spaces. Note that all of these conjectures have versions that apply to more general families of Kleinian groups.

The first two conjecture are from Bers' seminal paper [Ber70] which began the study of the the space $U(X)$.

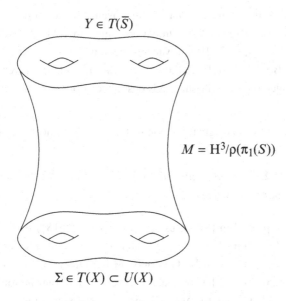

$$Y \in T(\overline{S})$$

$$M = \mathrm{H}^3/\rho(\pi_1(S))$$

$$\Sigma \in T(X) \subset U(X)$$

Figure 1: A quasifuchian manifold

The following conjecture is usually called the *Bers' density conjecture*:

Conjecture 6.2 ([Ber70]). $U(X) = \overline{T(X)}$.

A projective structure in $\partial T(X) = \overline{T(X)} \backslash T(X)$ is a *cusp* if the image of the holonomy representation contains cusps.

Conjecture 6.3 ([Ber70]). Cusps are dense in the boundary of $T(X)$.

The final conjecture we will state is Thurston's ending lamination conjecture. To do so we need to define an ending lamination. We will put off doing this till later and at this point simply state that to each $\Sigma \in U(X)$ we can define an *end invariant* which is determined by the hyperbolic manifold M.

Conjecture 6.4. An element of $U(X)$ is uniquely determined by its end-invariant.

We note that all three of these conjecture are now known to be true. In fact the ending lamination conjecture implies the previous two conjectures. Our purpose here is to describe how the deformation theory developed in this paper can be used to approach these conjectures. At present this approach still has significant gaps (at least for the first and third conjecture) but if completed it would provide new proofs of all three conjectures.

6.1. The Bers' density conjecture

In its most general form, the density conjecture states that every finitely generated Kleinian group is a limit of geometrically finite Kleinian groups. Very recently this complete version of the conjecture has been proven. To do so one needs to combine a number of results: the ending lamination conjecture ([Min03, BCM04]), tameness ([Bon86], [Ago04], [CG04]) and various theorems on limits of Kleinian groups ([Thu87, Ohs90, Bro00, KS02]). Although we will only address a very special case of the density conjecture here, the methods described apply in greater generality (see [BB04]).

Conjecture 6.2 was the original version of the density conjecture. In [Bro02] we proved the following result:

Theorem 6.5. *Let Σ be a projective structure in $U(X)$ such that the image of the holonomy has no parabolics. Then $\Sigma \in \overline{T(X)}$.*

The theorem is proved in two cases. Let $M = \mathbb{H}^3/\rho(\pi_1(S))$ be the quotient hyperbolic 3-manifold. Then M has *bounded geometry* if there is a lower bound on the length of any closed geodesic in M. Otherwise M has *unbounded geometry*. Minsky proved Theorem 6.5 when M has bounded geometry. Our contribution was the case when M has unbounded geometry. We will give a brief sketch of the proof, emphasizing those parts that use the deformation theory we have described in this paper.

The starting point is the following tameness theorem of Bonahon:

Theorem 6.6 ([Bon86]). *The manifold M is homeomorphic to $S \times (0,1)$.*

A simple closed curve γ in $S \times (0,1)$ is *unknotted* if it is isotopic to a simple closed curve on $S \times \{1/2\}$.

Theorem 6.7 ([Bro02]). *Let γ be an unknotted, simple closed geodesic in M and assume that the product structure is chosen such that γ lies on $S \times \{1/2\}$. Then there is a geometrically finite hyperbolic cone-manifold M_γ with the following properties:*

 (i) *The singular locus has a single component with cone angle 4π.*

 (ii) *The length of the singular locus in M_γ is equal to the length of γ in M. The tube radius of the singular locus in M_γ is greater than or equal to the tube radius of γ in M.*

 (iii) *M and M_γ are isometric on $S \times (0,1/2)$.*

The construction of M_γ is similar to the construction of grafting of complex projective structures. Although we will not go through it here, it is not difficult. The proof that M_γ is geometrically finite is more involved. For a proof in the above case see [Bro02]. A proof in a more general setting can be found in [BB04]. An expository account can be found in [BB03].

To apply Theorem 6.7 we use the following theorem of Otal:

Theorem 6.8 ([Ota95, Ota03]). *There exists an $\varepsilon_{unknot} > 0$ depending only on the genus of S such that if γ is a closed geodesic in M of length less than ε then γ is unknotted.*

Now assume that M has unbounded geometry. Then there exists a sequence of closed geodesics γ_i whose lengths limits to zero. In particular we can assume that $L_{\gamma_i}(M) \leq \min\{\varepsilon_{unknot}, \ell\}$ for all γ_i where ℓ is the constant in Theorem 4.3. Then for each γ_i, Theorem 6.7 gives us a cone-manifold M_i with cone-angle 4π. Furthermore, by (3) of Theorem 6.7 the component of the projective boundary on $S \times \{0\}$ of M_i will be the original projective structure Σ.

Next we apply Theorem 4.3 to decrease the cone-angle to 2π obtaining a quasi-fuschsian manifold M_i'. Let Σ_i be the $S \times \{0\}$ component of the projective boundary of M_i'. By Theorem 4.2 there exists a C such that $d(\Sigma, \Sigma_i) \leq CL_{\gamma_i}(M_i) = CL_{\gamma_i}(M)$ and therefore $\Sigma_i \to \Sigma$, as desired. This completes the sketch of the proof of Theorem 6.5 in the case of unbounded geometry.

What if M has bounded geometry? As we have already mentioned Minsky proved Theorem 6.5 in this case. He did so by proving the ending lamination conjecture for manifolds with bounded geometry. One might hope to find a direct approach to Conjecture 6.2 based on the methods outlined here.

There are two problems. First, there may not be a sequence of unknotted geodesics. Second, even if we are fortunate enough to have a sequence of unknotted geodesics the singular locus in the corresponding cone-manifolds will not be short and we won't be able to apply Theorem 4.3.

We can circumvent the first problem by lifting to a cover. The second problem is more serious. For a cone-manifold whose singular locus is not short to guarantee that the manifold can be deformed to cone angle zero we need to assume the the singular locus has a large tubular neighborhood. In particular we have the following theorem whose proof is beyond the scope of this paper:

Theorem 6.9. *Given any $\alpha, L > 0$ there exists an $R > 0$ such that the following holds. Let M_α be a geometrically finite hyperbolic cone-manifold with cone angle α and $L_C(\alpha) \leq L$ and assume that the singular locus has a tube radius $\geq R$. Then the one-parameter family M_t exists for all $t \in [0, \alpha]$.*

The next theorem allows us to circumvent both above problems by lifting to a cover. It is direct corollary of Theorems 2.1 and 4.3 of [FG01].

Theorem 6.10. *Let γ be a closed geodesic in a hyperbolic 3-manifold M with M homeophic $S \times (0,1)$. Given $R > 0$, M has a finite cover $\hat{M}$ for which γ has a homeomorphic lift $\hat{\gamma}$ that is unknotted and has a tubular neighborhood of radius $> R$.*

The tradeoff is that we are now working in a cover instead of with the original manifold. Because of this we can only prove that the projective structure lies in the boundary of the universal Teichmüller space. We now define this space. Let $P(1)$ be the space of bounded, holomorphic quadratic differentials on the unit disk Δ in $\widehat{\mathbb{C}}$. Given $\Phi \in P(1)$ there exists a locally conformal map $f : \Delta \longrightarrow \widehat{\mathbb{C}}$ with $Sf = \Phi$. This f is unique up to post-composition by elements of $PSL_2\mathbb{C}$. Let $U(1) \subset P(1)$ be those quadratic differentials where f is injective (or univalent in the language of complex analysis) and let $T(1) \subset U(1)$ be those quadratic differentials where f extends to a quasi-conformal homeomorphism of all of $\widehat{\mathbb{C}}$. The space $T(1)$ is usually called the *universal Teichmüller space* because all Teichmüller spaces $T(X)$ embed in $T(1)$. That is if Σ is a projective structure in $P(X)$ the the universal cover, $\tilde{\Sigma}$, is a projective structure in $P(1)$ and the map taking Σ to $\tilde{\Sigma}$ is an isometry.

Bers[2] made the following conjecture:

Conjecture 6.11. $U(1) = \overline{T(1)}$.

This conjecture is known to be false. Counterexamples where found by Gehring ([Geh78]) and later Thurston ([Thu86]). However, we can prove the following theorem

Theorem 6.12. *Let $\Sigma \in U(X)$ be a projective structure whose holonomy does not have parabolics. Then $\tilde{\Sigma} \in \overline{T(1)}$.*

Proof. In the course of proving Theorem 6.6, Bonahon shows that M has a sequence of closed geodesics γ_i with bounded length and $d(\gamma_i, \Sigma) \to \infty$. (Recall that $d(\gamma_i, \Sigma)$ is the distance from the component of the convex core boundary facing Σ to γ_i.) Now, for each γ_i, apply Theorem 6.10 to obtain a cover to which we can apply both Theorem 6.7 and Theorem 6.9. That is, in the cover, γ_i lifts to an unknotted geodesic $\hat{\gamma}_i$ along which we can graft to obtain a geometrically finite cone-manifold $\hat{M}_i$. The tube radius of singular locus will be sufficiently large so that we can decrease the cone angle to 2π and obtain a quasifuchsian manifold $\hat{M}_i'$.

Let $\hat{\Sigma}_i$ be the corresponding cover of the projective structure Σ. Then $\hat{\Sigma}_i$ is a component of the projective boundary of the cone-manifold $\hat{M}_i$. After the cone-manifold deformation this projective structure deforms to a projective structure $\hat{\Sigma}_i'$. By Theorem 5.2 we have

$$d(\hat{\Sigma}_i, \hat{\Sigma}_i') \le e^{-kd(\hat{\Sigma}_i, \hat{\gamma}_i)} L_{\gamma_i}(\hat{M}_i). \tag{6.1}$$

Now $d(\hat{\Sigma}_i, \hat{\gamma}_i) = d(\Sigma, \gamma)$ which limits to zero and $L_{\hat{\gamma}_i}(\hat{M}_i) = L_{\gamma_i}(M)$ is bounded so the left hand side of (6.1) limits to zero. Therefore $\tilde{\Sigma}_i \to \tilde{\Sigma} = \tilde{\Sigma}_i$ in $U(1)$ as desired. $\qquad\square$

---

[2]Conjectures 6.2 and 6.11 are labelled Conjectures II and I in [Ber70]. After stating Conjecture II Bers remarks "This would, of course, be a consequence of Conjecture I". This is not obvious to this author.

6.2. Cusps are dense

The conjugacy classes of parabolics in $\rho(\pi_1(S))$ correspond to disjoint simple closed curves on S. In particular there are at most $3g - 3$ conjugacy classes. A cusp whose holonomy has this maximal number of conjugacy classes of parabolics is called a *maximal cusp*. McMullen proved the following strong version of Conjecture 6.3.

Theorem 6.13 ([McM91]). *Maximal cusps are dense on the boundary of $T(X)$.*

Proof. Our proof will follow McMullen's except that we will replace his key estimate with Theorem 4.2. The part of the argument that we copy can be found on p. 221 of [McM91].

The first step is to note that projective structures whose holonomy does not have parabolics are dense in $\partial T(X)$. Let Σ be such a projective structure. To prove the theorem we need to show that Σ is approximated by maximal cusps. To do this McMullen finds projective structures $\Sigma_i \in T(X)$ that limit to Σ with the following property: The projective structures Σ_i correspond to conformal structures $Y_i \in T(\bar{S})$. For each Y_i there is a pants decompositions P_i such that $L_{P_i}(Y_i) \to 0$ where the length is measured in the unique hyperbolic metric on Y_i.

On Y_i we may assume that $L_{P_i}(Y_i) \leq \frac{1}{2}\min(\varepsilon_{unknot}, \ell)$ where ℓ is the constant in Theorem 4.3. By Bers' inequality ([Ber70]) this implies that $L_{P_i}(M_i) \leq \min(\varepsilon_{unknot}, \ell)$ where M_i is the quasifuchsian hyperbolic manifold determined by Σ_i. Now view M_i as a cone-manifold with singular locus P_i and cone angle 2π. Since $L_{P_i}(M_i) \leq \ell$ we can decrease the cone angle of M_i to zero to obtain a manifold M_i' with rank two cusps. The projective structure Σ_i is a component of the projective boundary of M_i and it deforms to a projective structure Σ_i'. By Theorem 4.2

$$d(\Sigma_i, \Sigma_i') \leq KL_{P_i}(M_i)$$

and therefore

$$\lim_{i\to\infty}\Sigma_i' = \lim_{i\to\infty}\Sigma_i = \Sigma.$$

What remains to show is that the Σ_i' are maximal cusps. Let $\hat{M}_i$ be the cover of M_i' corresponding to the boundary component Σ_i'. Since $L_{P_i}(M_i) \leq \varepsilon_{unknot}$, the geodesic representative of P_i is unknotted in M_i and therefore M_i' is homeomorphic to $S \times (0, 1)$ with the curves P_i removed from the halfway surface $S \times \{1/2\}$. Therefore $\hat{M}_i$ is homeomorphic to $S \times (0, 1)$ and every curve in P_i will be parabolic. This implies that Σ_i' is a maximal cusp. $\square$

There are versions of the density of cusps for more general quasiconformal deformation spaces in [CCHS03] and [CH04]. Both of these papers are generalizations of McMullen's methods. We note that our methods can also be used to prove these generalizations. See §8 of [Bro04a].

6.3. The ending lamination conjecture

The ending lamination conjecture is a classification of Kleinian groups isomorphic to a fixed group. The complete conjecture has recently be proven by Brock, Canary and Minsky ([Min03, BCM04]), completing a program of Minsky. In this section we will discuss an alternate approach to the conjecture. The approach is motivated by a theorem of R. Evans, which we will mention below. We also note, that this approach, if successful, uses some of Minsky's results in a key way and is heavily influenced by his ideas. The main difference is that we do not use the "model manifold".

The classifying objects are *end-invariants* which are objects associated to the surfaces that compactify the higher genus ends of the hyperbolic manifold. For groups without parabolics these invariants are either a conformal structure or a filling lamination on the surface compactifying the end. With parabolics the situation is more complicated. As usual, we will restrict to groups without parabolics.

For $\Sigma \in U(X)$ the corresponding manifolds M has two ends both compactified by S. On $S \times \{0\}$ the end-invariant is always the conformal structure X. If $\Sigma \in T(X)$ then the end-invariant for $S \times \{1\}$ will also be a conformal structure. In this case the conformal structure will be the image of Σ in $T(\bar{S})$ under the Bers' isomorphism. For $\Sigma \in U(X) \backslash T(X)$ the end-invariant is a lamination. To define it we recall that there is a sequence of closed geodesics, γ_i, whose length is bounded and such that $d(\gamma_i, \Sigma) \to 0$. Furthermore, Bonahon ([Bon86]) shows that these geodesics can be chosen to be homotopic to simple closed curves on $S \times \{1/2\}$. As simple closed curves on S, the γ_i will limit to a lamination λ. Most importantly this *ending lamination* will not depend on the initial choice of geodesics. This is also a result of Bonahon ([Bon86]).

The following theorem of Minsky shows the importance of the ending lamination. It is a combination of his proof of the ending lamination conjecture for bounded geometry manifolds and one of the first steps in the proof of the general conjecture.

Theorem 6.14 ([Min01]). *Let Σ and Σ' be projective structures in $U(X)$ and assume that the corresponding hyperbolic manifolds M and M' have the same end-invariant. Then either:*

(i) *M and M' are isometric.*

(ii) *M and M' both have unbounded geometry and there exists a sequence of simple closed curves γ_i so that both $L_{\gamma_i}(M)$ and $L_{\gamma_i}(M')$ limit to zero.*

We empasize that (1) and (2) are not mutually exclusive. In fact the goal is to show that (1) always holds. This is exactly the ending lamination conjecture.

For γ a simple closed curve on S, let $U(X,\gamma) \subset U(X)$ be those projective structures in $U(X)$ where the conjugacy class of γ is parabolic under the holonomy representation. Let $d_C(\gamma,X)$ be the distance between γ and X in the *curve complex*. That is $d_C(\gamma,X)$ is the minimum number k such that there exist $k+1$ essential simple closed curves, $\beta_0, \dots, \beta_k$ on S with $\beta_0 = \gamma$, β_i and β_{i+1} disjoint, and β_k a bounded length curve in the hyperbolic metric on X.

While we believe the following conjecture is interesting in its own right, as we will see below it also implies the ending lamination conjecture for manifolds in $U(X)$ with unbounded geometry.

Conjecture 6.15. There exists a constants C_1 and C_2, depending only on the genus of S, such that the diameter of $U(X,\gamma)$ in $P(X)$ is bounded by $C_1 e^{-C_2 d_C(X,\gamma)}$.

Our motivation for this conjecture is as follows. The distance $d_C(\gamma,X)$ gives a lower bound on the thickness of the convex core of every manifold in $U(X,\gamma)$. One then wants to combine this with geometric inflexibility to obtain the desired bound.

We now show how Theorem 6.14 and Conjecture 6.15 together imply the ending lamination conjecture for projective structures in $U(X)$ with unbounded geometry and no cusps.

Let Σ and Σ' be as in Theorem 6.14 and assume that M and M' have unbounded geometry. Let γ_i be the sequence given by (2) in Theorem 6.14. By Theorem 6.5 there exists a sequence Σ_i in $T(X)$ converging to Σ. Let M_i be the associated hyperbolic 3-manifolds. After passing to a subsequence we can assume that $L_{\gamma_i}(M_i) \to 0$. Now repeating the construction in the proof of Theorem 6.13, for each Σ_i we can find a cusp $\hat{\Sigma}_i$ in $U(X,\gamma_i)$ such that

$$d(\Sigma_i, \hat{\Sigma}_i) \leq CL_{\gamma_i}(M_i).$$

Therefore the sequence $\hat{\Sigma}_i$ converges to Σ. We similarly find a sequence $\hat{\Sigma}'_i$ converging to Σ' with each $\hat{\Sigma}'_i$ in $U(X,\gamma_i)$. Finally we note that the γ_i converge to the ending lamination so $d_C(\gamma_i,X)$ limits to infinity. Conjecture 6.15 then implies that both sequences $\hat{\Sigma}_i$ and $\hat{\Sigma}'_i$ have the same limit so $\Sigma = \Sigma'$.

Note that, in the above argument, if we replace the curves γ_i with pants decompositions P_i such that

$$\lim_{i \to \infty} L_{P_i}(M) = \lim_{i \to \infty} L_{P_i}(M') = 0$$

then $\hat{\Sigma}_i$ and $\hat{\Sigma}'_i$ will be maximal cusps. Since maximal cusps are uniquely determined by the pants decompositon $\hat{\Sigma}_i = \hat{\Sigma}'_i$ and the corresponding limits, Σ and Σ', are equal without appealing to Conjecture 6.15. This argument, due to Evans, leads to the following theorem that we mentioned at the begining of this section.

Theorem 6.16 ([Eva03]). *Let Σ and Σ' be projective structures in $U(X)$ with corresponding hyperbolic manifolds M and M'. Assume that there exist a sequence of pants decompositions P_i with*

$$\lim_{i\to\infty} L_{P_i}(M) = \lim_{i\to\infty} L_{P_i}(M') = 0.$$

Then $\Sigma = \Sigma'$.

We remark that having such a shrinking pants decomposition is not as restrictive as it may seem. In fact, the density of maximal cusps (Theorem 6.13) implies that there is a dense G_δ of such manifolds in $\partial T(X)$. Furthermore, if M has such a sequence of pants decompositions and M' has the same ending lamination as M then the lengths of the same sequence of pants will limit to zero in M'. This last statement is proven in [Min03] and is a large part of the proof of the ending lamination conjecture. Namely, in [Min03], Minsky constructs a model for M that is completely determined by combinatorial information coming from the ending lamination. He then shows that there is a Lipschitz map from this model to the hyperbolic manifold M. Furthermore, every sufficiently short curve in M will also be short in the model. Therefore, if M' has the same ending lamination as M then it will have the same model and the same short geodesics. The final step in Brock, Canary and Minsky's proof of the ending lamination conjecture is to show that this model is also bi-Lipschitz. This is done in [BCM04]. On the other hand, Theorem 6.16 completely avoids the work in [BCM04] which seems significant.

References

[Ago04]   I. Agol (2004). Tameness of hyperbolic 3-manifolds. Preprint, `math.GT/0405568v1`.

[BB]   J. Brock & K. Bromberg. Geometric inflexibility of hyperbolic 3-manifolds. In preparation.

[BB03]   J. F. Brock & K. W. Bromberg (2003). Cone-manifolds and the density conjecture. In *Kleinian Groups and Hyperbolic 3-Manifolds (Warwick, 2001)*, *London Math. Soc. Lecture Note Ser.*, volume 299, pp. 75–93. Cambridge Univ. Press, Cambridge.

[BB04]   J. F. Brock & K. W. Bromberg (2004). On the density of geometrically finite Kleinian groups. *Acta Math.* **192** (1), 33–93.

[BCM04]   J. Brock, R. Canary & Y. Minsky (2004). The classification of Kleinian surfaces groups II: the ending lamination conjecture. Preprint, `math.GT/0412006`.

[Ber60]   L. Bers (1960). Simultaneous uniformization. *Bull. Amer. Math. Soc.* **66**, 94–97.

[Ber70]   L. Bers (1970). On boundaries of Teichmüller spaces and on Kleinian groups. I. *Ann. of Math. (2)* **91**, 570–600.

[Bon86]   F. Bonahon (1986). Bouts des variétés hyperboliques de dimension 3. *Ann. of Math. (2)* **124** (1), 71–158.

[Bro00]   J. F. Brock (2000). Continuity of Thurston's length function. *Geom. Funct. Anal.* **10** (4), 741–797.

[Bro02]   K. Bromberg (2002). Projective structures with degenerate holonomy and the Bers density conjecture. Preprint, math.GT/0211402.

[Bro04a] K. Bromberg (2004). Hyperbolic cone-manifolds, short geodesics, and Schwarzian derivatives. *J. Amer. Math. Soc.* **17** (4), 783–826 (electronic).

[Bro04b] K. Bromberg (2004). Rigidity of geometrically finite hyperbolic cone-manifolds. *Geom. Dedicata* **105**, 143–170.

[CCHS03] R. D. Canary *et al.* (2003). Approximation by maximal cusps in boundaries of deformation spaces of Kleinian groups. *J. Differential Geom.* **64** (1), 57–109.

[CG04]   D. Calegari & D. Gabai (2004). Shrinkwrapping and the taming of hyperbolic 3-manifolds. Preprint, math.GT/0407161.

[CH04]   R. D. Canary & S. Hersonsky (2004). Ubiquity of geometric finiteness in boundaries of deformation spaces of hyperbolic 3-manifolds. *Amer. J. Math.* **126** (6), 1193–1220.

[Ear81]   C. J. Earle (1981). On variation of projective structures. In *Riemann Surfaces and Related Topics: Proceedings of the 1978 Stony Brook Conference (State Univ. New York, Stony Brook, N.Y., 1978), Ann. of Math. Stud.*, volume 97, pp. 87–99. Princeton Univ. Press, Princeton, N.J.

[Eva03]   R. Evans (2003). The ending lamination conjecture for super-slender hyperbolic 3-manifolds. Preprint.

[FG01]   L. Funar & S. Gadgil (2001). Topological geodesics and virtual rigidity. *Algebr. Geom. Topol.* **1**, 369–380 (electronic).

[Geh78]   F. W. Gehring (1978). Spirals and the universal Teichmüller space. *Acta Math.* **141** (1-2), 99–113.

[Hej75]   D. A. Hejhal (1975). Monodromy groups and linearly polymorphic functions. *Acta Math.* **135** (1), 1–55.

[HK]   C. Hodgson & S. Kerckhoff. The shape of hyperbolic Dehn surgery space. In preparation.

[HK98]   C. D. Hodgson & S. P. Kerckhoff (1998). Rigidity of hyperbolic cone-manifolds and hyperbolic Dehn surgery. *J. Differential Geom.* **48** (1), 1–59.

[HK02]   C. Hodgson & S. Kerckhoff (2002). Universal bounds for hyperbolic Dehn surgery. Preprint, math.GT/0204345.

[HK03]   C. D. Hodgson & S. P. Kerckhoff (2003). Harmonic deformations of hyperbolic 3-manifolds. In *Kleinian Groups and Hyperbolic 3-Manifolds (Warwick, 2001)*, London Math. Soc. Lecture Note Ser., volume 299, pp. 41–73. Cambridge Univ. Press, Cambridge.

[Hub81]   J. H. Hubbard (1981). The monodromy of projective structures. In *Riemann Surfaces and Related Topics: Proceedings of the 1978 Stony Brook Conference (State Univ. New York, Stony Brook, N.Y., 1978), Ann. of Math. Stud.*, volume 97, pp. 257–275. Princeton Univ. Press, Princeton, N.J.

[KS02]   G. Kleineidam & J. Souto (2002). Algebraic convergence of function groups. *Comment. Math. Helv.* **77** (2), 244–269.

[McM91]   C. McMullen (1991). Cusps are dense. *Ann. of Math. (2)* **133** (1), 217–247.

[McM96]   C. T. McMullen (1996). *Renormalization and 3-manifolds which Fiber over the Circle, Annals of Mathematics Studies*, volume 142. Princeton University Press, Princeton, NJ.

[Min01]   Y. N. Minsky (2001). Bounded geometry for Kleinian groups. *Invent. Math.* **146** (1), 143–192.

[Min03]   Y. Minsky (2003). The classification of Kleinian surface groups I: models and bounds. Preprint, math.GT/0302208.

[Ohs90]   K. Ohshika (1990). Ending laminations and boundaries for deformation spaces of Kleinian groups. *J. London Math. Soc. (2)* **42** (1), 111–121.

[Ota95]   J.-P. Otal (1995). Sur le nouage des géodésiques dans les variétés hyperboliques. *C. R. Acad. Sci. Paris Sér. I Math.* **320** (7), 847–852.

[Ota03]   J.-P. Otal (2003). Les géodésiques fermées d'une variété hyperbolique en tant que nœuds. In *Kleinian Groups and Hyperbolic 3-Manifolds (Warwick, 2001)*, *London Math. Soc. Lecture Note Ser.*, volume 299, pp. 95–104. Cambridge Univ. Press, Cambridge.

[Thu80]   W. P. Thurston (1980). The Geometry and Topology of 3-Manifolds. Lecture notes, Princeton University. Available at www.msri.org/publications/books/gt3m/.

[Thu86]   W. P. Thurston (1986). Zippers and univalent functions. In *The Bieberbach Conjecture (West Lafayette, Ind., 1985)*, *Math. Surveys Monogr.*, volume 21, pp. 185–197. Amer. Math. Soc., Providence, RI.

[Thu87]   W. P. Thurston (1987). Hyperbolic structures on 3-manifolds, II: Surface groups and 3-manifolds which fiber over the circle. Preprint; electronic version 1998, math.GT/9801045.

K. Bromberg

Department of Mathematics
University of Utah
155 S 1400 E, JWB 233
Salt Lake City, UT 84112
USA

bromberg@math.utah.edu

AMS Classification: 30F40, 37F30, 37F15

Keywords: Kleinian group, hyperbolic 3-manifold, cone-manifold

Spaces of Kleinian Groups
Lond. Math. Soc. Lec. Notes **329**, 29–48

Cambridge University Press
Y. Minsky, M. Sakuma & C. Series (Eds.)

© K. Ohshika & H. Miyachi, 2005

On topologically tame Kleinian groups with bounded geometry

Ken'ichi Ohshika and Hideki Miyachi

Abstract

We consider a generalisation of Minsky's rigidity theorem for freely inde-composable Kleinian groups with bounded geometry to topologically tame freely decomposable case in order to apply it to the following two results. The first is the uniqueness property for the problem of realising given end invariants by a group lying on the boundary of the quasi-conformal deformation space of a convex cocompact group. The second is a generalisation of Soma's result on the third bounded cohomology groups of closed surface groups to the case of free groups.

1. Introduction

A topologically tame hyperbolic 3-manifold has three pieces of information: the homeomorphism type, the conformal structures at infinity for geometrically finite ends (of non-cuspidal part), and the ending laminations for geometrically infinite ends. The ending lamination conjecture, due to Thurston, says that these pieces of information uniquely determine the isometry type of the manifold. Recently, Minsky proved this conjecture affirmatively, partially collaborating with Brock and Canary. Although the result is in the process of publication, the special case when manifolds have freely indecomposable fundamental group and bounded geometry, i.e., when the injectivity radii are bounded below by a positive constant, has been already published in [Min94], [Min00] and [Min01]. In the present paper, we shall explain how the argument of Minsky there can be generalised to the case of topologically tame manifolds with bounded geometry possibly with freely decomposable fundamental groups, and then show the following two kinds of its applications. It should be noted that the recent affirmative solution of Marden's tameness conjecture by Agol, Calegari–Gabai and Choi implies that the assumption of the topological tameness in our results can be removed.

The first result (Theorem 1.1) is about the uniqueness of hyperbolic 3-manifolds having given end invariants provided that the minimal arational geodesic laminations constituting the given end invariants satisfy the bounded geometry condition. The statement is as follows.

Theorem 1.1. *Let M be a compact hyperbolisable 3-manifold such that ∂M is non-empty and no component of ∂M is a torus. Denote by $S_1, \ldots, S_n$ the components of ∂M. On each S_i, let λ_i be either a marked hyperbolic structure or a minimal arational geodesic lamination with the bounded geometry condition (3.1) in Theorem 3.1, which supports a projective lamination contained in the Masur domain. We suppose furthermore that if M is a product I-bundle over a closed orientable surface, then λ_1 and λ_2 are not geodesic laminations homotopic in M, and that if M is a twisted I-bundle over a closed non-orientable surface, then λ_1 is not a lift of a geodesic lamination on the base surface to the boundary of M regarded as the double covering of the base surface. Then, a Kleinian group Γ satisfying the following conditions is unique up to conjugation.*

(1) $N_\Gamma = \mathbb{H}^3 / \Gamma$ *is homeomorphic to the interior of M and*

(2) *the end of N_Γ corresponding to S_i is either geometrically finite and the Teich-müller parameter at infinity is λ_i when λ_i is a marked hyperbolic structure, or geometrically infinite with its ending lamination equal to λ_i when λ_i is a geodesic lamination.*

The existence of such a Kleinian group Γ will be dealt with in Ohshika [Ohsa] and [Ohsc]. Our assumptions on given laminations are necessary to show this existence part.

Theorem 1.1, combined with some argument due to Namazi-Souto [NS] and Ohshika [Ohsc], implies a useful corollary as follows if we consider iterations of hyperbolic transformations of Teichmüller spaces. Let M be a compact hyperbolisable 3-manifold with non-empty ∂M containing no tori. The Teichmüller space $T(\partial M)$ of ∂M covers the quasiconformal deformation space $QH(M)$ of a convex cocompact hyperbolic structure of M. Let $AH(M)$ be the space of faithful and discrete representations of $\pi_1(M)$ to $PSL_2(\mathbb{C})$ modulo conjugacy. It is known that $QH(M)$ coincides with a component of the interior of $AH(M)$ (cf. Marden [Mar74] and Sullivan [Sul85]). The following will be proved in §5. Here a modular transformation is said to be *hyperbolic* when its restriction to each component is either the identity or pseudo-Anosov. Such a transformation is called *Masur* when each constituent pseudo-Anosov map has stable lamination in the Masur domain. When M is an I-bundle over a closed surface, we further assume that the stable laminations of the two pseudo-Anosov automorphisms of the coordinates are not homotopic if the bundle is trivial, and that the pseudo-Anosov map is not a lift of that of the base surface if the bundle is twisted.

Theorem 1.2. *Let M be as above. Let ω be a Masur hyperbolic modular transformation on the Teichmüller space $T(\partial M)$ of ∂M. Let X be a point in $T(\partial M)$ and $[\rho_n] \in QH(M)$ a convex cocompact representation uniformising $\omega^n(X)$. Then $\{[\rho_n]\}$*

converges to a unique $[\rho_\infty] \in \mathrm{AH}(M)$. *Furthermore, for two Masur hyperbolic modular transformations* ω_1, ω_2, *their limits as above coincide if and only if* ω_1 *is homotopic to* ω_2 *in M.*

The second application is about the third bounded cohomology groups of free groups. Soma showed in [Som97] that if we fix a positive number δ_0, then third bounded cohomology classes of a closed surface induced from the fundamental cohomology class of doubly degenerate hyperbolic 3-manifolds whose injectivity radii are bounded below by δ_0 cannot accumulate in the third bounded cohomology group. This implies as a corollary the third bounded cohomology group of a closed surface has dimension of cardinality of continuum. We shall show in this paper the same holds for free groups using topologically tame groups which are limits of Schottky groups constructed by the way of Theorem 1.2.

Acknowledgements. The authors would like to express their hearty gratitude to the organisers of the conference "Spaces of Kleinian Groups" for inviting us to this marvellous conference, to the Newton institute of Mathematical Sciences for its hospitality, and to Yair Minsky and the referee for useful comments.

2. Notation

2.1. Kleinian groups

A *Kleinian group* is a discrete subgroup of the group $\mathrm{PSL}_2(\mathbb{C}) = \mathrm{Isom}^+(\mathbb{H}^3)$ of orientation preserving isometries on the hyperbolic space $\mathbb{H}^3$. We always assume any Kleinian group in this paper to be *finitely generated* and *purely loxodromic* for simplicity. A Kleinian group Γ is said to be *topologically tame* if the quotient manifold $N_\Gamma := \mathbb{H}^3/\Gamma$ is almost compact, i.e., homeomorphic to the interior of a compact manifold. We say that a Kleinian group Γ or a manifold N_Γ has *bounded geometry* if the translation lengths of the elements in Γ are bounded below by a positive constant depending only on Γ. If Γ has bounded geometry, the injectivity radius at any point of N_Γ is bounded below by a constant depending only on the manifold.

2.2. Ends and End invariants

Let Γ be a Kleinian group. An *end* e of N_Γ is a collection $\{U_i\}_i$ of open subsets of N_Γ with the following three properties: (1) ∂U_i is compact but $\overline{U_i}$ is not, (2) for any i, j, some k satisfies $U_k \subset U_i \cap U_j$, and (3) $\{U_i\}_i$ is maximal under the conditions (1) and (2). Each element in e is called a *neighbourhood* of e. An end of N_Γ is said to be *geometrically finite* if it admits a neighbourhood which is disjoint from the convex core of N_Γ. Otherwise, it is called *geometrically infinite*. When Γ is topologically tame, the quotient manifold N_Γ contains a compact core C such that N_Γ is homeomorphic to the interior of C and the closure of the complement of C is homeomorphic to $\partial C \times [0, \infty)$. Furthermore, each component of $N_\Gamma - C$ is a neighbourhood of an end of N_Γ.

We define the *end invariant* $\nu(e)$ for each end e of N_Γ as follows. Recall that the quasiconformal deformation space of Γ is biholomorphically equivalent to the product $\prod_e T(S_e)/\text{Mod}_e$, where the product is taken over all geometrically finite ends of N_Γ, S_e is the surface at infinity corresponding to e, and Mod_e is a subgroup of the Teichmüller modular group acting on $T(S_e)$ which is determined by the topological structure of M (cf. Kra [Kra72] and Maskit [Mas71]).

When an end e is geometrically finite, we define the end invariant $\nu(e)$ of e to be the coordinate in the product space associated with the end e. Suppose that e is geometrically infinite. Fix a hyperbolic structure on the component S_e of ∂C facing e. A geodesic lamination λ is called the *ending lamination* for e if there exists a sequence $\{\gamma_i\}_i$ of simple closed curves on S_e such that their geodesic representatives γ_i^* in N_Γ tend to e and are homotopic to the γ_i in $N_\Gamma - C$, and their projective classes $[\gamma_i]$ converge to a projective measured lamination whose support is λ. As was shown in §10 of Canary [Can93], for any geometrically infinite end of topologically tame N_Γ, the ending lamination is uniquely defined. It should be also noted that an ending lamination is *minimal* and *arational*: that is, every leaf is dense and every complementary region is simply connected. We define the end invariant $\nu(e)$ of a geometrically infinite end e to be the ending lamination for e. For negatively curved manifolds, one can define ends and ending laminations in the same way as above.

3. Bounded Geometry Theorem

3.1. Bounded geometry condition

We recall the *bounded geometry condition* for geodesic laminations introduced by Minsky [Min00]. Let S be a compact orientable surface with negative Euler characteristic. Let $C(S)$ be the curve complex of S and $C_k(S)$ the k-skeleton of $C(S)$. Fix a hyperbolic structure on S of finite area and let λ denote a geodesic lamination on S. Let Y be an essential subsurface of S. Take a covering $\tilde{Y} \to S$ corresponding to $\pi_1(Y)$ and consider the preimage $\tilde{\lambda}$ on $\tilde{Y}$. If this preimage has leaves that are either non-peripheral closed curves or essential arcs that terminate in the boundary components of $\tilde{Y}$, these components determine a simplex of the complex of arcs $\mathcal{A}(\tilde{Y})$ (cf. Minsky [Min00] and [Min01]) and we define $\pi_Y(\lambda)$ to be its barycentre. If there are no such components, then we set $\pi_Y(\lambda) = \emptyset$. Let λ, ν be geodesic laminations on S

with non-empty $\pi_Y(\lambda)$ and $\pi_Y(\nu)$. We define their Y-distance by:

$$d_Y(\lambda,\nu) = d_{\mathcal{A}(\tilde{Y})}(\pi_Y(\lambda),\pi_Y(\nu)),$$

where $d_{\mathcal{A}(\tilde{Y})}$ is the canonical path metric on $\mathcal{A}(\tilde{Y})$.

Theorem 3.1 (Characterisation). *Let* Γ *be a purely loxodromic topologically tame Kleinian group and* C *a compact core such that* $N_\Gamma - C$ *is homeomorphic to* $\partial C \times [0,\infty)$. *Let* $e_1,\ldots,e_m$ *be the geometrically infinite ends of* N_Γ, *and* $S_1,\ldots,S_m$ *components of* ∂C *facing them. Let* λ_i *be the ending lamination for* e_i. *Then,* N_Γ *has bounded geometry if and only if for any pants decomposition* P_i *on* S_i *($i = 1,\ldots,m$), there is a constant* $K > 0$ *such that*

$$\sup_Y d_Y(P_i,\lambda_i) \le K, \tag{3.1}$$

where the supremum is taken over all essential subsurfaces for which $d_Y(P_i,\lambda_i)$ *is defined.*

Indeed, one can check that the characterisation above also holds for Kleinian groups with parabolic elements (if we modify the definition of Bounded geometry, see [Min01]). However, for simplicity, we only consider the case of Kleinian groups without parabolic elements.

This theorem is proved by the essentially same way as that of the case when Γ is freely indecomposable by Minsky (cf. [Min94] and [Min00]). For the sake of completeness, we shall give an outline of the proof for the case of topologically tame Kleinian groups.

Fix a topologically tame Kleinian group Γ. Set $N = N_\Gamma$ for simplicity. Then there exists a branched covering $pr : \tilde{N} \to N$ such that $\tilde{N}$ has a boundary-irreducible compact core $\tilde{C}$ and each end $\tilde{e}$ of $\tilde{N}$ has a neighbourhood $U_{\tilde{e}}$ which is mapped isometrically to a neighbourhood of an end of N by pr. Furthermore, each end of N has a neighbourhood which is isometrically lifted to a neighbourhood of an end of $\tilde{N}$ (see Canary [Can93]). Therefore, N has bounded geometry if and only if $\tilde{N}$ has.

We take $\tilde{C}$ such that $\tilde{N} - \tilde{C}$ is homeomorphic to $\partial\tilde{C} \times (0,\infty)$ and the restriction of pr to each component of $\tilde{N} - \tilde{C}$ is isometric. For an end e of N, we fix a component $U_{\tilde{e}}$ of $\tilde{N} - \tilde{C}$ such that $pr(U_{\tilde{e}})$ is a neighbourhood of e. Let $\tilde{e}$ be the end of $\tilde{N}$ which has $U_{\tilde{e}}$ as its neighbourhood and $S_{\tilde{e}}$ a component of $\partial\tilde{C}$ facing $\tilde{e}$.

Pleated surfaces Let σ be a hyperbolic metric on S and λ a geodesic lamination on (S,σ). A *pleated surface* $f : (S,\sigma) \to \tilde{N}$ with bending lamination λ is a π_1-injective continuous mapping such that both $f|_\lambda$ and $f_{S-\lambda}$ is locally isometric, and the path metric induced from $\tilde{N}$ coincides with σ. A geodesic lamination λ' is *realisable* if it is contained in the bending lamination of a pleated surface.

Analogously to the case of hyperbolic manifolds, we can confirm that the bounded diameter lemma (cf. [Bon86]), the uniform injectivity theorem (cf. Thurston [Thu86]), the short bridge arcs lemma (cf. [Min01]), and the efficiency of pleated surfaces (cf. [Min00]), the homotopy bound lemma (cf. Lemma 4.1 of [Min01]) hold in our case.

We shall briefly discuss these results and how to generalise them here. The bounded diameter lemma asserts that there is a uniform upper bound for the diameters of pleated surfaces modulo their thin parts, which depends only on the topological type of the surfaces. Since this derives from the Gauss-Bonnet formula, in our situation of negatively curved manifold with curvature ≤ -1, we get the same bound. The uniform injectivity theorem asserts that at any two points not too close each other in a thick part of any lamination realised by an incompressible pleated surface in any hyperbolic 3-manifold, the unit vectors tangent to the lamination at the two points are not mapped to vectors too close each other in the hyperbolic 3-manifold. This is proved by contradiction, supposing that there is a sequence of pairs of such tangent vectors mapped to vectors closer and closer, and using the Gromov limit of these pleated surfaces and the injectivity of the pleating locus of the limit. All of this machinery works also for our negatively curved manifolds. The short bridge arc lemma asserts that for any bridge arc on an incompressible pleated surface that either lies in the thick part or is primitive, if its endpoints lie on leaves that are nearly parallel in its image in the target hyperbolic 3-manifold, then the arc is short (independently of the pleated surface). Since this follows from the uniform injectivity, once we generalise the uniform injectivity, we can also generalise this lemma to our negatively curved manifolds. The efficiency of pleated surfaces asserts that on a doubly-incompressible pleated surface realising a finite-leaved maximal lamination whose compact leaves have lengths bounded away from 0, the difference between the length of any closed curve and that of the homotopic closed geodesic in the hyperbolic 3-manifold is bounded by the alternation number of the closed curve and the leaf. This is also derived from the uniform injectivity theorem; hence can be generalised to our setting. Finally, the homotopy bound lemma asserts that between two homotopic pleated surfaces realising the same system of simple closed curves, there is a good homotopy. This relies on the elementary hyperbolic geometry and some property of convex sets in the hyperbolic space (Lemma 8.6 in [Min01]). It is easy to see that convex sets in the universal cover of our negatively curved manifolds, where the sectional curvature is bounded above by -1, have the same property.

We can also show that any pleated surface takes only the ε_0-thin part of the surface into the ε_1-thin part of $\tilde{N}$ (cf. Lemma 3.1 of [Min00]). However, the constants appearing in all of them depend on $\tilde{N}$.

The following lemma is a corollary of the fact that any sequence of pleated surfaces realising multi-curves converging to an ending lamination of an end must tend to the same end. This fact can be proved by showing that there is a dichotomy for a measured lamination on a boundary component of a compact core: either it is realised by a pleated surface or any sequence of pleated surfaces realising multi-curves converging to the lamination tends to an end. (This is obtained from a generalisation of Bonahon's Proposition 5.1 in [Bon86] by Canary.)

Lemma 3.2. *Let S be a component of $\partial \tilde{C}$. Let $\tilde{e}$ and $\lambda_{\tilde{e}}$ be the end of $\tilde{N}$ facing S and the ending lamination of $\tilde{e}$, respectively. Then there is a neighbourhood of V of $\lambda_{\tilde{e}}$ in the geodesic laminations on S with the following properties.*

(1) *Any finite, leaved lamination in V is realisable. Furthermore, the image of any pleated surface realizing $\lambda \in V$ is contained in $U_{\tilde{e}}$.*

(2) *Let $P \to Q$ be an elementary move for two pants decompositions $P, Q \in V$. Then the internal lamination $\lambda_{P,Q}$ with respect to P and Q is realisable and the image of a pleated surface realising $\lambda_{P,Q}$ is contained in $U_{\tilde{e}}$.*

Next, we note the following.

Lemma 3.3. *There exist a positive number L_1 and a sequence $\{P_n\}_n$ of maximal curve systems on S such that*

(i) *the geodesic representative P_n^* exits the end $\tilde{e}$ as $n \to \infty$,*

(ii) *$0 < \ell_N(P_n^*) \le L_1$, and*

(iii) *the projective classes $[P_n]$ converge to a projective measured lamination whose support is $\lambda_{\tilde{e}}$.*

The constant L_1 depends only on $\tilde{N}$.

Proof. There exists a sequence of pleated surfaces $f_n : S \to \tilde{N}$ tending to the end $\tilde{e}$ which realise simple closed curves with projective classes converging to a projective measured lamination with support $\lambda_{\tilde{e}}$, by the definition of ending laminations. Let σ_n be the hyperbolic structure on S induced by f_n from $\tilde{N}$. Take a pants decomposition P_n on (S, σ_n) that has the smallest total length among all the pants decomposition. Then there is a positive constant L_1, i.e., the Bers constant, depending only on the topological type of S bounding the total lengths of the P_n above.

Let ε be a positive constant less than the Margulis constant for the hyperbolic 3-manifolds. Since $\tilde{N}$ has sectional curvature less than or equal to -1 everywhere, the ε-thin part of $\tilde{N}$ consists of Margulis tubes. Let P_n^* be the union of closed geodesics in $\tilde{N}$ freely homotopic to $f_n(P_n)$. By taking a subsequence, we can assume that either all the P_n^* have a component with length greater than or equal to ε or none of them have. First we consider the case when all the P_n^* have such a component. We denote by γ_n^* a component of P_n^* having length greater than or equal to ε. Let γ_n be the component of P_n corresponding to γ_n^*. Then the distance between $f_n(\gamma_n)$ and γ_n^* is bounded above by a constant depending only on ε since the length of $f_n(\gamma_n)$ is bounded above by L_1. Hence γ_n^* tends to the end $\tilde{e}$. Let $g_n : S \to \tilde{N}$ be a pleated surface realising P_n. Then the image of g_n contains γ_n^*. Since pleated surfaces cannot keep intersecting a compact set while having a part tending to an end, g_n also tends to the end $\tilde{e}$.

Next we consider the case when all the components of P_n^* have lengths less than ε. Let γ_n be a component of P_n, and γ_n^* the closed geodesic freely homotopic to $f_n(\gamma_n)$. There is an ε-Margulis tube V_n whose axis is γ_n^*. Since $f_n(\gamma_n)$ has length less than L_1, the distance between $f_n(\gamma_n)$ and V_n is bounded independently of n. If V_n does not tend to the end $\tilde{e}$, then either infinitely many tubes among the V_n are contained in a compact set or V_n tends to an end other than $\tilde{e}$ passing to a subsequence. The former cannot happen since a compact set can intersect only finitely many Margulis tubes. If the latter is the case, since the distance between V_n and γ_n^* tending to the end $\tilde{e}$ is bounded, V_n must intersect the core $\tilde{C}$. Again this is a contradiction. Thus V_n must tend to the end $\tilde{e}$ as $n \to \infty$. Therefore, as before, the pleated surface g_n realising P_n, which must contain the axis of V_n, also tends to $\tilde{e}$, and we have completed the proof. $\square$

Outline of proof of Characterisation   First, we prove "only if" part by reductio ad absurdum. Suppose that Γ has bounded geometry but there exist a component S of ∂C and a sequence $\{Y_n\}_n$ of essential subsurfaces of S such that

$$d_{Y_n}(P, \lambda_e) \to \infty \tag{3.2}$$

for some pants decomposition P and the end e of N facing S. Let $\tilde{e}$ be a lift of e. Since the projection $\tilde{N} \to N$ is isometric near ends, the divergence also holds for the end $\tilde{e}$ of $\tilde{N}$. Hence we may assume S is a component of $\tilde{C}$ (i.e., $S = S_{\tilde{e}}$ and $\lambda_{\tilde{e}} = \lambda_e$).

We may take such P with the property that all the components of P are realised as closed geodesics in U with length $\ell_{\tilde{N}}(P) \leq L_1$. Indeed, let P_1 and P_2 denote two pants decompositions on S. By applying Minsky's bounded geometry theorem for the quasi-Fuchsian group uniformising two surfaces on which the lengths of both P_1 and P_2 are bounded by L_1, we have for every incompressible subsurface Y on S that $d_Y(P_1, P_2)$ is bounded by a constant depending only on the associated quasi-Fuchsian manifold. This means that once some pants decomposition satisfies (3.2), any pants decomposition on S satisfies the same condition.

We here claim that projective classes of ∂Y_n ($n = 1, 2, \ldots$) converge to a projective measured lamination on S with support $\lambda_{\tilde{e}}$. Indeed, let $\rho : \pi_1(S) \to \mathrm{PSL}_2(\mathbb{C})$ denote a representation of a singly degenerate group such that the ending lamination of its geometrically infinite end is $\lambda_{\tilde{e}}$, and in the Teichmüller parameter of its geometrically finite end, P has length bounded by L_1. Since $\lambda_{\tilde{e}}$ is maximal, ρ admits no accidental parabolic transformations. Since $\pi_{Y_n}(P)$ is contained in $\pi_{Y_n}(C(\rho, L_1))$ and so is $\pi_{Y_n}(\lambda_{\tilde{e}})$ by Lemma 3.3, where $C(\rho, L_1) = \{\gamma \in C_0(S) \mid \ell_\rho(\gamma) \leq L_1\}$, by our assumption we have

$$\operatorname{diam}_{Y_n}(C(\rho, L_1)) = \operatorname{diam}_{\mathcal{A}(Y_n)}(\pi_{Y_n}(C(\rho, L_1))) \geq d_{Y_n}(P, \lambda_{\tilde{e}}) - 2 \to \infty.$$

Hence $\ell_\rho(\partial Y_n) \to 0$ as $n \to \infty$ by Theorem B of [Min00]. Therefore, the geodesic representatives of ∂Y_n must exit the geometrically infinite end of $\mathbb{H}^3/\rho(\pi_1(S))$. This implies that the projective classes of ∂Y_n converges to a projective measured lamination whose support is equal to $\lambda_{\tilde{e}}$.

Let us continue to prove "only if" part. We note that by the argument above, the geodesic representative of ∂Y_n exit the end $\tilde{e}$. By taking a subsequence of $\{Y_n\}_n$ if necessary, we may assume that either all or none of the Y_n are annuli. In either case, we shall be lead to a contradiction by an argument of Minsky in [Min00]. For the sake of simplicity, we only deal with the case where no Y_n is annulus.

The following lemmas can be proved in the same way as those in [Min00].

Lemma 3.4. *For $L > 0$ there is $D > 0$, depending only on L and the topological type of S, and $n_2 > 0$ for which the following statement holds:*

For any $n \geq n_2$ and any simple closed curve γ on S intersecting Y_n essentially with $\ell_N(\gamma) < L$, there exists a pleated surface $g_{\gamma,n} : S \to \tilde{N}$ with an induced metric σ_γ which is homotopic to the inclusion of S and maps ∂Y_n geodesically, such that for any shortest essential proper arc τ in (Y_n, σ_γ) we have

$$d_{Y_n}(\gamma, \tau) \leq D.$$

Lemma 3.5. *There exist D and $n_3 > 0$ such that the following holds:*

For $n \geq n_3$, let g_0 and g_1 be a pair of pleated surfaces both homotopic to the inclusion of S mapping ∂Y_n to closed geodesics. Let σ_0 and σ_1 be the induced metrics by g_0 and g_1 on S. If τ_0 and τ_1 are shortest essential proper arcs in (Y_n, σ_0) and (Y_n, σ_1), respectively, then

$$d_{Y_n}(\tau_0, \tau_1) \leq D.$$

We here notice that the length of the geodesic representative of any component of ∂Y_n is bounded below by a positive constant independent of n because Γ has bounded geometry (Compare Lemma 3.5 with Lemma 4.2 of [Min00]).

Fix a sequence of simple closed curves $\{\gamma_m\}_{m=1}^\infty$ on S such that the geodesic representatives γ_m^* exit $\tilde{e}$ and $\ell_{\tilde{N}}(\gamma_m^*) \leq L_1$. Take a sufficiently large $m > 0$ so that γ_m^*

lies deep inside $\tilde{e}$. Let γ denote a component of P. Consider two pleated surfaces $g_\gamma : (S, \sigma_\gamma) \to \tilde{N}$ and $g_{\gamma_m} : (S, \sigma_{\gamma_m}) \to \tilde{N}$ as Lemma 3.4. Let τ_γ and τ_{γ_m} denote the shortest proper essential arcs in Y_n with respect to the metrics σ_γ and σ_{γ_m} respectively. Then the propositions above imply

$$d_{Y_n}(\gamma, \gamma_m) \leq d_{Y_n}(\gamma, \tau_\gamma) + d_{Y_n}(\tau_\gamma, \tau_{\gamma_m}) + d_{Y_n}(\tau_{\gamma_m}, \gamma_m) \leq 3D$$

where $D > 0$ is independent of n and m. Thus, by letting $m \to \infty$, we deduce

$$d_{Y_n}(\gamma, \lambda_{\tilde{e}}) \leq 3D,$$

which contradicts (3.2) since γ is a component of P.

Next we shall give a sketch of the proof of "if" part. First of all, we shall show the quasiconvexity lemma for compressible boundary components. Let C be a subset of $C_0(S)$. We say that C is *B-quasiconvex at the end* $\tilde{e}$ if there exists a neighbourhood $U_{\tilde{e}}$ of $\tilde{e}$ with the following properties: Let $\{\beta_i\}_{i=0}^n$ be a geodesic in $C(S)$. When β_0, β_n lie in C and each geodesic representative β_i^* of β_i in $\tilde{N}$ is contained in $U_{\tilde{e}}$, the geodesic $\{\beta_i\}_{i=0}^n$ is contained in the B-neighbourhood of C.

For $L > 0$, let

$$C(S, L) := \{\gamma \in C(S) \mid \ell_{\tilde{N}}(\gamma^*) < L\}.$$

Then, we define the projection $\Pi_S : C(S) \to \mathcal{P}(C(S, L))$ as follows, where $\mathcal{P}(X)$ denotes the set of subsets of X: Given $\gamma \in C(S)$, let P_x be the curve/arc system associated to the smallest simplex containing x. If P_x contains vertices $\{v_i\}_i \in C_0(S)$ whose geodesic representatives are contained in no pleated surface in $\tilde{N}$, we define $\Pi_S(x) = \{v_i\}_i$. Otherwise, we define

$$\Pi_S(x) := \cup_f \mathbf{short}(\sigma_f),$$

where f ranges over all pleated surfaces realising P_x and $\mathbf{short}(\sigma_f)$ is a subset of $C_0(S)$ consisting of curves γ whose total lengths with respect to the hyperbolic structure σ_f induced by f are less than L_1.

Lemma 3.6. (Quasiconvexity at ends) *For $L \geq L_1$ there is $B > 0$ depending only on L and the topological type of S such that $C(S, L)$ is B-quasiconvex at the end $\tilde{e}$. Moreover, if $\{\beta_i\}_{i=0}^n$ is a geodesic in $C(S)$ with $\beta_0, \beta_n \in C(S, L)$ and the geodesic representative of a simple closed curve on S representing each β_i is contained in the neighbourhood $U_{\tilde{e}}$ of $\tilde{e}$ which appeared in the definition of quasiconvexity of the end, then*

$$d_{C(S)}(\beta_i, \Pi_S(\beta_i)) \leq B$$

for each $i = 0, \ldots, n$.

To show this lemma, we need the following proposition, which is proved by an argument similar to that of Lemma 3.2 of [Min01].

Proposition 3.7. (Coarse Projection) *The map Π_S satisfies the following conditions:*

(1) *(Coarse Lipschitz near ends) There exist a neighbourhood $U_{\tilde{e}}$ of $\tilde{e}$ and $b_0 > 0$ such that if $x, y \in \mathcal{A}(S)$ satisfy*

   (a) $d_{\mathcal{A}(S)}(x,y) \le 1$,

   (b) *neither* $\text{pleat}_S(P_x)$ *nor* $\text{pleat}_S(P_y)$ *is the empty set, and*

   (c) *the image of any pleated surface in* $\text{pleat}_S(P_x) \cup \text{pleat}_S(P_y)$ *is contained in* $U_{\tilde{e}}$,

   then

$$\text{diam}_{C(S)}(\Pi_S(x) \cup \Pi_S(y)) \le b_0.$$

(2) *(Coarse idempotence) If $x \in C(S, L_1)$ then*

$$d_{C(S)}(x, \Pi_S(x)) = 0.$$

Proof of Lemma 3.6. By Proposition 3.7, such $\{\beta_i\}_{i=0}^n$ are contained in the B-neighbourhood of $C(S, L)$. The last part of the assertion derives from lemma 3.3 of [Min01]. $\square$

We return to explaining the proof of the "if" part. Take a sequence of closed geodesics in $\{\gamma_k^*\}_k$ in $\tilde{N}$ such that

$$\ell_{\tilde{N}}(\gamma_k^*) \to \inf\{\ell_N(\gamma^*) \mid \gamma^* \text{ is a closed geodesic on } \tilde{N}\} \tag{3.3}$$

as $k \to \infty$. If infinitely many geodesics in the sequence touch a compact set of $\tilde{N}$, the length of γ_k^* cannot tend to zero. Hence Γ has bounded geometry and we are done.

Suppose $\{\gamma_k^*\}_{k=1}^\infty$ exits an end $\tilde{e}$ of $\tilde{N}$ and all the γ_k^* are contained in a neighbourhood $U_{\tilde{e}}$ of $\tilde{e}$. We have only to consider the case when $\ell_N(\gamma_k^*) \to 0$, which implies that γ_k^* is homotopic to a simple closed curve γ_k on S within $U_{\tilde{e}}$ by Otal's theorem [Ota95]. Taking $\tilde{C}$ large if necessary, we may assume that $U_{\tilde{e}}$ satisfies the conditions of Lemma 3.6 and Proposition 3.7. Let V denote a neighbourhood of $\lambda_{\tilde{e}}$ in the space of geodesic laminations on S such that the image of any pleated surface realising a finite lamination in V is contained in $U_{\tilde{e}}$ (Lemma 3.2). Fix a maximal curve system $P \in V$ with $\ell_{\tilde{N}}(P) \le L_1$. Since γ_k^* exits $\tilde{e}$, for any sufficiently large k, there is a maximal curve system $Q_k \in V$ with the following properties:

- $\ell_{\tilde{N}}(Q_k) \le L_1$ and

- for pleated surfaces f_+ and f_- realising P and Q_k, respectively, the ε_1-Margulis tube of γ_k^* is homologically encased by f_+ and f_-.

(cf. Lemma 3.3. See also Lemma 7.1 of [Min01]). Join P and Q_k with a resolution sequence

$$P = P_0 \to P_1 \to \cdots \to P_n = Q_k,$$

as Theorem 5.1 in [Min01]. If we choose $U_{\tilde{e}}$ appropriately again, we may assume that the image of the good homotopy (for the definition, see §4 of [Min01]) connecting pleated surfaces realising P_k and P_{k+1} is contained in $U_{\tilde{e}}$ (cf. Lemma 3.2). Thus, by applying Minsky's construction in §7 of [Min01] with Lemma 3.6, we have a model of the part of $\tilde{N}$ between f_+ and f_- which covers the ε_1-Margulis tube of γ_k^* in $U_{\tilde{e}}$. Since the number of blocks intersecting the Margulis tube of γ_k^* is bounded by a constant depending on the topology of S and the right-hand side of (3.1) (see (7.1) in [Min01]), by the definition of good homotopies, one can control the geometry of the intersection between the Margulis tube of γ_k^* and any block to conclude that the radius of the Margulis tube of γ_k^* is uniformly bounded, when γ_k is not a component of any P_j. The case when γ_k is a component of some P_j is also treated in a similar way to Minsky's method (see p.176–p.177 of [Min01]). Thus we get a lower bound of the lengths of all $\{\gamma_k^*\}_{k=1}^{\infty}$ which we desired (see (3.3)).

3.2. Ending Lamination Theorem for groups with bounded geometry

Now we can complete the proof of Theorem 1.1.

Theorem 3.8 (Uniqueness). *Let Γ_1 and Γ_2 be topologically tame Kleinian groups whose compact cores are mutually homeomorphic. Suppose that Γ_1 has bounded geometry. If the ending invariants of Γ_1 and Γ_2 coincide, then Γ_1 and Γ_2 are conjugate in* $\mathrm{PSL}_2(\mathbb{C})$.

Proof. Since Γ_2 has the same end invariants as those of Γ_1, by Theorem 3.1, Γ_2 also has bounded geometry. Thus, applying the ending lamination theorems by Minsky ([Min94]) and Ohshika ([Ohs96] and [Ohs98]), we have the assertion. □

4. Limit of Kleinian groups

To deduce Theorem 1.2 from Theorem 1.1, we need a convergence theorem proved in [Ohsa] and some argument in [Ohsc]. We shall review them in this section.

The following theorem was proved in Ohshika [Ohsa] applying the convergence theorem by Kleineidam-Souto [KS02] except for the condition (i) for the case when

M is a handlebody. When M is a handlebody, we need to use the result of Brock and Souto [BS04] to show that the limit given in [Ohsa] is topologically tame.

Theorem 4.1 (Existence of limit). *Let M be a hyperbolisable 3-manifold with non-empty ∂M without torus components, and $S_1,\ldots,S_m$ denote its boundary components. Suppose that arational measured laminations $\lambda_{j_1},\ldots,\lambda_{j_p}$ contained in the Masur domain are given on (at least one) boundary components $S_{j_1},\ldots,S_{j_p}$ among $S_1,\ldots,S_m$, and conformal structures $m_1,\ldots,m_q$ on the remaining boundary components $S_{i_1},\ldots,S_{i_q}$. When M is homeomorphic to $S_{j_1} \times I$ and $p = 2$, we further assume that λ_{j_1} and λ_{j_2} are not homotopic in M. When M is a twisted I-bundle over a non-orientable surface, we further assume that λ_{j_1} is not a lift of a measured lamination on the base surface. Let $\{X^n_k\}_{n=1}^\infty$ be a sequence in $T(S_k)$ which converges to either m_{i_l} if $k = i_l$ or a projective measured lamination whose support is λ_{j_1}, otherwise. Let $[\rho_n] \in QH(M)$ be a representation which uniformises $(X^n_1,\ldots,X^n_m)$. Then the sequence $\{[\rho_n]\}_{n=1}^\infty$ contains a subsequence which converges to $[\rho_\infty] \in AH(M)$. The limit group $\rho_\infty(\pi_1(M))$, which we denote by Γ, has the following properties:*

(i) *Γ is topologically tame,*

(ii) *a homotopy equivalence from M to $\mathbb{H}^3/\Gamma$ induced by ρ_∞ is homotopic to a homeomorphism h from M to a compact core C_Γ of $\mathbb{H}^3/\Gamma$,*

(iii) *the end e_k of $\mathbb{H}^3/\Gamma$ facing $h(S_k)$ is either geometrically finite with conformal structure $h_*(m_{i_l})$ at infinity when $k = i_l$, or geometrically infinite when $k = j_l$.*

(iv) *$h(\lambda_{j_l})$ is unrealisable by a pleated surface homotopic to $h|S_{j_l}$.*

In the theorem above, although we state that Γ is topologically tame and $h(\lambda_{j_l})$ is unrealisable, we do not know if it represents the ending lamination for the end facing $h(S_{j_l})$. We shall now show that if we add an extra condition that the length of λ_{j_l} with respect to $X^n_{j_l}$ is bounded as $n \to \infty$, the ending lamination facing $h(S_{j_l})$ is actually represented by $h(\lambda_{j_l})$. This was proved independently by Namazi-Souto [NS] and Ohshika [Ohsc]. We shall briefly review the gist of the argument in [Ohsc].

Proposition 4.2. *In the situation of Theorem 4.1, assume furthermore that the length of λ_{j_l} with respect to $X^n_{j_l}$ is bounded as $n \to \infty$. Then for each $k = j_l$, the ending lamination of the geometrically infinite end facing $h(S_k)$ is represented by $h(\lambda_k)$.*

Outline of Proof. We shall sketch the argument in [Ohsc]. Let e_k be the end facing $h(S_k)$ for $k = j_l$. We divide our argument into two cases: the first is when M is not a handlebody and the second is when M is a handlebody. We start with the first case, which is simpler.

Suppose that M is not a handlebody. Let $\{w_i\gamma_i\}$ be a sequence of weighted simple closed curves converging to λ_k. We can assume that $w_i\gamma_i$ is contained in the Masur domain of S_k. Then, there is a pleated surface $h_i : S_k \to \mathbb{H}^3/\Gamma$ homotopic to $h|S_k$ which realises γ_i as a closed geodesic γ_i^* tending to the end e_k. Since M is not a handlebody and S_k represents a non-trivial second homology class, as was shown in Ohshika [Ohsb], this sequence of pleated surfaces gives rise to a product structure $S_k \times \mathbb{R}$ of a neighbourhood of the end e_k such that h_i is homotopic to $S_k \times \{\mathrm{pt}\}$ within $S_k \times \mathbb{R}$ for large i.

On the other hand, since e_k is known to be topologically tame, we can assume that $h(S_k)$ cuts off $h(S_k) \times [0,\infty)$ containing e_k from $\mathbb{H}^3/\Gamma$ with $h(S_k) \times \{0\}$ identified with $h(S_k)$, and that there is a measured lamination μ in the Masur domain of S_k such that $h(\mu)$ represents the ending lamination of e_k. Let $r_i c_i$ be a sequence of simple closed curves on S_k converging to μ. Then the closed geodesic c_i^* is homotopic to $h(c_i)$ in $h(S_k) \times [0,\infty)$ for sufficiently large i. Since the product structure of a neighbourhood of e_k is homotopically unique, by isotoping h, we can assume that the product structures $S_k \times \mathbb{R}$ given by h_i and $h(S_k) \times \mathbb{R}$ which we have here coincide.

Bonahon showed that in this situation, if c_i^* and γ_i^* do not intersect thin Margulis tubes outside the axes, then we get $i(w_i\gamma_i, r_i c_i) \to 0$. By using the fact that both c_i and γ_i are simple and contained in the Masur domain, if we take sufficiently small $\varepsilon > 0$, we can assume, by changing $w_i\gamma_i$ and $r_i c_i$ preserving the supports of their limits, that the closed geodesics cannot intersect ε-Margulis tubes outside the axes. Thus we can show that $i(\lambda_k, \mu) = 0$, which means that $h(\lambda_k)$ also represents the ending lamination of e_k.

Next we consider the case when M is a handlebody, which is more difficult to deal with. We only state claims which constitute steps for the proof. In this case, we can first show that C_Γ is also a handlebody using the fact that the closed geodesic γ_i^* as above tends to e_k and Bonahon's intersection lemma. This implies that the convergence of $[\rho_n]$ to Γ is strong, which means that Γ is also a geometric limit of $[\rho_n]$. Then by pulling back the closed geodesics c_i^* by approximate isometries and using a version of Bonahon's intersection lemma generalised to deal with variable target hyperbolic 3-manifolds for $w_i c_i$ and λ_k in $\mathbb{H}^3/\rho_n(\pi_1(M))$, we see that $i(\lambda_k, \mu) = 0$. This means that $h(\lambda_k)$ also represents the ending lamination of e_k. $\square$

5. Iterations of hyperbolic actions

Let M be a hyperbolisable 3-manifold with $\partial M = \cup_{i=1}^m S_i \neq \emptyset$. Assume that no S_i is a torus. A holomorphic automorphism $\omega = (\omega_1, \ldots, \omega_n)$ of the Teichmüller space $T(\partial M)$ is said to be *hyperbolic* if each coordinate ω_i is either hyperbolic or the identity on $T(S_i)$. We say such a transformation is *Masur hyperbolic* when each hyperbolic transformation of a coordinate $T(S_i)$ has stable lamination in the Masur domain of S_i. We further assume that when M is a product I-bundle over a closed orientable surface, the stable laminations of the two coordinates of hyperbolic actions ω_1 and ω_2 are not mutually homotopic in M, and that when M is a twisted I-bundle over a closed non-orientable surface, ω_1 is not a lift of an automorphism of the base surface.

The quasi-conformal deformation space $QH(M)$ of a convex cocompact structure on M is identified with a quotient space $T(\partial M)/\mathrm{Mod}_0(M)$ of the Teichmüller space of ∂M (cf. Kra [Kra72]). Hence there exists the canonical projection $\Pi : T(\partial M) \to QF(M)$. If $[\rho] = \Pi(X)$ for some $X \in T(\partial M)$, we say that $[\rho]$ *uniformises* X.

Proof of Theorem 1.2. Set $X = (X^1, \ldots, X^m) \in T(\partial M)$ and let ω be a Masur hyperbolic modular action on $T(\partial M)$. Let $[\rho_n] = \Pi(\omega^n(X)) \in QF(M)$ ($n \geq 1$). We may assume that ω_i is hyperbolic if $i \leq k$ and the identity otherwise. Then for $i \leq k$, $\omega_i^n(X^i)$ converges to the stable measured lamination $[\mu_i]$ of the pseudo-Anosov mapping on S_i which induces ω_i.

By the extra-assumptions for the case of I-bundles and the arationality of the stable laminations, we can apply Theorem 4.1, and conclude that $\{[\rho_n]\}_n$ contains a subsequence converging to $[\rho_\infty] \in AH(M)$. The fact that the limit group has end invariants corresponding to $(|\mu_1|, \ldots, |\mu_k|, X^{k+1}, \ldots, X^m)$ follows from Proposition 4.2. Theorem 3.8 guarantees that the limit is unique. Consequently, $\{[\rho_n]\}_n$ converges to $[\rho_\infty]$.

Since two hyperbolic modular transformations are homotopic if and only if their stable laminations are homotopic in M, the latter part of the theorem also follows from Theorem 3.8. $\qquad\qquad\qquad\square$

6. Third bounded cohomology groups

As an application of the results above and those in Miyachi [Miy02], we shall consider a generalisation of Soma's theorems on the third bounded cohomology groups of surface groups to free groups. We recall basic definitions first.

For a topological space X, an n-dimensional cochain c is said to be *bounded* when $\sup\{|c(\sigma)| \| \sigma$ is an n-simplex in $X\}$ is finite. The bounded cochains constitute a subcomplex of the cochain groups $C^*(X)$. The n-th cohomology group of this complex of bounded cochains is the n-th bounded cohomology of X. Similarly, we can define the bounded cohomology groups for a group G, and they coincide with the bounded cohomology groups of $K(G, 1)$.

We consider a complete hyperbolic 3-manifold M homeomorphic to the interior of a handlebody of genus $g \geq 2$. The volume form Ω_M of M determines a bounded 3-cocycle, by setting the value on a singular 3-simplex σ to be the volume of the straightening of σ. The third bounded cohomology class represented by Ω_M this way is called the *fundamental class* on M and is denoted by ω_M.

Let G be a Fuchsian Schottky group of rank g, which is a free group of rank g if regarded as an abstract group. For $i = 1, 2$, let (Γ_i, ϕ_i) be an element in $AH(G)$, where $\phi_i : G \to \mathrm{PSL}_2(\mathbb{C})$ is a faithful discrete representation with its image Γ_i, which we assume to be geometrically infinite, topologically tame, purely loxodromic, and have bounded geometry. Pulling back $\omega_{\mathbb{H}^3/\Gamma_i}$ by ϕ_i, we get a third bounded cohomology class in $H_b^3(G, \mathbb{R})$.

For a bounded 3-cochain c of G, we define its norm by $\|c\| = \sup\{|c(\sigma)| \,|\, \sigma$ is a singular 3-simplex$\}$. This induces a pseudo-norm on $H_b^3(G, \mathbb{R})$ by $\|\gamma\| = \inf\{\|c\| \,|\, [c] = \gamma\}$.

The main result in this section is the following generalisation of Soma's result on the third bounded cohomology groups of surface groups.

Theorem 6.1. *Let $(\Gamma_i, \phi_i)(i = 1, 2)$ be purely loxodromic, geometrically infinite, and topologically tame groups in $AH(G)$ with bounded geometry as above. Suppose that the injectivity radii of points in $\mathbb{H}^3/\Gamma_1$ are bounded below by $\delta_0 > 0$. Then, there exists a constant ε depending only on δ_0 such that if $\|\phi_1^*(\omega_{\mathbb{H}^3/\Gamma_1}) - \phi_2^*(\omega_{\mathbb{H}^3/\Gamma_2})\| < \varepsilon$ in $H_b^3(G, \mathbb{R})$, then $(\Gamma_1, \phi_1) = (\Gamma_2, \phi_2)$ as elements in $AH(G)$.*

Before starting the proof of this theorem, we shall see its corollary. Let S be a closed surface of genus g, which is also regarded as the boundary of a handlebody H of genus g. Let f_1, f_2 be pseudo-Anosov automorphisms of S whose stable laminations are in the Masur domain of S. Take a point $X \in \mathcal{T}(S)$ uniformised by G. We consider the limit of $\rho_n(i) \in QH(G)$ uniformising $f_i^n(X)$ for $i = 1, 2$. By Theorem 1.2, we have limits $(\Gamma_i, \rho_\infty(i)) \in AH(G)$ for $i = 1, 2$, and $(\Gamma_1, \rho_\infty(1))$ coincides with $(\Gamma_2, \rho_\infty(2))$ if and only if ω_1 and ω_2 are homotopic in H. Having noted these, we now state a theorem of Soma [Som96] with an alternative proof obtained from our theorem above.

Corollary 6.2. *$H_b^3(G, \mathbb{R})$ has dimension of the cardinality of the continuum.*

Proof. Let (Γ, ϕ) be a limit of ρ_n uniformising $f^n(X)$ for a Masur hyperbolic f as above. Let λ be a stable lamination of f which is assumed to be contained in the Masur domain. Let c be a simple closed curve on S representing a non-trivial free homotopy class in H. Let D_k be the k-time right-hand Dehn twist on S along c. Regarding ϕ as a homotopy equivalence from H to $\mathbb{H}^3/\Gamma$, we consider an element $(\Gamma, \phi \circ D_k)$ in $AH(G)$. Since $f_k = D_k^{-1} f D_k$ has a stable lamination $D_k^{-1}(\lambda)$, Theorem 1.2 implies that the limit of elements in $QH(G)$ uniformising $(f_k)^n(X)$ as $n \to \infty$ coincides with $(\Gamma, \phi \circ D_k)$. Since D_k is not homotopic to the identity in H, these $(\Gamma, \phi \circ D_k)$ are all distinct. Let $\omega_k \in H_b^3(G, \mathbb{R})$ be $(\phi \circ D_k)^*(\omega_{\mathbb{H}^3/\Gamma})$. Then $\{\|\omega_k\|\}$ is totally bounded.

If $(\Gamma, \phi \circ D_k)$ spans a finite dimensional subspace, then it must have a subsequence converging to some element in $H_b^3(G, \mathbb{R})$ and contradicts Theorem 6.1. This implies that $HB^3(G, \mathbb{R}) = H_b^3(G, \mathbb{R}) / \{\gamma \in H_b^3(G, \mathbb{R}) | \|c\| = 0\}$ has infinite dimension. Then the same argument as in the surface group case works and we see that $H_b^3(G, \mathbb{R})$ has dimension of the cardinality of continuum. $\qquad \square$

Proof of Theorem 6.1. This can be proved by the same argument as the proof in Soma [Som96] for the case of closed surface groups. Soma's argument proceeds as follows: First, using the Cannon–Thurston map, he gets continuous maps from the limit set of G, which is $\hat{\mathbb{R}}$ in Soma's case, to the limit sets of Γ_1 and Γ_2, both of which are $\hat{\mathbb{C}}$ in Soma's case. This gives rise to an equivariant map from the complement of a measure 0-set of Λ_{Γ_1} to Λ_{Γ_2}. It is shown that this map can be extended to $\mathbb{H}^3$ equivariantly. The assumption of the bounded geometry implies that this map gives a conjugation from Γ_1 to Γ_2 which takes an ending lamination of $\mathbb{H}^3 / \Gamma_1$ to that of $\mathbb{H}^3 / \Gamma_2$. Minsky's ending lamination theorem for the bounded geometry case implies that this yields an isometry from $\mathbb{H}^3 / \Gamma_1$ to $\mathbb{H}^3 / \Gamma_2$. Now, we shall indicate a point where Soma's proof has to be changed to fit into our case.

Thus, the only point where our proof should be different from his is the construction of Cannon-Thurston maps for Γ_i and the negligible sets for Cannon-Thurston maps. We have to define a Cannon-Thurston map from the limit set of a Fuchsian Schottky group, whereas in Soma's case the map is defined on the entire circle at infinity. The problem caused by this difference can be avoided as follows: In [Miy02], we proved that there is a ϕ_i-equivariant continuous map F_i of $\hat{\mathbb{C}}$ for $i = 1, 2$. Furthermore, $F_i(x) = F_i(y)$ if and only if x and y are in either the same leaf or the same component of the complement of the lift of the ending lamination of Γ_i on $\Omega_G = \hat{\mathbb{C}} - \Lambda_G$, where Λ_G is the limit set of G. Let Λ_2 be the set of points in $\hat{\mathbb{C}}$ each of whose preimage with respect to F_1 consists of at least 2 points and set $\Lambda_1 = F_1^{-1}(\Lambda_2)$. Then we have a comparing map $F_2 \circ F_1^{-1} : \Lambda_{\Gamma_1} - \Lambda_2 \to \Lambda_{\Gamma_2}$.

Finally, we should check that Λ_1 and Λ_2 are negligible and F_1 is a measurable map from Λ_G to Λ_{Γ_1}. Indeed, by using the model manifold of $\mathbb{H}^3 / \Gamma_1$ (cf. [Ohs98]), one can see that Λ_2 consists of non-conical limit points. Hence, by Sullivan's ergodicity theorem [Sul81], Λ_2 has measure zero (cf. the discussion after the proof of Lemma 1 in [Som95]). Since Λ_1 consists of the endpoints of leaves of the ending lamination for ψ_1, by virtue of Birman and Series' theorem in [BS85], the Hausdorff dimension of Λ_1 is zero. Furthermore, since the Hausdorff dimension of Λ_{Γ_1} is positive and its Hausdorff measure is also positive (e.g. [Nic89]), F_1 is measurable. $\qquad \square$

References

[Bon86] F. Bonahon (1986). Bouts des variétés hyperboliques de dimension 3. *Ann. of Math. (2)* **124** (1), 71–158.

[BS85] J. S. Birman & C. Series (1985). Geodesics with bounded intersection number on surfaces are sparsely distributed. *Topology* **24** (2), 217–225.

[BS04] J. Brock & J. Souto (2004). Algebraic limits of geometrically finite manifolds are tame. Preprint.

[Can93] R. D. Canary (1993). Ends of hyperbolic 3-manifolds. *J. Amer. Math. Soc.* **6** (1), 1–35.

[Kra72] I. Kra (1972). On spaces of Kleinian groups. *Comment. Math. Helv.* **47**, 53–69.

[KS02] G. Kleineidam & J. Souto (2002). Algebraic convergence of function groups. *Comment. Math. Helv.* **77** (2), 244–269.

[Mar74] A. Marden (1974). The geometry of finitely generated Kleinian groups. *Ann. of Math. (2)* **99**, 383–462.

[Mas71] B. Maskit (1971). Self-maps on Kleinian groups. *Amer. J. Math.* **93**, 840–856.

[Min94] Y. N. Minsky (1994). On rigidity, limit sets, and end invariants of hyperbolic 3-manifolds. *J. Amer. Math. Soc.* **7** (3), 539–588.

[Min00] Y. Minsky (2000). Kleinian groups and the complex of curves. *Geometry and Topology* **4**, 117–148.

[Min01] Y. N. Minsky (2001). Bounded geometry for Kleinian groups. *Invent. Math.* **146** (1), 143–192.

[Miy02] H. Miyachi (2002). Semiconjugacy between topologically tame kleinian group with bounded geometry. Preprint.

[Nic89] P. J. Nicholls (1989). *The Ergodic Theory of Discrete Groups, London Mathematical Society Lecture Note Series*, volume 143. Cambridge University Press, Cambridge.

[NS] H. Namazi & J. Souto. Nonrealizability in handlebodies and ending laminations. In preparation.

[Ohsa] K. Ohshika. Constructing geometrically infinite groups on boundaries of deformation spaces. Submitted.

[Ohsb] K. Ohshika. Kleinian groups which are limits of geometrically finite groups. To appear, *Memoirs Amer. Math. Soc.* (2005).

[Ohsc] K. Ohshika. Realising end invariants by limits of minimally parabolic geometrically finite groups. Preprint, math.GT/0504546.

[Ohs96] K. Ohshika (1996). Topologically conjugate Kleinian groups. *Proc. Amer. Math. Soc.* **124** (3), 739–743.

[Ohs98] K. Ohshika (1998). Rigidity and topological conjugates of topologically tame Kleinian groups. *Trans. Amer. Math. Soc.* **350** (10), 3989–4022.

[Ota95] J.-P. Otal (1995). Sur le nouage des géodésiques dans les variétés hyperboliques. *C. R. Acad. Sci. Paris Sér. I Math.* **320** (7), 847–852.

[Som95] T. Soma (1995). Equivariant, almost homeomorphic maps between S^1 and S^2. *Proc. Amer. Math. Soc.* **123** (9), 2915–2920.

[Som96] T. Soma (1996). The third bounded cohomology and Kleinian groups. In *Topology and Teichmüller spaces (Katinkulta, 1995)*, pp. 265–277. World Sci. Publishing, River Edge, NJ.

[Som97] T. Soma (1997). Bounded cohomology of closed surfaces. *Topology* **36** (6), 1221–1246.

[Sul81] D. Sullivan (1981). On the ergodic theory at infinity of an arbitrary discrete group of hyperbolic motions. In *Riemann Surfaces and Related Topics: Proceedings of the 1978 Stony Brook Conference (State Univ. New York, Stony Brook, N.Y., 1978)*, Ann. of Math. Stud., volume 97, pp. 465–496. Princeton Univ. Press, Princeton, N.J.

[Sul85] D. Sullivan (1985). Quasiconformal homeomorphisms and dynamics. II. Structural stability implies hyperbolicity for Kleinian groups. *Acta Math.* **155** (3-4), 243–260.

[Thu86] W. P. Thurston (1986). Hyperbolic structures on 3-manifolds. I. Deformation of acylindrical manifolds. *Ann. of Math. (2)* **124** (2), 203–246.

Ken'ichi Ohshika

Graduate School of Science
Department of Mathematics
Osaka University
1-1, Machikaneyama-cho
Toyonaka
Osaka 560-0043
Japan

ohshika@math.wani.
            osaka-u.ac.jp

Hideki Miyachi

Dept. of Mathematical Sciences
Tokyo Denky University
Ishizaka, Hatoyama, Hiki
Saitama, 329-0394
Japan

miyachi@r.dendai.ac.jp

AMS Classification: 57M50, 30F40, 55N35

Keywords: Kleinian groups, end invariants, bounded cohomology

Spaces of Kleinian Groups
Lond. Math. Soc. Lec. Notes **329**, 49–73

Cambridge University Press
Y. Minsky, M. Sakuma & C. Series (Eds.)

An extension of the Masur domain

Cyril Lecuire

Abstract

The Masur domain is a subset of the space of projective measured geodesic laminations on the boundary of a 3-manifold M. This domain plays an important role in the study of the hyperbolic structures on the interior of M. In this paper, we define an extension of the Masur domain and explain that it shares a lot of properties with the Masur domain.

1. Introduction

A compression body is the connected sum along the boundary of a ball of I-bundles over closed surfaces and solid tori. Among the compression bodies are the handlebodies which are the connected sums along the boundary of solid tori $D^2 \times S^1$. If M is a compression body and if ∂M has negative Euler characteristic then, by Thurston hyperbolization theorem, its interior admits a hyperbolic structure. Namely there are discrete faithful representations $\rho : \pi_1(M) \to \text{Isom}(\mathbb{H}^3)$ such that $\mathbb{H}^3/\rho(\pi_1(M))$ is homeomorphic to the interior of M. If such a representation ρ is geometrically finite, it is said to uniformize M.

In [Mas86], H. Masur studied the space of projective measured foliations on the boundary of a handlebody. He described the limit set of the action of the modular group on this space and defined a subset of the space of projective measured foliations on which this action is properly discontinuous. In [Ota88], J.-P. Otal defined a similar subset O of the space of projective measured geodesic laminations on the boundaries of compression bodies. This set $O \subset \mathcal{PML}(\partial M)$ is called the Masur domain and J.-P. Otal showed that the action of the modular group on O is properly discontinuous. He also proved the following: if $\text{int}(M)$ is endowed with a convex cocompact hyperbolic metric, then any projective class of measured geodesic laminations lying in O is realized by a pleated surface. He also showed that the injectivity theorem of [Thu86] applies for such pleated surfaces.

Later it was shown that the projective classes of measured laminations in O are an analogous of what Thurston called binding laminations on I-bundles over closed surfaces. Namely if we have a sequence of geometrically finite representations $\rho_n : \pi_1(M) \to \text{Isom}(\mathbb{H}^3)$ uniformizing a compression body and a measured geodesic lamination $\lambda \in O$ such that $l_{\rho_n}(\lambda)$ is bounded, then the sequence (ρ_n) contains an algebraically converging subsequence. This property has been obtained for various cases in [Thu87], [Ota94], [Can93], [Ohs97] and the general statement comes from [KS02] and [KS03].

In this paper, we allow M to be any orientable 3-manifold with boundary satisfying the following: the Euler characteristic of ∂M is negative and the interior of M admits a complete hyperbolic metric. We will consider the following set :

$\mathcal{D}(M) = \{\lambda \in \mathcal{ML}(\partial M) | \exists \eta > 0$ such that $i(\lambda, \partial E) > \eta$ for any essential annulus or disc $E \subset M\}$.

First we will link this set $\mathcal{D}(M)$ with the result of [Lec02] and deduce from this that the support of a geodesic measured lamination lying in $\mathcal{D}(M)$ is also the support of a (in fact many) bending measured geodesic lamination of a representation uniformizing M. Using the continuity of the bending measure proved in [KS95] and [Bon98], we will show that $\mathcal{D}(M)$ is connected. We will also discuss the relationships between $\mathcal{D}(M)$ and the Masur domain.

After that, we will prove that the set $\mathcal{D}(M)$ has the following properties:

If $\text{int}(M)$ is endowed with a convex cocompact hyperbolic metric, any measured geodesic lamination lying in $\mathcal{D}(M)$ is realized by a pleated surface and such a pleated surface satisfies the injectivity theorem of [Thu86].

If ρ_n is a sequence of geometrically finite metrics uniformizing M and $\lambda \in \mathcal{D}(M)$ is a measured geodesic lamination such that $l_{\rho_n}(\lambda)$ is bounded, then the sequence (ρ_n) contains an algebraically converging subsequence.

We will also discuss the action of the modular group on $\mathcal{D}(M)$.

I would like to thank F. Bonahon, I. Kim, K. Ohshika and J.-P. Otal for fruitful discussions, and J. Souto who gave me the ideas of Proposition 4.2. I also thank the referee for pointing out some mistakes in a previous version.

2. Definitions

2.1. Geodesic Laminations

Let S be a closed surface endowed with a complete hyperbolic metric; a *geodesic lamination* on S is a compact subset that is the disjoint union of complete embedded geodesics. Using the fact that two complete hyperbolic metrics on S are quasi-isometric, this definition can be made independent of the chosen metric on S (see [Ota94] for example). A geodesic lamination whose leaves are all closed is called a *multi-curve.* If each half-leaf of a geodesic lamination L is dense in L, then L is *minimal*. Such a minimal geodesic lamination is either a simple closed curve or an *irrational lamination*. A leaf l of a geodesic lamination L is *recurrent* if it lies in a minimal geodesic lamination. Any geodesic lamination is the disjoint union of finitely many minimal laminations and non-recurrent leaves. A leaf is said to be an *isolated* leaf if it is either a non-recurrent leaf or a compact leaf without any leaf spiraling toward it.

Let L be a connected geodesic lamination which is not a simple closed curve and let us denote by $\bar{S}(L)$ the smallest surface with geodesic boundary containing L. Inside $\bar{S}(L)$ there are finitely many closed geodesics (including the components of $\partial \bar{S}(L)$) disjoint from L and these closed geodesics do not intersect each other (cf. [Lec02]); let us denote by $\partial' \bar{S}(L) \supset \partial \bar{S}(L)$ the union of these geodesics. Let us remove from $\bar{S}(L)$ a small tubular neighborhood of $\partial' \bar{S}(L)$ and let $S(L)$ be the resulting surface. We will call $S(L)$ the *surface embraced* by the geodesic lamination L and $\partial' \bar{S}(L)$ the *effective boundary* of $S(L)$. If L is a simple closed curve, let us define $S(L)$ to be an annular neighborhood of L and $\partial' \bar{S}(L) = L$. If L is not connected, $S(L)$ is the disjoint union of the surfaces embraced by the connected components of L and $\partial' \bar{S}(L) = \bigcup_{\{L^i \text{ is a component of } L\}} \partial' \bar{S}(L^i)$.

A *measured geodesic lamination* λ is a transverse measure for some geodesic lamination $|\lambda|$: any arc $k \approx [0,1]$ embedded in S transversely to $|\lambda|$, such that $\partial k \subset S - \lambda$, is endowed with an additive measure $d\lambda$ such that:

– the support of $d\lambda_{|k}$ is $|\lambda| \cap k$;

– if an arc k can be homotoped into k' by a homotopy respecting $|\lambda|$ then $\int_k d\lambda = \int_{k'} d\lambda$. We will denote by $\mathcal{ML}(S)$ the space of measured geodesic lamination topologised with the topology of weak* convergence. We will denote by $|\lambda|$ the support of a measured geodesic lamination λ.

Let γ be a weighted simple closed geodesic with support $|\gamma|$ and weight w and let λ be a measured geodesic lamination, the intersection number between γ and λ is defined by $i(\gamma, \lambda) = w \int_{|\gamma|} d\lambda$. The weighted simple closed curves are dense in $\mathcal{ML}(S)$ and this intersection number extends continuously to a function $i : \mathcal{ML}(S) \times \mathcal{ML}(S) \to \mathbb{R}$ (cf. [Bon86]). A measured geodesic lamination λ is *arational* if for any simple closed curve c, we have $i(c, \lambda) = \int_c d\lambda > 0$.

2.2. Real trees

An $\mathbb{R}$-tree $\mathcal{T}$ is a metric space such that any two points x, y can be joined by a unique simple arc. Let G be a group acting by isometries on an $\mathbb{R}$-tree $\mathcal{T}$; the action is *minimal* if there is no proper invariant subtree and *small* if the stabilizer of any non-degenerate arc is virtually Abelian.

A G-equivariant map ϕ between two $\mathbb{R}$-trees $\mathcal{T}$ and $\mathcal{T}'$ is a *morphism* if and only if every point $p \in \mathcal{T}$ lies in a non-degenerate segment $[a,b]$ (but p may be a vertex of $[a,b]$) such that the restriction $\phi_{|[a,b]}$ is an isometry. The point p is a *branching point* if there is no segment $[a,b]$ such that $\phi_{|[a,b]}$ is an isometry and that $p \in]a,b[$.

Let S be a connected hyperbolic surface and let $q : \mathbb{H}^2 \to S$ be the covering projection. Let $L \subset S$ be a geodesic lamination and let $\pi_1(M) \curvearrowright \mathcal{T}$ be a minimal action of $\pi_1(M)$ on an $\mathbb{R}$-tree $\mathcal{T}$; L is *realized* in $\mathcal{T}$ if there is a continuous equivariant map $\mathbb{H}^2 \to \mathcal{T}$ whose restriction to any lift of a leaf of L is injective.

Let $\lambda \in \mathcal{ML}(S)$ be a measured geodesic lamination; following [MO93], we will define the dual tree of λ. Consider the following metric space $pre\mathcal{T}_\lambda$: the points of $pre\mathcal{T}_\lambda$ are the complementary regions of $q^{-1}(\lambda)$ in $\mathbb{H}^2$, where $q : \mathbb{H}^2 \to S$ is the covering projection and the distance $d : \mathcal{T}_\lambda \times \mathcal{T}_\lambda \to \mathbb{R}$ is defined as follows. Let R_0 and R_1 be two complementary regions and choose a geodesic segment $k \subset \mathbb{H}^2$ whose vertices lie in R_0 and R_1; we set $d(R_0, R_1)$ to be the $q^{-1}(\lambda)$-measure of k. Then, there is a unique (up to isometry) $\mathbb{R}$-tree $\mathcal{T}_\lambda$ and an isometric embedding $e : pre\mathcal{T}_\lambda \to \mathcal{T}_\lambda$ such that **any point** of $\mathcal{T}_\lambda$ lies in a segment with endpoints in $e(pre\mathcal{T}_\lambda)$ (cf. [GS90]). The covering transformations yield an isometric action of $\pi_1(M)$ on $\mathcal{T}_\lambda$; if $\delta_\lambda(c)$ is the distance of translation of an isometry of $\mathcal{T}_\lambda$ corresponding to a simple closed curve c, we have $\delta_\lambda(c) = i(c, \lambda)$. This construction yields a natural projection $\mathbb{H}^2 - q^{-1}(\lambda) \to \mathcal{T}_\lambda$. If λ does not have closed leaves, this projection extends continuously to a map $\pi_\lambda : \mathbb{H}^2 \to \mathcal{T}_\lambda$. Otherwise, replacing closed leaves of λ by foliated annuli endowed with uniform transverse measures, we get also a continuous map $\pi_\lambda : \mathbb{H}^2 \to \mathcal{T}_\lambda$ (cf. [Ota94]).

2.3. Train tracks

A *train track* τ in S is the union of finitely many "rectangles" b_i called the *branches* and satisfying:

- any branch b_i is an imbedded rectangle $[0,1] \times [0,1]$ such that the preimage of the double points is a segment of $\{0\} \times [0,1]$ and a segment of $\{1\} \times [0,1]$;

- the intersection of two different branches is either empty or a non-degenerate segment lying in the vertical sides $\{0\} \times [0,1]$ and $\{1\} \times [0,1]$;

- any connected component of the union of the vertical sides is a simple arc embedded in $\partial_{\chi<0} M$.

A connected component of the union of the vertical sides is a *switch*. In each branch the segments $\{p\} \times [0,1]$ are the *ties* and the segments $[0,1] \times \{p\}$ are the *rails*.

A geodesic lamination L is *carried* by a train track τ when:

– L lies in τ;

– for each branch b_i of τ, $L \cap b_i$ is not empty, lies in the image of $[0,1] \times]0,1[$ and each leaf of L is transverse to the ties.

Notice that, in some papers, a geodesic lamination satisfying the above is said to be "minimally carried" by τ.

A measured geodesic lamination λ is carried by a train track τ if its support $|\lambda|$ is carried by τ.

Let S be a hyperbolic surface, let $\tau \subset S$ be a train track and let $\pi_1(M) \curvearrowright \mathcal{T}$ be a minimal action of $\pi_1(M)$ on an $\mathbb{R}$-tree $\mathcal{T}$. Let $q^{-1}(\tau) \subset \mathbb{H}^2$ be the preimage of τ under the covering projection; a *weak realization* of τ in $\mathcal{T}$, is a $\pi_1(M)$-equivariant continuous map $\pi : q^{-1}(\tau) \to \mathcal{T}$ such that π is constant on the ties of $q^{-1}(\tau)$, monotone and not constant on the rails and that the images of two adjacents branches lying on opposite sides of the same switch have disjoint interiors.

2.4. 3-manifolds

Let M be a 3-manifold, M is *irreducible* if any sphere embedded in M bounds a ball. We will say that M is a *hyperbolic manifold* if its interior can be endowed with a complete hyperbolic metric. Let Σ be a subsurface of ∂M; an *essential disc* in (M,Σ) is a disc D properly embedded in (M,Σ) that can not be mapped to ∂M by a homotopy fixing ∂D. The simple closed curve ∂D is a *meridian* curve. The manifold M is *boundary irreducible* if there is no essential disc in $(M,\partial M)$. An *essential annulus* in (M,Σ) is an incompressible annulus A properly embedded in (M,Σ) which can not be mapped to ∂M by a homotopy fixing ∂A. Let A be an essential annulus in M; if one component of ∂A lies in a toric component of ∂M we will call the other component of ∂A a *parabolic curve*.

Let $m \subset \partial M$ be a simple closed curve; a simple arc $k \subset \partial M$ such that $k \cap m = \partial k$ is an *m-wave* if there is an arc $k' \subset m$ such that $k' \cup k$ bounds an essential disc. A leaf $\tilde{l}$ of a geodesic lamination $\tilde{L} \subset \partial \tilde{M}$ is *homoclinic* if it contains two sequences of points (x_n) and (y_n) such that the distance between the points x_n and y_n measured on $\tilde{l}$ goes to ∞ whereas their distance measured in $\tilde{M}$ is bounded. A leaf l of a geodesic lamination $L \subset \partial M$ is *homoclinic* if a (any) lift of l to $\partial \tilde{M}$ is a homoclinic leaf. Notice that, with this definition, a meridian or a leaf spiraling around a meridian is homoclinic.

Let $\rho : \pi_1(M) \to \text{Isom}(\mathbb{H}^3)$ be a faithful discrete representation such that $\mathbb{H}^3/\rho(\pi_1(M))$ is homeomorphic to the interior of M. Let $L_\rho \subset S^2 = \partial \overline{\mathbb{H}}^3$ be the limit set of $\rho(\pi_1(M))$, let $C(\rho) \subset \mathbb{H}^3$ be the convex hull of L_ρ and let $C(\rho)^{ep}$ be the intersection of $C(\rho)$ with the preimage of the thick part of $\mathbb{H}^3/\rho(\pi_1(M))$. The quotient $N(\rho)$

of $C(\rho)$ by $\rho(\pi_1(M))$ is the convex core of ρ and ρ is said to be *geometrically finite* if $N(\rho)$ has finite volume. A geometrically finite representation $\rho : \pi_1(M) \to \mathrm{Isom}(\mathbb{H}^3)$ such that $\mathbb{H}^3/\rho(\pi_1(M))$ is homeomorphic to the interior of M is said to *uniformize M*. If ρ uniformize M, there is a natural homeomorphism (defined up to homotopy) $h : \tilde{M} \to C(\rho)^{ep}$ coming from the retraction map $S^2 - L_\rho \to C(\rho)^{ep}$. Let us choose a geometrically finite representation ρ with only rank 2 maximal parabolic subgroups (namely the maximal subgroups of $\rho(\pi_1(M))$ containing only parabolic isometries have rank 2). We will define the compactification $\overline{\tilde{M}}$ of $\tilde{M}$ as the closure of $h(\tilde{M}) = C(\rho)^{ep}$ in the usual unit ball compactification of $\mathbb{H}^3$. This compactification does not depend on the choice of the representation ρ (see [Lec02, section 2.1]). We will call this compactification the Floyd-Gromov compactification of $\tilde{M}$.

Let $\tilde{l}_+ \subset \partial\tilde{M}$ be a half-geodesic and let $\overline{\tilde{l}}_+$ be its closure in $\overline{\tilde{M}}$; we will say that $\tilde{l}_+$ has a *well defined endpoint* if $\overline{\tilde{l}}_+ - \tilde{l}_+$ contains one point. We will say that a geodesic $\tilde{l} \subset \partial\tilde{M}$ has two well defined endpoints if $\tilde{l}$ contains two disjoint half geodesics each having a well defined endpoint. Two distinct leaves $\tilde{l}_1$ and $\tilde{l}_2$ of a geodesic lamination $\tilde{L} \subset \partial\tilde{M}$ will be said to be *biasymptotic* if they both have two well defined endpoints in $\overline{\tilde{M}}$ and if the endpoints of $\tilde{l}_1$ are the same as the endpoints of $\tilde{l}_2$. A geodesic lamination $A \subset \partial M$ is *annular* if the preimage of A in $\partial\tilde{M}$ contains a pair of biasymptotic leaves.

2.5. Pleated surfaces

Let $\rho : \pi_1(M) \to \mathrm{Isom}(\mathbb{H}^3)$ be a discrete faithful representation and let $N = \mathbb{H}^3/\rho(\pi_1(M))$. A *pleated surface* in N is a map $f : S \to N$ from a surface S to N with the following properties:

- the metric s on S obtained by pulling back the metric induced on $f(S) \subset N$ by the length of rectifiable paths is a hyperbolic metric;

- every point in S lies in the interior of some s-geodesic arc that is mapped to a geodesic arc in N;

The *pleating locus* of a pleated surface is the set of points of S where the map fails to be a local isometry. The pleating locus of a pleated map is a geodesic lamination (cf. [Thu80]).

Let $\rho : \pi_1(M) \to \mathrm{Isom}(\mathbb{H}^3)$ be a discrete faithful representation such that there is a homeomorphism $h : \mathrm{int}(M) \to N = \mathbb{H}^3/\rho(\pi_1(M))$ and let $S \subset M$ be a properly embedded surface homeomorphic and homotopic to ∂M. A measured geodesic lamination $\lambda \in \mathcal{ML}(\partial M)$ is realized by a pleated surface in N if there is a pleated surface $f : S \to N$ homotopic to $h_{|S}$ such that the restriction of f to the support of λ is an isometry.

2.6. Masur domain

Let M be a compression body that is neither a solid torus nor an I-bundle over a closed surface; its boundary has a unique compressible component, the *exterior boundary* that we will denote by $\partial_e M$. Let $\mathcal{PML}(\partial_e M)$ be the space of projective measured geodesic laminations on $\partial_e M$ and let $\mathcal{M}'$ be the closure in $\mathcal{PML}(\partial_e M)$ of the set of projective classes of weighted meridians. The compression body M is said to be a *small compression body* if it is the connected sum along the boundary of an I-bundle over a closed surface and a solid torus or of two I-bundles over closed surfaces. It is said to be a *large compression body* otherwise. When M is a large compression body, the Masur domain is defined as follows:

$$O = \{\lambda \in \mathcal{PML}(\partial_e M) | i(\lambda, \mu) > 0 \text{ for any } \mu \in \mathcal{M}'\}.$$

When M is a small compression body, the definition is the following one:

$O = \{\lambda \in \mathcal{PML}(\partial_e M) | i(\lambda, \nu) > 0$ for any $\nu \in \mathcal{PML}(\partial_e BC)$ such that there is $\mu \in \mathcal{M}'$ with $i(\mu, \nu) = 0\}$.

We will denote by $\hat{O} \subset \mathcal{ML}(\partial M)$ the set of measured geodesic laminations whose projective class lies in O.

Let M be an orientable hyperbolic 3-manifold such that ∂M has negative Euler characteristic. We will say that a measured geodesic lamination $\lambda \in \mathcal{ML}(\partial M)$ is *doubly incompressible* if and only if

$$\exists \eta > 0 \text{ such that } i(\lambda, \partial E) \geq \eta \text{ for any essential annulus or disc } E.$$

We will denote by $\mathcal{D}(M) \subset \mathcal{ML}(\partial M)$ the set of doubly incompressible measured geodesic laminations.

Doubly incompressible multi-curve were first introduced by W. Thurston in [Thu] and we have the following equivalence: $(\partial M, |\gamma|, \subset)$ is doubly incompressible (in the sense of [Thu]) if and only if there is a weighted multi-curve $\gamma \subset \mathcal{ML}(\partial M)$ with support $|\gamma|$ satisfying the condition above except in the following situation (in which γ lies in $\mathcal{D}(M)$ but $(\partial M, |\gamma|, \subset)$ is not doubly incompressible in Thurston's sense):

$(-)$ there is a homeomorphism between M and an I-bundle over a pair of pants P such that $|\gamma|$ is mapped to a section of the bundle over ∂P.

The set $\mathcal{D}(M)$ of doubly incompressible measured geodesic laminations is the extension of Masur domain we will study in this paper.

3. Relations between $O(M)$, $\mathcal{D}(M)$ and $\mathcal{P}(M)$

When a statement deals with the Masur domain, it means that we have assumed that M is a compression body but neither a solid torus nor an I-bundle over a closed surface.

Lemma 3.1. *The set $\hat{O}$ is a subset of $\mathcal{D}(M)$.*

Proof. Let $\lambda \notin \mathcal{D}(M)$ be a measured geodesic lamination. We will show, using the following lemma of [Ota88], that $\lambda \notin \hat{O}$.

Lemma 3.2 ([Ota88]). *Let E be an essential annulus in a large compression body M; then there is a projective measured geodesic lamination $\mu \in \mathcal{M}'$ with support lying in ∂E.*

Proof. Since [Ota88] is not published, we will write the details of the proof. The boundary of ∂M has only one compressible component $\partial_e M$ called the exterior boundary. Let us choose a complete hyperbolic metric on $\partial_e M$.

Claim 3.3. *Let $c \subset \partial_e M$ be a simple closed curve that is disjoint from one non separating meridian or from two separating meridians. Then there is a projective measured geodesic lamination $\mu \in \mathcal{M}'$ whose support is c.*

Proof. Let us first consider that there is a non separating meridian m disjoint from c. Let D be an essential disc bounded by m. Since m does not separate ∂M, there is a sequence of simple closed curves (c_i) that approximates c, namely the sequence (c_i) converges to c in $\mathcal{PML}(\partial M)$, such that each c_i intersects m in one point. Consider a small neighborhood $\mathcal{V}_i$ of $D \cup c_i$ in M. The closure of $\partial \mathcal{V}_i - \partial M$ is an essential disc D_i and the sequence (∂D_i) converges to c in $\mathcal{PML}(\partial M)$.

Let us now assume that there are two disjoint separating meridians m_1 and m_2 that do not intersect c. Let D_1 and D_2 be two essential discs bounded by m_1 and m_2 respectively. Let N be the closure of the connected component of $M - (D_1 \cup D_2)$ whose boundary contains c. If N intersects D_1 and D_2, we can approximate c by a sequence of arcs k_i joining m_1 to m_2. Let $\mathcal{V}_i$ be a small neighborhood of $D_1 \cup k_i \cup D_2$. The closure of $\partial \mathcal{V}_i - \partial M$ is an essential disc Δ_i and the sequence $(\partial \Delta_i)$ converges to c in $\mathcal{PML}(\partial M)$.

If N intersects only one disc D_1 or D_2, by considering an arc in $\partial M - N$ joining D_1 to D_2, we can construct an essential disc D_3 such that one component of $M - (D_1 \cup D_3)$ or of $M - (D_2 \cup D_3)$ contains c and intersects D_1 and D_3 or D_2 and D_3. Thus we are in the previous case and we can conclude as above. $\square$

To prove Lemma 3.2, it remains to consider the case where there is at most one meridian disjoint from E and this meridian separates M.

Let us assume that the two components of ∂E are not homotopic in ∂M. Since M is a large compression body, E intersects a meridian m. Let us choose an orientation for E and let $\psi : M \rightarrow M$ be the Dehn twist along E. The curve $\psi^n(m)$ is a meridian. The restriction of ψ to ∂M is a Dehn twist along ∂E. It follows that the sequence $(\psi^n(m))$ tends to a projective measured geodesic lamination $\mu \in \mathcal{M}'$ with $|\mu| \subset \partial E$.

Consider now that there is an annulus $E' \subset \partial M$ with $\partial E' = \partial E$. By cutting M along an essential disc disjoint from E (if there is one), we can assume that E intersects any essential disc in M. Since there is at most one meridian disjoint from E, the part we have removed (if there is one) has incompressible boundary (so it is an I-bundle over a closed surface). The resulting manifold M is still a compression body but it could be an I-bundle over a closed surface or a solid torus. In particular, M is atoroidal and $E \cup E'$ bounds a solid torus $T \subset M$. Since the components of ∂E are isotopic on ∂M, each component of ∂E represents a non primitive element in $\pi_1(M)$. It follows that M can not be an I-bundle over a closed surface. Especially M contains an essential disc D and D intersects ∂E transversely. Choose D so that the geometric intersection between ∂D and ∂E is minimal in the homotopy class of ∂D. Let D' be an outermost component of $D - E$, namely the closure of D' is bounded by an arc $k \subset \partial D - \partial E$ and by an arc $k' \subset E$. If k' is homotopic in E relatively to its endpoints to an arc lying in ∂E, then D' is homotopic to an essential disc that is disjoint from E. This contradicts the fact that ∂E intersects any essential disc. So k' is not homotopic in E relatively to its endpoints to an arc lying in ∂E. Consider a small neighborhood $\mathcal{V}$ of $E \cup D'$. The closure of $\partial \mathcal{V} - \partial M$ is the union of an essential annulus parallel to E and a properly embedded disc Δ that does not intersect ∂E. The manifold $W = T \cup \mathcal{V}$ is a solid torus and the closure of $\partial W - \partial M$ is Δ'. Since Δ' does not intersect ∂E, it is not an essential disc. So Δ is homotopic to a disc $\Delta' \subset \partial M$. The sphere $\Delta \cup \Delta' \subset M$ bounds a ball and M is the union of W and this ball. It follows that M is a solid torus. Recalling that we may have cut M along an essential disc, we conclude that M was originally the connected sum along the boundary of a solid torus and an I-bundle over a closed surface. This contradicts our assumption that M is a large compression body. □

Let λ be a measured geodesic lamination such that $\lambda \notin \mathcal{D}(M)$. Then there is a sequence of essential discs or annuli $E_n \subset M$ such that $i(\lambda, \partial E_n) \longrightarrow 0$. We will show that $\lambda \notin \hat{O}$.

We will first assume that M is a large compression body. By Lemma 3.2, there is a sequence of multi-curves (e_n) such that $e_n \subset \partial E_n$ and that $e_n \in \mathcal{M}'$. Let $\varepsilon > 0$ and let εe_n be the weighted multi-curve obtained by endowing each leaf of e_n with a Dirac mass with weight ε. Up to extracting a subsequence, there is a sequence (ε_n) converging to 0 such that the sequence $(\varepsilon_n e_n)$ converges to some measured geodesic lamination α. Since $\varepsilon_n e_n \in \mathcal{M}'$ for any n, then $\alpha \in \mathcal{M}'$. Since we have $\varepsilon_n \longrightarrow 0$ and $i(\lambda, e_n) = 0$, we have $i(\lambda, \alpha) = 0$ hence $\lambda \notin \hat{O}$.

Let us assume now that M is a small compression body. By the proof of Lemma 3.2, for each n, either E_n is disjoint from an essential disc or there is a projective measured geodesic lamination $\mu \in \mathcal{M}'$ with support lying in ∂E_n. Especially, for any n, there is $\mu_n \in \mathcal{M}'$ with $i(\mu_n, \partial E_n) = 0$. Furthermore, we can choose the μ_n such that a subsequence of (μ_n) converges in $\mathcal{M}L(\partial M)$ to a measured geodesic lamination $\mu \in \mathcal{M}'$. Let $e_n \subset \partial E_n \cap \partial_e M$ be a simple closed curve and choose $\varepsilon_n \longrightarrow 0$ such that the sequence $(\varepsilon_n e_n)$ converges to some measured geodesic lamination α. We have then $i(\alpha, \mu) = 0$ and $i(\alpha, \lambda) = 0$ hence $\lambda \notin \hat{O}$.

Thus we have shown that if $\lambda \notin \mathcal{D}(M)$, then $\lambda \notin \hat{O}$. $\square$

The opposite is not true but we have the following:

Lemma 3.4. *Let $\lambda \in \mathcal{D}(M)$ be an arational measured geodesic lamination; then λ lies in $\hat{O}$.*

Proof. Let us assume the contrary.

If M is a large compression body, there is $\mu \in \mathcal{M}'$ such that $i(\mu, \lambda) = 0$. It follows from the assumption that λ is arational that λ and μ share the same support $|\mu|$. Since $\mu \in \mathcal{M}'$, there is a sequence of meridians $c_n \subset \partial M$ and a sequence $\varepsilon_n \longrightarrow 0$ such that $\varepsilon_n c_n$ converges to μ in the topology of $\mathcal{M}L(\partial M)$. Up to extracting a subsequence, (c_n) converges in the Hausdorff topology to a geodesic lamination L and we have $|\mu| \subset L$. By Casson's criterion (cf. [Ota88], [Lec02, Theorem B.1] or [Lec04b]), L contains a homoclinic leaf l. Since $|\mu| \subset L$ is the support of λ, l does not intersect λ transversely. This contradicts Lemma 3.6 below.

If M is a small compression body, there are $\mu \in \mathcal{M}'$ and $\alpha \in \mathcal{M}L(\partial_e M)$ such that $i(\mu, \alpha) = i(\alpha, \lambda) = 0$. Since λ is arational, λ, α and μ share the same support. Using the fact that λ and μ have the same support, we can finish the proof in the same way as in the case of a large compression body. $\square$

In [Lec02] (see also [Lec04b]), one studied the subset $\mathcal{P}(M)$ of $\mathcal{M}L(\partial M)$ defined as follows. Let $\lambda \in \mathcal{M}L(\partial M)$ be a measured geodesic lamination; then $\lambda \in \mathcal{P}(M)$ if and only if:

(a) no closed leaf of λ has a weight greater than π;

(b) $\exists \eta > 0$ such that, for any essential annulus E, $i(\partial E, \lambda) \geq \eta$;

(c) $i(\lambda, \partial D) > 2\pi$ for any essential disc D.

Let $\rho : \pi_1(M) \to \text{Isom}(\mathbb{H}^3)$ be a geometrically finite representation uniformizing M and let h be an isotopy class of homeomorphisms $M \to N(\rho)^{ep}$ homotopic to the identity; we will denote by $\mathcal{G}\mathcal{F}(M)$ the set of such pairs (ρ, h). There is a well defined map

$b: \mathcal{GF}(M) \to \mathcal{ML}(\partial M)$ which to a pair (ρ, h) associates the preimage under h of the bending measured geodesic lamination of $N(\rho)$, let us call this map the bending map. It is shown in [BO] and [Lec02] that $\mathcal{P}(M)$ is the image of b.

In [Lec02], it was proved that a measured geodesic lamination lying in $\mathcal{P}(M)$ intersects transversely all the homoclinic leaves and all the annular laminations. In order to get the same property for the laminations lying in $\mathcal{D}(M)$, we will discuss the relationships between $\mathcal{P}(M)$ and $\mathcal{D}(M)$.

We clearly have $\mathcal{P}(M) \subset \mathcal{D}(M)$, conversely, we have:

Lemma 3.5. *Let $\lambda \in \mathcal{D}(M)$ be a measured geodesic lamination not satisfying condition $(-)$, then there is a measured geodesic lamination $\alpha \in \mathcal{P}(M)$ with the same support as λ.*

Proof. Since $\lambda \in \mathcal{D}(M)$, $\exists \eta > 0$ such that $i(\partial E, \lambda) > \eta$ for any essential annulus or disc E. Let $\frac{2\pi}{\eta}\lambda$ be the measured geodesic lamination obtained by multiplying the measure λ by $\frac{2\pi}{\eta}$; then $\frac{2\pi}{\eta}\lambda$ satisfies the properties $b)$ and $c)$ above. Let $\lambda^{(p)}$ be the union of the leaves of $\frac{2\pi}{\eta}\lambda$ with a weight greater than π and let α be the measured geodesic lamination obtained from $\frac{2\pi}{\eta}\lambda$ by decreasing the weight of the leaves of $\lambda^{(p)}$ to π. This measured geodesic lamination α satisfies $a)$ and $b)$, let us show that it satisfies also $c)$.

Let $D \subset M$ be an essential disc; then $i(\frac{2\pi}{\eta}\lambda, \partial D) > 2\pi$. If ∂D does not intersect $\lambda^{(p)}$ transversely, we have $i(\lambda', \partial D) = i(\frac{2\pi}{\eta}\lambda, \partial D) > 2\pi$.

If ∂D intersects $\lambda^{(p)}$ in one point x, let c be the leaf of $\lambda^{(p)}$ containing x. Let $\mathcal{V}$ be a small neighborhood of $c \cup D$; $\mathcal{V}$ is a solid torus. Let D' be the closure of $\partial \mathcal{V} - \partial M$, D' is a disc properly embedded in M not intersecting $\lambda^{(p)}$ hence $i(\partial D', \alpha) = i(\partial D', \frac{2\pi}{\eta}\lambda)$. If D' is not an essential disc, then $\partial D'$ bounds a disc $D'' \subset \partial M$. Since M is irreducible, $D' \cup D''$ bounds a ball $B \subset M$ and $M = B \cup \mathcal{V}$ is a solid torus. By assumption, M is not a solid torus hence D' is an essential disc and $i(\partial D', \frac{2\pi}{\eta}\lambda) > 2\pi$. By construction, we have $i(\partial D', \alpha) \leq 2(i(\partial D, \alpha) - \pi)$, therefore we have $i(\partial D, \alpha) \geq \frac{i(\partial D', \alpha)}{2} + \pi = \frac{i(\partial D', \frac{2\pi}{\eta}\lambda)}{2} + \pi > 2\pi$.

If ∂D intersects $\lambda^{(p)}$ in two points x and y, we have $i(\alpha, \partial D) = 2\pi + i(\frac{2\pi}{\eta}\lambda - \lambda^{(p)}, \partial D)$. Hence we just have to show that $i(\lambda - \lambda^{(p)}, \partial D) > 0$. Assuming the contrary, we have $\lambda \cap \partial D = \{x, y\}$. If x and y lie in two distinct leaves $c \subset |\lambda|$ and $d \subset |\lambda|$, let $\mathcal{V}$ be a small neighborhood of $c \cup d \cup D$; $\mathcal{V}$ is an I-bundle over a pair of pants. The closure of $\partial \mathcal{V} - \partial M$ is an annulus with boundary not intersecting $|\lambda|$. By condition $b)$, this annulus is not essential. It follows that M is an I-bundle over a pair of pants P and that $|\lambda|$ lies in a section of the bundle over ∂P. This contradicts our assumptions hence x and y lie in the same leaf c of $\lambda^{(p)}$.

Let $\mathcal{V}$ be a small neighborhood of $c \cup D$; it is again an I-bundle over a pair of pants. If the tangents vectors $\frac{dc}{dt}|_x$ and $\frac{dc}{dt}|_y$ do not point to the same side of ∂D, the closure of $\partial \mathcal{V} - \partial M$ is the union of two annuli with boundaries not intersecting λ. This yields the same contradiction as above.

Next let us consider the case where $\frac{dc}{dt}|_x$ and $\frac{dc}{dt}|_y$ point to the same side of ∂D. Let k be a connected component of $c - \{x, y\}$ and let $\mathcal{V}'$ be a small neighborhood of $k \cup D$; the closure of $\partial \mathcal{V}' - \partial M$ is an essential disc D'. Replacing D by D', we are in the situation of the previous paragraph and get the same contradiction.

If ∂D and $\lambda^{(p)}$ intersect each other in more than 2 points, $i(\lambda', \partial D) \geq 3\pi$.   □

Combining Lemma 3.5 and results of [Lec02] (see also [Lec04b]) we get the following:

Lemma 3.6. *A measured geodesic lamination* $\lambda \in \mathcal{D}(M)$ *not satisfying condition* $(-)$ *has the following property:*

λ *intersects transversely any annular lamination and any geodesic lamination containing a homoclinic leaf.*

Remark 3.7. Let us add a few comments about the case where λ satisfies condition $(-)$. Any homoclinic leaf l intersects λ at least once. If an annular geodesic lamination A does not intersect λ transversely, then A contains two disjoint half-leaves both spiraling in the same direction toward the same leaf of λ. This can not happen for a Hausdorff limit of multi-curves. Therefore λ has the property above if we consider only annular laminations that are Hausdorff limits of multi-curves.

4. Topological properties of $\mathcal{D}(M)$

Lemma 4.1. *The set* $\mathcal{D}(M)$ *is an open set.*

Proof. Let us assume the contrary. Then there are $\lambda \in \mathcal{D}(M)$ and a sequence of measured geodesic laminations $\lambda_n \notin \mathcal{D}(M)$ converging to λ. Therefore there is a sequence of essential discs or annuli E_n such that $i(\lambda_n, \partial E_n) \longrightarrow 0$. Let us extract a subsequence such that ∂E_n converge in the Hausdorff topology to a geodesic lamination A. Then A does not intersect λ transversely. By [Lec02] (see also [Lec04b]) either A contains a homoclinic leaf ([Lec02, Theorem B1]) or A is annular ([Lec02, Lemma C2]), both contradicting Lemma 3.6.   □

A train track τ carrying a measured geodesic lamination is *complete* if it is not a subtrack of a train track carrying a measured geodesic lamination (cf. [PH92]).

Any measured geodesic lamination λ is carried by some (maybe many) complete train track τ. The weight system on a complete train track gives rise to a coordinate system for a simplex of the piecewise linear manifold $\mathcal{M}L(\partial M)$. The *rational depth* of a measured geodesic lamination λ is the dimension of the rational vector space of linear functions with rational coefficients (from the simplex previously defined to $\mathbb{R}$) vanishing on the coordinates of λ. Let us denote by $I(\partial M)$ the set of measured geodesic laminations with rational depth equal to 0. If a measured geodesic lamination λ lies in I, then λ is arational (cf. [Thu80, Proposition 9.5.12]).The set I is a dense open subset of $\mathcal{M}L(\partial M)$ (cf. [Thu80, chap 9]).

Proposition 4.2. *The set $\mathcal{D}(M)$ is path-connected.*

Proof. Let $\lambda_1, \lambda_2 \in \mathcal{D}(M)$; since I is a dense subset of $\mathcal{M}L(\partial M)$ and since $\mathcal{D}(M)$ is open, there are α_1 and $\alpha_2 \in \mathcal{D}(M) \cap I$ such that λ_j is connected to α_j by a path $k_j \subset \mathcal{D}(M)$.

Since $\alpha_j \in \mathcal{D}(M)$ there is $\eta > 0$ such that $i(\alpha_j, \partial E) > \eta$ for any essential disc or annulus $E \subset M$. Since $\alpha_j \in I$, it has no closed leaf and by the proof of Lemma 3.5, $\frac{2\pi}{\eta}\alpha_j \in \mathcal{P}(M)$. Let $CC(M) \subset \mathcal{GF}(M)$ be the set of hyperbolic metrics uniformizing M and having only rank 2 cusps; by results of Ahlfors-Bers ([Ber60]), $CC(M)$ is homeomorphic to the Cartesian product of the Teichmüller spaces of the connected components of $\partial_{\chi<0}M$, indeed $CC(M)$ is path-connected. Let $\mathcal{P}_{nc}(M)$ be the set of measured geodesic laminations lying in $\mathcal{P}(M)$ and having no closed leaves with weight π. By [Lec02] (see also [Lec04b]) $\mathcal{P}_{nc}(M)$ is the image of $CC(M)$ under the bending map. By [KS95] and [Bon98], the bending map is continuous on $CC(M)$ hence $\mathcal{P}_{nc}(M)$ is path-connected. Since $\frac{2\pi}{\eta}\alpha_j$ has no closed leaf, $\frac{2\pi}{\eta}\alpha_j \in \mathcal{P}_{nc}(M)$. Therefore there is a path $\alpha : [0,1] \to \mathcal{P}(M) \subset \mathcal{D}(M)$ such that $\alpha(0) = \frac{2\pi}{\eta}\alpha_1$ and that $\alpha(1) = \frac{2\pi}{\eta}\alpha_2$. Let $\kappa_j : [0,1] \to \mathcal{D}(M)$ be the path defined by $\kappa_j(t) = (1 - t + t\frac{2\pi}{\eta})\alpha_j$. The union of the paths k_j, κ_j for $j = 1,2$ and of the path $\alpha([0,1])$ is a path lying in $\mathcal{D}(M)$ joining λ_1 to λ_2.

Thus we get that $\mathcal{D}(M)$ is path-connected. $\square$

Let us discuss now the connectedness of $\hat{O}$. Assume $\lambda_1, \lambda_2 \in \hat{O}$ and follow the proof of Proposition 4.2 replacing $\mathcal{D}(M)$ by $\hat{O}$. The only problem is that we do not have $\mathcal{P}(M) \subset \hat{O}(M)$. But by Lemmas 3.5 and 3.4, if a measured geodesic lamination lying in $\mathcal{P}(M)$ is arationnal, it lies in $\hat{O}$. So if we can choose α so that $\alpha(t)$ is arational for any $t \in [0,1]$, we have $\alpha([0,1]) \subset \hat{O}(M)$. Especially if the set of arational measured geodesic laminations is path-connected, the set $\hat{O}$ is path-connected.

5. Pleated surfaces

Theorem 5.1. *Let M be an orientable 3-manifold, let $\rho : \pi_1(M) \to \mathrm{Isom}(\mathbb{H}^3)$ be a geometrically finite representation uniformizing N and having only rank 2 maximal parabolic subgroups and let $h : N = \mathbb{H}^3/\rho(\pi_1(M)) \to \mathrm{int}(M)$ be a homeomorphism; then any measured geodesic lamination $\lambda \in \mathcal{D}(M)$ is realized by a pleated surface in N.*

Proof. If M is a compression body and λ is arational, then λ lies in the Masur domain and the theorem has been proved by Otal ([Ota88]). If M is boundary irreducible, then any geodesic lamination is realized in N (see [CEG87, chap. 5]). In order to prove our general statement, we will follow the main lines of Otal's proof.

Lemma 5.2. *Let $\lambda \in \mathcal{D}(M)$ be a weighted multi-curve, then λ is realized by a pleated surface in N.*

Proof. Let us extend $|\lambda|$ to a geodesic lamination L (namely $|\lambda| \subset L$) such that all the components of $\partial M - L$ are triangles and that L has finitely many leaves. Since $\lambda \in \mathcal{D}(M)$ and since ρ has only rank 2 cusps, any closed leaf of L is homotopic to a closed geodesic in N. Let $S \subset M$ be a properly embedded surface homeomorphic and homotopic to ∂M and let us change the restriction of h to S by a homotopy in order to get a map $f : S \to N$ mapping the closed leaves of L into closed geodesics. For each connected component of S, let us lift this to a map $\hat{f} : \mathbb{H}^2 \to \mathbb{H}^3$; this map $\hat{f}$ defines a map from the endpoints of the lifts of the leaves of L to L_ρ. Furthermore, if $\hat{l} \in \mathbb{H}^2$ is a lift of a leaf of L, by Lemma 3.6, the images of its two endpoints are distincts. Following [CEG87, Theorem 5.3.6], this allows us to construct a pleated surface realizing L. $\square$

Now let us consider the general case. Let $\lambda \in \mathcal{D}(M)$ be a measured geodesic lamination; let λ_n be a sequence of weighted multi-curves such that $\lambda_n \longrightarrow \lambda$ in $\mathcal{M}L(\partial M)$ and that $|\lambda_n| \to |\lambda|$ in the Hausdorff topology. Since $\mathcal{D}(M)$ is open, $\lambda_n \in \mathcal{D}(M)$ for large n. Let γ be a weighted multi-curve with a maximal number of leaves such that $i(\lambda, \gamma) = 0$; since $\lambda_n \in \mathcal{D}(M)$ for large n, $\lambda_n \cup \gamma$ is also a measured geodesic lamination lying in $\mathcal{D}(M)$. By the previous lemma, $\lambda_n \cup \gamma$ is realized by a pleated surface $f_n : S \to N$. We will show that a subsequence of (f_n) converges to a pleated surfaces realizing λ.

Let us denote by s_n the metric on S induced by the map $f_n : S \to N$ and let us show that (s_n) contains a converging subsequence. First we will prove that the sequence of metrics (s_n) is bounded in the modular space. By Mumford's Lemma, it is sufficient to prove that the injectivity radius of s_n is bounded from below.

Claim 5.3. *Let* (c_n) *be a sequence of curves such that* $l_{s_n}(c_n) \longrightarrow 0$ *and let us extract a subsequence* (c_n) *which converges in the Hausdorff topology to a geodesic lamination* C; *then* C *does not intersects* λ *transversely.*

Proof. Let assume the contrary and let c be a leaf of C intersecting λ transversely. Since λ is recurrent, we can consider a segment $k = k([0,1])$ of $|\lambda|$ such that $k \cap C = \partial k$ and that $\frac{dk}{dt}(0)$ is close (for some reference metric on S) to $-\frac{dk}{dt}(1)$ and a short segment κ of c joining the ends of k so that we get a closed curve $d = k \cup \kappa$. Since $\lambda_n \longrightarrow \lambda$ and $c_n \longrightarrow C$, there exist arcs $k_n \subset \lambda_n$ and $\kappa_n \subset c_n$ near k and κ such that $d_n = k_n \cup \kappa_n$ is homotopic on S to d. Since $l_{s_n}(c_n) \longrightarrow 0$, c_n is the core of a very deep Margulis tube and $l_{s_n}(k_n) \longrightarrow \infty$. Since $l_{s_n}(\kappa_n) \leq l_{s_n}(c_n) \longrightarrow 0$ and $f_n(k_n) \subset f_n(\lambda_n)$ is a geodesic arc, $f_n(d_n) = f_n(k_n \cup \kappa_n)$ is a quasi-geodesic and is very close to the geodesic d_n^* of N in its homotopy class. This implies that $l_\rho(d_n^*) \longrightarrow \infty$ but d_n is homotopic to d so $d_n^* = d^*$ giving the expected contradiction. $\qquad\square$

Let (c_n) be a sequence of curves such that $l_{s_n}(c_n) \longrightarrow 0$. If we can extract a converging (in the Hausdorff topology) subsequence such that all the c_n are meridians then, by Casson's criterion (cf. [Ota88], [Lec02, Theorem B.1]), the limit contains a homoclinic leaf. By Lemma 3.6 such a homoclinic leaf intersects λ transversely contradicting Claim 5.3. This implies that for large n, the c_n are not meridians. If we can extract a converging subsequence such that all the c_n are parabolic curves, then $i(c_n, \lambda) > \eta$ for any n, leading to the same contradiction.

It follows that, for large n, each $f_n(c_n)$ is homotopic to a closed geodesic c_n^* of N. But this would mean that $l_\rho(c_n^*) \longrightarrow 0$ and since N is geometrically finite, there is a uniform lower bound for the length of a closed geodesic. We get then from Mumford's Lemma ([CEG87, Proposition 3.2.13]):

Claim 5.4. *The sequence* (s_n) *is bounded in the moduli space.*

Let us now show that (s_n) is bounded in the Teichmüller space. By the previous claim, there exists a sequence (φ_n) of diffeomorphisms such that, up to extracting a subsequence, $(\varphi_n^* s_n)$ converges in the Teichmüller space to a metric s'_∞. By construction $l_{\varphi_n^* s_n}(\varphi_n^{-1}(\gamma)) = l_{s_n}(\gamma) = l_\rho(\gamma)$, therefore the s'_∞-length of the multi-curve $\varphi_n^{-1}(\gamma)$ is bounded. This implies that we can choose some n_0 and a subsequence such that any diffeomorphism $(\varphi_n^{-1} \circ \varphi_{n_0})$ preserves this multi-curve, component by component.

For large n, λ_n intersects transversely all the parabolic curves. Therefore λ_n intersects the thick part of N which is compact. It follows that all the $f_n(S)$ intersect the same compact subset of N. Using Ascoli's theorem we can choose a subsequence of (φ_n) such that the sequence of pleated surfaces $(f_n \circ \varphi_n)$ converges. This implies that the maps $f_n \circ \varphi_n$ are homotopic for n sufficiently large. Thus, up to changing

n_0, the diffeomorphisms $\psi_n = \varphi_n^{-1} \circ \varphi_{n_0}$ are homotopic in M to the identity. Let R be a complementary region of γ. If the map $i^* : \pi_1(R) \to \pi_1(M)$ induced by the inclusion is injective, then by [Wal68], $\psi_{n|R}$ is isotopic to the identity in S. If the map $i^* : \pi_1(R) \to \pi_1(M)$ is not injective, R contains a meridian. Since $\lambda \in \mathcal{D}(M)$, R must contain a component λ^i of λ and since γ has a maximal number of components, λ^i must be arational in R. Let us call r_n the restriction of s_n to R and suppose that the sequence (r_n) is not bounded in Teichmüller space. Since the length of ∂R is bounded, we can use Thurston's compactification and assume that (r_n) tends to a measured geodesic lamination ν. Since $l_{r_n}(\lambda_n \cap R) = l_\rho(\lambda_n \cap R) \leq l_{r_{n_0}}(\lambda_n \cap R) \to l_{r_{n_0}}(\lambda^i)$, $i(\nu, \lambda^i) = 0$ and ν and λ^i share the same support.

Let $m \subset R$ be a meridian. Then $m_n = \psi_n(m)$ is homotopic to m and therefore (m_n) is a sequence of meridians. We can assume that (m_n) converges in $\mathcal{PML}$ to a projective measured lamination represented by μ. Since $(\psi_n^* s_n)$ converges, then $l_{s_n}(m_n) = l_{\psi_n^* s_n}(\psi_n^{-1}(m_n)) = l_{\psi_n^* s_n}(m)$ converges and therefore $i(\mu, \nu) = 0$. Since ν and λ^i have the same support and since λ^i is arational in R, this implies that μ and λ^i have the same support. But Casson's criterion (c.f. [Ota88], [Lec02, Theorem B.1]) says that there exists a simple geodesic $l \subset R$ which is homoclinic and does not intersect μ transversely. This contradicts Lemma 3.6 and proves that the sequence (r_n) is bounded.

This applies to each component of $\partial M - \gamma$. It follows that we can choose the ψ_n such that each one is the composition of Dehn twists along the leaves of γ. We have seen above that the ψ_n are homotopic to the identity; by [Wal68], each ψ_n can be extended to a homeomorphism of the whole manifold M. Let $\mathcal{V} \subset S$ be a small neighbourhood of γ; since $\lambda \subset \mathcal{D}(M)$, $\mathcal{V}$ does not contain the boundary of any essential annulus. It follows then from [Joh79, Proposition 27.1] that, up to isotopy, each ψ_n has finite order. Since the ψ_n are compositions of Dehn twists along disjoint curves, they can not have finite order except when they are isotopic to the identity. We get from [CEG87] that a subsequence of (f_n) converges to a pleated surface realizing λ. $\square$

Let $f : S \to N$ be a pleated surface realizing a geodesic lamination L. Let $\mathbb{P}(N)$ be the tangent line bundle of N. We define a map $\mathbb{P}f$ from L to $\mathbb{P}(N)$ by mapping a point $x \in L$ to the direction of the unit vector tangent to $f(L)$ at $f(x)$.

The following injectivity theorem has been proved by Thurston ([Thu86]) when M is boundary irreducible and by Otal ([Ota88]) when M is a compression body and $\lambda \in \hat{O}$.

Theorem 5.5. *Let $\lambda \in \mathcal{D}(M)$ be a measured geodesic lamination not satisfying the condition $(-)$, let L be a geodesic lamination containing the support of λ and let $f : \partial M \to N$ be a pleated surface realizing L. Then the map $\mathbb{P}f : L \to \mathbb{P}(N)$ is a* *homeomorphism into its image.*

Proof. Since the map f reduces the length, it is easy to see that $\mathbb{P}f$ is a continuous map and since L is compact, we need only to show that $\mathbb{P}f$ is injective.

Let us assume the contrary, there are two points u and $v \subset L$ such that $\mathbb{P}f(u) = \mathbb{P}f(v)$; let $\hat{f} : \mathbb{H}^2 \to \mathbb{H}^3$ be a lift of f and let $\hat{u}$ and $\hat{v}$ be lifts of u and v such that $\mathbb{P}\hat{f}(\hat{u}) = \mathbb{P}\hat{f}(\hat{v})$. Since $\hat{f}$ is an isometry on the preimage of L, it is injective on each leaf of the preimage of L. Therefore $\hat{u}$ and $\hat{v}$ lie in two different leaves $\hat{l}_1$ and $\hat{l}_2$ of the preimage of L. Since $\mathbb{P}\hat{f}(\hat{u}) = \mathbb{P}\hat{f}(\hat{v})$, then $\hat{f}(\hat{l}_1) = \hat{f}(\hat{l}_2)$. It follows that L is an annular lamination and since L does not intersect $\lambda \in \mathcal{D}(M)$ transversely, this contradicts Lemma 3.6. $\qquad\square$

Remark 5.6. If λ satisfies the condition $(-)$, the same is true for λ but not for any geodesic lamination containing λ.

6. Action on $\mathbb{R}$-trees

We will prove the following:

Proposition 6.1. *Let $\mathcal{T}$ be a real tree, let $\pi_1(M) \times \mathcal{T} \to \mathcal{T}$ be a small minimal action and let $\lambda \in \mathcal{D}(M)$ be a measured geodesic geodesic lamination. Then at least one connected component of λ is realized in $\mathcal{T}$.*

Proof. Let us first notice that this result has been proved by G. Kleineidam and J. Souto ([KS02] and [KS03]) when M is a compression body and λ lies in the Masur domain. The general case need just a reorganization of the proof of [Lec02, Proposition 6]. Here we will sketch the proof which consists essentially in putting together ideas of [BO] and of [KS02].

If λ satisfies the condition $(-)$ then the elements of $\pi_1(M)$ corresponding to the leaves of λ form a generating subset of $\pi_1(M)$. In this case Proposition 6.1 is a straightforward consequence of [MS84].

Let us assume that λ does not satisfies the condition $(-)$. For $c \in \pi_1(M)$ let us denote by $\delta_\mathcal{T}(c)$ the distance of translation of c on $\mathcal{T}$. Let S be a connected component of ∂M with $\chi(S) < 0$; the inclusion $i_* : \pi_1(S) \to \pi_1(M)$ provides us with an action of $\pi_1(S)$ on $\mathcal{T}$. By [MO93], there exist a measured geodesic lamination $\beta \in \mathcal{ML}(S)$ and a morphism $\phi : \mathcal{T}_\beta \to \mathcal{T}_S$ from the dual tree of β to the minimal subtree of $\mathcal{T}$ that is invariant under the action of $\pi_1(S)$. Since the action of $\pi_1(S)$ is not a priori small, ϕ is not, a priori, an isomorphism and there might be many laminations β with this property. We will consider such a lamination β which is adapted to our problem.

Let (λ_n) be a sequence of weighted multi-curves converging to λ in $\mathcal{ML}(\partial M)$ such that $(|\lambda_n|)$ converges to $|\lambda|$ in the Hausdorff topology. For each irrational sublamination λ^i of λ let us denote by $S(\lambda^i)$ the surface embraced by $|\lambda^i|$. For n large enough such that $|\lambda_n|$ does not intersect $\partial'\bar{S}(\lambda)$ transversely, let us add simple closed curves to $\partial'\bar{S}(\lambda) \cup |\lambda_n|$ in order to obtain a multi-curve L_n whose complementary regions are pairs of pants. By [MO93], there are measured geodesic laminations $\beta_n \in \mathcal{ML}(\partial M)$ and equivariant morphisms $\phi_n : \mathcal{T}_{\beta_n} \to \mathcal{T}$ such that for any leaf l_n of L_n, either $\delta_{\mathcal{T}}(l_n) > 0$ and the restriction of ϕ_n to the axis of l_n is an isometry or $\delta_{\mathcal{T}}(l_n) = 0$ and $i(l_n, \beta_n) = 0$, see [Lec02, §4.1] for more details.

Extract a subsequence such that $(|\beta_n|)$ converges to a geodesic lamination B in the Hausdorff topology. The first step of the proof is to show that B intersects $|\lambda|$ transversely, this will allow us to follow [KS02] by using a realization of a train track carrying λ to prove the proposition.

Lemma 6.2. *The geodesic lamination B intersects $|\lambda|$ transversely.*

Proof. The proof is done by contradiction; let us assume that $|\lambda|$ does not intersect B transversely.

If B is a multi-curve, then for large n, $|\beta_n| = B$ and β_n does not intersect λ transversely. By the definition of $\mathcal{D}(M)$, a small neighbourhood of B does not contain any essential disk, annulus or Moebius band. By [MS84, Corollary IV 1.3], this implies that the action of $\pi_1(M)$ fixes a point of $\mathcal{T}$. This would contradict the assumption that this action is minimal.

Let us now consider the case where B is not a multi-curve. The first step in this case is to prove that $S(B)$ is incompressible for any connected component B^i of B. This will imply that a subsequence of $(|\beta_n|)$ is constant.

Claim 6.3. *If B does not intersect $|\lambda|$ transversely, then for any connected component B^i of B, the surface $S(B^i)$ is incompressible.*

Proof. Since we have assumed that B does not intersect $|\lambda|$ transversely, if B^i is a closed curve, the claim follows from the definition of $\mathcal{D}(M)$.

Let B^i be a component of B which is not a closed curve and let us assume that $S(B^i)$ contains a meridian. It follows from the ideas of [KS02], that $S(B^i)$ contains a homoclinic leaf h which does not intersect B^i transversely (see [Lec02, Lemma 4.3] for details). Since we have assumed that B does not intersect λ transversely, then $|\lambda| \cap S(B^i) \subset B^i$. Especially, h does not intersect λ transversely, contradicting Lemma 3.6. □

Let us explain how Claim 6.3 implies that for large n the support of β_n does not depend on n. Let B^i be a connected component of B; if B^i is a closed leaf then for large n, $B^i \subset |\beta_n|$. Let us next assume that B^i is not a closed leaf; by claim 6.3, $S(B^i)$ is incompressible, hence the action of $i_*(\pi_1(S(B^i)))$ on its minimal subtree $\mathcal{T}_{S(B^i)} \subset \mathcal{T}$ is small. Since B does not intersect $\partial'\bar{S}(B^i)$, for large n, β_n does not intersect $\partial'\bar{S}(B^i)$. It follows that for each component d of $\partial'\bar{S}(B^i)$, the action of $i_*(d)$ has a fixed point in $\mathcal{T}_{S(B^i)}$. This allows us to apply Skora's theorem [Sko90] which says that $\beta_n^i = \beta_n \cap S(B^i)$ is dual to the action of $i_*(\pi_1(S(B^i)))$ on $\mathcal{T}_{S(B^i)}$. Doing this for each component of B, we obtain that, for large n, $|\beta_n|$ does not depend on n. Let us endow B with the measure of one of the β_n and let us call β the measured geodesic lamination thus obtained.

The last step in the proof of Lemma 6.2 is to show that $|\beta| = B$ is annular. Since we have assumed that B does not intersect $|\lambda|$ transversely, this will contradict the fact that $\lambda \in \mathcal{D}(M)$ (Lemma 3.6).

Claim 6.4. *The measured geodesic lamination β is annular*

Proof. By hypothesis β does not intersect λ transversely hence $S(\beta) \cap |\lambda| \subset |\beta|$.

Since $S(\beta)$ is incompressible, we might consider a characteristic submanifold W of $(M, S(\beta))$ (cf. [Joh79] and [JS79]). Such a characteristic submanifold is a union of essential I-bundles and Seifert fibered manifolds such that any essential annulus in $(M, S(\beta))$ can be homotoped in W. For each component Σ of $\partial M - S(B)$, $i_*(\Sigma)$ fixes a point in $\mathcal{T}$, hence by [Thu] (see also [MS88, theorem IV 1.2]) W can be isotoped in such a way that we have $\beta \subset W \cap \partial M$.

We are considering the case where β is not a multi-curve, therefore it contains an irrational sublamination β^1. Since the Seifert fibered manifolds composing W intersect ∂M in annuli, $|\beta^1|$ lies in a component W^1 of W which is an essential I-bundle over a compact surface F: $W^1 = F \times I$. Let us denote by $p : F \times \partial I \to F$ the projection along the fibers. By Skora's theorem [Sko90], for any component Σ of $W^1 \cap \partial M$, $\Sigma \cap \beta$ is dual to the action of $i_*(\pi_1(S))$ on $\mathcal{T}_\Sigma$. Since this action factorizes through the action of $\pi_1(W^1) = \pi_1(F)$, there is a measured geodesic lamination $\beta' \in \mathcal{ML}(F)$ such that $\beta \cap \partial W^1 \supset p^{-1}(\beta')$. Since the lamination $p^{-1}(\beta')$ is annular, β is annular (compare with [BO, Lemma 14]). □

This claim concludes the proof of Lemma 6.2. □

Let us now complete the proof of Proposition 6.1. Let λ^i be a connected component of λ that intersects B transversely. Let us denote by $\pi_{\beta_n} : \mathbb{H}^2 \to \mathcal{T}_{\beta_n}$ the projection associated to the dual tree of β_n (as defined in §2.2). Since B intersects λ^i transversely, the construction in [Ota88, chap 3] yields a train track τ^i such that for large n, π_{β_n} is a weak realization of τ^i in $\mathcal{T}_{\beta_n}$.

Let l_n be a component of $L_n \cap S(\lambda^i)$. Up to extracting a subsequence, l_n converge in the Hausdorff topology to a geodesic lamination $L' \subset S(\lambda^i)$ that does not intersect λ^i transversely (by the choice of L_n). Therefore $|\lambda^i| \subset L'$. If up to extracting a subsequence, $i_*(l_n)$ has a fixed point in $\mathcal{T}$; then $i(\beta_n, l_n) = 0$. Letting n tend to ∞, we would get that B does not intersect $|\lambda^i|$ transversely, contradicting our choice of λ^i.

It follows from the previous paragraph that the restriction of ϕ_n to l_n is an isometry. For large n, each branch of $\hat{\tau}$ intersects transversely a lift of l_n. The fact that the restriction of ϕ_n to the axis of l_n is an isometry implies that $\phi_n \circ \pi_{\beta_n}$ is a weak realization of τ^i in $\mathcal{T}$ (compare with [KS02, Lemma 11]). By [Ota88] this map $\phi_n \circ \pi_{\beta_n}$ is homotopic to a realization of λ^i in $\mathcal{T}$. $\square$

Let $\rho_n : \pi_1(M) \to \mathrm{Isom}(\mathbb{H}^3)$ be a sequence of representations containing no converging subsequence; in [MS84], J. Morgan and P. Shalen described a way to associate a small minimal action of $\pi_1(M)$ on an $\mathbb{R}$-tree to some subsequence of (ρ_n). This can be stated in the following way: the sequence (ρ_n) tends to the action $\pi_1(M) \curvearrowright \mathcal{T}$ in the sense of Morgan and Shalen if there is a sequence $\varepsilon_n \longrightarrow 0$ such that for any $a \in \pi_1(M)$, $\varepsilon_n \delta_{\rho_n}(a) \longrightarrow \delta_\tau(a)$. In [Ota94], J.-P. Otal described, in the special case of handlebodies, the behavior of the length of measured geodesic laminations which are realized in $\mathcal{T}$. A careful look at the proof yields the following statement.

Theorem 6.5 (Continuity Theorem [Ota94]). *Let (ρ_n) be a sequence of discrete and faithful representations of $\pi_1(M)$ tending in the sense of Morgan and Shalen to a small minimal action of $\pi_1(M)$ on an $\mathbb{R}$-tree $\mathcal{T}$. Let $\varepsilon_n \longrightarrow 0$ be such that $\forall g \in \pi_1(M)$, $\varepsilon_n \delta_{\rho_n}(g) \longrightarrow \delta_{\mathcal{T}}(g)$ and let $L \subset \partial M$ be a geodesic lamination which is realized in $\mathcal{T}$. Then there exists a neighbourhood $\mathcal{V}(L)$ of L, and constants K, n_0 such that for any simple closed curve $c \subset \mathcal{V}(L)$ and for any $n \geq n_0$,*

$$\varepsilon_n l_{\rho_n}(c^*) \geq K l_{s_0}(c).$$

In the preceding statement s_0 is a fixed complete hyperbolic metric on $\partial_{\chi<0} M$. Using this and Proposition 6.1, we get the following

Theorem 6.6. *Let ρ_n be a sequence of faithful representations of $\pi_1(M)$ such that $\mathbb{H}^3/\rho_n(\pi_1(M))$ is homeomorphic to $\mathrm{int}(M)$, let $\lambda \in \mathcal{D}(M)$ and let λ_n be a sequence of measured geodesic laminations such that:*

– the sequence λ_n converges to λ in $\mathcal{ML}(\partial M)$;

– the sequence $|\lambda_n|$ converges to $|\lambda|$ in the Hausdorff topology;

– the sequence $l_{\rho_n}(\lambda_n)$ is bounded.

Then (ρ_n) *contains a converging subsequence.*

Proof. Approximating each λ_n by weighted multi-curves, we produce a sequence of multi-curves also satisfying the hypothesis of the theorem. Let us assume that (ρ_n) does not contain an algebraically converging subsequence, then by [MS84], a subsequence of (ρ_n) tends to a small minimal action of $\pi_1(M)$ on an $\mathbb{R}$-tree $\mathcal{T}$. By Proposition 6.1, λ is realized in $\mathcal{T}$ and it follows from Theorem 6.5 that $l_{\rho_n}(\gamma_n) \longrightarrow \infty$ giving us the desired contradiction. $\qquad\qquad\qquad\qquad\qquad\qquad\qquad\qquad\qquad\qquad\qquad\quad\square$

Remark 6.7. When M is an I-bundle over a closed surface, the proof of this theorem can be found in [Thu86]; this result has been extended to manifolds with incompressible boundary in [Ohs89]. When M is a compression body and $\lambda \in \hat{O}$, this result has been proved in [KS02] and [KS03].

7. Conclusion

To complete this paper, we should also mention the action of $\mathrm{Mod}(M)$ on $\mathcal{D}(M)$. The following result is proved in [Lec04b] using some properness properties of the bending map. The proof of these properties is long and is the subject of [Lec04a]. Here we will only give an outline of the proof, the reader interested in a complete proof should refer to [Lec04b] or to [Lec04a].

Proposition 7.1. *If M is not a genus 2 handlebody, the action of* $\mathrm{Mod}(M)$ *on* $\mathcal{D}(M)$ *is properly discontinuous.*

Outline of the proof. Here $\mathrm{Mod}(M)$ is the group of isotopy classes of diffeomorphisms $M \to M$.

Let us assume that Proposition 7.1 is not true. There are measured geodesic laminations $\lambda \in \mathcal{D}(M)$, $(\lambda_n) \in \mathcal{D}(M)$ and diffeomorphisms $(\phi_n) \in \mathrm{Mod}(M)$ such that (λ_n) and $(\phi_n(\lambda_n))$ converge to λ in $\mathcal{ML}(\partial M)$ and that for any $n \neq m$, ϕ_n is not isotopic to ϕ_m. Since $\lambda \in \mathcal{D}(M)$, $\exists \eta > 0$ such that $i(\lambda, \partial D) > \eta$ for any essential disc D. Let $\frac{2\pi}{\eta}\lambda$ be the measured geodesic lamination obtained by rescaling the measure of λ by $\frac{2\pi}{\eta}$. Let λ^i be a compact leaf of $\frac{2\pi}{\eta}\lambda$ with a weight greater than or equal to π; if, up to extracting a subsequence, λ^i is a compact leaf of all the measured geodesic laminations λ_n, let us replace, in $\frac{2\pi}{\eta}\lambda$ and in all $\frac{2\pi}{\eta}\lambda_n$, λ^i by the same leaf with weight π. Let λ'_∞ and λ'_n be the measured geodesic laminations obtained by doing the same for all the leaves of $\frac{2\pi}{\eta}\lambda$ with a weight greater than π; let us remark that λ'_∞ may have some leaves with a weight greater than π but that for n large enough, the compact leaves of λ'_n have a weight less than or equal to π. Let us also remark that (λ'_n) and $(\phi_n(\lambda'_n))$ converge to λ'_∞ in $\mathcal{ML}(\partial M)$. By Lemma 3.5, λ'_∞ and λ'_n satisfy the conditions b), c). For n large enough, the λ'_n also satisfy the condition a) hence, by [Lec02] (see also [Lec04b]), there is a geometrically finite metric ρ_n on the interior of M whose bending measured lamination is (λ'_n); here a geometrically finite metric is a geometrically finite representation $\rho : \pi_1(M) \to \mathrm{Isom}(\mathbb{H}^3)$ together with an isotopy class of homeomorphisms $M \to N^{ep}(\rho)$. The bending measured geodesic lamination of $\phi_{n*}(\rho_n)$ is $\phi_n(\lambda'_n)$ and by construction $\phi_n(\lambda'_n) \longrightarrow \lambda'_\infty$. It is at this point that we need the properness property of the bending map mentioned before the statement of Proposition 7.1: it follows from [Lec02] that there is a subsequence such that (ρ_n) and $(\phi_{n*}(\rho_n))$ converge to some geometrically finite metrics.

The conclusion comes from the fact that the action of $\mathrm{Mod}(M)$ on the space of isotopy classes of geometrically finite metrics (see [Lec04a] for a definition) on the interior of M is properly discontinuous. This fact can be shown by using the arguments of the proof of the properness properties mentioned above (cf. [Lec04a]).   □

As has been mentioned throughout this paper, almost all the above results have been already proved when $\lambda \in \hat{O}$. In an attempt to convince the reader of the interest of this paper we will give some examples of laminations lying in $\mathcal{D}$ but not in $\hat{O}$.

Let M be an I-bundle over a compact surface S with boundary; this manifold M is a handlebody. Let $(\gamma, \alpha) \in \mathcal{ML}(S)$ be a pair of binding measured geodesic laminations, namely for any measured geodesic lamination $\beta \in \mathcal{ML}(S)$, $i(\beta, \gamma) + i(\beta, \alpha) > 0$. Such a pair of binding measured geodesic laminations has the following property: $\exists \eta > 0$ such that $i(c, \gamma) + i(c, \alpha) \geq \eta$ for any closed curve $c \subset S$. Let us defined a measured geodesic lamination $\lambda \in \mathcal{ML}(\partial M)$ as follows: on one component $\{0\} \times S$ of $\partial I \times S$, $\lambda \cap (\{0\} \times S)$ is γ, on the other component, $\lambda \cap (\{1\} \times S)$ is α and on the remaining part $I \times \partial S$ of the boundary, $\lambda \cap (I \times \partial S)$ is $\{p\} \times \partial S$ for some $p \in]0, 1[$ endowed with a Dirac mass η.

For any essential disc $D \subset M$, ∂D intersects $\{p\} \times \partial S$, hence $i(\partial D, \lambda) \geq \eta$. If A is an essential annulus, either ∂A intersects $\{p\} \times \partial S$ and $i(\partial A, \lambda) \geq \eta$, or A can be homotoped to a vertical annulus $c \times I \subset I \times S$ with c being a simple closed curve. In the second case, we have $i(\partial A, \lambda) = i(c, \gamma) + i(c, \alpha) \geq \eta$. We have thus proved that $\lambda \in \mathcal{D}(M)$. By [KS02] the measured geodesic laminations $\lambda \cap \{0\} \times S$ and $\lambda \cap \{1\} \times S$ have the same supports as some measured laminations lying in $\mathcal{M}'$ hence $\lambda \notin \hat{O}$.

References

[Ber60]   L. Bers (1960). Simultaneous uniformization. *Bull. Amer. Math. Soc.* **66**, 94–97.

[BO]   F. Bonahon & J.-P. Otal. Laminations mesurées de plissage des variétés hyperboliques de dimension 3. To appear, *Ann. of Math.*

[Bon86] F. Bonahon (1986). Bouts des variétés hyperboliques de dimension 3. *Ann. of Math. (2)* **124** (1), 71–158.

[Bon98] F. Bonahon (1998). Variations of the boundary geometry of 3-dimensional hyperbolic convex cores. *J. Differential Geom.* **50** (1), 1–24.

[Can93] R. D. Canary (1993). Algebraic convergence of Schottky groups. *Trans. Amer. Math. Soc.* **337** (1), 235–258.

[CEG87] R. D. Canary, D. B. A. Epstein & P. Green (1987). Notes on notes of Thurston. In *Analytical and Geometric Aspects of Hyperbolic Space (Coventry/Durham, 1984), London Math. Soc. Lecture Note Ser.*, volume 111, pp. 3–92. Cambridge Univ. Press, Cambridge.

[GS90] H. Gillet & P. B. Shalen (1990). Dendrology of groups in low **Q**-ranks. *J. Differential Geom.* **32** (3), 605–712.

[Joh79] K. Johannson (1979). *Homotopy Equivalences of 3-Manifolds with Boundaries, Lecture Notes in Mathematics*, volume 761. Springer, Berlin.

[JS79] W. H. Jaco & P. B. Shalen (1979). Seifert fibered spaces in 3-manifolds. *Mem. Amer. Math. Soc.* **220**.

[KS95] L. Keen & C. Series (1995). Continuity of convex hull boundaries. *Pacific J. Math.* **168** (1), 183–206.

[KS02] G. Kleineidam & J. Souto (2002). Algebraic convergence of function groups. *Comment. Math. Helv.* **77** (2), 244–269.

[KS03] G. Kleineidam & J. Souto (2003). Ending laminations in the Masur domain. In *Kleinian Groups and Hyperbolic 3-Manifolds (Warwick, 2001), London Math. Soc. Lecture Note Ser.*, volume 299, pp. 105–129. Cambridge Univ. Press, Cambridge.

[Lec02] C. Lecuire (2002). Plissage des variétés hyperboliques de dimension 3. Preprint.

[Lec04a] C. Lecuire (2004). Bending map and strong convergence. Preprint.

[Lec04b] C. Lecuire (2004). *Structures hyperboliques convexes sur les variétés de dimension 3*. Ph.D. thesis, ENS Lyon.

[Mas86] H. Masur (1986). Measured foliations and handlebodies. *Ergodic Theory Dynam. Systems* **6** (1), 99–116.

[MO93]   J. W. Morgan & J.-P. Otal (1993). Relative growth rates of closed geodesics on a surface under varying hyperbolic structures. *Comment. Math. Helv.* **68** (2), 171–208.

[MS84]   J. W. Morgan & P. B. Shalen (1984). Valuations, trees, and degenerations of hyperbolic structures. I. *Ann. of Math. (2)* **120** (3), 401–476.

[MS88]   J. W. Morgan & P. B. Shalen (1988). Degenerations of hyperbolic structures. III. Actions of 3-manifold groups on trees and Thurston's compactness theorem. *Ann. of Math. (2)* **127** (3), 457–519.

[Ohs89]   K. Ohshika (1989). On limits of quasi-conformal deformations of Kleinian groups. *Math. Z.* **201** (2), 167–176.

[Ohs97]   K. Ohshika (1997). A convergence theorem for Kleinian groups which are free products. *Math. Ann.* **309** (1), 53–70.

[Ota88]   J.-P. Otal (1988). *Courants géodésiques et produits libres*. Thèse d'état, Université Paris-Sud, Orsay.

[Ota94]   J.-P. Otal (1994). Sur la dégénérescence des groupes de Schottky. *Duke Math. J.* **74** (3), 777–792.

[PH92]   R. C. Penner & J. L. Harer (1992). *Combinatorics of Train Tracks*. Princeton University Press, Princeton, NJ.

[Sko90]   R. K. Skora (1990). Splittings of surfaces. *Bull. Amer. Math. Soc. (N.S.)* **23** (1), 85–90.

[Thu]   W. P. Thurston. Hyperbolic structures on 3-manifolds, III: Deformations of 3-manifolds with incompressible boundary. Preprint, math.GT/9801058.

[Thu80]   W. P. Thurston (1980). The Geometry and Topology of 3-Manifolds. Lecture notes, Princeton University. Available at www.msri.org/publications/books/gt3m/.

[Thu86]   W. P. Thurston (1986). Hyperbolic structures on 3-manifolds. I. Deformation of acylindrical manifolds. *Ann. of Math. (2)* **124** (2), 203–246.

[Thu87]   W. P. Thurston (1987). Hyperbolic structures on 3-manifolds, II: Surface groups and 3-manifolds which fiber over the circle. Preprint; electronic version 1998, math.GT/9801045.

[Wal68]   F. Waldhausen (1968). On irreducible 3-manifolds which are sufficiently large. *Ann. of Math. (2)* **87**, 56–88.

Cyril Lecuire

Mathematics Institute
University of Warwick
Coventry, CV4 7AL
U.K.

clecuire@maths.warwick.ac.uk

AMS Classification: 30F40, 20H10, 20E08

Keywords: Geodesic laminations, Kleinian groups

Spaces of Kleinian Groups
Lond. Math. Soc. Lec. Notes **329**, 75–89

Cambridge University Press
Y. Minsky, M. Sakuma & C. Series (Eds.)

Thurston's bending measure conjecture for once punctured torus groups

Caroline Series

Abstract

We prove Thurston's bending measure conjecture for quasifuchsian once punctured torus groups. The conjecture states that the bending measures of the two components of the convex hull boundary uniquely determine the group.

1. Introduction

Thurston conjectured that a convex hyperbolic structure on a 3-manifold with boundary is uniquely determined by the bending measure on the boundary. In this paper we prove the conjecture in one of the simplest possible examples, namely when the manifold is an interval bundle over a once punctured torus.

In [KS04], the author and L. Keen studied the convex hull boundary of this class of hyperbolic 3-manifolds in great detail, without however addressing Thurston's conjecture *per se*. Since the question came up several times during the Newton Institute programme, it seemed worthwhile to investigate, even though the once punctured torus is a very special case. Given the current state of knowledge, the basic idea of the present proof is rather simple. It does however build on a large number of rather deep results from [KS04] and elsewhere.

Let $G \subset \mathrm{PSL}(2,\mathbb{C})$ be quasifuchsian, so that the corresponding quotient 3-manifold $\mathbb{H}^3/G$ is an interval bundle over a surface S. The boundary of the convex core of $\mathbb{H}^3/G$ has two components $\partial C^{\pm}$ each of which are pleated surfaces bent along measured laminations $\beta^{\pm}$ on S called the *bending measures*. The claim of the bending measure conjecture is that $\beta^{\pm}$ uniquely determine G, up to conjugation in $\mathrm{PSL}(2,\mathbb{C})$. If the underlying supports of $\beta^{\pm}$ are closed curves, then a result of Bonahon and Otal [BO] (which applies to general Kleinian groups not just the quasifuchsian case under discussion here) asserts this is indeed the case. They also gave necessary and sufficient conditions for existence, but not uniqueness, of quasifuchsian groups for which the bending measures $\beta^{\pm}$ are any given pair of bending measures μ, ν. If S is a once punctured torus, these conditions reduce to $i(\mu, \nu) > 0$ and the proviso that the weight of any closed curve is less than π. In a recent preprint [Bon02], Bonahon has also shown uniqueness for general quasifuchsian groups which are sufficiently close to being Fuchsian.

The object of [KS04] was to describe the space $Q\mathcal{F}$ of quasifuchsian groups when the surface S is a once punctured torus in terms of the geometry of $\partial C^{\pm}$. This was done by analysing the *pleating planes* $\mathcal{P}(\mu, \nu)$ consisting of all groups whose bending laminations lie in a particular pair of projective classes $[\mu], [\nu]$, see Section 2. In place of the bending measures, we concentrated on the lengths l_μ and l_ν of μ and ν, which can be extended to well-defined holomorphic functions, the *complex lengths* λ_μ and λ_ν, on $Q\mathcal{F}$, see Section 3. We showed, see Theorem 3.1, that λ_μ and λ_ν are local holomorphic coordinates for $Q\mathcal{F}$ in a neighbourhood of $\mathcal{P}(\mu, \nu)$ whose restrictions to $\mathcal{P}(\mu, \nu)$ are real valued. In this way we obtained a diffeomorphism of $\mathcal{P}(\mu, \nu)$ with a certain open subset in $\mathbb{R}^+ \times \mathbb{R}^+$. In particular, the lengths l_μ and l_ν uniquely determine the group.

Given this background, to prove the bending measure conjecture we just need to show that, restricted to a given pleating variety, the map from lengths to bending measures is injective. (Here by bending measure, we really mean the scale factors which relate $\beta^{\pm}$ to a fixed choice of laminations $\mu \in [\mu], \nu \in [\nu]$.) For rational laminations, as mentioned above, we already know by [BO] that angles determine the group. In [CS], again in the context of general Kleinian groups, we showed that for rational bending laminations the map from bending angles to lengths is injective and moreover that its Jacobian is symmetric and negative definite. If S is a once punctured torus, it follows (see Section 5) that if both bending laminations are rational and if we fix the length of the bending line on say ∂C^+, then the bending angle on ∂C^- is a monotonic function of the length on ∂C^-.

Now keeping the bending line on ∂C^+ fixed and of fixed length, we take limits as the the bending laminations on ∂C^- converge projectively to an arbitrary irrational lamination ν. Using the fact that the limit of monotone functions is monotone, we deduce the scale factor of the bending measure on ν is still a monotone function of the length l_ν. Since this scale factor is real analytic, it is either strictly monotonic or constant; a global geometrical argument rules out constancy. An elaboration of the argument in Sections 6 and 7 then allows us relax the requirement that the bending lamination and length on ∂C^+ remain fixed and prove the conjecture.

In [KS04] we also showed that the rational pleating varieties for which the supports of the bending laminations are simple closed curves are dense. This is not enough for the present proof: to compare our monotone functions properly, we need the "limit pleating theorem" of [KS04], see Theorem 3.3. This is a deep result closely related to the "lemme de fermeture" in [BO]. Roughly, it asserts the existence of an algebraic limit in $\mathcal{P}(\mu, \nu)$ for any sequence along which the pleating lengths l_μ, l_ν remain bounded.

A simpler version of the same analysis, not spelled out here, would show that the bending angle also uniquely determines the group for the so-called Riley slice of Schottky space, see [KS94].

I would like to thank in particular Cyril Lecuire and Pete Storm whose questioning fixed this problem in my mind.

2. Preliminaries

Throughout the paper, S will denote a fixed topological once punctured torus. A representation $\pi_1(S) \to \mathrm{PSL}(2,\mathbb{C})$ is called quasifuchsian if the image is geometrically finite, torsion free, if the image of a simple loop around the puncture is parabolic, and if there are no other parabolics. Then in particular the quotient hyperbolic manifold $M = \mathbb{H}^3/G$ is homeomorphic to $S \times (-1,1)$. It is Fuchsian if it is quasifuchsian and if in addition the representation is conjugate to a representation into $\mathrm{PSL}(2,\mathbb{R})$. Let $\mathcal{F}$ and $Q\mathcal{F}$ denote the spaces of Fuchsian and quasifuchsian representations respectively, modulo conjugation in $\mathrm{PSL}(2,\mathbb{C})$. The space $Q\mathcal{F}$ is a smooth complex manifold of dimension 2, with natural holomorphic structure induced from $\mathrm{PSL}(2,\mathbb{C})$.

2.1. Measured laminations

We assume the reader is familiar with geodesic laminations, see for example [Thu80, Ota96]. A measured lamination μ on S consists of a geodesic lamination, called the support of μ and denoted $|\mu|$, together with a transverse measure, also denoted μ. We topologise the space $\mathcal{ML}$ of all measured laminations on S with the topology of weak convergence of transverse measures, that is, two laminations are close if the measures they assign to any finite set of transversals are close. We write $[\mu]$ for the projective class of $\mu \in \mathcal{ML}(S)$ and denote the set of projective equivalence classes on S by $\mathcal{PML}$.

Let $\mathcal{S}$ be the set of simple closed curves on S. We call $\mu \in \mathcal{ML}$ *rational* if $|\mu| \in \mathcal{S}$. (On a more general surface, we say a lamination is rational if its support is a disjoint union of simple closed curves.) Equivalently, μ is rational if $\mu = c\delta_\gamma$ where $\gamma \in \mathcal{S}$, $c \geq 0$ and δ_γ is the measured lamination which assigns unit mass to each intersection with γ. The set of all rational measured laminations on S is denoted $\mathcal{ML}_\mathbb{Q}$; this set is dense in $\mathcal{ML}$.

The geometric intersection number $i(\gamma,\gamma')$ of two geodesics $\gamma,\gamma' \in \mathcal{S}$ extends to a jointly continuous function $i(\mu,\nu)$ on $\mathcal{ML}$. It is special to the once punctured torus that $i(\mu,\nu) > 0$ is equivalent to $[\mu] \neq [\nu]$, moreover (since all laminations on S are uniquely ergodic) if $\mu,\nu \in \mathcal{ML}$ with $|\mu| = |\nu|$, then $[\mu] = [\nu]$.

For general surfaces, convergence of laminations in the topology of $\mathcal{ML}$ does not imply Hausdorff convergence of their supports. However:

Lemma 2.1. *([KS04] Lemma 1) Suppose S is a once punctured torus. Suppose that $\mu \notin \mathcal{ML}_\mathbb{Q}$ and that $\mu_n \to \mu$ in the topology of weak convergence on $\mathcal{ML}$. Then $|\mu_n| \to |\mu|$ in the Hausdorff topology on the set of closed subsets of S.*

2.2. Lengths

Given a hyperbolic structure on S (associated to a Fuchsian representation of $\pi_1(S)$) and $\gamma \in \pi_1(S)$, we define the length l_γ to be the hyperbolic length of the unique geodesic freely homotopic to γ. This definition can be extended to general laminations $\mu \in \mathcal{ML}$. The following theorem summarizes the results of [Ker85], Lemma 2.4 and [Ker83], Theorem 1:

Proposition 2.2. *The function $(c\delta_\gamma, p) \mapsto cl_{\delta_\gamma}(p)$ from $\mathcal{ML}_\mathbb{Q} \times \mathcal{F}$ to $\mathbb{R}^+$ extends to a real analytic function $(\mu, p) \mapsto l_\mu(p)$ from $\mathcal{ML} \times \mathcal{F}$ to $\mathbb{R}^+$. If $\mu_n \in \mathcal{ML}_\mathbb{Q}$, $\mu_n \to \mu$ then $l_{\mu_n}(p) \to l_\mu(p)$ uniformly on compact subsets of $\mathcal{F}$.*

We showed in [KS04] that for $\mu \in \mathcal{ML}$, the length function l_μ on $\mathcal{F}$ extends to a non-constant holomorphic function λ_μ, called the *complex length* of μ, on $Q\mathcal{F}$. The extension is done in such a way that $\lambda_{c\mu} = c\lambda_\mu$ for $c > 0$, and such that if $\mu = \delta_\gamma$ then $q \mapsto \lambda_\mu(q)$ is a well-defined branch of the complex length $2\cosh^{-1}\text{Tr}\rho(\gamma)/2$, where $\rho : \pi_1(S) \to PLS(2, \mathbb{C})$ represents $q \in Q\mathcal{F}$.

Proposition 2.3. *([KS04, Theorem 20]) The family $\{\lambda_\mu\}$ is uniformly bounded and equicontinuous on compact subsets of $\mathcal{ML} \times Q\mathcal{F}$, in particular if $\mu_n \to \mu \in \mathcal{ML}$ and $q_n \to q \in Q\mathcal{F}$ then $\lambda_{\mu_n}(q_n) \to \lambda_\mu(q)$.*

The real part l_μ of the complex length λ_μ is the lamination length of μ in $\mathbb{H}^3/G$ as defined in [Thu80] p. 9.21.

2.3. The Thurston boundary

We recall the fundamental inequality which governs Thurston's compatification of Teichmüller space $\mathcal{F}$ with projective measured lamination space $\mathcal{PML}$, see [FLP79] Lemme II.1 and [Thu87] Theorem 2.2:

Proposition 2.4. *Suppose $\sigma_n \to [\xi] \in \mathcal{PML}$. Then there exist $C > 0$, $d_n \to \infty$ and $\xi_n \to \xi$ in $\mathcal{ML}$ such that*

$$d_n i(\xi_n, \zeta) \le l_\zeta(\sigma_n) \le d_n i(\xi_n, \zeta) + Cl_\zeta(\sigma_0)$$

for all $\zeta \in \mathcal{ML}$.

Corollary 2.5. *Suppose given a sequence of measured laminations μ_n on the once punctured torus S such that $\mu_n \to \mu$ in $\mathcal{ML}$, and suppose that $\sigma_n \in \mathcal{F}$ is a sequence of metrics such that $l_{\mu_n}(\sigma_n) \to 0$. Then $\sigma_n \to [\mu] \in \mathcal{PML}$.*

Proof. If σ_n converged to a point σ_∞ in $\mathcal{F}$ then $l_\mu(\sigma_\infty) = \lim_n l_{\mu_n}(\sigma_n) = 0$ which is impossible. Thus σ_n converges to some point $[\xi] \in \mathcal{PML}$. Substituting in the fundamental inequality we obtain $\xi_n \to \xi$ in $\mathcal{ML}$ and $d_n \to \infty$ such that:

$$d_n i(\xi_n, \mu_n) \leq l_{\mu_n}(\sigma_n) \leq d_n i(\xi_n, \mu_n) + C l_{\mu_n}(\sigma_0).$$

Since $i(\xi_n, \mu_n) \to i(\xi, \mu)$, we see that $l_{\mu_n}(\sigma_n)/d_n \to i(\xi, \mu)$. We conclude that $i(\xi, \mu) = 0$, and hence, since S is a punctured torus, that $[\xi] = [\mu]$.   □

2.4. Bending measures

Let $q \in \mathcal{QF}$ and let $G = G(q)$ be a group representing q. The convex core C of $\mathbb{H}^3/G$ is the smallest closed set containing all closed geodesics; it is the projection to the quotient of the convex hull of the limit set and has non-zero volume if and only if $G \in \mathcal{QF} - \mathcal{F}$. In this case its boundary has two connected components $\partial C^\pm$ each of which is homeomorphic to S. The metric induced on $\partial C^\pm$ from $\mathbb{H}^3/G$ makes each a pleated surface. The bending laminations of these surfaces carry natural transverse measures, the *bending measures* $\beta^\pm$, see [EM87, KS95]. The underlying geodesic laminations $|\beta^\pm|$ are called the *pleating loci* of G. Since the same geodesic lamination cannot be the pleating locus on both sides, we have $i(\beta^+, \beta^-) > 0$, see [KS04, BO]. If β^+ is rational with support $\gamma \in S$, then $\beta^+ = \theta_\gamma \delta_\gamma$ where θ_γ is the bending angle along γ. In this case we have the obvious constraint $\theta_\gamma < \pi$. If G is Fuchsian we define $\beta^\pm = 0$. One of the main results of [KS95] is that $\beta^\pm$ are continuous functions on $\mathcal{QF}$.

The following central existence result is a special case of Bonahon and Otal's "Lemme de fermeture":

Theorem 2.6. *([BO] Proposition 8) Suppose $\mu, \nu \in \mathcal{ML}$ with $i(\mu, \nu) > 0$. If $\mu = c\delta_\gamma$ is rational assume also that $c < \pi$, and similarly for ν. Then there exists a quasifuchsian group $G \in \mathcal{QF}$ such that $\beta^+ = \mu$ and $\beta^- = \nu$. If μ, ν are rational, then G is unique.*

The object of this paper is to prove the uniqueness part of this statement for arbitrary μ and ν.

3. Pleating varieties

In this section we review the results we shall need from [KS04].

Fix $\mu, \nu \in \mathcal{ML}$. The (μ, ν)-*pleating variety* is the set $\mathcal{P}(\mu, \nu) \subset \mathcal{QF}$ such that $\beta^+ \in$ $[\mu]$ and $\beta^- \in [\nu]$ with $\beta^\pm \neq 0$. (The last condition is equivalent to $\mathcal{P}(\mu, \nu) \cap \mathcal{F} = \emptyset$.) Notice that $\mathcal{P}(\mu, \nu)$ actually only depends on the projective classes $[\mu]$ and $[\nu]$. By Theorem 2.6 or alternatively [KS04] Theorem 2, $\mathcal{P}(\mu, \nu)$ is non-empty if and only if $i(\mu, \nu) > 0$. By [KS04] Proposition 22, the complex length λ_μ is real valued whenever the projective class of β^+ (or β^-) is $[\mu]$. In this case, $l_\mu = \Re \lambda_\mu$ is both the lamination length in $\mathbb{H}^3/G$ and the length of μ in the hyperbolic structure on ∂C^+ (or ∂C^-).

Let (μ, ν) be measured laminations on S with $i(\mu, \nu) > 0$. Then by [Ker92], the function $l_\mu + l_\nu$ has a unique minimum on $\mathcal{F}$. Moreover as c varies in $(0, \infty)$, the minima of $l_\mu + l_{c\nu}$ form a line $\mathcal{L}(\mu, \nu)$; it is easy to see that this line only depends on the projective classes $[\mu]$ and $[\nu]$. The lines $\mathcal{L}(\mu, \nu)$ vary continuously with μ and ν, see [Ker92]. (In [KS04] we gave a slightly different description of $\mathcal{L}(\mu, \nu)$ in terms of the minimum of l_μ on an earthquake path along ν, namely, $\mathcal{L}(\mu, \nu)$ is the locus where $\frac{\partial l_\mu}{\partial t_\nu}$ vanishes, where t_ν denotes the earthquake along ν. This definition only works for the once punctured torus; for the equivalence of the two definitions, see [KS04] Lemma 6.) We proved in [Ser05] Theorem 1.7 (in the context of arbitrary hyperbolisable surfaces S) that the closure of $\mathcal{P}(\mu, \nu)$ meets $\mathcal{F}$ exactly in $\mathcal{L}(\mu, \nu)$.

For each $c > 0$, the line $\mathcal{L}(\mu, \nu)$ meets the horocycle $l_\mu = c$ in $\mathcal{F}$ in exactly one point. Let $f_{\mu, \nu} : \mathbb{R}^+ \to \mathbb{R}^+$ be the function $f_{\mu, \nu}(c) = l_\nu(d)$ where $d = \mathcal{L}(\mu, \nu) \cap l_\mu^{-1}(c)$. We showed ([KS04] Lemma 6) that $f_{\mu, \nu}$ is surjective and strictly monotone decreasing. We denote by $\mathcal{R}(\mu, \nu)$ the open region in $\mathbb{R}^+ \times \mathbb{R}^+$ bounded between the two coordinate axes and the graph of $f_{\mu, \nu}$. The following description of $\mathcal{P}(\mu, \nu)$ is one of the main results of [KS04]:

Theorem 3.1. *([KS04] Theorem 2) Let $\mu, \nu \in \mathcal{ML}$ be measured laminations with $i(\mu, \nu) > 0$. Then the function $\lambda_\mu \times \lambda_\nu$ is locally injective in a neighbourhood of $\mathcal{P}(\mu, \nu)$, moreover its restriction to $\mathcal{P}(\mu, \nu)$ is a diffeomorphism to $\mathcal{R}(\mu, \nu)$.*

This means that $\mathcal{P}(\mu, \nu)$ is totally real, in other words, there are local holomorphic coordinates such that a neighbourhood of $x \in \mathcal{P}(\mu, \nu) \hookrightarrow \mathcal{QF}$ is identified with a neighbourhood of $0 \in \mathbb{R}^2 \hookrightarrow \mathbb{C}^2$. Moreover $\mathcal{P}(\mu, \nu)$ is a connected real 2-manifold, on which we can take (l_μ, l_ν) as global real analytic coordinates, where as above we define $l_\mu = \Re \lambda_\mu, l_\nu = \Re \lambda_\nu$.

Corollary 3.2. *The set $L_c = l_\nu^{-1}(c) \subset \mathcal{P}(\mu, \nu)$ is connected and hence can be regarded as a line in $\mathcal{R}(\mu, \nu)$ parameterised by l_μ.*

Proof. This is immediate from the fact that $f_{\mu, \nu}$ is strictly monotonic.   □

A crucial ingredient of Theorem 3.1 was the following, which we call the *limit pleating theorem*:

Theorem 3.3. *([KS04] Theorem 5.1) Suppose $q_n \in \mathcal{P}(\mu, \nu)$ is a sequence such that the lengths $l_\mu(q_n), l_\nu(q_n)$ are uniformly bounded above. Then there is a subsequence of q_n such that the corresponding groups $G(q_n)$ converge algebraically to a group G_∞. If both $\lim_n l_\mu(q_n)$ and $\lim_n l_\nu(q_n)$ are strictly positive, then $G_\infty \in \mathcal{F} \cup \mathcal{P}(\mu, \nu)$ and the convergence is strong.*

It is convenient to have a somewhat stronger version of this theorem, in which we allow the bending loci to vary also:

Theorem 3.4. *Suppose that $\mu, \nu \notin \mathcal{ML}_{\mathbb{Q}}$ and that $\mu_n \to \mu$ and $\nu_n \to \nu$. Suppose $q_n \in \mathcal{P}(\mu_n, \nu_n)$ and that l_{μ_n}, l_{ν_n} are uniformly bounded above. Then there is a subsequence of q_n such that $G(q_n)$ converges algebraically to a group G_∞. If both $\lim_n l_{\mu_n}(q_n)$ and $\lim_n l_{\nu_n}(q_n)$ are strictly positive, then $G_\infty \in \mathcal{F} \cup \mathcal{P}(\mu, \nu)$ and the convergence is strong.*

The proof is almost identical to that in [KS04]. We make the assumption that $\mu, \nu \notin \mathcal{ML}_{\mathbb{Q}}$ since we need that the convergence of μ_n to μ be Hausdorff; this follows from Lemma 2.1.

Very briefly, the proof of Theorem 3.3 goes as follows. The bounds on l_μ and l_ν allow one to use Thurston [Thu87] Theorem 3.3 (as in the proof of the double limit theorem) to conclude the existence of an algebraic limit G_∞. The main part of the work is to show that G_∞ is quasifuchsian. One uses continuity of the lamination length to show that the laminations μ and ν are realised in the limit 3-manifold $\mathbb{H}^3/G_\infty$. By carefully using the full force of the algebraic convergence, one can then show that, in the universal cover $\mathbb{H}^3$, the lifts of the geodesic laminations $|\mu|$ and $|\nu|$ from $\mathbb{H}^3/G(q_n)$ approach the lifts of the realisations of $|\mu|$ and $|\nu|$ from $\mathbb{H}^3/G_\infty$ in the Hausdorff topology. From this it is not hard to deduce that the laminations $|\mu|$ and $|\nu|$ lie in boundary of the convex core of $\mathbb{H}^3/G_\infty$. One deduces that this boundary has two components (or one two-sided component if G_∞ is Fuchsian) from which it follows that $G_\infty \in \mathcal{QF}$. Inspection shows that exactly the same proof gives the more general version Theorem 3.4, provided we have Hausdorff convergence of $|\mu_n|$ to $|\mu|$ and $|\nu_n|$ to $|\nu|$.

Alternatively, one can modify the proof of the Lemme de fermeture [BO] Proposition 8. They make an assumption on the limit bending laminations (which in particular rules out that the limit group is Fuchsian) but since the main use of this assumption is to get a bound on the lengths $l_{\beta^\pm}$, their proof can be adapted relatively easily to our case.

3.1. Scaling functions

For fixed $\mu, \nu \in \mathcal{ML}$ we define the *scaling functions* ξ_μ, ξ_ν by $\beta^+ = \xi_\mu \mu$ and $\beta^- = \xi_\nu \nu$.

Proposition 3.5. *The scaling function ξ_μ is real analytic on $\mathcal{P}(\mu, \nu)$.*

Proof. Suppose first that $\mu \in \mathcal{ML}_{\mathbb{Q}}$ and let $\gamma = |\mu|$. Let $\lambda_\gamma, \tau_\gamma$ be complex Fenchel Nielsen coordinates for $Q\mathcal{F}$ with respect to γ. Here λ_γ is the complex length of γ and τ_γ is the complex twist along γ, see for example [KS04] or [KS97]. The quasifuchsian group G can be described up to conjugation by these parameters, which are holomorphic functions on $Q\mathcal{F}$. As explained in detail in [KS04], if $\beta^+ \in [\mu]$ then $\beta^+ = \theta_\gamma \delta_\gamma$ and $\theta_\gamma = \Im\tau_\gamma$. This easily gives the result.

For general μ, we can obtain any group for which $\beta^+ \in [\mu]$ by a complex earthquake along μ. The scale factor ξ_μ is easily seen to be $\Im\tau_\mu$. This function is again holomorphic, see Section 7.3 of [KS04]. One can also obtain this result by using the shear-bend coordinates developed by Bonahon in [Bon96], see also [Bon02].   □

In the special case under discussion the bending measure conjecture can thus be viewed as the assertion that the map $\mathcal{P}(\mu, \nu) \to \mathbb{R}^+ \times \mathbb{R}^+$, $(l_\mu, l_\nu) \mapsto (\xi_\mu, \xi_\nu)$ is injective.

4. Global geometry

We shall need various results which control the behaviour of bending angles versus bending lengths. The following basic inequality is due to Bridgeman:

Proposition 4.1. *([Bri98] Proposition 2) There exists a universal constant $K > 0$ such that if $\beta^+ \in [\mu] \in \mathcal{PML}$, then $l_\mu \xi_\mu \leq K$.*

Say $G \in Q\mathcal{F}$. If $\gamma \in \pi_1(S)$ we let γ^* be the geodesic representative of γ in $\mathbb{H}^3/G$ and we write $\gamma^\pm$ for the geodesic representative in the hyperbolic structure of $\partial C^\pm$. Denote the respective lengths by l_{γ^*} and $l_{\gamma^\pm}$. We recall that $l_{\gamma^*} = \Re\lambda_{\gamma^*}$ where λ_{γ^*} is the complex length defined in Section 3. The following result, a simplified version of Lemma A.1 of [Lec02], see also Proposition 5.1 in [Ser05], gives a good comparison between the lengths of γ^+ and γ^* when the intersection of γ with the bending measure is reasonably small.

Proposition 4.2. *Let β^+ be the bending measure of the component $\partial C^+/G$ of the convex hull boundary of the manifold $\mathbb{H}^3/G$. Let γ be a simple closed curve with geodesic representatives γ^* in $\mathbb{H}^3/G$ and γ^+ in ∂C^+. Then there exist $A, B > 0$ such that if $i(\gamma, \beta^+) < \pi/12$, then $l_{\gamma^+} < Al_{\gamma^*} + B$.*

Now we can prove a result which will give the global control we need.

Proposition 4.3. *Fix $a > 0$. Then for any $K > 0$, there exists $\varepsilon > 0$ such that whenever $q \in \mathcal{P}(\mu, \nu)$ and $l_\mu(q) < \varepsilon$ with $\xi_\mu(q) = a$, then $l_\nu(q) > K$.*

Proof. If the result is false, then there is a sequence $q_m \in \mathcal{P}(\mu, \nu)$ for which $\xi_\mu(q_m) = a$ and $l_\nu(q_m) \leq K$ but $l_\mu(q_m) \to 0$. By Theorem 3.3, under these conditions the groups $G(q_m)$ have an algebraic limit so that in particular, the hyperbolic lengths $l_{\gamma^*}(q_m) = \Re \lambda_{\gamma^*}(q_m)$ are uniformly bounded above for any simple closed curve $\gamma \in \pi_1(S)$.

Suppose first that $\mu \in \mathcal{ML}_\mathbb{Q}$, $\mu = c\delta_\eta$ for $\eta \in S$. Then the bending angle θ_η is $c\xi_\mu$, where necessarily $\theta_\eta < \pi$. Since S is a punctured torus, we can choose a simple closed curve γ such that $i(\gamma, \eta) = 1$. As on p.195 in [PS95], for any $G \in \mathcal{QF}$, we have the equation

$$\cosh \tau_\eta / 2 = \cosh \lambda_{\gamma^*} / 2 \, \tanh \lambda_{\eta^*} / 2$$

where τ_η is the complex Fenchel Nielsen twist along η. Since η is assumed to be the bending locus of ∂C^+, we have $\tau_\eta = t_\eta + i\theta_\eta$ (see [KS04] or [KS97] for detailed discussion). Noting that $\lambda_{\eta^*} = l_{\eta^*} \in \mathbb{R}$ and taking real parts we see that if $\theta_\eta = ac < \pi$ is fixed, then $l_{\eta^*} \to 0$ implies that $l_{\gamma^*} = \Re \lambda_{\gamma^*} \to \infty$. This contradicts the uniform upper bound to the lengths $l_{\gamma^*}(q_m)$ from the first paragraph.

Now suppose that $\mu \notin \mathcal{ML}_\mathbb{Q}$ and that $\xi_\mu = a$ is fixed. By following a leaf of $|\mu|$ which returns sufficiently close to itself along some transversal, we can choose a simple closed curve γ such that $i(\gamma, \beta^+) = ai(\gamma, \mu) < \pi/12$. Since $a\mu$ is the bending measure of $\partial C^+(q_m)$, the lengths l_{μ^*} and l_{μ^+} of the lamination μ in $\mathbb{H}^3/G(q_m)$ and on the pleated surface $\partial C^+(q_m)$ coincide. Since $l_{\mu^+}(q_m) \to 0$, by Corollary 2.5 the hyperbolic structures of the pleated surfaces $\partial C^+(q_m)$ tend to $[\mu] \in \mathcal{PML}$. Since $i(\gamma, \mu) > 0$, it follows that $l_{\gamma^+}(q_m) \to \infty$. By Proposition 4.2, we have $l_{\gamma^*}(q_m) \to \infty$, and the same contradiction as before completes the proof. $\qquad\square$

5. Monotonicity of angle for fixed length

Suppose that the bending laminations μ, ν are rational, supported by simple closed curves $\gamma, \delta \in S$. Theorem 3.1 asserts that the pleating variety $\mathcal{P}(\mu, \nu) = \mathcal{P}(\gamma, \delta)$ is an open real 2-manifold parameterised by the lengths l_γ, l_δ. Theorem 2.6 shows on the other hand that the bending angles $\theta_\gamma, \theta_\delta$ are global coordinates for $\mathcal{P}(\gamma, \delta)$.

In [CS] we studied the relationship between bending angle and length for general hyperbolic 3-manifolds with boundary, where the bending laminations were rational. In particular, we showed in Proposition 7.1, that if we regard the lengths l_i of the bending lines as functions of the bending angles θ_i, then the Jacobian matrix $\left(\frac{\partial l_i}{\partial \theta_j} \right)$ evaluated at any point in the relevant pleating variety is negative definite and symmetric. (The main point of [CS] is to establish that in general lengths are parameters. Given that both lengths and angles are parameters, so that the Jacobian is non-singular, the fact that it is symmetric and negative definite follows easily from the Schläfli formula for variation of volume of the convex core and the symmetry of second derivatives; see [CS].)

Since the inverse of a negative definite symmetric matrix is also negative definite and symmetric, and since the diagonal entries of a negative definite matrix cannnot vanish, we deduce in our special case that $\frac{\partial \theta_\gamma}{\partial l_\gamma}(q) < 0$ for $q \in \mathcal{P}(\gamma, \delta)$ (where the partial derivative is taken keeping l_δ fixed). Recall from Corollary 3.2 that $L_c = l_\nu^{-1}(c) \subset \mathcal{P}(\mu, \nu)$ can be regarded as a line parameterised by l_μ. Our discussion proves:

Proposition 5.1. *Let $\mu, \nu \in \mathcal{ML}_\mathbb{Q}$. Fix $c > 0$. Then the scaled bending angle ξ_μ is a strictly monotone function of l_μ on the line L_c.*

We want to take limits to prove the same result for general μ, ν. We use the following simple fact about analytic functions:

Lemma 5.2. *Let f_n be a sequence of real valued monotonic functions on $(a, b) \subset \mathbb{R}$ which converges pointwise to a real analytic function $f : (a, b) \to \mathbb{R}$. Then f is either strictly monotonic or constant.*

Notice that a *real analytic* function f may be strictly monotonic even though its derivative vanishes at some points. (The sequence $f_n(x) = x^3 + x/n$ is an instructive example.) Thus in the following result we only claim monotonicity and not necessarily that $\frac{\partial \xi_\mu}{\partial l_\mu} < 0$.

Proposition 5.3. *Let $\mu, \nu \in \mathcal{ML}$. Fix $c > 0$. Let $L_c = l_\nu^{-1}(c) \subset \mathcal{P}(\mu, \nu)$. Then the function ξ_μ is a strictly monotone function of l_μ on the line L_c.*

Proof. Choose $\mu_n, \nu_n \in \mathcal{ML}_\mathbb{Q}$ with $\mu_n \to \mu$ and $\nu_n \to \nu$ in $\mathcal{ML}$. If either μ or ν is in $\mathcal{ML}_\mathbb{Q}$ then choose the corresponding sequence to be constant. By Lemma 2.1, both sequences of laminations also converge in the Hausdorff topology.

Fix $b, c > 0$ such that $(b, c) \in \mathcal{R}(\mu, \nu)$. By Theorem 3.1, there is a unique point $w(b, c)$ with $l_\mu = b$ and $l_\nu = c$ in $\mathcal{P}(\mu, \nu)$. Since the lines of minima vary continuously with the laminations, for sufficently large n, $\mathcal{R}(\mu_n, \nu_n)$ is close to $\mathcal{R}(\mu, \nu)$ and hence there is a unique point $w_n(b, c)$ in $\mathcal{P}(\mu_n, \nu_n)$ for which $l_{\mu_n} = b$ and $l_{\nu_n} = c$. By the strengthened version of the limit pleating theorem 3.4, up to extracting a subsequence these points converge to a point in $\mathcal{P}(\mu, \nu)$. At this point $l_\mu = b$ and $l_\nu = c$, hence by Theorem 3.1 the limit point is unique and must equal $w(b, c)$. Now keep c fixed and vary b. For each n, we know from Proposition 5.1 that $\xi_{\mu_n}(w_n(b, c))$ is a strictly monotone decreasing function of b. By continuity we have $\xi_{\mu_n}(w_n(b, c)) \to \xi_\mu(w(b, c))$. Hence for c fixed, $\xi_\mu(w(b, c))$ is monotone decreasing. Moreover for fixed $l_\nu = c$, the function ξ_μ is real analytic in l_μ. Hence it is either strictly decreasing or constant. However ξ_μ is certainly not constant since $\xi_\mu \to 0$ as we approach Fuchsian space along L_c. The result follows. $\qquad\square$

We remark that a somewhat simpler version of the above proof would show monotonicity of angle in the Riley slice of Schottky space, proving the bending measure conjecture for a genus two handlebody with both handles pinched (so that the boundary is a sphere with four punctures).

6. The constant angle variety

Fix $a > 0$ and let $V_a = \{q \in \mathcal{P}(\mu, \nu) | \xi_\mu = a\}$. Notice that $V_a \neq \emptyset$ by Theorem 2.6.

Lemma 6.1. V_a *contains no isolated points.*

Proof. Suppose the contrary, and let $q \in \mathcal{P}(\mu, \nu)$ be an isolated point of V_a. Then there exists an open disk D containing q with $V_a \cap D = \{q\}$. The image of $D - \{q\}$ under ξ_μ is connected, and it follows that a is a local maximum or minimum for the function ξ_μ restricted to the line $L_c = l_\nu^{-1}(c)$, $c = l_\nu(q)$. This contradicts the strict monotonicity of ξ_μ on L_c from Proposition 5.3. $\square$

Lemma 6.2. V_a *is a real analytic variety of dimension* 1.

Proof. By definition, V_a has dimension $2 - r$ where r is the maximum rank of $\left(\frac{\partial \xi_\mu}{\partial l_\mu}, \frac{\partial \xi_\mu}{\partial l_\nu}\right)$ on V_a. Clearly $r \leq 1$. If $r = 0$ then the Jacobian vanishes identically on V_a, from which we deduce that all partial derivatives of ξ_μ vanish identically. This would mean the real analytic function ξ_μ was constant in a neighbourhood of V in $\mathcal{P}(\mu, \nu)$, contradicting Proposition 5.3. $\square$

We can now use the classical local description of real analytic varieties, see for example [Loj91] p.173 (or [Mil68] Lemma 3.3 for the analogous real algebraic version):

Theorem 6.3. *Let* $P = (x_0, y_0)$ *be a non-isolated point in a real one dimensional algebraic variety* $V \subset \mathbb{R}^2$. *Then there is a neighbourhood of* P *in which* V *consists of a finite number of branches, each of which is homeomorphic to an open interval on* $\mathbb{R}$. *After interchanging the two coordinates if necessary, we may assume this homeomorphism has the form* $t \mapsto (x_0 + t^k, y_0 + \sum_{r=1}^{\infty} b_r t^r)$ *where the highest common factor of* k *and the indices of the non-vanishing coefficients* b_r *is* 1.

Proposition 6.4. V_a *consists of a unique connected component with no branch points which intersects each line* L_c *in a unique point.*

Proof. We already know by Proposition 5.3 that $V_a \cap L_c$ consists of at most one point. In other words, the restriction of l_ν to V_a is injective, hence in particular, has no local maxima or minima. Denote this restriction by f. It is easy to see that injectivity of f implies that V_a has no branch points.

Let W be a connected component of V_a. We claim that the image $f(W)$ is $(0,\infty)$. It follows from Theorem 6.3 and the above observation about local maxima and minima that $f(W)$ is open. Now we show that f is proper. Choose $w_n \in W$ with $f(w_n) \to b$ where $0 < b < \infty$, so that by definition the lengths $l_v(w_n)$ are bounded above. Since $\xi_\mu = a$ on V_a, it follows from Bridgeman's inequality that the lengths $l_\mu(w_n)$ are also uniformly bounded above. By Theorem 3.3, the corresponding sequence of groups $G(w_n)$ converge algebraically to a group G and since G is clearly not Fuchsian, either $G \in \mathcal{P}(\mu,v)$, or $f(w_n) = l_v(w_n) \to 0$ or $l_\mu(w_n) \to 0$. By Proposition 4.3, if $l_\mu(w_n) \to 0$ then $l_v(w_n) \to \infty$, ruling out the last possibility. If $G \in \mathcal{P}(\mu,v)$ then by continuity G corresponds to a point $w \in W$ with $f(w) = b$. This shows that f is proper and hence that $f(W) = (0,\infty)$ as claimed.

Thus $W \cap L_c$ is non-empty for each $c \in (0,\infty)$. If V_a had any other connected component, then for some c the intersection $V_a \cap L_c$ would contain more than one point, which is impossible. The result follows. $\qquad\square$

7. Proof of the bending measure conjecture

Suppose $a, b > 0$. Any quasifuchsian group with $\beta^+ = a\mu, \beta^- = bv$ necessarily lies in the variety V_a. The following result therefore concludes the proof of the bending measure conjecture:

Proposition 7.1. *Fix $a > 0$. The angle ξ_v is strictly monotonic on V_a.*

Proof. Let $q(c)$ denote the point $V_a \cap L_c$. Choose sequences $v_n \to v, \mu_n \to \mu$, where as usual if $\mu \in \mathcal{ML}_\mathbb{Q}$ we assume that the sequence μ_n is constant. Let $q_n = q_n(c) \in \mathcal{P}(\mu_n, v_n)$ be the unique point for which $l_{v_n} = c$ and $\xi_{\mu_n} = a$; this exists for large n since $\mathcal{R}(\mu_n, v_n)$ is close to $\mathcal{R}(\mu, v)$. By Bridgeman's inquality the lengths $l_{v_n}(q_n)$ are uniformly bounded above. Hence by the limit pleating theorem, we can extract a subsequence for which $G(q_n)$ converges algebraically to a group G.

We claim that $l_{\mu_n}(q_n)$ does not tend to zero. If $\mu \in \mathcal{ML}_\mathbb{Q}$ we can follow exactly the argument in the second paragraph of the proof of Proposition 4.3. If $\mu \notin \mathcal{ML}_\mathbb{Q}$, we proceed as in the third paragraph of that proof. Now $[\mu_n]$ is the projective class of the bending measure of $\partial C^+(q_n)$ so that $l_{\mu_n^+}(q_n)$ and $l_{\mu_n^*}(q_n)$ coincide. By Corollary 2.5, if $l_{\mu_n}(q_n) \to 0$ then the hyperbolic structures $\partial C^+(q_n)$ converge to $[\mu]$. This leads to a contradiction to the existence of the algebraic limit of the groups $G(q_n)$ exactly as in Proposition 4.3.

We conclude from Theorem 3.3 that G is represented by a point $q' \in \mathcal{P}(\mu,v)$. Moreover since $\xi_{v_n}(q_n) \to \xi_v(q')$ and $l_{v_n}(q_n) \to l_v(q')$, we deduce that $q' = q(c)$, so that the limit is independent of the subsequence.

By definition $q_n(c)$ is the point $V_a^n \cap L_c^n$. We already know that the angle function $\xi_{v_n}(q_n(c))$ is monotonic in c. We have just shown that $q_n(c) \to q(c)$ and so $\xi_{v_n}(q_n(c)) \to \xi_v(q(c))$. We deduce that $\xi_v(q(c))$ is monotonic in c.

It remains to show that $\xi_v(q(c))$ is strictly monotonic. Now $\xi_v(q(c))$ is a real analytic function of the parameters l_μ, l_v for $\mathcal{P}(\mu, v)$. By Theorem 6.3, on the real analytic variety V_a, each length function is a real analytic function of some $t \in \mathbb{R}$. Thus so is $\xi_v(q(c))$. We deduce that $\xi_v(q(c))$ is either strictly monotonic or constant on V_a. To rule out the second possibility, notice that if ξ_v is constant on V_a, then Bridgeman's inequality gives a uniform upper bound to l_v on V_a, contradicting Proposition 6.4. $\quad\square$

References

[BO]   F. Bonahon & J.-P. Otal. Laminations mesurées de plissage des variétés hyperboliques de dimension 3. To appear, *Ann. of Math.*

[Bon96] F. Bonahon (1996). Shearing hyperbolic surfaces, bending pleated surfaces and Thurston's symplectic form. *Ann. Fac. Sci. Toulouse Math. (6)* **5** (2), 233–297.

[Bon02] F. Bonahon (2002). Kleinian groups which are almost fuchsian. Preprint, math.DG/0210233v1.

[Bri98] M. Bridgeman (1998). Average bending of convex pleated planes in hyperbolic three-space. *Invent. Math.* **132** (2), 381–391.

[CS]   Y. E. Choi & C. Series. Lengths are coordinates for convex structures. Preprint, University of Warwick, math.GT/0406257.

[EM87] D. B. A. Epstein & A. Marden (1987). Convex hulls in hyperbolic space, a theorem of Sullivan, and measured pleated surfaces. In *Analytical and Geometric Aspects of Hyperbolic Space (Coventry/Durham, 1984)*, London Math. Soc. Lecture Note Ser., volume 111, pp. 113–253. Cambridge Univ. Press, Cambridge.

[FLP79] A. Fathi, F. Laudenbach & V. Poenaru (1979). Travaux de Thurston sur les surfaces. *Astérisque* **66–67**.

[Ker83] S. P. Kerckhoff (1983). The Nielsen realization problem. *Ann. of Math. (2)* **117** (2), 235–265.

[Ker85] S. P. Kerckhoff (1985). Earthquakes are analytic. *Comment. Math. Helv.* **60** (1), 17–30.

[Ker92]   S. P. Kerckhoff (1992). Lines of minima in Teichmüller space. *Duke Math. J.* **65** (2), 187–213.

[KS94]   L. Keen & C. Series (1994). The Riley slice of Schottky space. *Proc. London Math. Soc. (3)* **69** (1), 72–90.

[KS95]   L. Keen & C. Series (1995). Continuity of convex hull boundaries. *Pacific J. Math.* **168** (1), 183–206.

[KS97]   L. Keen & C. Series (1997). How to bend pairs of punctured tori. In *Lipa's Legacy (New York, 1995)*, *Contemp. Math.*, volume 211, pp. 359–387. Amer. Math. Soc., Providence, RI.

[KS04]   L. Keen & C. Series (2004). Pleating invariants for punctured torus groups. *Topology* **43** (2), 447–491.

[Lec02]   C. Lecuire (2002). Plissage des variétés hyperboliques de dimension 3. Preprint.

[Loj91]   S. Lojasiewicz (1991). *Introduction to Complex Analytic Geometry*. Birkhäuser Verlag, Basel.

[Mil68]   J. Milnor (1968). *Singular Points of Complex Hypersurfaces*. Annals of Mathematics Studies, No. 61. Princeton University Press, Princeton, N.J.

[Ota96]   J.-P. Otal (1996). Le théorème d'hyperbolisation pour les variétés fibrées de dimension 3. *Astérisque* **235**.

[PS95]   J. R. Parker & C. Series (1995). Bending formulae for convex hull boundaries. *J. Anal. Math.* **67**, 165–198.

[Ser05]   C. Series (2005). Limits of quasi-fuchsian groups with small bending. *Duke Math. J.* **128** (2), 285–329.

[Thu80]   W. P. Thurston (1980). The Geometry and Topology of 3-Manifolds. Lecture notes, Princeton University. Available at www.msri.org/publications/books/gt3m/.

[Thu87]   W. P. Thurston (1987). Hyperbolic structures on 3-manifolds, II: Surface groups and 3-manifolds which fiber over the circle. Preprint; electronic version 1998, math.GT/9801045.

Caroline Series

Mathematics Institute
University of Warwick
Coventry, CV4 7AL
U.K.

cms@maths.warwick.ac.uk

AMS Classification: 30F40, 20H10, 57M50

Keywords: Kleinian group, once punctured torus, convex hull, bending measure

Spaces of Kleinian Groups
Lond. Math. Soc. Lec. Notes **329**, 91–120

Cambridge University Press
Y. Minsky, M. Sakuma & C. Series (Eds.)

Complexity of 3-manifolds

Bruno Martelli[1]

Abstract

We give a summary of known results on Matveev's complexity of compact
3-manifolds. The only relevant new result is the classification of all closed orientable irreducible 3-manifolds of complexity 10.

1. Introduction

In 3-dimensional topology, various quantities are defined, that measure how complicated a compact 3-manifold M is. Among them, we find the Heegaard genus, the
minimum number of tetrahedra in a triangulation, and Gromov's norm (which equals
the volume when M is hyperbolic). Both Heegaard genus and Gromov norm are additive on connected sums, and behave well with respect to other common cut-and-paste
operations, but it is hard to classify all manifolds with a given genus or norm. On
the other hand, triangulations with n tetrahedra are more suitable for computational
purposes, since they are finite in number and can be easily listed using a computer, but
the minimum number of tetrahedra is a quantity which does not behave well with any
cut-and-paste operation on 3-manifolds. (Moreover, it is not clear what is meant by
"triangulation": do the tetrahedra need to be embedded? Are ideal vertices admitted
when M has boundary?)

In 1988, Matveev introduced [Mat88] for any compact 3-manifold M a non-negative
integer $c(M)$, which he called the *complexity* of M, defined as the minimum number of
vertices of a *simple spine* of M. The function c is finite-to-one on the most interesting
sets of compact 3-manifolds, and it behaves well with respect to the most important
cut-and-paste operations. Its main properties are listed below.

additivity $c(M \# M') = c(M) + c(M')$;

finiteness for any n there is a finite number of closed $\mathbb{P}^2$-irreducible M's with $c(M) = n$, and a finite number of hyperbolic N's with $c(N) = n$;

monotonicity $c(M_F) \leqslant c(M)$ for any incompressible $F \subset M$ cutting M into M_F.

---

[1] Supported by the INTAS project "CalcoMet-GT" 03-51-3663.

We recall some definitions used throughout the paper. Let M be a compact 3-manifold, possibly with boundary. We say that M is *hyperbolic* if it admits (after removing all tori and Klein bottles from the boundary) a complete hyperbolic metric of finite volume (possibly with cusps and geodesic boundary). Such a metric is unique by Mostow's theorem (see [McM90] for a proof). A surface in M is *essential* if it is incompressible, ∂-incompressible, and not ∂-parallel. Thurston's Hyperbolicity Theorem for Haken manifolds ensures that a compact M with boundary is hyperbolic if and only if every component of ∂M has $\chi \leqslant 0$, and M does not contain essential surfaces with $\chi \geqslant 0$. The complexity satisfies also the following strict inequalities.

filling every closed hyperbolic M is a Dehn filling of some hyperbolic N with $c(N) < c(M)$;

strict monotonicity $c(M_F) < c(M)$ if F is essential and M is closed $\mathbb{P}^2$-irreducible or hyperbolic;

Some results in complexity zero already show that the finiteness property does not hold for all compact 3-manifolds.

complexity zero the closed $\mathbb{P}^2$-irreducible manifolds with $c = 0$ are $S^3, \mathbb{RP}^3$, and $L(3,1)$. We also have $c(S^2 \times S^1) = c(S^2 \mathbin{\tilde{\times}} S^1) = 0$. Interval bundles over surfaces and handlebodies also have $c = 0$.

The ball and the solid torus have therefore complexity zero. Moreover, the additivity property actually also holds for ∂-connected sums. These two facts together imply the following.

stability The complexity of M does not change when adding 1-handles to M or removing interior balls from it.

Note that both such operations that not affect c are "invertible" and hence topologically inessential. In what follows, a simplicial face-pairing T of some tetrahedra is a *triangulation* of a closed 3-manifold M when $M = |T|$. Tetrahedra are therefore not necessarily embedded in M. A simplicial pairing T is an *ideal triangulation* of a compact M with boundary if M is $|T|$ minus open stars of all the vertices. The finiteness property above follows easily from the following.

naturality if M is closed $\mathbb{P}^2$-irreducible and not $S^3, \mathbb{RP}^3$, or $L(3,1)$, then $c(M)$ is the minimum number of tetrahedra in a triangulation of M. If N is hyperbolic with boundary, then $c(N)$ is the minimum number of tetrahedra in an ideal triangulation of N.

The beauty of Matveev's complexity theory relies on the fact that simple spines are more flexible than triangulations: for instance spines can often be simplified by puncturing faces, and can always be cut along normal surfaces. In particular, we have the following result. An (ideal) triangulation T of M is *minimal* when M cannot be (ideally) triangulated with fewer tetrahedra. A *normal surface* in T is one intersecting the tetrahedra in normal triangles and squares, see [Hem76].

normal surfaces let T be a minimal (ideal) triangulation of a closed $\mathbb{P}^2$-irreducible (hyperbolic with boundary) manifold M different from S^3, $\mathbb{RP}^3$, and $L(3,1)$. If F is a normal surface in T containing some squares, then $c(M_F) < c(M)$.

As an application of the previous properties, the following result was implicit in Matveev's paper [Mat90].

Corollary 1.1. *Let T be a minimal triangulation of a closed $\mathbb{P}^2$-irreducible 3-manifold M different from $S^3, \mathbb{RP}^3, L(3,1)$. Then T has one vertex only, and it contains no normal spheres, except the vertex-linking one.*

Computers can easily handle spines and triangulations, and manifolds of low complexity have been classified by various authors. Closed orientable irreducible manifolds with $c \leqslant 6$ were classified by Matveev [Mat88] in 1988. Those with $c = 7$ were then classified in 1997 by Ovchinnikov [Ovc97, Mat03a], and those with $c = 8, 9$ in 2001 by Martelli and Petronio [MP01a]. We present here the results we recently found for $c = 10$. The list of all manifolds with $c = 10$ has also been computed independently by Matveev [Mat03b], and the two tables (each consisting of 3078 manifolds) coincide. The closed $\mathbb{P}^2$-irreducible non-orientable manifolds with $c \leqslant 7$ have been listed independently by Amendola and Martelli [AM05], and Burton [Bur03].

Hyperbolic manifolds with cusps and without geodesic boundary were listed for all $c \leqslant 3$ in the orientable case by Matveev and Fomenko [MF88] in 1988, and for all $c \leqslant 7$ by Callahan, Hildebrand, and Weeks [CHW99] in 1999. Orientable hyperbolic manifolds with geodesic boundary (and possibly some cusps) were listed for $c \leqslant 2$ by Fujii [Fuj90] in 1990, and for $c \leqslant 4$ by Frigerio, Martelli, and Petronio [FMP04] in 2002.

All properties listed above were proved by Matveev in [Mat90], and extended when necessary to the non-orientable case by Martelli and Petronio in [MP02b], except the filling property, which is a new result proved below in Subsection 2.3. The only other new results contained in this paper are the complexity-10 closed census (also constructed independently by Matveev [Mat03b]), and the following counterexample (derived from that census) of a conjecture of Matveev and Fomenko [MF88] stated in Subsection 5.3.

Proposition 1.2. *There are two closed hyperbolic fillings M and M′ of the same cusped hyperbolic N with c(M) < c(M′) and* Vol(M) > Vol(M′).

We mention the most important discovery of our census.

Proposition 1.3. *There are 25 closed hyperbolic manifolds with c = 10 (while none with c ⩽ 8 and four with c = 9).*

This paper is structured as follows: the complexity of a 3-manifold is defined in Section 2. We then collect in Section 3 and 4 the censuses of closed and hyperbolic 3-manifolds described above, together with the new results in complexity 10. Relations between complexity and volume of hyperbolic manifolds are studied in Section 5. Lower bounds for the complexity, together with some infinite families of hyperbolic manifolds with boundary for which the complexity is known, are described in Section 6. The algorithm and tools usually employed to produce a census are described in Section 7. Finally, we describe the decomposition of a manifold into *bricks* introduced by Martelli and Petronio in [MP01a, MP02b], necessary for our closed census with c = 10, in Section 8. All sections may be read independently, except that Sections 7 and 8 need the definitions contained in Section 2.

Acknowledgements

The author warmly thanks Carlo Petronio and Sergej Matveev for their continuous support. He also thanks the referee for her/his suggestions.

2. The complexity of a 3-manifold

We define here simple and special spines, and the complexity of a 3-manifold. We then show a nice relation between spines without vertices and Riemannian geometry, found by Alexander and Bishop [AB00]. Finally, we prove the filling property stated in the Introduction.

2.1. Definitions

We start with the following definition. A compact 2-dimensional polyhedron P is *simple* if the link of every point in P is contained in the graph ⊗. Alternatively, P is simple if it is locally contained in the polyhedron shown in Fig. 1-(3). A point, a compact graph, a compact surface are therefore simple. The polyhedron given by two orthogonal discs intersecting in their diameter is not simple. Three important possible kinds of neighborhoods of points are shown in Fig. 1. A point having the whole of ⊗

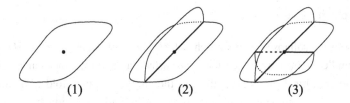

Figure 1: Neighborhoods of points in a special polyhedron.

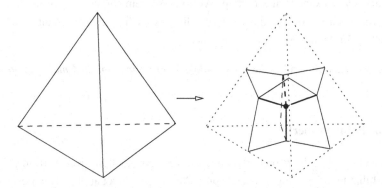

Figure 2: A special spine of M is dual to a triangulation, which is ideal or 1-vertex, depending on whether M has boundary or not.

as a link is called a *vertex*, and its regular neighborhood is shown in Fig. 1-(3). The set $V(P)$ of the vertices of P consists of isolated points, so it is finite. Note that points, graphs, and surfaces do not contain vertices.

A compact polyhedron $P \subset M$ is a *spine* of a compact manifold M with boundary if M collapses onto P. When M is closed, we say that $P \subset M$ is a *spine* if $M \setminus P$ is an open ball. The *complexity* $c(M)$ of a compact 3-manifold M is the minimal number of vertices of a simple spine of M. As an example, a point is a spine of S^3, and therefore $c(S^3) = 0$. A simple polyhedron is *special* when every point has a neighborhood of one of the types (1)-(3) shown in Fig. 1, and the sets of such points induce a cellularization of P. That is, defining $S(P)$ as the set of points of type (2) or (3), the components of $P \setminus S(P)$ should be open discs – the *faces* – and the components of $S(P) \setminus V(P)$ should be open segments – the *edges*.

Remark 2.1. A special spine of a compact M with boundary is dual to an ideal triangulation of M, and a special spine of a closed M is dual to a 1-vertex triangulation of M, as suggested by Fig. 2. In particular, a special spine is a spine of a unique manifold. Therefore the naturality property of c may be read as follows: every closed irreducible or hyperbolic manifold distinct from $S^3, \mathbb{RP}^3$, and $L(3,1)$ has a special spine with $c(M)$ vertices. Such a special spine is then called *minimal*.

2.2. Complexity zero

A handlebody M collapses onto a graph, which has no vertices, hence $c(M) = 0$. An interval bundle M over a surface has that surface as a spine, and hence $c(M) = 0$ again. Note that, by shrinking the fibers of the bundle, the manifold M admits product metrics with arbitrarily small injectivity radius and uniformly bounded curvature. This is a particular case of a relation between spines and Riemannian geometry found by Alexander and Bishop [AB00]. A Riemannian 3-manifold M is *thin* when its curvature-normalized injectivity radius is less than some constant $a_2 \approx 0.075$, see [AB00] for details. We have the following.

Proposition 2.2 (Alexander–Bishop [AB00]). *A thin Riemannian 3-manifold has complexity zero.*

2.3. The filling property

We prove here the filling property, stated in the Introduction. Recall from [Mat90, Mat03a] that by thickening a special spine P of M we get a handle decomposition ξ_P of the same M. Normal surfaces in ξ_P correspond to normal surfaces in the (possibly ideal) triangulation dual to P.

Theorem 2.3. *Every closed hyperbolic manifold M is a Dehn filling of some hyperbolic N with $c(N) < c(M)$.*

Proof. Let P be a minimal special spine of M, which exists by Remark 2.1. Take a face f of P. By puncturing f and collapsing the resulting polyhedron as much as possible, we get a simple spine Q of some N obtained by drilling M along a curve. Since P is special, f is incident to at least one vertex. During the collapse, all vertices adjacent to f have disappeared, hence Q has less vertices than P. This gives $c(N) < c(M)$.

If N is hyperbolic we are done. Suppose it is not. Then it is reducible, Seifert, or toroidal. If N is reducible, the drilled solid torus is contained in a ball of M and we get $N = M\#M'$ for some M', hence $c(M) \leqslant c(N) < c(M)$ by the additivity property. Then N is irreducible. Moreover ∂N is incompressible in N (because M is not a lens space). Then the 1-dimensional portion of Q can be removed, and we can suppose $Q \subset P$ is a spine of N having only points of the type of Fig. 1.

Our N cannot be Seifert (because M is hyperbolic), hence its JSJ decompostion consists of some tori $T_1, \ldots, T_k$. Each T_i is essential in N and compressible in M. Each T_i can be isotoped in normal position with respect to ξ_Q. Since $Q \subset P$, every normal surface in ξ_Q is normal also in ξ_P. The only normal surface in ξ_P not containing squares is the vertex-linking sphere, therefore we have $c(M_{T_i}) < c(M)$ for all i by the normal surfaces property. Each T_i is compressible in M, hence either it bounds a solid

	0	1	2	3	4	5	6	7	8	9	10
					orientable						
lens spaces	3	2	3	6	10	20	36	72	136	272	528
other elliptic	.	.	1	1	4	11	25	45	78	142	270
flat	.	.	.	.	.	.	6	.	.	.	.
Nil	.	.	.	.	.	.	7	10	14	15	15
$SL_2\mathbb{R}$	.	.	.	.	.	.	.	39	162	513	1416
Sol	.	.	.	.	.	.	.	5	9	23	39
$\mathbb{H}^2 \times \mathbb{R}$	.	.	.	.	.	.	.	.	2	.	8
hyperbolic	.	.	.	.	.	.	.	.	.	4	25
not geometric	.	.	.	.	.	.	.	4	35	185	777
total orientable	**3**	**2**	**4**	**7**	**14**	**31**	**74**	**175**	**436**	**1154**	**3078**
					non-orientable						
flat	.	.	.	.	.	.	4	.			
$\mathbb{H}^2 \times \mathbb{R}$	.	.	.	.	.	.	.	2			
Sol	.	.	.	.	.	.	1	1			
total non-orientable	.	.	.	.	.	.	**5**	**3**			

Table 1: The number of closed $\mathbb{P}^2$-irreducible manifolds of given complexity (up to 10 in the orientable case, and up to 7 in the non-orientable one) and geometry. Recall that there is no $\mathbb{P}^2$-irreducible manifold of type $S^2 \times \mathbb{R}$, and no non-orientable one of type S^3, Nil, and $SL_2\mathbb{R}$.

torus or is contained in a ball. The latter case is excluded, otherwise M_{T_i} is the union of $M\#M'$ and a solid torus, and $c(M) \leqslant c(M_{T_i}) < c(M)$.

Therefore each T_i bounds a solid torus in M. Each solid torus contains the drilled curve, hence they all intersect, and there is a solid torus H bounded by a T_i containing all the others. Therefore $M_{T_i} = N' \cup H$ where N' is a block of the JSJ decomposition, which cannot be Seifert, hence it is hyperbolic. We have $c(N') = c(M_{T_i}) < c(M)$, and M is obtained by filling N', as required. □

Remark 2.4. The proof Theorem 2.3 is also valid for M *hyperbolike*, i.e. irreducible, atoroidal, and not Seifert.

3. Closed census

We describe here the closed orientable irreducible manifolds with $c \leqslant 10$, and the closed non-orientable $\mathbb{P}^2$-irreducible ones with $c \leqslant 7$. Such manifolds are collected in terms of their geometry, if any, in Table 1. The complete list of manifolds can be downloaded from [Pet].

	$\chi^{\text{orb}} > 0$	$\chi^{\text{orb}} = 0$	$\chi^{\text{orb}} < 0$
$e = 0$	$S^2 \times \mathbb{R}$	E^3	$H^2 \times \mathbb{R}$
$e \neq 0$	S^3	Nil	$\widetilde{SL_2\mathbb{R}}$

Table 2: The six Seifert geometries.

3.1. The first 7 geometries

We recall [Sco83] that there are eight important 3-dimensional geometries, six of them concerning Seifert manifolds. A Seifert fibration is described via its *normalized parameters* $(F, (p_1, q_1), \ldots, (p_k, q_k), t)$, where F is a closed surface, $p_i > q_i > 0$ for all i, and $t \geqslant -k/2$ (obtained by reversing orientation if necessary). The Euler characteristic χ^{orb} of the base orbifold and the Euler number e of the fibration are given respectively by

$$\chi^{\text{orb}} = \chi(F) - \sum_{i=1}^{k} \left(1 - \frac{1}{p_i}\right), \qquad e = t + \sum_{i=1}^{k} \frac{q_i}{p_i}$$

and they determine the geometry of the Seifert manifold (which could have different fibrations) according to Table 2. The two non-Seifert geometries are the Sol and the hyperbolic ones [Sco83].

The following result shows how to compute the complexity (when $c \leqslant 10$) of most manifolds belonging to the first 7 geometries. It is proved for $c \leqslant 9$ in [MP04], and completed for $c = 10$ here in Subsection 8.7. We define the norm $|p, q|$ of two coprime non-negative integers inductively by setting $|1, 0| = |0, 1| = |1, 1| = 0$ and $|p+q, q| = |p, q+p| = |p, q| + 1$. A norm $\|A\|$ on matrices $A \in \text{GL}_2(\mathbb{Z})$ is also defined in [MP04].

Theorem 3.1. *Let M be a geometric non-hyperbolic manifold with $c(M) \leqslant 10$:*

(i) *if M is a lens space $L(p, q)$, then $c(M) = |p, q| - 2$;*

(ii) *if M is a torus bundle with monodromy A then $c(M) = \min\{\|A\| + 5, 6\}$.*

(iii) *if $M = \left(S^2, (2, 1), (3, 1), (m, 1), -1\right)$ with $m \geqslant 5$, we have $c(M) = m$;*

(iv) *if $M = \left(S^2, (2, 1), (n, 1), (m, 1), -1\right)$ is not of the type above, we have $c(M) = n + m - 2$;*

(v) *if $M = \left(S^2, (2, 1), (3, 1), (p, q), -1\right)$ with $p/q > 5$ is not of the types above, we have $c(M) = |p, q| + 2$;*

(vi) *if $M = \left(F, (p_1, q_1), \ldots, (p_k, q_k), t\right)$ is not of the types above, then*

$$c(M) = \max\left\{0, t - 1 + \chi(F)\right\} + 6\left(1 - \chi(F)\right) + \sum_{i=1}^{k} \left(|p_i, q_i| + 2\right).$$

The symmetries of this link act transitively on the components, in such a way that to define the $(p/q, r/s, t/u)$-surgery we do not need to associate a component to each parameter.

Figure 3: The chain link with 3 components.

Note from Table 1 that a Seifert manifold with $c < 6$ has $\chi^{orb} > 0$ and one with $c \leqslant 6$ has $\chi^{orb} \geqslant 0$, whereas for higher c most Seifert manifolds have $\chi^{orb} < 0$.

Remark 3.2. Theorem 3.1, together with analogous formulas for some non-geometric graph manifolds, follows from the decomposition of closed manifolds into bricks, introduced in Section 8. The lists of all non-hyperbolic manifolds with $c \leqslant 10$ is then computed from such formulas by a computer program, available from [Pet]. A mistake in that program produced in [MP01a] for $c = 9$ a list of 1156 manifolds instead of 1154 (two graph manifolds with distinct parameters were counted twice). Using Turaev-Viro invariants, Matveev has also recently checked that all the listed closed manifolds with $c \leqslant 10$ are distinct [Mat03b].

3.2. Hyperbolic manifolds

Table 3 shows all closed hyperbolic manifolds with $c \leqslant 10$. Each such manifold is a Dehn surgery on the chain link with 3 components shown in Fig. 3, with parameters shown in the table.

It is proved in [MF88] that every closed 3-manifold with $c \leqslant 8$ is a graph manifold, and that the first closed hyperbolic manifolds arise with $c = 9$. The hyperbolic manifolds with $c = 9$ then turned out [MP01a] to be the 4 smallest ones known. The most interesting question about those with $c = 10$ is then whether they are also among the smallest ones known, for instance comparing them with the closed census [HW94] also used by SnapPea [Wee]. As explained in [DT03], the manifolds in that census have all geodesics bigger than .3, and therefore some manifolds having $c = 10$ are not present there (namely, those in Table 3 corresponding to N. 16, 21, 24). We have therefore used SnapPea (in the python version) to compute a list of many surgeries on the chain link with 3 components (avoiding the non-hyperbolic ones, listed in [MP02a]), available from [Pet], which contains many closed manifolds of volume smaller than 2 that are not present in SnapPea's closed census. The entry "N." in Table 3 tells the position of the manifold in our table from [Pet]. The first 10 manifolds of the two lists nevertheless coincide and are also fully described in [HW94], and they all have $c \leqslant 10$, as Table 3 shows.

surgery parameters	N.	volume	shortest geod	homology
		complexity 9		
$1, -4, -3/2$	1	0.942707362	0.5846	$\mathbb{Z}_5 + \mathbb{Z}_5$
$1, -4, 2$	2	0.981368828	0.5780	$\mathbb{Z}_5$
$1, -5, -1/2$	3	1.014941606	0.8314	$\mathbb{Z}_3 + \mathbb{Z}_6$
$1, -3/2, -3/2$	4	1.263709238	0.5750	$\mathbb{Z}_5 + \mathbb{Z}_5$
		complexity 10		
$1, -5, 2$	5	1.284485300	0.4803	$\mathbb{Z}_6$
$1, 2, 1/2$	6	1.398508884	0.3661	trivial
$1, -5, 1/2$	7	1.414061044	0.7941	$\mathbb{Z}_6$
$1, -4, 3$	8	1.414061044	0.3648	$\mathbb{Z}_{10}$
$1, -4, -4/3$	9	1.423611900	0.3523	$\mathbb{Z}_{35}$
$1, 2, -1/2$	10	1.440699006	0.3615	$\mathbb{Z}_3$
$1, 2, -3/2$	12	1.529477329	0.3359	$\mathbb{Z}_5$
$1, -4, -5/2$	13	1.543568911	0.3353	$\mathbb{Z}_{35}$
$1, -1/2, -5/2$	14	1.543568911	0.5780	$\mathbb{Z}_{21}$
$1, -4, -5/3$	16	1.583166660	0.2788	$\mathbb{Z}_{40}$
$1, -6, -1/2$	17	1.583166660	0.5577	$\mathbb{Z}_{21}$
$1, -1/2, -7/2$	18	1.583166660	0.7774	$\mathbb{Z}_3 + \mathbb{Z}_9$
$2, -3/2, -3/2$	19	1.588646639	0.3046	$\mathbb{Z}_{30}$
$1, -5, -3/2$	20	1.588646639	0.5345	$\mathbb{Z}_{30}$
$1, -4, 3/2$	21	1.610469711	0.2499	$\mathbb{Z}_5$
$1, 2, -5/2$	24	1.649609715	0.2627	$\mathbb{Z}_7$
$1, -1/2, -3/2$	25	1.649609715	0.5087	$\mathbb{Z}_{15}$
$1, 1/2, -6$	34	1.757126029	0.7053	$\mathbb{Z}_7$
$1, -1/2, -1/2$	49	1.824344322	0.4680	$\mathbb{Z}_3 + \mathbb{Z}_3$
$1, -5, -1/3$	55	1.831931188	0.5306	$\mathbb{Z}_2 + \mathbb{Z}_{12}$
$1, -3/2, -5/3$	74	1.885414725	0.3970	$\mathbb{Z}_{40}$
$1, -5/2, -5/2$	76	1.885414725	0.5846	$\mathbb{Z}_7 + \mathbb{Z}_7$
$-5/2, -1/2, -1/2$	77	1.885414725	0.5846	$\mathbb{Z}_{39}$
$1, -5, -2/3$	91	1.910843793	0.4421	$\mathbb{Z}_{30}$
$1, -4/3, -3/2$	139	1.953708315	0.3535	$\mathbb{Z}_{35}$

Table 3: The hyperbolic manifolds of complexity 9 and 10. Each such manifold is described as the surgery on the chain link with some parameters.

	0	1	2	3	4	5	6	7	8	9	10
					geometric						
lens spaces	3	2	3	6	10	20	36	72	136	272	528
$S^2,3$	.	.	1	1	4	11	31*	84	226	586	1477
$S^2,4$	.	.	.	.	.	.	2	4	14	40	120
$S^2,5$	.	.	.	.	.	.	.	.	.	2	5
$\mathbb{RP}^2,2$	.	.	.	.	.	.	2	4	14	34	90
$\mathbb{RP}^2,3$	.	.	.	.	.	.	.	.	.	2	5
T or K	.	.	.	.	.	4*	2	2	2	2	
$T,1$ or $K,1$	.	.	.	.	.	.	.	.	4	10	
T-fiberings over S^1	.	.	.	.	.	.	.	2	2	6	6
T-fiberings over I	.	.	.	.	.	.	.	3	7	17	33
				non-geometric							
$D,2 - D,2$	.	.	.	.	.	.	.	4	35	168	674
$A,1$	.	.	.	.	.	.	.	.	8	24	
$D,2 - D,3$	.	.	.	.	.	.	.	.	3	24	
$S,1 - D,2$	.	.	.	.	.	.	.	.	3	24	
$D,2 - A,1 - D,2$	.	.	.	.	.	.	.	.	3	31	
total	3	2	4	7	14	31	74	175	436	1150	3053

Table 4: The type of graph manifolds of given complexity, up to 10. Here, I, D, S, A, T, K denote respectively the closed interval, the disc, the Möbius strip, the annulus, the torus, and the Klein bottle. We denote by X, n a block with base space the surface X and n exceptional fibers. We write X for $X, 0$. We have counted as T-fiberings only the Sol manifolds, not the manifolds also admitting a Seifert structure. There is a flat manifold with $c = 6$ counted twice, since it has two different fibrations, corresponding to the asterisks.

3.3. Non-geometric manifolds

Every non-hyperbolic orientable manifold with $c \leqslant 10$ is a graph manifold, i.e. its JSJ decomposition consists of Seifert or Sol blocks. A non-geometric orientable manifold whose decomposition contains a hyperbolic block with $c \leqslant 11$ is constructed in [AM03], and from our census now it follows that it cannot have $c \leqslant 10$. Therefore we have proved the following.

Theorem 3.3. *The first closed orientable irreducible manifold with non-trivial JSJ decomposition containing hyperbolic blocks has $c = 11$.*

All graph manifolds with $c \leqslant 10$ are collected in Table 4 according to their JSJ decomposition into fibering pieces, and to the type of fiberings of each piece.

3.4. The simplest manifolds

As the following discussion shows, in most geometries, the manifolds with lowest complexity are the "simplest" ones.

3.4.1. Elliptic

The elliptic manifolds of smallest complexity are $S^3, \mathbb{RP}^3$, and $L(3,1)$, having $c = 0$. The first manifold which is not a lens space is $\left(S^2, (2,1), (2,1), (2,1), -1\right)$ and has $c = 2$. It is the elliptic manifold with smallest non-cyclic fundamental group, having order 8 [Mat03a].

3.4.2. Flat

Every (orientable or not) flat manifold has $c = 6$. A typical way to obtain some flat 3-manifold M is from a face-pairing of the cube: by taking a triangulation of the cube with 6 tetrahedra matching along the face-pairing, we get a minimal triangulation of M.

3.4.3. $\mathbb{H}^2 \times \mathbb{R}$

The first manifolds of type $\mathbb{H}^2 \times \mathbb{R}$ are non-orientable and have $c = 7$, and are also the manifolds of that geometry with smallest base orbifold [AM05], having volume $-2\pi\chi^{\text{orb}} = \pi/3$.

3.4.4. Sol

The first manifold of type Sol is also non-orientable and has $c = 6$, and it is the unique filling of the Gieseking manifold, the cusped hyperbolic manifold with smallest volume $1.0149\ldots$ [Ada87] and smallest complexity 1 [CHW99]. It is also the unique torus fibering whose monodromy $A = \left(\begin{smallmatrix} 0 & 1 \\ 1 & 1 \end{smallmatrix}\right)$ is hyperbolic with $|\text{tr}\, A| < 2$ [AM05].

3.4.5. Hyperbolic

As we said above, the first orientable hyperbolic manifolds are the smallest ones known. It would be interesting to know the complexity of the first non-orientable closed hyperbolic manifold, whose volume is probably considerably bigger than in the orientable case, see [HW94].

4. Census of hyperbolic manifolds

We describe here the compact hyperbolic manifolds with boundary with $\chi = 0$ and $c \leqslant 7$, and the orientable ones with $\chi < 0$ and $c \leqslant 4$.

topological boundary	0	1	2	3	4	5	6	7
orientable								
T	.	.	2	9	52	223	913	3388
T,T	.	.	.	.	4	11	48	162
T,T,T	.	.	.	.	.	.	1	2
total orientable	.	.	**2**	**9**	**56**	**234**	**962**	**3552**
non-orientable								
K	.	1	1	5	14	52	171	617
K,K	.	.	1	2	9	23	68	208
K,K,K	.	.	.	.	.	.	3	6
K,K,K,K	.	.	.	.	.	.	1	.
T	.	.	.	.	1	1	4	19
T,T	.	.	.	.	.	.	1	.
K,T	.	.	.	.	1	2	8	31
K,K,T	.	.	.	.	1	.	3	6
total non-orientable	.	**1**	**2**	**7**	**26**	**78**	**259**	**887**

Table 5: The number of cusped hyperbolic manifolds of given complexity, up to 7. The "topological boundary" indicates the tori T and Klein bottles K present as cusps.

4.1. Manifolds with $\chi = 0$

Recall that we define a compact M to be hyperbolic when it admits a complete metric of finite volume and geodesic boundary, after removing all boundary components with $\chi = 0$. Therefore, hyperbolic manifolds M with $\chi(M) = 0$ have some cusps based on tori or Klein bottles, and those with $\chi(M) < 0$ have geodesic boundary and possibly some cusps. To avoid confusion, we define the *topological boundary* of M to be the union of the geodesic boundary and the cusps.

Hyperbolic manifolds with $\chi(M) = 0$ and $c \leqslant 7$ were listed by Hodgson and Weeks in [CHW99] and form the cusped census used by SnapPea. They are collected, according to their topological boundary, in Table 5. Hyperbolicity of each manifold was checked by solving Thurston's equations, and all manifolds were distinguished computing their Epstein-Penner *canonical decomposition* [EP88]. In practice, volume, homology, and the length of the shortest geodesic are usually enough to distinguish two such manifolds.

4.2. Manifolds with $\chi < 0$

Equations analogous to Thurston's were constructed by Frigerio and Petronio in [FP04] for an ideal triangulation T of a manifold M with $\chi(M) < 0$. A solution of such equations gives a realization of the hyperbolic structure of M via partially truncated hyper-

topological boundary	0	1	2	3	4
2	.	.	8	76	628
3	.	.	.	74	2034
4	.	.	.	.	2340
2,0	.	.	.	1	18
3,0	.	.	.	.	12
2,0,0	.	.	.	.	1
total	.	.	**8**	**151**	**5033**

Table 6: The number of orientable hyperbolic manifolds with non-empty geodesic boundary of given complexity, up to 4. The "topological boundary" indicates the genera of the boundary components, with zeroes correspond to cusps.

bolic tetrahedra. One such tetrahedron is parametrized by its 6 interior dihedral angles $\alpha_1, \ldots, \alpha_6$. The sum of the 3 of them incident to a given vertex must be less or equal than π, and the vertex is truncated if the sum is less than π, or ideal if it is π. The compatibility equations ensure that identified edges all have the same length and that dihedral angles sum to 2π around each resulting edge. These equations, together with others checking the completeness of the cusps, realize the hyperbolic structure for M. Then Kojima's *canonical decomposition* [Koj90], analogous to Epstein-Penner's, is a complete invariant which allows one to distinguish manifolds. In contrast with the case $\chi = 0$, there are plenty of manifolds having the same complexity that are not distinguished by volume, homology, Turaev-Viro invariants, and the canonical decomposition seems to be the only available tool, see Subsection 6.2. The results from [FMP04] are summarized in Table 6.

Remark 4.1. The two censuses of hyperbolic manifolds described in this Section have a slightly more experimental nature than the closed census of Section 3, since solving hyperbolicity equations and calculating the canonical decomposition involve numerical calculations with truncated digits.

5. Complexity and volume of hyperbolic manifolds

We describe here some relations between the complexity and the volume of a hyperbolic 3-manifold.

5.1. Ideal tetrahedra and octahedra

As Theorem 5.1 below shows, there is a constant K such that $\mathrm{Vol}(M) < K \cdot c(M)$ for any hyperbolic M. Let $v_T = 1.0149\ldots$ and $v_O = 3.6638\ldots$ be the volumes respectively of the regular ideal hyperbolic tetrahedron and octahedron.

Theorem 5.1. *Let M be hyperbolic, with or without boundary. If $\chi(M) = 0$ we have* $\mathrm{Vol}(M) \leqslant v_T \cdot c(M)$. *If $\chi(M) < 0$ we have* $\mathrm{Vol}(M) < v_O \cdot c(M)$.

Proof. First, note that by the naturality property of the complexity $c(M)$ is the minimum number of tetrahedra in an (ideal) triangulation. If M is closed, take a minimal triangulation T and straighten it. Tetrahedra may overlap or collapse to low-dimensional objects, having volume zero. Since geodesic tetrahedra have volume less than v_T, we get the inequality.

If M is not closed, let T be an ideal triangulation for M with $c(M)$ tetrahedra. We can realize topologically M with its boundary tori removed, by partially truncating each tetrahedron in T (i.e. removing the vertex only in presence of a cusp, and an open star of it in presence of true boundary). Then we can straighten every truncated tetrahedron with respect to the hyperbolic structure in M. As above, tetrahedra may overlap or collapse. In any case, the volume of each such will be at most v_T if there is no boundary, and strictly less than v_O in general, since any ideal tetrahedron has volume at most equal to v_T, and any partially truncated tetrahedron has volume strictly less than v_O [Ush03]. □

The constants v_T and v_O are the best possible ones, see Remark 6.9. A converse result of type $c(M) < K' \cdot \mathrm{Vol}(M)$ is impossible, because for big C's there are a finite number of hyperbolic manifolds with complexity less than C, and an infinite number of such with volume less than C.

5.2. First segments of c and Vol

Complexity and volume give two partial orderings on the set $\mathcal{H}$ of all hyperbolic 3-manifolds. By what was just said, they are globally qualitatively very different. Nevertheless, as noted in [MF88], they might have similar behaviours on some subsets of $\mathcal{H}$. We propose the following conjecture.

Conjecture 5.2. Among hyperbolic manifolds with the same topological boundary, the ones with smallest complexity have volume smaller than the other ones.

The conjecture is stated more precisely as follows: let $\mathcal{M}_\Sigma$ be the set of hyperbolic manifolds having some fixed topological boundary Σ. Suppose $M \in \mathcal{M}_\Sigma$ is so that $c(M') \geqslant c(M)$ for all $M' \in \mathcal{M}_\Sigma$. We conjecture that $\mathrm{Vol}(M') > \mathrm{Vol}(M)$ for all $M' \in \mathcal{M}_\Sigma$ having $c(M') > c(M)$. We now discuss our conjecture.

5.2.1. Closed case

The closed hyperbolic manifolds with smallest $c = 9$ are the four having smallest volume known, see Table 3. Therefore Conjecture 5.2 claims that these four are actually the ones having smallest volumes among all closed hyperbolic manifolds.

5.2.2. Connected topological boundary

In this case, Conjecture 5.2 is true, as the following shows.

Theorem 5.3. *Among hyperbolic manifolds whose topological boundary is a connected surface, the ones with smallest volume are the ones with smallest complexity.*

Proof. Among manifolds having one toric cusp, the figure-8 knot complement and its sibling are those with minimal volume $2 \cdot v_T$ [CM01] and minimal complexity 2. Among those with a Klein bottle cusp, the Gieseking manifold is the one with minimal volume v_T [Ada87] and minimal complexity 1. Our assertion restricted to orientable 3-manifolds bounded by a connected surface of higher genus is proved in [FMP03a] combining the naturality property of the complexity with Miyamoto's description [Miy94] of all such manifolds with minimal volume. The same proof also works in the general case. $\square$

5.2.3. Experimental data

Conjecture 5.2 is true when restricted to the manifolds of Tables 3, 5, and 6, for all the boundary types involved (see [CHW99], [Wee], and [FMP04]). One sees from Table 3 that the manifolds of type (K,K), (T,T), (K,T), (K,K,T), (T,T,T), (K,K,K), and (K,K,K,K) with smallest complexity have respectively $c = 2,4,4,4,6,6$, and 6. The manifolds with $c = 2$ are constructed with two regular ideal tetrahedra, and hence have volume $2 \cdot v_T$. Those with $c = 4$ are constructed either with 4 regular ideal tetrahedra, hence having volume $4 \cdot v_T = 4.05976\ldots$, or with one regular ideal octahedron, of volume $v_O = 3.6638\ldots$ (therefore Conjecture 5.2 claims that every other M with the same topological boundary has volume bigger than $4 \cdot v_T$). Those with $c = 6$ have volume $2 \cdot v_D = 5.3334\ldots$, where $v_D = 2.6667\ldots$ is the volume of the "triangular ideal drum" used by Thurston [Thu80] to construct the complement of the chain link of Fig. 3, which is the only orientable manifold among them.

Problem 5.4. Classify the hyperbolic (orientable) manifolds of smallest complexity among those having $\chi = 0$ and k toric cusps, and compute their volume, for each k.

5.3. Matveev-Fomenko conjecture

As we mentioned above, the orderings given by c and Vol are qualitatively different on the whole set $\mathcal{M}$ of hyperbolic manifolds, but might be similar on some subsets of $\mathcal{M}$. The following conjecture was proposed by Matveev and Fomenko in [MF88].

Conjecture 5.5 (Matveev-Fomenko [MF88]). Let M be a hyperbolic manifold with one cusp. Among Dehn fillings N and N' of M, if $c(N) < c(N')$ then $\mathrm{Vol}(N) < \mathrm{Vol}(N')$.

The complexity-10 closed census produces a counterexample to Conjecture 5.5.

Proposition 5.6. *Let $N(p/q)$ be the p/q-surgery on the figure-8 knot. We have*

$$\mathrm{Vol}\big(N(5/2)\big) = 1.5294773\ldots \qquad c\big(N(5/2)\big) = 11$$
$$\mathrm{Vol}\big(N(7)\big) = 1.4637766\ldots \qquad c\big(N(7)\big) > 11$$

Proof. We first note that $N(p/q) = N(-p/q)$ is the $(1,2,1-p/q)$-surgery on the chain link. The manifold $N(7)$ does not belong to Table 3 (it is the manifold labeled as N.11 in our census of surgeries on the chain link of [Pet]), and hence has $c > 11$, whereas $N(5/2)$ is the manifold N.12 and has $c = 11$. $\qquad\square$

6. Lower bounds

Providing upper bounds for the complexity of a given manifold M is relatively easy: from any combinatorial description of M one recovers a spine of M with n vertices, and certainly $c(M) \leqslant n$. Finding lower bounds is a much more difficult task. The only ∂-irreducible manifolds whose complexity is known are those listed in the censuses of Sections 3 and 4, and some infinite families of hyperbolic manifolds with bundary described below. In particular, for a closed irreducible M, the value $c(M)$ is only known when $c(M) \leqslant 10$, i.e. for a finite number of manifolds.

6.1. The closed case

The only available lower bound for closed irreducible orientable manifolds is the following one, due to Matveev and Pervova. We denote by $|\mathrm{Tor}(H_1(M))|$ the order of the torsion subgroup of $H_1(M)$, while b_1 is the rank of the free part, i.e. the fist Betti number of M.

Theorem 6.1 (Matveev-Pervova [MP01b]). *Let M be a closed orientable irreducible manifold different from $L(3,1)$. Then $c(M) \geqslant 2 \cdot \log_5 |\mathrm{Tor}(H_1(M))| + b_1 - 1$.*

Recall that Theorem 3.1 holds only for $c \leqslant 10$. Actually, the same formulas in the statement give an upper bound for $c(M)$. Some such upper bounds for lens spaces, torus bundles, and simple Seifert manifolds were previously found by Matveev and Anisov, who proposed the following conjectures.

Conjecture 6.2 (Matveev [Mat03a]). We have

$$c\big(L(p,q)\big) = |p,q| - 2 \quad \text{and} \quad c\big(S^2, (2,1), (2,1), (m,1), -1\big) = m$$

Conjecture 6.3 (Anisov [Ani01]). The complexity of a torus bundle M over S^1 with monodromy $A \in \mathrm{GL}_2(\mathbb{Z})$ is $c(M) = \min\{\|A\| + 5, 6\}$.

6.2. Families of hyperbolic manifolds with boundary of known complexity

The following corollaries of Theorem 5.1 were first noted by Anisov.

Corollary 6.4 (Anisov [Ani02]). *The complexity of a hyperbolic manifold decomposing into n ideal regular tetrahedra is n.*

Corollary 6.5 (Anisov [Ani02]). *The punctured torus bundle with monodromy $\begin{pmatrix} 2 & 1 \\ 1 & 1 \end{pmatrix}^n$ is a hyperbolic manifold of complexity $2n$.*

For each $n \geqslant 2$, Frigerio, Martelli, and Petronio defined [FMP03a] the family $\mathcal{M}_n$ of all orientable compact manifolds admitting an ideal triangulation with one edge and n tetrahedra.

Theorem 6.6 (Frigerio–Martelli–Petronio [FMP03a]). *Let $M \in \mathcal{M}_n$. Then M is hyperbolic with a genus-n surface as geodesic boundary, and without cusps. It has complexity n. Its homology, volume, Heegaard genus, and Turaev-Viro invariants also depend only on n.*

The manifolds in $\mathcal{M}_n$ are distinguished by their Kojima's canonical decomposition (see Subsection 4.2), which is precisely the triangulation with one edge defining them. Therefore combinatorially different such triangulations give different manifolds.

Theorem 6.7 (Frigerio–Martelli–Petronio [FMP03a, FMP03b]). *Manifolds in $\mathcal{M}_n$ correspond bijectively to triangulations with one edge and n tetrahedra. The cardinality $\#\mathcal{M}_n$ grows as n^n.*

We say that a sequence a_n *grows as* n^n when there exist constants $0 < k < K$ such that $n^{k \cdot n} < a_n < n^{K \cdot n}$ for all $n \gg 0$.

Corollary 6.8 (Frigerio–Martelli–Petronio [FMP03b]). *The number of hyperbolic manifolds of complexity n grows as n^n.*

Remark 6.9. From the families introduced here we see that the inequalities of Theorem 5.1 cannot be strengthened. The torus bundles M above have $\mathrm{Vol}(M) = v_{\mathrm{T}} \cdot c(M)$, and the manifolds in $\mathcal{M}_n$ have $\mathrm{Vol}(M) = v_n \cdot c(M)$, with v_n equals to the volume of a truncated tetrahedron with all angles $\pi/(3n)$, so that $v_n \to v_{\mathrm{O}}$ for $n \to \infty$.

The set $\mathcal{M}_n$ is also the set mentioned in Theorem 5.3 of all manifolds having both minimal complexity and minimal volume among those with a genus-n surface as boundary. We therefore get from Table 6 that $\#\mathcal{M}_n$ is $8, 74, 2340$ for $n = 2, 3, 4$.

The class $\mathcal{M}_n$ is actually contained as $\mathcal{M}_n = \mathcal{M}_{n,0}$ in a bigger family $\mathcal{M}_{g,k}$, defined in [FMP03b]. The set $\mathcal{M}_{g,k}$ consists of all orientable hyperbolic manifolds of complexity $g + k$ with connected geodesic boundary of genus g and k cusps. Theorems 6.6 and 6.7 hold similarly for all such sets. For any fixed g and k, $\mathcal{M}_{g,k}$ is the set of all manifolds with minimum complexity among those with that topological boundary. Therefore Conjecture 5.2 would imply the following.

Conjecture 6.10 (Frigerio–Martelli–Petronio [FMP03b]). The manifolds of smallest volume among those with a genus-g geodesic surface as boundary and k cusps are those in $\mathcal{M}_{g,k}$.

7. Minimal spines

We describe here some known results about minimal spines, which are crucial for computing the censuses of Sections 3 and 4.

7.1. The algorithm

The algorithm used to classify all manifolds with increasing complexity n typically works as follows:

(1) list all special spines with n vertices (or triangulations with n tetrahedra);

(2) remove from the list the many spines that are easily seen to be non-minimal, or not to thicken to an irreducible (or hyperbolic) manifold;

(3) try to recognize the manifolds obtained from thickening the remaining spines;

(4) eliminate from that list of manifolds the duplicates, and the manifolds that have already been found previously in some complexity-n' census for some $n' < n$.

Typically, step (1) produces a huge list of spines, 99.99... % of which are canceled via some quick criterion of non-minimality during step (2), and one is left with a much smaller list, so that steps (3) and (4) can be done by hand.

7.2. Cutting dead branches

Step (1) of the algorithm above needs a huge amount of computer time already for $c = 5$, due to the very big number of spines listed. Therefore one actually uses the non-minimality criteria (step (2)) *while* listing the special spines with n vertices (step (1)), to cut many "dead branches". Step (1) remains the most expensive one in terms of computer time, so it is worth describing it with some details.

A special spine or its dual (possibly ideal) triangulation T (see Remark 2.1) with n tetrahedra can be encoded roughly as follows. Take the face-pairing 4-valent graph G of the tetrahedra in T. It has n vertices and $2n$ edges. After fixing a simplex on each vertex, a label in S_3 on each (oriented) edge of G encodes how the faces are glued. We therefore get 6^{2n} gluings (the same combinatorial T is usually realized by many distinct gluings). Point (1) in the algorithm consists of two steps:

(1a) classify all 4-valent graphs G with n vertices;

(1b) for each graph G, fix a simplex on each vertex, and try the 6^{2n} possible labelings on edges.

Step (1b) is by far the most expensive one, because it contains many "dead branches"; most of them are cut as follows: a partial labeling of some k of the $2n$ edges defines a partial gluing of the tetrahedra. If such partial gluing already fulfills some local non-minimality criterion, we can forget about every labeling containing this partial one.

Remark 7.1. A spine of an *orientable* manifold can be encoded more efficiently by fixing an immersion of the graph G in $\mathbb{R}^2$, and assigning a colour in $\mathbb{Z}_2$ to each vertex and a colour in $\mathbb{Z}_3$ to each edge [BP95].

Local non-minimality criteria used to cut the branches are listed in Subsection 7.3. We discuss in Subsection 7.4 another powerful tool, which works in the closed case only: it turns out that most 4-valent graphs G can be quickly checked *a priori* not to give rise to any minimal spine (of closed manifolds).

7.3. Local non-minimality criteria

We start with the following results.

Proposition 7.2 (Matveev [Mat90]). *Let P be a minimal special spine of a 3-manifold M. Then P contains no embedded face with at most 3 edges.*

Proposition 7.3 (Matveev [Mat90]). *Let P be a minimal special spine of a closed orientable 3-manifold M. Let e be an edge of P. A face f cannot be incident 3 times to e, and it cannot run twice on e with opposite directions.*

In the orientable setting, both Propositions 7.2 and 7.3 are special cases of the following. Recall that $S(P)$ is the subset of a special spine P consisting of all points of type (2) and (3) shown in Fig. 1.

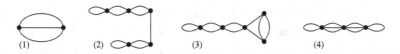

Figure 4: Portions of graphs forbidden for minimal triangulations/spines of closed manifolds.

n	3	4	5	6	7	8	9	10	11
useful	2	4	12	39	138	638	3366	20751	143829
all	4	10	28	97	359	1635	8296	48432	316520

Table 7: Useful graphs among all 4-valent graphs with $n \leqslant 11$ vertices.

Proposition 7.4 (Martelli–Petronio [MP01a]). *Let P be a minimal spine of a closed orientable 3-manifold M. Every simple closed curve $\gamma \subset P$ bounding a disc in the ball $M \setminus P$ and intersecting $S(P)$ transversely in at most 3 points is contained in a small neighborhood of a point of P.*

Analogous results in the possibly non-orientable setting are proved by Burton [Bur04].

7.4. Four-valent graphs

Quite surprisingly, some 4-valents graphs can be checked *a priori* not to give any minimal special spine of closed 3-manifold.

Remark 7.5. The face-pairing graph of a (possibly ideal) triangulation is also the set $S(P)$ in the dual special spine P.

Proposition 7.6 (Burton [Bur04]). *The face-pairing graph G of a minimal triangulation with at least 3 tetrahedra does not contain any portions of the types shown in Fig. 4-(1,2,3), except if G itself is as in Fig. 4-(4).*

A portion of G is of type shown in Fig. 4-(2,3,4) when it is as in that picture, with chains of arbitrary length. In the algorithm of Subsection 7.2, step (1b) can be therefore restricted to the *useful* 4-valent graphs, i.e. the ones that do not contain the portions forbidden by Proposition 7.6. Table 7, taken from [Bur04], shows that some 40 % of the graphs are useful.

8. Bricks

As shown in Sections 2 and 7, classifying all closed $\mathbb{P}^2$-irreducible manifolds with complexity n reduces to listing all minimal special spines of such manifolds with n

vertices. Non-minimality criteria as those listed in Section 7 are then crucial to eliminate the many non-minimal spines (by cutting "dead branches") and gain a lot of computer time. Actually, closed manifolds often have many minimal spines, and it is not necessary to list them all: a criterion that eliminates some, but not all, minimal spines of the same manifold is also suitable for us. This is the basic idea which underlies the decomposition of closed $\mathbb{P}^2$-irreducible manifolds into *bricks*, introduced by Martelli and Petronio in [MP01a], and described in the orientable case in this Section. (For the nonorientable one, see [MP02b].)

8.1. A quick introduction

The theory is roughly described as follows: every closed irreducible manifold M decomposes along tori into pieces on which the complexity is additive. Each torus is marked with a θ-graph in it, and the complexity of each piece is not the usual one, because it depends on that graphs. A manifold M which does not decompose is a *brick*. Every closed irreducible manifold decomposes into bricks. The decomposition is not unique, but there can be only a finite number of such. In order to classify all manifolds with $c \leqslant 10$, one classifies all bricks with $c \leqslant 10$, and then assemble them in all possible (finite) ways to recover the manifolds.

For $c \leqslant 10$, bricks are atoroidal, hence either Seifert or hyperbolic. And the decomposition into bricks is tipically a mixture of the JSJ, the graph-manifolds decomposition, and the thick-thin decomposition for hyperbolic manifolds. Very few closed manifolds do not decompose, i.e. are themselves bricks.

Proposition 8.1. *There are 25 closed bricks with $c \leqslant 10$. They are: 24 Seifert manifolds of type $\left(S^2, (2,1), (m,1), (n,1), -1\right)$, and the hyperbolic manifold N.34 of Table 3.*

Among closed bricks, we have Poincaré sphere $\left(S^2, (2,1), (3,1), (5,1), -1\right)$.

Proposition 8.2. *There are 25 non-closed bricks with $c \leqslant 10$.*

There are 4978 closed irreducible manifolds with $c \leqslant 10$, see Table 1. Therefore $4953 = 4978 - 25$ such manifolds are obtained with the 25 non-closed bricks above.

Before giving precise definitions, we note that the *layered triangulations* [Bur03, JR03] of the solid torus H are particular decompositions of H into bricks. Our experimental results show the following.

Proposition 8.3. *Every closed irreducible atoroidal manifold with $c \leqslant 10$ has a minimal triangulation containing a (possibly degenerate [Bur03]) layered triangulation, except for some $\left(S^2, (2,1), (m,1), (n,1), -1\right)$ and the hyperbolic N.34 of Table 3.*

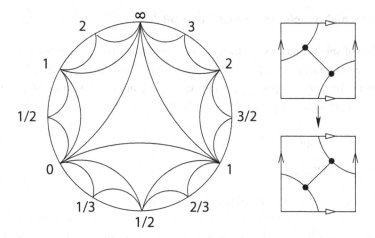

Figure 5: The Farey tesselation of the Poincaré disc into ideal triangles (left) and a flip (right).

8.2. θ-graphs in the torus

In this paper, a θ-*graph* θ in the torus T is a graph with two vertices and three edges inside T, having an open disc as a complement. That is, it is a trivalent spine of T. Dually, this is a one-vertex triangulation of T.

The set of all θ-graphs in T up to isotopy can be described as follows. After choosing a meridian and a longitude, every *slope* on T (i.e. isotopy class of simple closed essential curves) is determined by a number $p/q \in \mathbb{Q} \cup \{\infty\}$. Those numbers are the ideal vertices of the Farey tesselation of the Poincaré disc sketched in Fig. 5-left. A θ-graph contains three slopes, which are the vertices of an ideal triangle of the tesselation. This gives a correspondence between the θ-graphs in T and the triangles of the tesselation. Two θ-graphs correspond to two adjacent triangles when they share two slopes, i.e. when they are related by a *flip*, shown in Fig. 5-right.

8.3. Manifolds with marked boundary

Let M be a connected compact 3-manifold with (possibly empty) boundary consisting of tori. By associating to each torus component of ∂M a θ-graph, we get a *manifold with marked boundary*.

Let M and M' be two marked manifolds, and $T \subset \partial M, T' \subset \partial M'$ be two boundary tori. A homeomorphism $\psi : T \to T'$ sending the marking of T to the one of T' is an *assembling* of M and M'. The result is a new marked manifold $N = M \cup_\psi M'$. We define analogously a *self-assembling* of M along two tori $T, T' \subset \partial M$, the only difference is that for some technical reason we allow the map to send one $\theta \subset T$ either to $\theta' \subset T$ itself or to one of the 3 other θ-graphs obtained from θ' via a flip.

8.4. Spines and complexity for marked manifolds

The notion of spine extends from the class of closed manifold to the class of manifolds with marked boundary. Recall from Subsection 2.1 that a compact polyhedron is *simple* when the link of each point is contained in ⊗. A sub-polyhedron P of a manifold with marked boundary M is called a *spine* of M if:

- $P \cup \partial M$ is simple,

- $M \setminus (P \cup \partial M)$ is an open ball,

- $P \cap \partial M$ is a graph contained in the marking of ∂M.

Note that P is not in general a spine of M in the usual sense[2]. The *complexity* of a 3-manifold with marked boundary M is of course defined as the minimal number of vertices of a simple spine of M. Three fundamental properties extend from the closed case to the case with marked boundary: complexity is still additive under connected sums, it is finite-to-one on orientable irreducible manifolds, and every orientable irreducible M with $c(M) > 0$ has a minimal special spine [MP01a]. (Here, a spine $P \subset M$ is *special* when $P \cup \partial M$ is: the spine P is actually a *special spine with boundary*, with $\partial P = \partial M \cap P$ consisting of all the θ-graphs in ∂M.)

8.5. Bricks

An important easy fact is that if M is obtained by assembling M_1 and M_2, and P_i is a spine of M_i, then $P_1 \cup P_2$ is a spine of M. This implies the first part of the following result.

Proposition 8.4 (Martelli–Petronio [MP01a]). *If M is obtained by assembling M_1 and M_2, we have $c(M) \leqslant c(M_1) + c(M_2)$. If M is obtained by self-assembling N, we have $c(M) \leqslant c(N) + 6$.*

When $c(M) = c(M_1) + c(M_2)$ or $c(M) = c(N) + 6$, and the manifolds involved are irreducible[3], the (self-)assembling is called *sharp*. An orientable irreducible marked manifold M is a *brick* when it is not the result of any sharp (self-)assembling.

Theorem 8.5 (Martelli–Petronio [MP01a]). *Every closed orientable irreducible M is obtained from some bricks via a combination of sharp (self-)assemblings.*

There are only a finite number of such combinations giving the same M.

---

[2]To avoid confusion, the term *skeleton* was used in [MP01a].

[3]This hypothesis is actually determinant only in one case, see [MP01a].

Figure 6: If 4 edges disconnect G, then one of the two pieces is of one of these types.

8.6. The algorithm that finds the bricks

The algorithm described in Subsection 7.2 also works for classifying all bricks of increasing complexity, with some modifications, which we now sketch. As we said above, every brick with $c > 0$ has a minimal spine P such that $P \cup \partial M$ is special. The 4-valent graph $H = S(P \cup \partial M)$ contains the θ-graphs marking the boundary ∂M. By substituting (i.e. identifying) in H each θ-graph with a point, we get a simpler 4-valent graph G. We mark the edges of G containing that new points with a symbol $\star$. It is then possible to encode the whole P by assigning labels in S_3 on the remaining edges of G, as in Subsection 7.2. The spine P is uniquely determined by such data.

Every edge of G can have a label in $S_3 \cup \{\star\}$, giving 7^{2n} possibilities to analyze during step (1b) of the algorithm (actually, they are $2^n(3+1)^{2n}$ by Remark 7.1). Although there are more possibilities to analyze than in the closed case (7^{2n} against 6^{2n}), the non-minimality criteria for bricks listed below are so powerful, that step (1b) is actually experimentally much quicker for bricks than for closed manifolds. This should be related with the experimental fact that there are much more manifolds than bricks.

Proposition 8.6 (Martelli–Petronio [MP01a]). *Let P be a minimal special spine of a brick with $c > 3$. The 3 faces incident to an edge e of P are all distinct. A face can be incident to at most one θ-graph in ∂P.*

Theorem 8.7 (Martelli–Petronio [MP01a]). *Let G be the 4-valent graph associated to a minimal special spine of a brick with $c > 3$. Then:*

(i) *no pair of edges disconnects G;*

(ii) *if $c \leqslant 10$ and a quadruple of edges disconnects G, one of the two resulting components must be of one of the forms shown in Fig. 6.*

Point (ii) of Theorem 8.7 is proved for $c \leqslant 9$ in [MP01a] and conjectured there to be true for all c: its extension to the case $c = 10$ needed here is technical and we omit it. We can restrict step (1b) of the algorithm to the *useful* 4-valent graphs, i.e. the ones that are not forbidden by Theorem 8.7. Table 8 shows that only 2.1 % of the graphs are useful for $c = 10, 11$.

n	3	4	5	6	7	8	9	10	11
useful	1	2	4	11	27	57	205	1008	6549
all	4	10	28	97	359	1635	8296	48432	316520

Table 8: Useful graphs among all 4-valent graphs with $n \leqslant 11$ vertices.

8.7. Bricks with $c \leqslant 10$.

We list here the bricks found. There are two kinds of bricks: the closed ones, and the ones with boundary. The closed ones correspond to the closed irreducible 3-manifolds that do not decompose.

Theorem 8.8. *The closed bricks having $c \leqslant 10$ are:*

- $\left(S^2,(2,1),(3,1),(m,1),-1\right)$ *with $m \geqslant 5, m \neq 6$, having $c = m$;*

- $\left(S^2,(2,1),(n,1),(m,1),-1\right)$ *not of the type above and with $\{n,m\} \neq \{3,6\},\{4,4\}$, having $c = n+m-2$;*

- *the closed hyperbolic manifold N.34 from Table 3, with volume $1.75712\ldots$ and homology $\mathbb{Z}_7$, obtained as a $(1,-5,-3/2)$-surgery on the chain link, having $c = 10$.*

Remark 8.9. The manifolds $\left(S^2,(2,1),(n,1),(m,1),-1\right)$ with $\{n,m\} = \{3,6\}$ or $\{4,4\}$ are not bricks. Actually, they are flat torus bundles, whereas every other such manifold is atoroidal.

In the following statement, we denote by $N(\alpha,\beta,\gamma)$ the following marked manifold: take the chain link of Fig. 3; if $\alpha \in \mathbb{Q}$, perform an α-surgery on one component, and if $\alpha = \theta^{(i)}$, drill that component and mark the new torus with the θ-graph containing the slopes ∞, i, and $i+1$. Do the same for β and γ (the choice of the components does not matter, see Fig. 3).

Theorem 8.10. *The bricks with boundary having $c \leqslant 10$ are:*

$c = 0$: *one marked $T \times [0,1]$ and two marked solid tori;*

$c = 1$: *one marked $T \times [0,1]$;*

$c = 3$: *one marked (pair of pants)$\times S^1$;*

$c = 8$: *one marked $\left(D,(2,1),(3,1)\right)$, and $N(1,-4,\theta^{(-1)})$;*

The complement M of this link is a hyperbolic manifold. On each cusp, there are two shortest loops of equal length, and hence two preferred θ-graphs, the ones containing both loops. Up to symmetries of M, there are only 3 marked M's with such preferred θ-graphs, and these are the ones with $c = 10$.

Figure 7: The complement of a chain link with 4 components.

$c = 9$: *four bricks of type* $N(\alpha, \beta, \gamma)$, *with* (α, β, γ) *being one of the following:*

$$(1, -5, \theta^{(-1)}), \ (1, \theta^{(-2)}, \theta^{(-2)}), \ (\theta^{(-3)}, \theta^{(-2)}, \theta^{(-2)}), \ (\theta^{(-2)}, \theta^{(-2)}, \theta^{(-2)});$$

$c = 10$: *eleven bricks of type* $N(\alpha, \beta, \gamma)$, *with* (α, β, γ) *being one of the following:*

$$(1, 2, \theta^{(i)}) \ \text{with} \ i \in \{-3, -2, -1, 0\}, \quad (1, -6, \theta^{(-1)}),$$

$$(-5, \theta^{(-2)}, \theta^{(-1)}), \quad (-5, \theta^{(-1)}, \theta^{(-1)}), \quad (1, \theta^{(-1)}, \theta^{(-1)}),$$

$$(1, \theta^{(-4)}, \theta^{(-1)}), \quad (2, \theta^{(-2)}, \theta^{(-2)}), \quad (\theta^{(-3)}, \theta^{(-1)}, \theta^{(-1)}),$$

and three marked complements of the same link, shown in Fig. 7.

Remark 8.11. Using the bricks with $c \leqslant 1$, one constructs every marked solid torus. This construction is the *layered solid torus* decomposition [Bur03, JR03]. An atoroidal manifold with $c \leqslant 10$ is either itself a brick, or it decomposes into one brick B of Theorem 8.10 and some layered solid tori.

Remark 8.12. The generic graph manifold decomposes into some Seifert bricks with $c \leqslant 3$. As Theorem 3.1 suggests, the only exceptions with $c \leqslant 10$ are the closed bricks listed by Theorem 8.8, and some surgeries of the Seifert brick with $c = 8$.

Remark 8.13. Table 3 is deduced from Theorems 8.8 and 8.10, using SnapPea via a python script available from [Pet].

Remark 8.14. The proof of Theorem 3.1 from [MP04] extends to $c = 10$. One has to check that the new hyperbolic bricks with $c = 10$ do not contribute to the complexity of non-hyperbolic manifolds, at least for $c = 10$: we omit this discussion.

We end this Section with a conjecture, motivated by our experimental results, which implies that the decomposition into bricks is always finer than the JSJ.

Conjecture 8.15. Every brick is atoroidal.

References

[AB00] S. B. Alexander & R. L. Bishop (2000). Spines and homology of thin Riemannian manifolds with boundary. *Adv. Math.* **155** (1), 23–48.

[Ada87] C. C. Adams (1987). The noncompact hyperbolic 3-manifold of minimal volume. *Proc. Amer. Math. Soc.* **100** (4), 601–606.

[AM03] G. Amendola & B. Martelli (2003). Non-orientable 3-manifolds of small complexity. *Topology Appl.* **133** (2), 157–178.

[AM05] G. Amendola & B. Martelli (2005). Non-orientable 3-manifolds of complexity up to 7. *Topology Appl.* **150** (2), 179–195.

[Ani01] S. Anisov (2001). Towards lower bounds for complexity of 3-manifolds: a program. Preprint, math.GT/0103169.

[Ani02] S. Anisov (2002). Complexity of torus bundles over the circle. Preprint, math.GT/0203215.

[BP95] R. Benedetti & C. Petronio (1995). A finite graphic calculus for 3-manifolds. *Manuscripta Math.* **88** (3), 291–310.

[Bur03] B. Burton (2003). Structures of small closed non-orientable 3-manifold triangulations. Preprint, math.GT/0311113.

[Bur04] B. A. Burton (2004). Face pairing graphs and 3-manifold enumeration. *J. Knot Theory Ramifications* **13** (8), 1057–1101.

[CHW99] P. J. Callahan, M. V. Hildebrand & J. R. Weeks (1999). A census of cusped hyperbolic 3-manifolds. *Math. Comp.* **68** (225), 321–332.

[CM01] C. Cao & G. R. Meyerhoff (2001). The orientable cusped hyperbolic 3-manifolds of minimum volume. *Invent. Math.* **146** (3), 451–478.

[DT03] N. M. Dunfield & W. P. Thurston (2003). The virtual Haken conjecture: experiments and examples. *Geom. Topol.* **7**, 399–441 (electronic).

[EP88] D. B. A. Epstein & R. C. Penner (1988). Euclidean decompositions of noncompact hyperbolic manifolds. *J. Differential Geom.* **27** (1), 67–80.

[FMP03a] R. Frigerio, B. Martelli & C. Petronio (2003). Complexity and Heegaard genus of an infinite class of compact 3-manifolds. *Pacific J. Math.* **210** (2), 283–297.

[FMP03b] R. Frigerio, B. Martelli & C. Petronio (2003). Dehn filling of cusped hyperbolic 3-manifolds with geodesic boundary. *J. Differential Geom.* **64** (3), 425–455.

[FMP04]   R. Frigerio, B. Martelli & C. Petronio (2004).   Small hyperbolic 3-manifolds with geodesic boundary. *Experiment. Math.* **13** (2), 171–184.

[FP04]   R. Frigerio & C. Petronio (2004).   Construction and recognition of hyperbolic 3-manifolds with geodesic boundary. *Trans. Amer. Math. Soc.* **356** (8), 3243–3282 (electronic).

[Fuj90]   M. Fujii (1990).   Hyperbolic 3-manifolds with totally geodesic boundary which are decomposed into hyperbolic truncated tetrahedra. *Tokyo J. Math.* **13** (2), 353–373.

[Hem76]   J. Hempel (1976).   *3-Manifolds.* Princeton University Press, Princeton, N.J.

[HW94]   C. D. Hodgson & J. R. Weeks (1994).   Symmetries, isometries and length spectra of closed hyperbolic three-manifolds. *Experiment. Math.* **3** (4), 261–274.

[JR03]   W. Jaco & J. H. Rubinstein (2003).   0-efficient triangulations of 3-manifolds. *J. Differential Geom.* **65** (1), 61–168.

[Koj90]   S. Kojima (1990). Polyhedral decomposition of hyperbolic manifolds with boundary. *Proc. Work. Pure Math.* **10**, 37–57.

[Mat88]   S. V. Matveev (1988). The theory of the complexity of three-dimensional manifolds. Akad. Nauk Ukrain. SSR Inst. Mat. Preprint, **13**.

[Mat90]   S. V. Matveev (1990). Complexity theory of three-dimensional manifolds. *Acta Appl. Math.* **19** (2), 101–130.

[Mat03a]   S. Matveev (2003).   *Algorithmic Topology and Classification of 3-manifolds, Algorithms and Computation in Mathematics*, volume 9. Springer-Verlag, Berlin.

[Mat03b]   S. V. Matveev (2003). Private communication.

[McM90]   C. McMullen (1990). Iteration on Teichmüller space. *Invent. Math.* **99** (2), 425–454.

[MF88]   S. V. Matveev & A. T. Fomenko (1988). Isoenergetic surfaces of Hamiltonian systems, the enumeration of three-dimensional manifolds in order of growth of their complexity, and the calculation of the volumes of closed hyperbolic manifolds. *Uspekhi Mat. Nauk* **43** (1(259)), 5–22, 247.

[Miy94]   Y. Miyamoto (1994).   Volumes of hyperbolic manifolds with geodesic boundary. *Topology* **33** (4), 613–629.

[MP01a]　B. Martelli & C. Petronio (2001). Three-manifolds having complexity at most 9. *Experiment. Math.* **10** (2), 207–236.

[MP01b]　S. V. Matveev & E. L. Pervova (2001). Lower bounds for the complexity of three-dimensional manifolds. *Dokl. Akad. Nauk* **378** (2), 151–152.

[MP02a]　B. Martelli & C. Petronio (2002). Dehn fillings of the "magic" 3-manifold. Preprint, math.GT/0204228.

[MP02b]　B. Martelli & C. Petronio (2002). A new decomposition theorem for 3-manifolds. *Illinois J. Math.* **46** (3), 755–780.

[MP04]　B. Martelli & C. Petronio (2004). Complexity of geometric three-manifolds. *Geom. Dedicata* **108**, 15–69.

[Ovc97]　M. Ovchinnikov (1997). A table of closed orientable prime 3-manifolds of complexity 7. Preprint, Chelyabinsk University.

[Pet]　C. Petronio. Home page. www.dm.unipi.it/~petronio/.

[Sco83]　P. Scott (1983). The geometries of 3-manifolds. *Bull. London Math. Soc.* **15** (5), 401–487.

[Thu80]　W. P. Thurston (1980). The Geometry and Topology of 3-Manifolds. Lecture notes, Princeton University. Available at www.msri.org/publications/books/gt3m/.

[Ush03]　A. Ushijima (2003). A volume formula for generalized hyperbolic tetrahedra. Preprint, math.GT/0309216.

[Wee]　J. R. Weeks. *SnapPea*: a computer program for creating and studying hyperbolic 3-manifolds. Available from www.northnet.org/weeks.

Bruno Martelli

Dipartimento di Matematica
Università di Pisa
Via F. Buonarroti 2
56127 Pisa, Italy

martelli@dm.unipi.it

AMS Classification: 57M27 (primary), 57M20, 57M50 (secondary)

Keywords: 3-manifolds, spines, complexity, enumeration

Spaces of Kleinian Groups
Lond. Math. Soc. Lec. Notes **329**, 121–149

Cambridge University Press
Y. Minsky, M. Sakuma & C. Series (Eds.)

Moduli of continuity of Cannon–Thurston maps

Hideki Miyachi

Abstract

In this paper, we deal with the moduli of continuity of Cannon–Thurston maps for Kleinian 2-orbifold groups with bounded geometry. As an application, we establish the continuity of Cannon–Thurston maps when the corresponding representations vary. We also obtain an estimate for a fractal measure of a set of endpoints of leaves of a certain geodesic lamination.

1. Introduction

A *Cannon–Thurston map* is, by definition, a continuous map which is equivariant under group actions (cf. §2.2). A typical example is a quasiconformal mapping which appears in deforming a Kleinian group. In this paper, we will mainly concentrate on Cannon–Thurston maps for geometrically infinite representations of Fuchsian groups.

Historically, J. Cannon and W. Thurston first described the existence of Cannon–Thurston maps for geometrically infinite representations as follows.

Theorem ([CT89]). *Let S be a closed surface of genus $\mathfrak{g}(S) \geq 2$ and let N denote a hyperbolic manifold which is an S bundle over the circle. Let $\widetilde{S} \to S$ and $\widetilde{N} \to N$ denote their universal covering spaces. Then, a lift of the inclusion $S \hookrightarrow N$ of the fibre is extended as a continuous map from $S^1 = \partial \widetilde{S}$ to $S^2 = \partial \widetilde{N}$.*

Since the action of the fibre subgroup of $\pi_1(N)$ on S^2 is naturally semiconjugate to that of $\pi_1(S)$ on S^1 via the extension, the extension gives a sphere-filling curve equivariant under the group actions (and hence, it is a Cannon–Thurston map). In addition, they also dealt with a hyperbolic manifold with one geometrically infinite end whose associated ending lamination is the support of the stable lamination of a pseudo-Anosov homeomorphism on S. In this case, the limit set of the corresponding Kleinian group is a continuous image of S^1. As a consequence, they found it locally connected. In [Min94], Y. Minsky developed their works and proved the existence of Cannon–Thurston maps for Kleinian surface groups with bounded geometry (see §4).

Recently, many people observed several intriguing properties of the limit sets of Kleinian groups with bounded geometry (for instance, see [BJ97], [Min94] and [Miy03]), and verified the existence of Cannon–Thurston maps for various classes of Kleinian groups (cf. [Bow02], [Flo80], [Kla99], [McM01], [Mit98] and [Miy02]).

Throughout this paper, we shall abbreviate Cannon–Thurston map by CT-map for the sake of simplicity.

1.1. Main Theorem

The aim of this paper is to give an analytic property of CT-maps for Kleinian 2-orbifold groups with bounded geometry. Let Γ be a Kleinian group and $L(\Gamma)$ the infimum of the translation lengths of elements of infinite order in Γ. We say that Γ has *bounded geometry* if $L(\Gamma) > 0$. Denote by d_e the spherical distance on $\widehat{\mathbb{C}}$.

The main theorem in this paper is the following:

Theorem 1.1. *Let G be a finitely generated Fuchsian group of the first kind without parabolic elements acting on $\mathbb{D} = \{z \in \mathbb{C} \,|\, |z| < 1\}$, and let ρ denote a discrete faithful representation of G to $\mathrm{PSL}_2(\mathbb{C})$. If $\rho(G)$ has bounded geometry, the CT-map F for ρ satisfies*

$$d_e(F(x), F(y)) \leq A \left\{ \log \left(\frac{3}{|x - y|} \right) \right\}^{-B} \quad \text{for } x, y \in \partial \mathbb{D},$$

where $B > 0$ depends only on $L(\rho(G))$ and the topology of $\mathbb{D}/G$. Furthermore, when $\rho(G)$ is geometrically infinite, F is not Hölder continuous.

We note that $A > 0$ in Theorem 1.1 is determined by $L(\rho(G))$, the topology of $\mathbb{D}/G$, the hyperbolic distance between $O = (0,0,1) \in \mathbb{H}^3$ and the convex hull $\mathrm{CH}(\rho(G))$ of the limit set of $\rho(G)$, and the *distortion* $\mathrm{dis}(G, \rho)$ of G *in terms of* ρ, where the last constant will be defined in §3. In fact, one can take A as

$$A = A_1 e^{-d_{\mathbb{H}^3}(O, \mathrm{CH}(\rho(G)))} \mathrm{dis}(G, \rho)^B \tag{1.1}$$

where A_1 is dependent only on $L(\rho(G))$ and the topology of $\mathbb{D}/G$ (see §5.2.1, §5.2.2, §5.2.3, and Remark 5.6).

Modulus of continuity   The *modulus of continuity* mod_F of a continuous map F between metric spaces (X, d_X) and (Y, d_Y) is a function on $\mathbb{R}_{\geq 0}$ defined by

$$\mathrm{mod}_F(t) := \sup\{d_Y(F(x_1), F(x_2)) : x_1, x_2 \in X, \ d_X(x_1, x_2) \leq t\}.$$

We here regard CT-maps as continuous maps between two metric spaces $(\partial \mathbb{D}, |\cdot|)$ and $(\widehat{\mathbb{C}}, d_e)$. Then, by Theorem 1.1, we establish an estimate of the moduli of continuity of CT-maps as follows:

Corollary 1.2 (Modulus of Continuity). *Let F be the CT-map for a faithful discrete representation with bounded geometry. Then,*

$$\mathrm{mod}_F(t) = O\big(|\log t|^{-B}\big),$$

where B is the constant in Theorem 1.1.

1.2. Corollaries

In this section, we collect together some results derived from Theorem 1.1.

1.2.1. Families of Cannon–Thurston maps

We will give a rigorous proof for the following folklore in §6.1.

Corollary 1.3 (Continuity). *Let $\{\rho_n\}_{n \in \mathbb{N}}$ be a sequence of discrete faithful representations of G to $\mathrm{PSL}_2(\mathbb{C})$ with $\mathcal{L}(\rho_n(G)) \geq \varepsilon_0$ for some $\varepsilon_0 > 0$. Let F_n denote the CT-map for ρ_n. If ρ_n converges to a representation ρ_∞ algebraically, then F_n converges uniformly to the CT-map for ρ_∞.*

We note that, under the assumption in Corollary 1.3, the limit set of $\rho_n(G)$ converges to that of $\rho(G)$ in the Hausdorff topology (see [AC96] and [AC00]). However, this convergence does not immediately lead to the convergence of CT-maps.

The condition on injectivity radii in Corollary 1.3 can not be entirely removed. In fact, there exists a sequence of quasifuchsian groups such that its algebraic limit is a geometrically finite group with parabolic elements and the limit sets of the quasifuchsian groups do not converge to that of the limit group in the Hausdorff topology (cf.[McM99]). Meanwhile, the author believes that the conclusion of Corollary 1.3 is also valid without the bounded geometry condition when the limit group contains no (accidental) parabolic elements and admits a CT-map.

In §6.3, we also deal with a family of representations defined by iterations of a Teichmüller modular transformation induced by a pseudo-Anosov homeomorphism.

1.2.2. Approximating Peano curves by Polygonal curves

Let $F : \partial\mathbb{D} \to \widehat{\mathbb{C}}$ denote a CT-map as above. For a partition $\mathbf{x} = \{x_k\}_{k=0}^{n+1}$ on $\partial\mathbb{D}$ with $x_{n+1} = x_0$, we define a continuous map $F(\cdot;\mathbf{x}) : \partial\mathbb{D} \to \widehat{\mathbb{C}}$ such that for $k = 0, \ldots, n$, $F(\cdot;\mathbf{x})$ maps monotonically the interval $[x_k, x_{k+1}]$ to a spherical geodesic from $F(x_k)$ to $F(x_{k+1})$ at constant speed. Then, we obtain a guarantee of quality of a polygonal curve $F(\cdot;\mathbf{x})$ as follows (cf. §7. See also §16 of [CT89]).

Corollary 1.4 (Approximating Peano curves). *The following holds:*

$$\max_{y \in \partial\mathbb{D}} d_e(F(y;\mathbf{x}), F(y)) = O(|\log \mathcal{D}|^{-B}),$$

where $\mathcal{D} = \max_{0 \leq k \leq n} |x_{k+1} - x_k|$ and $B > 0$ is as in Theorem 1.1.

1.2.3. Generalised Hausdorff measures of sets

W. Dicks and J. Porti [DP02] and T. Soma [Som95] studied measurements of certain subsets of S^2 related to the actions of Kleinian groups. Using our main theorem, we will deduce the following in §8.

Corollary 1.5. *Let B be as above. For $E \subset \partial \mathbb{D}$, the following statements hold.*

(1) *If the Hausdorff dimension $\dim(F(E))$ of $F(E)$ is positive, then the generalised Hausdorff measure of E with the gauge function $|\log t|^{-b}$ is infinite for $b < B \dim(F(E))$.*

(2) *The Hausdorff dimension of $F(E)$ is zero, when the generalised Hausdorff measure of E with the gauge function $|\log \log(1/|t|)|^{-1}$ vanishes.*

Let λ be a geodesic lamination on a hyperbolic surface of finite area, and $\tilde{\lambda}$ the lift of λ on the universal cover $\mathbb{D}$. Using J. Birman and C. Series' result in [BS85], we deduce that

$$E(\lambda) := \cup\{\partial l \subset \partial \mathbb{D} : l \text{ is a leaf of } \tilde{\lambda}\}$$

has Hausdorff dimension zero (cf. §5.1). From Corollary 1.5 and Proposition 5.1, we obtain an (opposite) estimate for its fractal measure as follows.

Corollary 1.6. *Let λ_e denote the ending lamination of a singly degenerate group with bounded geometry. Then the generalised Hausdorff measure of $E(\lambda_e)$ with the gauge function $|\log t|^{-b}$ is infinite for $b < B$, where $B > 0$ is the constant in Theorem 1.1.*

Acknowledgments   First of all, the author would like to express his gratitude to the organisers of the conference "Spaces of Kleinian Groups and Hyperbolic 3-Manifolds" for their invitation and their great organisation, and the Newton Institute of Mathematical Sciences for their hospitality and financial support.

The author would like to thank Prof. Caroline Series for her constant encouragement, and the Mathematics Institute of the University of Warwick for their hospitality while the author was visiting from August to November in 2001. Almost all of this work was done during the visit. He is very grateful to Prof. Yair N. Minsky for useful comments about the author's many questions.

The author thanks Prof. Ken'ichi Ohshika and Prof. Makoto Sakuma for profitable discussions, Dr. Hirotaka Akiyoshi for pointing out a crucial error in the previous version, and Dr. Yûsuke Okuyama for kind and useful comments on the previous version of this paper. He also thanks Osaka University for financial support from October 2003 to March 2004. Finally, he thanks the referee for useful suggestions and comments.

2. Notation

2.1. Kleinian groups.

A *Kleinian group* is a discrete subgroup of the group of isometries of the hyperbolic 3-space ($\mathbb{H}^3, d_{\mathbb{H}^3}$). We mainly discuss the upper half-space model of the hyperbolic 3-space in this paper. The *limit set* Λ_Γ of a Kleinian group Γ is the set of accumulation points of the orbit of a point in $\mathbb{H}^3$. A *Fuchsian group* is, by definition, a Kleinian group whose limit set is contained in a round circle, and is said to be *of the first kind* if its limit set coincides with the whole circle. By a *Kleinian 2-orbifold group*, we mean a Kleinian group isomorphic to a finitely generated Fuchsian group of the first kind with no parabolic elements. A torsion-free Kleinian 2-orbifold group is called a *Kleinian surface group*. An end of a hyperbolic manifold N is said to be *geometrically infinite* if all of its neighbourhoods intersect the convex core of N. A finitely generated Kleinian group Γ is said to be *geometrically infinite* if some end of $\mathbb{H}^3/\Gamma$ is geometrically infinite. Otherwise, Γ is *geometrically finite*.

Let Γ be a Kleinian surface group and let S denote a closed surface of genus $\mathfrak{g}(S) \geq 2$ such that $\pi_1(S)$ is isomorphic to Γ. By Bonahon's tameness theorem ([Bon86]), any isomorphism ρ from $\pi_1(S)$ to Γ is induced by a homeomorphism from $S \times \mathbb{R}$ to $N_\Gamma := \mathbb{H}^3/\Gamma$. An end E of N_Γ is said to be *positive* (resp. *negative*) if the image of $S \times \mathbb{R}_{>0}$ (resp. $S \times \mathbb{R}_{<0}$) under the homeomorphism is a neighbourhood of E.

By a *degenerate group*, we mean a geometrically infinite Kleinian 2-orbifold group without parabolic elements. A degenerate group is said to be *singly (resp. doubly) degenerate* if its limit set is not the whole sphere (resp. is the whole sphere). If we fix a hyperbolic structure σ and an isomorphism ρ of $\pi_1(S)$ to a torsion-free degenerate group Γ, we associate a geodesic lamination on (S, σ) for any end of N_Γ, which is called the *ending lamination* of the end (cf. [Bon86] and [Min94]).

2.2. Cannon–Thurston maps and their basic properties

Let G be a Kleinian group and ρ a representation of G to $\mathrm{PSL}_2(\mathbb{C})$. By a *Cannon–Thurston map* (we abbreviate by *CT-map*) for ρ, we mean a continuous map $F : \Lambda_G \to \widehat{\mathbb{C}}$ with the equivariance:

$$\rho(g) \circ F(z) = F \circ g(z) \qquad (2.1)$$

for all $g \in G$ and $z \in \Lambda_G$ (compare[1] [Bow02], [Kla99], and [Mit98]).

The following lemma might be well-known. However, we shall sketch a proof for the sake of completeness.

---

[1] In discussing CT-maps, we ordinary recognise G as an abstract (hyperbolic) group and Λ_G as its Gromov boundary. However, in this paper, we adopt G as a subgroup of $\mathrm{PSL}_2(\mathbb{C})$ and Λ_G as its limit set in $\widehat{\mathbb{C}}$. The author hopes these differences make no confusion for readers.

Lemma 2.1. *Let G be a finitely generated non-elementary Kleinian group and ρ : $G \to \mathrm{PSL}_2(\mathbb{C})$ a discrete faithful representation. Then, the following statements hold.*

(1) *If two continuous maps h_1, h_2 on Λ_G satisfy (2.1), then $h_1 = h_2$.*

(2) *Let H be a normal subgroup of G. If $\rho \mid_H$ admits a CT-map, so does ρ.*

Proof. (1)   Let $g \in G$ be a loxodromic element and let $z \in \Lambda_G$ denote its attracting fixed point. By the discreteness of $\rho(G)$, $\rho(g)$ is either loxodromic or parabolic. From (2.1), one can check that $h_i(z)$ is either the attracting fixed point of $\rho(g)$ if $\rho(g)$ is loxodromic, or the fixed point of $\rho(g)$ otherwise. In any case, we have $h_1(z) = h_2(z)$ and hence $h_1 = h_2$ by the density of fixed points in Λ_G.

(2)   Let F denote a CT-map for $\rho \mid_H$. Since $\Lambda_G = \Lambda_H$ (cf. Lemma 2.22 of [MT98]), it suffices to show that for all $g \in G$, $F_g := \rho(g^{-1}) \circ F \circ g$ coincides with F on Λ_G. Indeed, for $h \in H$, we set $h' := ghg^{-1} \in H$. Then

$$
\begin{aligned}
F_g \circ h = \rho(g^{-1}) \circ F \circ (gh) &= \rho(g^{-1}) \circ F \circ (h'g) \\
&= \rho(g^{-1}) \circ \rho(h') \circ F \circ g \\
&= \rho(hg^{-1}) \circ F \circ g = \rho(h) \circ F_g.
\end{aligned}
$$

From (1), we have $F_g = F$ on Λ_G. $\square$

By Minsky's theorem in [Min94] (cf. §4.3) and (2) of Lemma 2.1, we have

Corollary 2.2. *Every isomorphism from a Fuchsian 2-orbifold group to a Kleinian 2-orbifold group with bounded geometry admits a unique CT-map.*

Remark 2.3. Throughout this paper, G will be a Fuchsian surface group or an orbifold group and therefore its limit set Λ_G is always the same as $\partial \mathbb{D}$, which is a circle. Thus, all CT maps will be maps from a circle to the sphere.

3. Distortions

In this section, we will define the distortion of a fuchsian group G in terms of a representation ρ of G to $\mathrm{PSL}_2(\mathbb{C})$. We first define it in the case of Fuchsian surface groups (§3.1.2). Afterwards, we treat it in the case of Fuchsian orbifold groups (§3.2). We will study the behaviour of distortions for a convergent sequence of representations (§3.1.3).

3.1. Distortions: The case of Fuchsian surface groups

3.1.1. Pleated surfaces

Let G be a Fuchsian surface group acting on $\mathbb{D}$. Denote by S the underlying surface of $\mathbb{D}/G$ and by proj_S the projection $\mathbb{D} \to S$. Set $\mathbf{p} = \text{proj}_S(0)$. Then we have a canonical identification

$$\xi_0 : G \to \pi_1(S, \mathbf{p}), \tag{3.1}$$

from the covering theory.

Let $\rho : G \to \text{PSL}_2(\mathbb{C})$ be a discrete faithful representation and $\Gamma = \rho(G)$. A *pleated surface* of N_Γ is a map $f : S \to N_\Gamma$ together with a hyperbolic metric σ_f, called the *induced metric*, and σ_f-geodesic lamination λ on S, so that the following holds: f is length-preserving on paths, maps leaves of λ to geodesics, and is totally geodesic on the complement of λ.

Proposition 3.1 (Bonahon and Thurston). *Suppose Γ as above[2] Then any point in the convex core of N_Γ is contained in a B_0 neighbourhood of the image of a pleated surface, where $B_0 > 0$ is a universal constant.*

For more details, see [Bon86], [CEG87], and [Min01].

We denote by $\text{np}(\rho)$ the nearest point to $O = (0,0,1) \in \mathbb{H}^3$ in the convex hull of Λ_Γ. Notice $\text{np}(\rho) = O$ when O is in the convex hull. Let $\text{pleat}(\rho, R)$ denote the set of all pleated surfaces of N_Γ inducing ρ such that each of their R neighbourhoods contains the projection of $\text{np}(\rho)$. By Proposition 3.1, $\text{pleat}(\rho, b) \neq \emptyset$ for $b \geq B_0$.

3.1.2. Definition of Distortions

We first note that for a π_1-injective continuous map $f : S \to N_\Gamma$ with $f_* = \rho$, there is a unique lift $\widetilde{f} : \mathbb{D} \to \mathbb{H}^3$ of f with $\rho(g) \circ \widetilde{f} = \widetilde{f} \circ g$ for $g \in G$, since the centre of Γ is trivial. Set $\text{bc}(f) := \widetilde{f}(0)$. We next notice that for a diffeomorphism ψ of S isotopic to the identity and a pleated surface f of N_Γ, $f \circ \psi$ is also a pleated surface with same image as f and its induced hyperbolic metric is equal to the pull-back of σ_f via ψ.

Let f be a pleated surface in $\text{pleat}(\rho, B_0)$ with

$$d_{\mathbb{H}^3}(\text{np}(\rho), \text{bc}(f)) \leq B_0 + 1. \tag{3.2}$$

Such a pleated surface with (3.2) in $\text{pleat}(\rho, B_0)$ exists by virtue of Proposition 3.1 and the observation above. Denote by $\text{pr}_{\sigma_f} : \mathbb{D} \to (S, \sigma_f)$ the hyperbolic universal covering

---

[2]Notice here that Γ is isomorphic to a closed surface group. Proposition 3.1 is also valid in the case where S admits punctures. However in such case, we has to impose an additional condition on ρ that it preserves every peripheral parabolic subgroups.

projection with $\text{pr}_{\sigma_f}(0) = \mathbf{p}$ and by G_f its covering transformation group. Then, there is a canonical identification $\pi_1(S, \mathbf{p}) \cong G_f$. Therefore, we obtain an isomorphism

$$\xi_f : G \to G_f \tag{3.3}$$

by composing (3.1) and the identification above. Note that a quasiconformal mapping $w_0 : (\mathbb{D}/G, \mathbf{p}) \to ((S, \sigma_f), \mathbf{p})$ isotopic to the identity rel $\mathbf{p}$ induces a quasiconformal mapping w of $\mathbb{D}$ with $w(0) = 0$ and $\xi_f(g) \circ w = w \circ g$ for $g \in G$:

$$
\begin{array}{ccc}
\mathbb{D} & \xrightarrow[w(0)=0]{w} & \mathbb{D} \\
{\scriptstyle \text{proj}_S} \downarrow & & \downarrow {\scriptstyle \text{pr}_{\sigma_f}} \\
\mathbb{D}/G & \xrightarrow[w_0(\mathbf{p})=\mathbf{p}]{w_0} & (S, \sigma_f).
\end{array}
\tag{3.4}
$$

Let $\text{dis}_f(G, \rho)$ denote the infimum of the maximal dilatations of quasiconformal mappings $w : \mathbb{D} \to \mathbb{D}$ with $w(0) = 0$ and $\xi_f(g) \circ w = w \circ g$ for $g \in G$. We define the *distortion* $\text{dis}(G, \rho)$ of G in terms of ρ by

$$\text{dis}(G, \rho) = \inf_f \text{dis}_f(G, \rho), \tag{3.5}$$

where f runs over all pleated surfaces in $\text{pleat}(\rho, B_0)$ satisfying (3.2).

The following proposition follows from the definition of distortions.

Proposition 3.2. *There exist a pleated surface* $f : (S, \sigma_f) \to N_\Gamma$ *in* $\text{pleat}(\rho, B_0)$ *and a* $2\text{dis}(G, \rho)$ *quasiconformal mapping* w *of* $\mathbb{D}$ *such that*

$$d_{\mathbb{H}^3}(\text{np}(\rho), \text{bc}(f)) \leq B_0 + 1,$$

$w(0) = 0$, *and* $\xi_f(g) \circ w = w \circ g$ *for* $g \in G$.

3.1.3. A Property of Distortions

We will need the following proposition to show the continuity of Cannon–Thurston maps when representations vary (cf. §6.1).

Proposition 3.3. *Let* $\{\rho_n\}_{n=1}^{\infty}$ *be a sequence of faithful discrete type-preserving representations of* G. *When* ρ_n *converges to a type-preserving representation* ρ_∞ *algebraically, distortions* $\text{dis}(G, \rho_n)$ *are uniformly bounded.*

Proof. Take a pleated surface $f_n : (S, \sigma_n) \to N_{\rho_n(G)}$ as in Proposition 3.2 for all n. Denote by G_n the covering transformation group associated to f_n as above. Let $\widetilde{f_n}$:

$\mathbb{D} \to \mathbb{H}^3$ denote the lift of f_n with $\rho_n \circ \xi_{f_n}^{-1}(g) \circ \widetilde{f_n} = \widetilde{f_n} \circ g$ for all $g \in G_n$. By definition, $\mathrm{bc}(f_n) = \widetilde{f_n}(0)$. Since the convergence $\rho_n \to \rho_\infty$ is type-preserving, it is strong (cf. [AC96] and [AC00], see also [Thu80] and [Ohs92]). In particular, by (3.2),

$$d_{\mathbb{H}^3}(\widetilde{f_n}(0), O) \le M_0 \tag{3.6}$$

where M_0 is independent of n. Thus, $f_n(\mathbf{p})$ is in the projection of the hyperbolic ball of centre O and radius M_0.

We here claim the following.

Claim 3.4. *A sequence of pointed hyperbolic surfaces* $\{((S, \sigma_n), \mathbf{p})\}_{n=1}^\infty$ *contains a subsequence which converges to some* $((S, \sigma_\infty), \mathbf{p})$ *geometrically.*

Proof. Fix a constant $\varepsilon > 0$ which is at most the half of the Margulis constant. From (3.6), by taking ε sufficiently small, we may suppose that $\mathbf{p}$ is in the ε-thick part of (S, σ_n) for all n. Let S_n denote the component of $2\varepsilon/3$-thick part of (S, σ_n) which contains $\mathbf{p}$. Since S_n has uniform diameter and $\rho_n \to \rho_\infty$ strongly, by choosing a subsequence in an appropriate way, we have an $M_1 > 0$ which is independent of n such that the lengths of geodesics in $\mathbb{H}^3/\rho_\infty(G)$ corresponding to curves in ∂S_n are at most ε for sufficiently large n, and their ε-Margulis tubes intersect the projection of the hyperbolic ball of centre O and radius M_1.

Thus, when (S, σ_n) contains a curve γ_n whose length tends to 0 as $n \to \infty$, $\rho_\infty(G)$ has to involve an accidental parabolic element, which contradicts our assumption that ρ_∞ is type-preserving. Consequently, $\{((S, \sigma_n), \mathbf{p})\}_{n=1}^\infty$ is precompact in the geometric topology. $\square$

We next claim:

Claim 3.5. $\{\xi_{f_n}\}_{n=1}^\infty$ *contains a subsequence which converges to a representation of* G.

Proof. Let $\zeta_n : ((S, \sigma_\infty), \mathbf{p}) \to ((S, \sigma_n), \mathbf{p})$ denote an ε_n-approximation with $\varepsilon_n \to 0$.

Let G be a system of generators of $\pi_1(S, \mathbf{p})$. Take a positive constant ℓ so that for all elements $\gamma \in G$, the length of the geodesic representative of γ in (S, σ_∞) with base point $\mathbf{p}$ is at most ℓ. Let $h_{\gamma,n}$ denote an element in G_n corresponding to $(\zeta_n)_*(\gamma)$. Since $d_{\mathbb{D}}(0, h_{\gamma,n}(0)) \le (1+\varepsilon_n)\ell$, $\{h_{\gamma,n}\}_{n=1}^\infty$ is precompact in $\mathrm{Aut}(\mathbb{D})$. Furthermore, because $\widetilde{f_n}$ is 1-Lipschitz, by (3.6),

$$d_{\mathbb{H}^3}(O, \rho_n \circ \xi_{f_n}^{-1}(h_{\gamma,n})(O)) \le 2M_0 + d_{\mathbb{H}^3}(\widetilde{f_n}(0), \widetilde{f_n}(h_{\gamma,n}(0)))$$
$$\le 2M_0 + d_{\mathbb{D}}(0, h_{\gamma,n}(0))$$
$$\le 2M_0 + (1+\varepsilon_n)\ell.$$

Hence $\{\rho_n \circ \xi_{f_n}^{-1}(h_{\gamma,n})\}_{n=1}^{\infty}$ is also precompact in $\mathrm{PSL}_2(\mathbb{C})$.

For the sake of simplicity, we set $g_{\gamma,n} := \xi_{f_n}^{-1}(h_{\gamma,n}) \in G$. Suppose $\{\rho_n(g_{\gamma,n})\}_{n=1}^{\infty}$ converges to $T_\infty \in \mathrm{PSL}_2(\mathbb{C})$. Since $\rho_n \to \rho_\infty$ strongly, there is $g_{\gamma,\infty} \in G$ such that $T_\infty = \rho_\infty(g_{\gamma,\infty})$. On the other hand, since $\rho_n(g_{\gamma,\infty}) \to \rho_\infty(g_{\gamma,\infty})$ as $n \to \infty$, $\rho_n(g_{\gamma,n} \circ g_{\gamma,\infty}^{-1})$ converges to the identity in $\mathrm{PSL}_2(\mathbb{C})$ as $n \to \infty$. By applying Jørgensen's inequality to $\rho_n(g_{\gamma,n} \circ g_{\gamma,\infty}^{-1})$ and $\rho_n(g_{\gamma,\infty})$, we see that $\rho_n(g_{\gamma,n})$ and $\rho_n(g_{\gamma,\infty})$ generate an elementary group for sufficiently large n. Since any elementary subgroup in G is cyclic and ρ_n is an isomorphism, there exist an element $g_0 \in G$ and $m_n, m_\infty \in \mathbb{Z}$ such that $g_{\gamma,n} = g_0^{m_n}$ and $g_{\gamma,\infty} = g_0^{m_\infty}$.

Let $\lg_n$ be the complex translation length of $\rho_n(g_0)$. Then the square of the trace of $\rho_n(g_{\gamma,n})$ satisfies

$$\left| \mathrm{Tr}^2 \rho_n(g_{\gamma,n}) \right| \geq 4\cosh^2(m_n \mathrm{Re}(\lg_n)/2). \tag{3.7}$$

Since ρ_∞ is type-preserving and $\rho_n(g_0) \to \rho_\infty(g_0)$, $\rho_\infty(g_0)$ is not parabolic and $\mathrm{Re}(\lg_n)$ is bounded below by a positive constant independent of n. Thus, m_n should be uniformly bounded because $\rho_n(g_{\gamma,n})$ converges to $\rho_\infty(g_{\gamma,\infty})$.

We now see that $\{m_n\}_{n=1}^{\infty}$ is bounded. After passing to a subsequence if necessary, we can suppose that all m_n are equal. Then $g_0^{m_n} = g_0^{m_{n'}}$ for $n, n' \geq 1$ and $\rho_n(g_0^{m_n - m_\infty}) = \rho_n(g_{\gamma,n} \circ g_{\gamma,\infty}^{-1})$ converges to the identity. Hence we deduce that $m_n = m_\infty$ and $g_{\gamma,n} = g_{\gamma,\infty}$, equivalently $\xi_{f_n}(g_{\gamma,\infty}) = h_{\gamma,n}$, for $\gamma \in G$ and sufficiently large n. Since $\{h_{\gamma,n}\}_{\gamma \in G}$ generates G_n, $\{g_{\gamma,\infty}\}_{\gamma \in G}$ also does G. Thus, Claim 3.5 derives from the precompactness of $\{h_{\gamma,n}\}_{n=1}^{\infty}$ for all $\gamma \in G$. $\square$

We now return to the proof of Proposition 3.3. By Claim 3.5, we may suppose that $\mathrm{dis}(G, \rho_n)$ tends to $\sup_{n \in \mathbb{N}} \mathrm{dis}(G, \rho_n)$ and ξ_{f_n} converges to a representation ξ_∞ of G as $n \to \infty$. We denote by w_n the extremal quasiconformal mapping on $\mathbb{D}$ which commutes with ξ_{f_n} (see [IT92]). Since $\xi_{f_n} \to \xi_\infty$, the Teichmüller classes of $\{w_n\}_{n=1}^{\infty}$ are precompact in the Teichmüller space of S. Hence the maximal dilatation of w_n is uniformly bounded and (after passing to a subsequence, if necessary) w_n converges to a quasiconformal mapping w_∞ which commutes with ξ_∞. In particular, the set $\{w_n(0)\}_{n=1}^{\infty}$ is hyperbolically bounded in $\mathbb{D}$ from the origin. Therefore, one can find a quasiconformal mapping W_n of $\mathbb{D}$ such that $W_n \circ g(z) = g \circ W_n(z)$ for $z \in \mathbb{D}$ and $g \in G_n$, $W_n(w_n(0)) = 0$, and its maximal dilatation is uniformly bounded. Since $\mathrm{dis}(G, \rho_n)$ is at most the maximal dilatation of $W_n \circ w_n$, the distortions are uniformly bounded. $\square$

Remark 3.6. The following is a typical example of representations whose distortions diverge: Fix a faithful discrete representation ρ of G and let $g \in G$ be a hyperbolic element. Let $\xi_n : G \to G$ be an isomorphism defined by $\xi_n(h) = g^n h g^{-n}$. Then $\{\rho_n\}_{n=1}^{\infty} := \{\rho \circ \xi_n\}_{n=1}^{\infty}$ is a desired sequence. Indeed, in this case, the corresponding quasiconformal mapping w_n for ξ_n as §3.1 satisfies $w_n(0) = 0$ and coincides with g^n on $\partial \mathbb{D}$ modulo composition with quasiconformal mappings of uniformly bounded maximal dilatations. One can check that the maximal dilatation of w_n diverges as $n \to \infty$, and hence so does $\mathrm{dis}(G, \rho_n)$. We note that CT-maps for $\{\rho_n\}_{n=1}^{\infty}$ are not convergent, since the CT-map for ρ_n is $F \circ g^n$.

3.2. Distortions : The case of orbifold groups

Let G be a Fuchsian 2-orbifold group with presentation

$$G = \left\langle a_1, \cdots, a_g, b_1, \cdots, b_g, e_1, \cdots, e_m \mid \prod_{i=1}^{g} [a_i, b_i] \prod_{i=1}^{m} e_i = e_1^{n_1} = \cdots = e_m^{n_m} = 1 \right\rangle$$

Under the notation, we set

$$\mathrm{Index}(G) := \begin{cases} 2 \cdot LCM(n_1, \cdots, n_m) & \text{if } G \text{ includes torsion, i.e. } m \geq 1, \\ 1 & \text{otherwise.} \end{cases}$$

According to [EEK82], G contains a Fuchsian surface group of index at most $\mathrm{Index}(G)$. Then, for a faithful discrete representation ρ of G to $\mathrm{PSL}_2(\mathbb{C})$, we define the *distortion* $\mathrm{dis}(G, \rho)$ of G in terms of ρ by

$$\mathrm{dis}(G, \rho) := \inf_{H} \mathrm{dis}(H, \rho \mid_H), \tag{3.8}$$

where H runs over all Fuchsian surface subgroups in G with $[G : H] \leq \mathrm{Index}(G)$.

4. Model manifolds and Cannon–Thurston maps.

Following Minsky [Min94], we shall recall briefly the definition and relevant properties of model manifolds of hyperbolic manifolds associated with Kleinian surface groups with bounded geometry.

4.1. Model manifolds

Let Γ be a Kleinian surface group with bounded geometry and ρ an isomorphism $\rho : G \to \Gamma$. We identify G with $\pi_1(S, \mathbf{p})$ by (3.1). Notice that the injectivity radius ε_0 of N_Γ is half of $\mathcal{L}(\Gamma)$. Here, we assume that the projection of O is in the convex core of N_Γ and fix a pleated surface $f_0 : (S, \sigma_{f_0}) \to N_\Gamma$ as Proposition 3.2. Let e_+ (resp. e_-) denote the positive (resp. negative) end of N_Γ.

• **Case of quasifuchsian groups** For $s = \pm$, let $f_s : (S, \sigma_s) \to N_\Gamma$ be the pleated surface corresponding to the component of the boundary of the convex core of M_Γ

facing the end e_s with $(f_s)_* = \rho$. Consider a Teichmüller geodesic $L \subset \mathrm{Teich}(S)$ from $[\sigma_-]$ to $[\sigma_+]$, where $[\cdot]$ expresses the Teichmüller class (cf. [IT92]).

By Theorem A of [Min93], there is $[\sigma_0] \in L$ such that $d_T([\sigma_0], [\sigma_{f_0}]) \leq A_0$, where $A_0 = A_0(\varepsilon_0, \mathfrak{g}(S)) > 0$ and d_T is the Teichmüller distance on $\mathrm{Teich}(S)$. Set $d_{\pm} := d_T([\sigma_0], [\sigma_{\pm}])$ and let Φ_0 denote a holomorphic quadratic differential on (S, σ_0) associated to L with L^1-norm one. We here notice that there is an ambiguity of choosing a representative σ_0. In fact, it will be taken appropriately in §5.2 (see also §5.2.3).

We define a singular metric ds_Γ^2 on $S \times \mathbb{R}$ by

$$
ds_\Gamma^2 = \begin{cases}
e^{|2t|-2d_-}(e^{-2d_-}dx^2 + e^{2d_-}dy^2) + dt^2 & (t \leq -d_-) \\
e^{2t}dx^2 + e^{-2t}dy^2 + dt^2 & (-d_- \leq t \leq d_+) \\
e^{2t-2d_+}(e^{2d_+}dx^2 + e^{-2d_+}dy^2) + dt^2 & (t \geq d_+)
\end{cases}
$$

away from the zeros of Φ_0, where dx (resp. dy) is the measure in the horizontal (resp. vertical) direction of Φ_0.

• **Case of degenerate groups**   Without loss of generality, we may assume that e_+ is geometrically infinite. If Γ is singly degenerate, we denote by $f_- : (S, \sigma_-) \to N_\Gamma$ the pleated surface corresponding to the boundary of the convex core.

By applying Theorem A of [Min93] again, there is a geodesic ray (if Γ is singly degenerate) or a geodesic (if Γ is doubly degenerate) L such that for any pleated surface f with $f_* = \rho$, there is $[\sigma] \in L$ such that $d_T([\sigma], [\sigma_f]) \leq A_0$. Let $[\sigma_0] \in L$ with $d_T([\sigma_0], [\sigma_{f_0}]) \leq A_0$ and Φ_0 as above. We note that L is defined to emanate from $[\sigma_-]$ (see [Min93]). Set $d_- := d_T([\sigma_-], [\sigma_0])$

The singular metric ds_Γ^2 on $S \times \mathbb{R}$ is defined by

$$
ds_\Gamma^2 = \begin{cases}
e^{|2t|-2d_-}(e^{-2d_-}dx^2 + e^{2d_-}dy^2) + dt^2 & (t \leq -d_-) \\
e^{2t}dx^2 + e^{-2t}dy^2 + dt^2 & (-d_- \leq t < \infty)
\end{cases}
$$

when Γ is singly degenerate, or by

$$
ds_\Gamma^2 = e^{2t}dx^2 + e^{-2t}dy^2 + dt^2 \qquad (t \in \mathbb{R})
$$

otherwise, where dx and dy are taken as the case of quasifuchsian groups.

• **Model manifolds.**   In any case, let M_Γ denote the path-metric space $(S \times \mathbb{R}, ds_\Gamma^2)$.

Theorem 4.1 (Minsky [Min94]). *There is a liftable (K, κ) quasi-isometry $f : M_\Gamma \to N_\Gamma$ such that $f_* = \rho$ and $f|_{S \times \{0\}} = f_0$, where K and κ depend only on ε_0 and $\mathfrak{g}(S)$.*

4.2. Universal covering space of M_Γ

Next we describe the universal covering space of M_Γ. Let G_0 be the Fuchsian group acting on $\mathbb{D}$ which uniformises (S, σ_0) as in §3.1. Then the universal covering projection $\mathrm{pr}_{\sigma_0} : \mathbb{D} \to (S, \sigma_0)$ (cf. §3.1) induces the covering map $\Pi : \mathbb{D} \times \mathbb{R} \to S \times \mathbb{R}$. By pulling back the model metric ds_Γ^2 we get a path-metric space $\widetilde{M} = (\mathbb{D} \times \mathbb{R}, d\widetilde{s}_\Gamma^2)$ and recognise Π as a locally isometric projection from $\widetilde{M}$ to M_Γ. Furthermore, the covering transformation group is

$$\{(z, t) \mapsto (g(z), t) : g \in G_0\}.$$

4.3. Cannon–Thurston maps

We here think of ρ as an isomorphism from G_0 to Γ by precomposition with $\xi_{f_0}^{-1}$ (see (3.3)). By Theorem 4.1, we have a (K, κ)-quasi-isometry $F : \widetilde{M} \to \mathbb{H}^3$ satisfying $f \circ \Pi = \mathrm{proj} \circ F$ and $F(g(z), t) = \rho(g) \circ F(z, t)$ for all $g \in G_0$, where $\mathrm{proj} : \mathbb{H}^3 \to N_\Gamma$ is the covering projection. Since $f|_{S \times \{0\}} = f_0$ (but not path-isometric with respect to the model metric), by (3.2),

$$d_{\mathbb{H}^3}(F(q_0), O) = d_{\mathbb{H}^3}(\mathrm{bc}(f_0), \mathrm{np}(\rho)) \leq B_1 + 1, \tag{4.1}$$

where $q_0 := (0, 0) \in \widetilde{M}$.

In [Min94], Y. Minsky showed that F is extended continuously from $\partial(\mathbb{D} \times \mathbb{R})$ onto $\widehat{\mathbb{C}}$. Furthermore, he also obtained that the restriction of the extension to $\partial\mathbb{D} \times \{0\}$ gives the CT-map from $\partial\mathbb{D}$ $(= \Lambda_{G_0})$ to the limit set Λ_Γ of Γ. By abuse of notation, we denote this CT-map by F.

4.4. Geometry of $\widetilde{M}$ and F

Let γ be a closed geodesic with respect to the $|\Phi_0|$-metric on S. We take a lift $\widetilde{\gamma}$ of γ on $\mathbb{D}$. Then $\widetilde{\gamma}$ is a simple curve on $\mathbb{D}$.

Lemma 4.2. $\widetilde{\gamma} \times \mathbb{R} \subset \widetilde{M}$ *is K_1-quasiconvex, where K_1 depends only on ε_0, $\mathfrak{g}(S)$ and the $|\Phi_0|$-length of γ.*

A set C is said to be K_1-quasiconvex if every geodesic connecting any two points of C is in a K_1 neighbourhood of C.

Proof. Notice from Theorem 4.1 that $\widetilde{M}$ is $\Delta = \Delta(\varepsilon_0, \mathfrak{g}(S))$-hyperbolic. Let l_0 be the $|\Phi_0|$-length of γ. Because any vertical line $\{p\} \times \mathbb{R}$ is a geodesic in $\widetilde{M}$, it suffices to verify that $\widetilde{\gamma} \times \{0\}$ is $K_1 = K_1(\varepsilon_0, \mathfrak{g}(S), l_0)$-quasiconvex in $\widetilde{M}$. Indeed, let $p_1 =$

$(x_1, t_1), p_2 = (x_2, t_2) \in \widetilde{\gamma} \times \mathbb{R}$ and $p_1 p_2$ be a geodesic in $\widetilde{M}$ connecting p_1 and p_2. Let $q_1 = (x_1, 0)$ and $q_2 = (x_2, 0)$. Since $\widetilde{M}$ is Δ-hyperbolic, $p_1 p_2$ is contained in an $R' = R'(\Delta)$-neighbourhood of $p_1 q_1 \cup q_1 q_2 \cup q_2 p_2$. Since $p_1 q_1, q_2 p_2 \subset \widetilde{\gamma} \times \mathbb{R}$, when $\widetilde{\gamma} \times \{0\}$ is K_1-quasiconvex, the geodesic $p_1 p_2$ is contained in an $R' + K_1$-neighbourhood of $\widetilde{\gamma} \times \mathbb{R}$.

We shall check the quasiconvexity of $\widetilde{\gamma} \times \{0\}$. Let n_0 be an integer with $K\kappa/\varepsilon_0 + 1 \leq n_0 < K\kappa/\varepsilon_0 + 2$ and g an element in G_0 corresponding to γ. Fix a point $x_0 \in \widetilde{\gamma}$ and set $x_n := g^n(x_0)$. Then

$$d_{\widetilde{M}}((x_0, 0), (x_{kn_0}, 0)) \geq \frac{1}{K} d_{\mathbb{H}^3}(F(x_0, 0), F(x_{kn_0}, 0)) - \kappa$$

$$= \frac{1}{K} d_{\mathbb{H}^3}(F(x_0, 0), \rho(g)^{kn_0} \circ F(x_0, 0)) - \kappa$$

$$\geq |k| n_0 \varepsilon_0 / K - \kappa$$

$$\geq |k|(K\kappa + \varepsilon_0)/K - \kappa = |k| \varepsilon_0/K + (|k| - 1)\kappa$$

for all $k \in \mathbb{Z}$.

Let $\widetilde{\gamma}_0 : \mathbb{R} \to \mathbb{D}$ denote the $|\Phi_0|$-length parametrisation of $\widetilde{\gamma}$ with $\widetilde{\gamma}_0(0) = x_0$. We think of $\widetilde{\gamma}_0$ as a distance non-increasing map $\mathbb{R} \to \widetilde{M}$ with $\widetilde{\gamma}_0(\mathbb{R}) = \widetilde{\gamma} \times \{0\}$. Let $p_1, p_2 \in \widetilde{\gamma} \times \{0\}$. Take $k_i \in \mathbb{Z}$ such that p_i is contained in the part of $\widetilde{\gamma} \times \{0\}$ between $(x_{k_i n_0}, 0)$ and $(x_{(k_i+1)n_0}, 0)$. Let $s_i \in \mathbb{R}$ with $\widetilde{\gamma}_0(s_i) = p_i$. Then

$$d_{\widetilde{M}}(\widetilde{\gamma}_0(s_1), \widetilde{\gamma}_0(s_2)) \geq d_{\widetilde{M}}(\widetilde{\gamma}_0(k_1 n_0 l_0), \widetilde{\gamma}_0(k_2 n_0 l_0))$$

$$- d_{\widetilde{M}}(\widetilde{\gamma}_0(k_1 n_0 l_0), \widetilde{\gamma}_0(s_1)) - d_{\widetilde{M}}(\widetilde{\gamma}_0(k_2 n_0 l_0), \widetilde{\gamma}_0(s_2))$$

$$\geq d_{\widetilde{M}}((x_{k_1 n_0}, 0), (x_{k_2 n_0}, 0)) - n_0 l_0 - n_0 l_0$$

$$\geq |k_1 - k_2| \varepsilon_0 / K - 2 n_0 l_0.$$

Since $k_i n_0 l_0 \leq s_i \leq (k_i + 1) n_0 l_0$,

$$|s_1 - s_2| \leq |k_1 n_0 l_0 - k_2 n_0 l_0| + 2 n_0 l_0 \leq n_0 l_0(|k_1 - k_2| + 2).$$

Therefore, we obtain

$$d_{\widetilde{M}}(\widetilde{\gamma}_0(s_1), \widetilde{\gamma}_0(s_2)) \geq \frac{\varepsilon_0}{n_0 l_0 K} |s_1 - s_2| - 2(\varepsilon_0/K + n_0 l_0)$$

$$\geq \frac{\varepsilon_0}{(K\kappa/\varepsilon_0 + 1) l_0 K} |s_1 - s_2| - 2(\varepsilon_0/K + (K\kappa/\varepsilon_0 + 2)l_0),$$

which means that $\widetilde{\gamma} \times \{0\}$ is a pseudo-geodesic in the sense of Bowditch [Bow91]. Hence, by Proposition 4.9 of [Bow91], $\widetilde{\gamma} \times \{0\}$ is K_1-quasiconvex, where $K_1 > 0$ is dependent only on K, κ, Δ, ε_0, and l_0. Since all constants K, κ, and Δ are dependent only on ε_0 and $\mathfrak{g}(S)$, K_1 is the desired constant. $\square$

Assume that $\widetilde{\gamma}$ does not pass through the origin $0 \in \mathbb{D}$. Denote by $H_{\widetilde{\gamma}}$ the component of $\mathbb{D} \times \mathbb{R} - \widetilde{\gamma} \times \mathbb{R}$ with $q_0 = (0,0) \notin H_{\widetilde{\gamma}}$. Let $\widetilde{\Phi}_0$ be the lift of Φ_0 on $\mathbb{D}$. We will use the following estimate in the next lemma: Any geodesic segment γ in $\mathbb{H}^3$ satisfies

$$\operatorname{diam}_e(\gamma) \leq 2/\sinh r \qquad (4.2)$$

when $d_{\mathbb{H}^3}(O, \gamma) > r$, where $\operatorname{diam}_e(\cdot)$ denotes the spherical diameter which is measured in the unit ball model of $\mathbb{H}^3$ (under the identification O corresponds to the origin). Indeed, we may observe (4.2) for $\mathbb{D}$ instead of the unit 3-ball by cutting with the hyperplane in $\mathbb{R}^3$ passing through the origin and the endpoints of γ. Let $z_0 \in \gamma$ is the nearest point to $0 \in \mathbb{D}$. By the convexity of γ, there is a disk D_0 such that ∂D_0 intersects $\partial \mathbb{D}$ at two points orthogonally, $\gamma \subset \overline{D_0}$ and $z_0 \in \partial D_0$. One can take D_0 such that its Euclidean radius is $1/\sinh d_{\mathbb{D}}(0, z_0)$, which is what we wanted.

Lemma 4.3. *There exist positive constants C_0 and R_1 dependent only on ε_0, $\mathfrak{g}(S)$, and the $|\Phi_0|$-length l_0 of γ such that when $d_{|\widetilde{\Phi}_0|}(0, \widetilde{\gamma}) \geq R \geq R_1$, the spherical diameter $\operatorname{diam}_e(F(H_{\widetilde{\gamma}}))$ is at most $C_0 R^{-1/4K}$.*

Proof. From the proof of Lemma 7.3 of [Min94], we know

$$d_{\widetilde{M}}(q_0, H_{\widetilde{\gamma}}) \geq \min\left\{ (\log R)/4, \sqrt{R} \right\} = (\log R)/4$$

when $d_{|\widetilde{\Phi}_0|}(0, \widetilde{\gamma}) > R$. Therefore, by (4.1), we have

$$d_{\mathbb{H}^3}(O, F(H_{\widetilde{\gamma}})) \geq (\log R)/4K - (\kappa + B_1 + 1)$$

for $R > 0$. Since $F(H_{\widetilde{\gamma}})$ is $K_1 = K_1(K, \kappa, l_0)$-quasiconvex in $\mathbb{H}^3$ by 4.2, any geodesic γ connecting two points in $F(H_{\widetilde{\gamma}})$ is at least $(\log R)/4K - (\kappa + B_1 + 1) - K_1$ far from the origin O. By (4.2), we conclude

$$\operatorname{diam}_e(F(H_{\widetilde{\gamma}})) \leq \frac{2}{\sinh\left((\log R)/4K - (\kappa + B_1 + 1 + K_1)\right)} \leq C_0 R^{-1/4K}$$

for $R \geq R_1$, where C_0 and R_1 are dependent only on ε_0, $\mathfrak{g}(S)$, and l_0. $\qquad \square$

5. Proof of Main Theorem

Let G be a Fuchsian 2-orbifold group acting on $\mathbb{D}$ and ρ a faithful discrete representation of G to $\mathrm{PSL}_2(\mathbb{C})$ such that $\Gamma = \rho(G)$ has bounded geometry. Let F be the CT-map for ρ.

5.1. Cannon–Thurston maps are not Hölder continuous

We check that the CT-map for a geometrically infinite representation is not Hölder continuous. To show this, we can suppose that G is a Fuchsian surface group because G contains a torsion free subgroup with finite index. Let λ be a geodesic lamination on $\mathbb{D}/G$ and $\widetilde{\lambda}$ its lift on $\mathbb{D}$. Recall from §1 that $E(\lambda) \subset \partial\mathbb{D}$ denotes the set of endpoints of all leaves of $\widetilde{\lambda}$.

Let λ_e denote the one of the ending laminations for ρ. Since the image $F(E(\lambda_e))$ contains infinitely many quasi-arcs[3] (cf. [Miy03]), we have

Proposition 5.1. *The Hausdorff dimension of $F(E(\lambda_e))$ is at least 1.*

By virtue of Birman and Series' theorem in [BS85], the set of endpoints of leaves of $\widetilde{\lambda}_e$ in $\partial\mathbb{D} \times \partial\mathbb{D}$ has Hausdorff dimension zero. Since $E(\lambda_e)$ is its Lipschitz image (via the projections to the coordinates), $E(\lambda_e)$ also has Hausdorff dimension zero. Hence, we conclude the following.

Corollary 5.2. *The CT-map is not Hölder continuous if ρ is geometrically infinite.*

5.2. Reductions

In this section, we shall explain how the special case in Theorem 1.1 (Proposition 5.5) induces the general case. This consists of three sections (§5.2.1, §5.2.2 and §5.2.3) and one remark (Remark 5.6 in §5.3). In any case, we are mainly concerned with how the dependence (1.1) of the constant A in Theorem 1.1 is derived.

5.2.1. Reduction: The first stage

We first note that we may assume that G is a Fuchsian surface group. Indeed, when G contains torsion, we may consider a torsion free subgroup H with $[G:H] \leq \mathrm{Index}(G)$ and $\mathrm{dis}(H, \rho \mid_H) \leq 2\mathrm{dis}(G, \rho)$ instead of G for our purpose (see (3.8)).

5.2.2. Reduction: The second stage

We may assume that $O \in \mathrm{CH}(\rho(G))$. Indeed, to obtain the dependence (1.1) of constants in Theorem 1.1 under the condition $O \notin \mathrm{CH}(\rho(G))$, we claim the following.

Claim 5.3. *Let E be a set in $\widehat{\mathbb{C}}$ such that its convex hull $\mathrm{CH}(E)$ (in $\mathbb{H}^3$) does not contain O and set $d = d_{\mathbb{H}^3}(O, \mathrm{CH}(E))$. Then there is a Möbius transformation T such that $d_e(z_1, z_2) \leq C_6 e^{-d} d_e(T(z_1), T(z_2))$ for $z_1, z_2 \in E$ and $\mathrm{CH}(T(E)) \ni O$, where C_6 is a universal constant.*

---

[3] In [Miy03], the author stated this result only the case when Γ is singly degenerate. However the proof is also available for the case of doubly degenerated groups.

Before proving this claim, we explain the reduction in this case. Assume the convex core $CH(\rho(G))$ does not contain O. Take T as in Claim 5.3 for $F(\partial\mathbb{D}) = \Lambda_{\rho(G)}$ and consider the composition $T \circ F$. Notice that $T \circ F$ is the CT-map for a new representation $g \mapsto T \circ \rho(g) \circ T^{-1}$ and $O \in CH(T \circ F(\partial\mathbb{D}))$. Thus, when Theorem 1.1 and the dependence (1.1) in the case where O is in the convex hull are correct, so are the case of $O \notin CH(F(\partial\mathbb{D}))$ because

$$d_e(F(z_1), F(z_2)) \le C_6 e^{-d_{\mathbb{H}^3}(O, CH(\rho(G)))} d_e(T \circ F(z_1), T \circ F(z_2))$$

for $z_1, z_2 \in \partial\mathbb{D}$.

Proof of Claim 5.3. Let $p_0 \in CH(F(\partial\mathbb{D}))$ be the nearest point to O. Because of the convexity of $CH(F(\partial\mathbb{D}))$, there are half-spaces $\widetilde{H}_0$ and $\widetilde{H}_1$ including $CH(F(\partial\mathbb{D}))$ such that $O \in \partial\widetilde{H}_0$, $np(\rho) \in \partial\widetilde{H}_1$, and each $\partial\widetilde{H}_i$ is perpendicular to the geodesic passing through O and p_0. After composing a rigid motion T_0 of $(\widehat{\mathbb{C}}, d_e)$ to F, we may suppose that $\widetilde{H}_0 \cap \mathbb{C} = \overline{\mathbb{D}}$ and the centre of the disc $\widetilde{H}_1 \cap \mathbb{C}$ is the origin $0 \in \mathbb{C}$. Let $T_1(z) = e^d z$. Then $T := T_0^{-1} \circ T_1 \circ T_0$ satisfies the desired property. $\square$

5.2.3. Reduction: The third stage

To explain the third reduction, we shall fix notations. Take $f_0 \in \text{pleat}(\rho, B_0)$ as Proposition 3.2. Let $\xi_{f_0} : G \to G_{f_0}$ be the natural isomorphism as in §3.1. Then, there is a quasiconformal mapping w_0 of $\mathbb{D}$ such that $\xi_{f_0}(g) \circ w_0 = w_0 \circ g$ for $g \in G$, $w_0(0) = 0$ and the maximal dilatation of w_0 is at most $2\text{dis}(G, \rho)$. Take $[\sigma_0] \in \text{Teich}(S)$ with $d_T([\sigma_0], [\sigma_{f_0}]) \le A_0$ (see §4.1). Let σ_0 be a representative of $[\sigma_0]$ such that the identity mapping $\text{id} : S \to S$ is a $2A_0$-quasiconformal mapping from (S, σ_{f_0}) to (S, σ_0). Let G_0 be the Deck transformation group of $\text{pr}_{\sigma_0} : \mathbb{D} \to (S, \sigma_0)$ as §4.2. Then $\text{dis}(G_0, \rho) \le 2A_0$ since the lift w_1 of the identity mapping is a $2A_0$-quasiconformal mapping of $\mathbb{D}$ which commutes with an isomorphism $\xi_0 : G_{f_0} \to G_0$, and satisfies $w_1(0) = 0$.

We recall Mori's 16-theorem (cf. [Mor56]):

Theorem 5.4 (Mori). *Let w be a K-quasiconformal homeomorphism of $\mathbb{D}$ with $w(0) = 0$. Then w satisfies*

$$|w(x) - w(y)| \le 16|x - y|^{1/K}$$

for $x, y \in \overline{\mathbb{D}}$.

Since $w := w_1 \circ w_0$ and $\xi := \xi_0 \circ \xi_{f_0}$ are a $4A_0\text{dis}(G, \rho)$-quasiconformal mapping and an isomorphism from $G \to G_0$ which satisfy that $\xi(g) \circ w = w \circ g$ for $g \in G$ and $w(0) = 0$, w satisfies

$$|w(x) - w(y)| \le 16|x - y|^{1/(4A_0\text{dis}(G, \rho))} \tag{5.1}$$

by Theorem 5.4.

We are now ready to discuss the third reduction: We claim here that it suffices to prove Theorem 1.1 only for G_0, $\rho \circ \xi^{-1}$ and $F \circ w^{-1}|_{\partial \mathbb{D}}$ instead of general G, ρ and F. Indeed, when we consider G_0, $F \circ w^{-1}|_{\partial \mathbb{D}}$ and $\rho \circ \xi^{-1}$ instead of G, ρ and F, the constants A and B in Theorem 1.1 are dependent only on ε_0 and $\mathfrak{g}(S)$ (see Proposition 5.5 and Remark 5.6). By (5.1), we conclude

$$d_e(F(x), F(y)) = d_e(F \circ w^{-1}(w(x)), F \circ w^{-1}(w(y)))$$

$$\leq A \left\{ \log \left(\frac{3}{|w(x) - w(y)|} \right) \right\}^{-B}$$

$$\leq A_2 \mathrm{dis}(G, \rho)^B \left\{ \log \left(\frac{3}{|x - y|} \right) \right\}^{-B}$$

for $|x - y| \leq (3/16)^{4A_0 \mathrm{dis}(G, \rho)} =: \delta_1$, where A, B, and A_2 depend only on ε_0 and $\mathfrak{g}(S)$.

We next check the dependence of the constants. Consider a function

$$\partial \mathbb{D} \times \partial \mathbb{D} \ni (x, y) \mapsto d_e(F(x), F(y)) \, |\log (3/|x - y|)|^B \in \mathbb{R}. \tag{5.2}$$

The function (5.2) does not exceed $A_2 \mathrm{dis}(G, \rho)$ if $|x - y| \leq \delta_1$. When $|x - y| \geq \delta_1$,

$$d_e(F(x), F(y)) \, |\log (3/|x - y|)|^B \leq \mathrm{diam}_e(\widehat{\mathbb{C}}) \times |\log (3/\delta_1)|^B$$

$$\leq A_3 \mathrm{dis}(G, \rho)^B,$$

where $A_3 = A_3(\varepsilon_0, \mathfrak{g}(S)) > 0$. Thus, by adopting

$$\max\{A_2, A_3\} \mathrm{dis}(G, \rho)^B \tag{5.3}$$

instead of A, we establish the desired inequality.

5.3. Estimation of modulus of continuity

From §5.2.1 and §5.2.2, we may assume that G is a Fuchsian surface group isomorphic to $\pi_1(S)$ and the convex hull $\mathrm{CH}(\rho(G))$ of $\rho(G)$ contains O. Furthermore, from §5.2.3, we may also suppose that G, ρ and F are equal to G_0, $\rho \circ \xi^{-1}$ and $F \circ w^{-1}|_{\partial \mathbb{D}}$, respectively. Let σ_0 denote a hyperbolic structure on S as in §5.2.3.

For any interval $I \subset \partial \mathbb{D}$, we denote by $|I|$ its arclength. When $|I| \leq \pi$, the hyperbolic distance d between $0 \in \mathbb{D}$ and a geodesic connecting endpoints of I satisfies (cf. [Bus92])

$$\cosh d = 1/\sin(|I|/2).$$

This equality induces that

$$d \geq -\log|I| + \log 2 \tag{5.4}$$

since $e^d \geq \cosh d = 1/\sin(|I|/2) \geq 2/|I|$. Let $[x,y]$ denote a shortest interval between $x, y \in \partial\mathbb{D}$.

Our main theorem follows from the following proposition.

Proposition 5.5. *There exist positive constants A, B, and δ_2 such that*

$$d_e(F(x), F(y)) \leq A \left| \log\left(\frac{\pi}{|[x,y]|}\right)\right|^{-B} \tag{5.5}$$

for $x, y \in \partial\mathbb{D}$ with $|[x,y]| \leq \delta_2$, where A, B, and δ_2 are dependent only on ε_0 and $\mathfrak{g}(S)$.

Remark 5.6 (Reduction : The final stage). By a similar argument to that in §5.2.3, we can get rid of the assumption $|x-y| < \delta_2$ from Proposition 5.5 with an appropriate constant A. Indeed, without loss of generality, we may assume that $\delta_2 \leq 2$, because the diameter of $\mathbb{D}$ is equal to 2. Since $|x-y| \leq |[x,y]| \leq \pi|x-y|/2$, by (5.5), the function (5.2) is at most

$$A \sup_{0 < |[x,y]| \leq \delta_2} \left(\frac{\log 3 - \log|x-y|}{\log \pi - \log|[x,y]|}\right)^B =: A_4(\varepsilon_0, \mathfrak{g}(S)) < \infty$$

if $|[x,y]| \leq \delta_2$. Otherwise, we have

$$d_e(F(x), F(y)) \left|\log(3/|x-y|)\right|^B \leq \operatorname{diam}_e(\widehat{\mathbb{C}}) \times \left|\log(6/\pi\delta_2')\right|^B =: A_5(\varepsilon_0, \mathfrak{g}(S)).$$

Thus, we establish

$$d_e(F(x), F(y)) \left|\log(3/|x-y|)\right|^B \leq \max\{A_4, A_5\} \tag{5.6}$$

for all $x, y \in \partial\mathbb{D}$.

5.3.1. A lemma from hyperbolic geometry

Let β be a proper path in $\mathbb{D}$ with distinct endpoints in $\partial\mathbb{D}$ and $I(\beta)$ a shortest interval in $\partial\mathbb{D}$ connecting the endpoints of β.

Lemma 5.7. *For any $\theta \in (0, \pi/2]$, there is a constant $C_1 = C_1(\theta) > 0$ with the following property: Let $z \in \partial\mathbb{D}$ and l a geodesic ray from 0 to z. Let β be a complete geodesic in $\mathbb{D}$. If β intersects l with angle at least θ, $I(\beta)$ contains an interval J with centre z and $|J| \geq C_1|I(\beta)|$.*

Proof. Let J be the maximal subinterval of $I(\beta)$ with centre z and let β_J denote a complete geodesic with the endpoints of J. Let d (resp. d_J) be the distance between 0

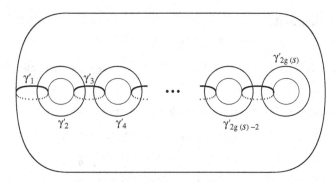

Figure 1: Curves $\{\gamma_i'\}_{i=1}^{2\mathfrak{g}(S)}$

and β (resp. β_J). Then, our assertion derives from a geometric observation that $d - d_J$ is at most the length of $\cosh^{-1}(1/\sin\theta)$ (cf. Theorem 2.2.2 of [Bus92]).   □

5.3.2. Geometries on surfaces

Let γ be a simple closed curve on (S, σ_0) and $\widetilde{\gamma}$ a lift of γ. We denote by $I(\gamma)$ the set of intervals $I(\widetilde{\gamma})$ for all such lifts of γ. For a system $\{\gamma_i\}_i$ of simple closed curves on S, we define $I(\{\gamma_i\}_i) = \cup_i I(\gamma_i)$.

Proposition 5.8. *There exist constants* δ_3, C_2, α_1, N, $\ell_1 > 0$ *and a system* $\{\gamma_i\}_i$ *of simple closed curves on* (S, σ_0) *such that*

(1) *the* $|\Phi_0|$-*lengths of all* γ_i *are at most* ℓ_1,

(2) $I(\{\gamma_i\}_i)$ *covers* $\partial\mathbb{D}$, *and*

(3) *for any interval* J *with* $|J| \leq \delta_3$, *there exist at most* N *intervals* $\{I_m\}_m$ *in* $I(\{\gamma_i\}_i)$ *which cover* J *and* $|I_m| \leq C_2|J|^{\alpha_1}$ *for all* m,

where all constants above are dependent only on $\mathfrak{g}(S)$ *and* ε_0.

Proof. Consider a system of simple closed curves $\{\gamma_i'\}_i$ on S as Figure 1 (cf. Section 3 of [Ser86]). We note that by Lemma 3.3 in [Min94], the identity mapping on S is a liftable quasi-isometry between $(S, |\Phi_0|)$ and (S, σ_0). Therefore, by the compactness of thick surfaces in the moduli space of S, there are positive constants ε_1, ℓ_1 and θ_1 depending on $\mathfrak{g}(S)$ and ε_0, a point $p_1 \in P$, and a diffeomorphism φ on S such that

- P contains an ε_1-disc with centre p_1,

- the hyperbolic geodesic representative γ_i of $\varphi(\gamma_i')$ has $|\Phi_0|$-length at most ℓ_1, and

- the angle between γ_i and γ_j at $\gamma_i \cap \gamma_j$ is at least θ_1.

Let $P = S - \cup_i \gamma_i$. Notice that the union of all lifts of P forms an equivariant tessellation under the action of G. Let P_0 denote the lift of P whose closure contains 0. Since Möbius transformations acting on $\mathbb{D}$ are Lipschitz on $\partial \mathbb{D}$, we may suppose that p_1 is under $0 \in \mathbb{D}$. Since the diameter of P_0 depends only on $\mathfrak{g}(S)$ and ℓ_1, the Lipschitz constant depends only on $\mathfrak{g}(S)$ and ε_0.

For a tile Q of the tessellation, we set

$$\widehat{Q} := \cup \{ g(\overline{P_0}) : g \in H_0 \text{ with } g(\overline{P_0}) \cap \overline{Q} \neq \emptyset \}.$$

Let e be an edge of $\widehat{Q}$ and denote by $\widetilde{\gamma}(e)$ the lift of a curve $\{\gamma_i\}_i$ with $e \subset \widetilde{\gamma}(e)$. By the compactness of the set of geodesic rays intersecting $\overline{P_0}$, there are $\ell_2, \ell_3, \theta_2 > 0$ such that for any geodesic l with $l \cap \overline{P_0} \neq \emptyset$, the length of $\widehat{P_0} \cap l$ is in the interval $[\ell_2, \ell_3]$, and the angle at $\partial \widehat{P_0} \cap l$ between l and $\widetilde{\gamma}(e)$ is in $[\theta_2, \pi/2 - \theta_2]$, where e is an edge of $\partial \widehat{P_0}$ with $e \cap l \subset \partial \widehat{P_0} \cap l$. Using the compactness of thick surfaces in the moduli space again, one can check that the constants ℓ_2, ℓ_3, θ_2 are dependent only on $\mathfrak{g}(S)$ and ε_0.

Let $z \in \partial \mathbb{D}$ and l a geodesic ray connecting 0 and z. Then, it follows from the argument above that there exist a sequence $\{P_n\}_n$ of tiles, an edge e_n of $\widehat{P_n}$ intersecting l such that

$$n\ell_4 \leq d_{\mathbb{D}}(0, \widetilde{\gamma}(e_n)) \leq n\ell_5,$$

where ℓ_4 and ℓ_5 are dependent only on ℓ_2 and ℓ_3 (and hence, on $\mathfrak{g}(S)$ and ε_0). Since $\theta_2 = \theta_2(\mathfrak{g}(S), \varepsilon_0)$, e_n satisfies $z \in I(\widetilde{\gamma}(e_n))$ and

$$D_1^{-n} \leq |I(\widetilde{\gamma}(e_n))| \leq D_2^{-n}, \tag{5.7}$$

where $D_i = D_i(\mathfrak{g}(S), \varepsilon_0) > 1$ $(i = 1, 2)$.

Take C_1 as Lemma 5.7 for θ_2. Let J be an interval in $\partial \mathbb{D}$ with $|J| \leq \delta_3 := D_1^{-1}$ and take $n \in \mathbb{N}$ with $D_1^{-(n+1)} \leq |J| < D_1^{-n}$. Divide J into (at most) $N = N(D_1, C_1)$ components $\{J_m\}_m$ such that each J_m has the length at most $D_1^{-(n+1)}/C_1$. By Lemma 5.7 and (5.7), any J_m is included in an interval $I_m := I(\widetilde{\gamma}_m)$ with $D_1^{-(n+1)} \leq |I_m| \leq D_2^{-(n+1)}$. Thus, at most N intervals $\{I_m\}_m$ cover J and $|I_m| \leq C_2|J|^{\alpha_1}$ for all m with $\alpha_1 = \log D_2 / \log D_1$. $\qquad \square$

5.3.3. Proof of Proposition 5.5

Let us prove Proposition 5.5. From §5.2.2, we suppose that the convex core of $N_\Gamma = \mathbb{H}^3/\Gamma$ contains the projection of $O = (0, 0, 1) \in \mathbb{H}^3$. In this case, the CT-map F for ρ is the restriction of the model map for N_Γ to $\widehat{\mathbb{C}}$ (cf. §4.3). Take $\{\gamma_i\}_i$, δ_3, C_2, α_1 and N as in Proposition 5.8. We may assume $\delta_3 < (\pi/C_2)^{1/\alpha_1}$.

Let $x, y \in \partial\mathbb{D}$ with $|[x,y]| \leq \delta_3$. By Proposition 5.8, there are (at most) N intervals $\{I_k\}_k$ with $|I_k| \leq C_2|[x,y]|^{\alpha_1}$ and $[x,y] \subset \cup_k I_k$. By (5.4), the hyperbolic geodesic β_k with the endpoints of I_k satisfies

$$d_{\mathbb{D}}(0, \beta_k) \geq -\log|I_k| + \log 2 \geq -\alpha_1 \log|[x,y]| - \log C_2 + \log 2. \tag{5.8}$$

Let $\widetilde{\gamma}_k$ denote a $|\widetilde{\Phi}_0|$-geodesic sharing endpoints with β_k. By Lemma 3.3 of [Min94] and (5.8), there exist positive constants C_3 and C_4 such that

$$d_{|\widetilde{\Phi}|}(0, \widetilde{\gamma}_k) \geq C_3 \log(\pi/|[x,y]|) - C_4.$$

Therefore, by Lemma 4.3, we deduce

$$d_e(F(x), F(y)) \leq \mathrm{diam}_e(F([x,y]) \leq \sum_{k=1}^{N} \mathrm{diam}_e(F(I_k))$$

$$\leq NC_0 \left(C_3 \log \frac{\pi}{|[x,y]|} - C_4 \right)^{-1/4K}$$

$$= NC_0 C_3^{-1/4K} \left(\log \frac{\pi}{|[x,y]|} - C_3^{-1}C_4 \right)^{-1/4K}$$

whenever $C_3 \log(\pi/|[x,y]|) - C_4 \geq R_1$. Thus, we conclude

$$d_e(F(x), F(y)) \leq C_5 \left| \log \frac{\pi}{|[x,y]|} \right|^{-1/4K} \tag{5.9}$$

for all $x, y \in \partial\mathbb{D}$ with $|[x,y]| \leq \delta_2 := \pi e^{-(R_1+C_4)/C_3}$, where $C_5 > 0$ is dependent only on ε_0 and $\mathfrak{g}(S)$.

6. Families of Cannon–Thurston maps

In this section, we treat the behaviour of the limit sets when representations vary.

6.1. Continuity of Cannon–Thurston maps

This section gives a proof of Corollary 1.3 in §1.2.1. To this end, we first note (without proof) the following proposition which immediately follows from the combination of Theorem 1.1 and Proposition 3.3 (see also (1.1)).

Proposition 6.1. *Let $\{\rho_n\}_{n=1}^{\infty}$ be a sequence of discrete faithful representations of G which converges to a representation of G. Suppose that there is a constant $\varepsilon_0 > 0$ such that $\mathcal{L}(\rho_n(G)) \geq \varepsilon_0$ for $n \in \mathbb{N}$. Then, the family $\{F_n\}_{n=1}^{\infty}$ of CT-maps for $\{\rho_n\}_{n=1}^{\infty}$ is equicontinuous.*

Proof of Corollary 1.3. By Proposition 6.1 and the Ascoli's theorem, $\{F_n\}_{n=1}^{\infty}$ contains a subsequence which converges to a continuous map F uniformly. Since $\rho_n(g) \circ F_n = F_n \circ g$ for all $g \in G$ and $\rho_n \to \rho_\infty$, $\rho_\infty(g) \circ F = F \circ g$ for all $g \in G$. This implies that F is the CT-map for ρ_∞. Therefore, the limit of any convergent subsequence of $\{F_n\}_{n=1}^{\infty}$ is equal to F by (1) of Lemma 2.1. This means that the original sequence converges to the CT-map for ρ_∞ uniformly. □

6.2. Motion of limit sets on the representation space

Let G be a Fuchsian 2-orbifold group. Denote by $\mathrm{AH}(G)$ the set of equivalence classes of discrete faithful representations from G to $\mathrm{PSL}_2(\mathbb{C})$. The equivalence relation is defined by conjugations in $\mathrm{PSL}_2(\mathbb{C})$. We topologise $\mathrm{AH}(G)$ by the algebraic topology. For $\varepsilon > 0$, let $\mathrm{AH}_\varepsilon(G)$ be the set of $[\rho] \in \mathrm{AH}(G)$ with $\mathcal{L}(\rho(G)) \geq \varepsilon$.

Theorem 6.2. *There is a continuous map*

$$\hat{F}_\varepsilon : \mathrm{AH}_\varepsilon(G) \times \Lambda_G \to \widehat{\mathbb{C}}$$

such that for $[\rho] \in \mathrm{AH}_\varepsilon(G)$, $\hat{F}_\varepsilon([\rho], \cdot)$ *is the CT-map for some* $\rho \in [\rho]$.

Proof. Fix non-commutative elements $g_1, g_2 \in G$. We choose a representative ρ of $[\rho] \in \mathrm{AH}_\varepsilon(G)$ such that the attracting fixed points of $\rho(g_i)$ are 0 and 1, respectively and the repelling fixed point of $\rho(g_1)$ is ∞. Then one can check that when $[\rho_n] \to [\rho_\infty]$ in $\mathrm{AH}_\varepsilon(G)$, the representatives $\{\rho_n\}_{n=1}^{\infty}$ converge to ρ_∞ algebraically.

Let $F_{[\rho]}$ denote the CT-map for the representative ρ. Then

$$\hat{F}_\varepsilon : \mathrm{AH}_\varepsilon(G) \times \Lambda_G \ni ([\rho], x) \longrightarrow F_{[\rho]}(x) \in \widehat{\mathbb{C}}$$

is continuous by Corollary 1.3. □

6.3. Iterations under pseudo-Anosov actions

Suppose that G is isomorphic to $\pi_1(S)$. Let X, Y be points in the Teichmüller space $\mathrm{Teich}(S)$ of S. In this section, we adopt the model of the Teichmüller space such that any point is represented by a pair (X_0, f_0), where X_0 is a hyperbolic surface and $f_0 : S \to X_0$ is an orientation preserving homeomorphism (see [IT92]). Denote by $\rho_{(X,Y)} : G \to \mathrm{PSL}_2(\mathbb{C})$ a quasifuchsian representation which uniformise X and Y (cf. [Ber60]). Let φ_* be a Teichmüller modular transformation acting on $\mathrm{Teich}(S)$ induced by a pseudo-Anosov homeomorphism φ on S. Namely, when $X = (X_0, f_0)$, we set $\varphi_*(X) = (X_0, f_0 \circ \varphi^{-1})$. Let $X_n := \varphi_*^n(X)$ and $Y_n := \varphi_*^n(Y)$. Thurston showed that a sequence $\{[\rho_{(X_n, Y_{-n})}]\}_{n=1}^{\infty}$ converges to $[\rho_\infty] \in \mathrm{AH}(G)$ which represents the $\mathbb{Z}$-cover of the mapping torus with monodromy φ (cf. [Thu87], see also [McM96] and [Min01]). Let ρ_n denote a representative of $\left[\rho_{(X_n, Y_{-n})}\right]$ which converges to some $\rho_\infty \in [\rho_\infty]$.

Theorem 6.3. *The CT-map for ρ_n converges uniformly to that for ρ_∞.*

Proof. By Corollary 1.3, it suffices to show the existence of a positive constant ε_0 satisfying $\mathcal{L}(\rho_n(G)) \geq \varepsilon_0$ for all $n > 0$. Let P_X (resp. P_Y) be a pants decomposition on S such that the length of any component of P_X on X (resp. P_Y on Y) is less than the *Bers' constant* (cf. [Bus92]), which is a constant depending only on $\mathfrak{g}(S)$.

Before proving the theorem, we claim the existence of a positive constant K_0 satisfying that

$$d_Z\left(\pi_Z(\varphi^n(P_X)), \pi_Z(\varphi^{-n}(P_Y))\right) \leq K_0 \qquad (6.1)$$

for all subsurface Z of S and $n > 0$, where d_Z is the distance on the arc complex $\mathcal{A}(Z)$ for Z and $\pi_Z : \mathcal{GL}(S) \to \mathcal{A}(Z)$ is the projection from the set $\mathcal{GL}(S)$ of geodesic laminations (with fixing a hyperbolic structure on S) for S to $\mathcal{A}(Z)$ (for the definitions, see [Min01]). Indeed, let λ_+ be the support of the stable lamination of φ and $Z_n = \varphi^n(Z)$. Then

$$\begin{aligned}
d_Z\left(\pi_Z(\varphi^n(P_X)), \pi_Z(\varphi^{-n}(P_Y))\right) &\leq d_Z\left(\pi_Z(\varphi^n(P_X)), \pi_Z(\lambda_+)\right) \\
&\quad + d_Z\left(\pi_Z(\lambda_+), \pi_Z(\varphi^{-n}(P_Y))\right) \\
&= d_{Z_{-n}}\left(\pi_{Z_{-n}}(P_X), \pi_{Z_{-n}}(\varphi^{-n}(\lambda_+))\right) \\
&\quad + d_{Z_n}\left(\pi_{Z_n}(\varphi^{-n}(\lambda_+)), \pi_{Z_n}(P_Y)\right) \\
&= d_{Z_{-n}}\left(\pi_{Z_{-n}}(P_X), \pi_{Z_{-n}}(\lambda_+)\right) \\
&\quad + d_{Z_n}\left(\pi_{Z_n}(\lambda_+), \pi_{Z_n}(P_Y)\right).
\end{aligned}$$

The last term is dominated by a constant independent of n and Z, because the hyperbolic manifold with one geometrically infinite end whose ending lamination is λ_+ has bounded geometry (e.g. an example in p.121 of [Min00]). Therefore, (6.1) holds.

We now return to the proof of Theorem 6.3. Suppose to the contrary that there is a sequence $\{\gamma_j\}_{j=1}^\infty$ of simple closed curves on S with $\ell(\rho_{n_j}(\gamma_j)) \to 0$ as $j \to \infty$. Since external short curves in the sense of Minsky for ρ_n are in $\varphi^n(P_X) \cup \varphi^{-n}(P_Y)$, from (6.1) and Minsky's Bounded geometry theorem, if γ is a closed curve on S with $\gamma \notin \varphi^n(P_X) \cup \varphi^{-n}(P_Y)$, the translation length $\ell(\rho_n(\gamma))$ of $\rho_n(\gamma)$ is bounded below by a constant depending only on K_0 and $\mathfrak{g}(S)$ (see p.144 and p.150 of [Min01]). Thus, without loss of generality, we may assume $\gamma_j = \varphi^{n_j}(\gamma)$ for some $\gamma \in P_X$.

By [Ber83], after taking subsequence if necessary, we deduce that $\left[\rho_{(X, Y_{-2n_j})}\right]$ converges to a singly degenerate group $[\eta_\infty] \in \mathrm{AH}(G)$ (see also the discussion after this proof). On the other hand, since

$$\ell(\rho_{(X,Y_{-2n_j})}(\gamma)) = \ell(\rho_{n_j}(\varphi^{n_j}(\gamma))) = \ell(\rho_{n_j}(\gamma_j)) \to 0$$

as $n_j \to \infty$, $\eta_\infty(G)$ has to contain a parabolic element, which contradicts that η_∞ is singly degenerate. $\qquad\square$

C. McMullen proved that a sequence $\{[\rho_{(X,Y_n)}]\}_n$ converges to $[\eta_\infty] \in \mathrm{AH}(G)$ and the image $\eta_\infty(G)$ is a singly degenerate group whose end invariants consist of X and the support of the stable lamination of φ (cf. [McM96]). We choose a representative $\eta_n \in [\rho_{(X,Y_n)}]$ for each $n \geq 1$ such that η_n converges to η_∞. In a similar way to the proof of Theorem 6.3, we also conclude the following.

Theorem 6.4. *The CT-map for η_n converges uniformly to that for η_∞.*

7. Approximations to Peano curves

In this section we treat an approximation of a group-equivariant Peano curve or dendrite by a spherical polygonal curve, which is stated in Corollary 1.4.

Let F be the CT-map for a representation of a Kleinian 2-orbifold group with bounded geometry. Take a partition $\mathbf{x} = \{x_k\}_{k=0}^{n+1}$ ($x_{n+1} = x_0$) of $\partial\mathbb{D}$ and let $F(\cdot;\mathbf{x})$ denote the spherical polygonal curve defined as §1.2.2 for $\mathbf{x}$. Set $\mathcal{D} := \max_k |x_{k+1} - x_k|$. Then, by Theorem 1.1,

$$\mathrm{diam}_e(F([x_k, x_{k+1}])) \leq A|\log \mathcal{D}|^{-B} \quad (k = 0, \cdots, n).$$

Thus, we obtain

$$d_e(F(x;\mathbf{x}), F(x)) \leq d_e(F(x;\mathbf{x}), F(x_k)) + d_e(F(x_k), F(x)) \leq 2A|\log \mathcal{D}|^{-B}$$

for $x \in [x_k, x_{k+1}]$ and $k = 0, \cdots, n$.

8. Generalised Hausdorff measures

Before showing Corollary 1.5 in §1.2.3, we recall the definition and basic properties of generalised Hausdorff measures on a metric space (see e.g. §4.9 of [Mat95]).

Let Ψ be an increasing function on $[0,\infty)$ with $\Psi(0) = 0$. For $\varepsilon > 0$, we set

$$\mathcal{H}_{\Psi,\varepsilon}(E) := \inf_{\{U_i\}_i} \left\{ \sum_i \Psi(\mathrm{diam}(U_i)) : \mathrm{diam}(U_i) < \varepsilon, \quad E \subset \cup_i U_i \right\}$$

where $\{U_i\}_i$ runs over all the possible coverings of E by sets of diameter less than ε. As ε decreases, $\mathcal{H}_{\Psi,\varepsilon}(E)$ increases and we define the *generalised Hausdorff measure* $\mathcal{H}_\Psi(E)$ with the gauge function Ψ of E by

$$\mathcal{H}_\Psi(E) = \lim_{\varepsilon \to 0} \mathcal{H}_{\Psi,\varepsilon}(E) = \sup_{\varepsilon > 0} \mathcal{H}_{\Psi,\varepsilon}(E).$$

When $\Psi(t) = t^\alpha$, we call $\mathcal{H}_\alpha(E)$ the α-*dimensional Hausdorff measure* and denote it by $\mathcal{H}_\alpha(E)$ for simplicity. It is known that there is a unique constant $\dim(E) \geq 0$, called the *Hausdorff dimension* of E, such that $\mathcal{H}_{\dim(E)}(E) = \infty$ for $\alpha < \dim(E)$ and $\mathcal{H}_{\dim(E)}(E) = 0$ for $\alpha > \dim(E)$.

Proposition 8.1. *Let* $F : (X, d_X) \to (Y, d_Y)$ *be a continuous map between metric spaces and* $\Psi(t) = \mathrm{mod}_F(t)$. *Then* $\mathcal{H}_\alpha(F(E)) \leq \mathcal{H}_{\Psi^\alpha}(E)$ *for* $E \subset X$.

Proof. Fix $\varepsilon, \varepsilon' > 0$ and take $\varepsilon'' > 0$ such that $\Psi(\varepsilon'') < \varepsilon$. Let $\{U_i\}_i$ be a covering of E such that $\mathrm{diam}_X(U_i) < \varepsilon''$ and $\mathcal{H}_{\Psi^\alpha,\varepsilon''}(E) > \sum_i \Psi(\mathrm{diam}_X(U_i))^\alpha - \varepsilon'$. Then $\{F(U_i)\}_i$ is a covering of $F(E)$ of diameter less than ε and

$$\mathcal{H}_{\alpha,\varepsilon}(F(E)) \leq \sum_i \mathrm{diam}_Y(F(U_i))^\alpha \leq \sum_i \Psi(\mathrm{diam}_X(U_i))^\alpha$$
$$\leq \mathcal{H}_{\Psi^\alpha,\varepsilon''}(E) + \varepsilon' \leq \mathcal{H}_{\Psi^\alpha}(E) + \varepsilon'.$$

Letting $\varepsilon, \varepsilon' \to 0$, we have the assertion.   □

The following corollary follows immediately from Proposition 8.1.

Corollary 8.2. *Let* $F : (X, d_X) \to (Y, d_Y)$ *be a continuous map between metric spaces and* $\Psi(t) = \mathrm{mod}_F(t)$. *Then the following statements hold for* $E \subset X$.

(1) If $\dim(F(E)) > 0$, *then* $\mathcal{H}_{\Psi^\alpha}(E) = \infty$ *for all* $\alpha < \dim(E)$.

(2) Let Ψ_0 *be an increasing continuous function of* $[0, \infty)$ *with* $\Psi_0(0) = 0$ *such that* $\Psi(t)^\alpha = O(\Psi_0(t))$ *as* $t \to 0$ *for all* $\alpha > 0$. *If* $\mathcal{H}_{\Psi_0}(E) = 0$, *then* $\dim(F(E)) = 0$.

We finish this paper with the proof of Corollary 1.5 as follows.

Proof of Corollary 1.5. (1)   This follows from Corollary 1.2 and (1) of Corollary 8.2.

(2)   Since $|\log t|^{-b} = O(|\log|\log t||^{-1})$ as $t \to 0$ for all $b > 0$, by (2) of Corollary 8.2, we conclude the assertion.   □

References

[AC96] J. W. Anderson & R. D. Canary (1996). Cores of hyperbolic 3-manifolds and limits of Kleinian groups. *Amer. J. Math.* **118** (4), 745–779.

[AC00] J. W. Anderson & R. D. Canary (2000). Cores of hyperbolic 3-manifolds and limits of Kleinian groups. II. *J. London Math. Soc. (2)* **61** (2), 489–505.

[Ber60] L. Bers (1960). Simultaneous uniformization. *Bull. Amer. Math. Soc.* **66**, 94–97.

[Ber83] L. Bers (1983). On iterates of hyperbolic transformations of Teichmüller space. *Amer. J. Math.* **105** (1), 1–11.

[BJ97] C. J. Bishop & P. W. Jones (1997). The law of the iterated logarithm for Kleinian groups. In *Lipa's Legacy (New York, 1995), Contemp. Math.*, volume 211, pp. 17–50. Amer. Math. Soc., Providence, RI.

[Bon86] F. Bonahon (1986). Bouts des variétés hyperboliques de dimension 3. *Ann. of Math. (2)* **124** (1), 71–158.

[Bow91] B. H. Bowditch (1991). Notes on Gromov's hyperbolicity criterion for path-metric spaces. In *Group Theory from a Geometrical Viewpoint (Trieste, 1990)*, pp. 64–167. World Sci. Publishing, River Edge, NJ.

[Bow02] B. Bowditch (2002). The cannon–thurston map for punctured-surface groups. Preprint.

[BS85] J. S. Birman & C. Series (1985). Geodesics with bounded intersection number on surfaces are sparsely distributed. *Topology* **24** (2), 217–225.

[Bus92] P. Buser (1992). *Geometry and Spectra of Compact Riemann Surfaces, Progress in Mathematics*, volume 106. Birkhäuser Boston Inc., Boston, MA.

[CEG87] R. D. Canary, D. B. A. Epstein & P. Green (1987). Notes on notes of Thurston. In *Analytical and Geometric Aspects of Hyperbolic Space (Coventry/Durham, 1984), London Math. Soc. Lecture Note Ser.*, volume 111, pp. 3–92. Cambridge Univ. Press, Cambridge.

[CT89] J. Cannon & W. Thurston (1989). Group invariant Peano curves. Preprint (Version 1.7).

[DP02] W. Dicks & J. Porti (2002). On the Hausdorff dimension of the Gieseking fractal. *Topology Appl.* **126** (1-2), 169–186.

[EEK82]   A. L. Edmonds, J. H. Ewing & R. S. Kulkarni (1982). Torsion free sub-groups of Fuchsian groups and tessellations of surfaces. *Invent. Math.* **69** (3), 331–346.

[Flo80]   W. J. Floyd (1980). Group completions and limit sets of Kleinian groups. *Invent. Math.* **57** (3), 205–218.

[IT92]   Y. Imayoshi & M. Taniguchi (1992). *An Introduction to Teichmüller spaces*. Springer-Verlag, Tokyo.

[Kla99]   E. Klarreich (1999). Semiconjugacies between Kleinian group actions on the Riemann sphere. *Amer. J. Math.* **121** (5), 1031–1078.

[Mat95]   P. Mattila (1995). *Geometry of Sets and Measures in Euclidean Spaces, Cambridge Studies in Advanced Mathematics*, volume 44. Cambridge University Press, Cambridge.

[McM96]   C. T. McMullen (1996). *Renormalization and 3-manifolds which Fiber over the Circle, Annals of Mathematics Studies*, volume 142. Princeton University Press, Princeton, NJ.

[McM99]   C. T. McMullen (1999). Hausdorff dimension and conformal dynamics. I. Strong convergence of Kleinian groups. *J. Differential Geom.* **51** (3), 471–515.

[McM01]   C. T. McMullen (2001). Local connectivity, Kleinian groups and geodesics on the blowup of the torus. *Invent. Math.* **146** (1), 35–91.

[Min93]   Y. N. Minsky (1993). Teichmüller geodesics and ends of hyperbolic 3-manifolds. *Topology* **32** (3), 625–647.

[Min94]   Y. N. Minsky (1994). On rigidity, limit sets, and end invariants of hyperbolic 3-manifolds. *J. Amer. Math. Soc.* **7** (3), 539–588.

[Min00]   Y. Minsky (2000). Kleinian groups and the complex of curves. *Geometry and Topology* **4**, 117–148.

[Min01]   Y. N. Minsky (2001). Bounded geometry for Kleinian groups. *Invent. Math.* **146** (1), 143–192.

[Mit98]   M. Mitra (1998). Cannon-Thurston maps for trees of hyperbolic metric spaces. *J. Differential Geom.* **48** (1), 135–164.

[Miy02]   H. Miyachi (2002). Semiconjugacies between actions of topologically tame Kleinian groups. Preprint.

[Miy03]   H. Miyachi (2003). Quasi-arcs in the limit set of a singly degenerate group with bounded geometry. In *Kleinian Groups and Hyperbolic 3-Manifolds (Warwick, 2001), London Math. Soc. Lecture Note Ser.*, volume 299, pp. 131–144. Cambridge Univ. Press, Cambridge.

[Mor56]   A. Mori (1956). On an absolute constant in the theory of quasi-conformal mappings. *J. Math. Soc. Japan* **8**, 156–166.

[MT98]   K. Matsuzaki & M. Taniguchi (1998). *Hyperbolic Manifolds and Kleinian Groups*. Oxford Mathematical Monographs. The Clarendon Press Oxford University Press, New York.

[Ohs92]   K. Ohshika (1992). Geometric behaviour of Kleinian groups on boundaries for deformation spaces. *Quart. J. Math. Oxford Ser. (2)* **43** (169), 97–111.

[Ser86]   C. Series (1986). Geometrical Markov coding of geodesics on surfaces of constant negative curvature. *Ergodic Theory Dynam. Systems* **6** (4), 601–625.

[Som95]   T. Soma (1995). Equivariant, almost homeomorphic maps between S^1 and S^2. *Proc. Amer. Math. Soc.* **123** (9), 2915–2920.

[Thu80]   W. P. Thurston (1980). The Geometry and Topology of 3-Manifolds. Lecture notes, Princeton University. Available at `www.msri.org/publications/books/gt3m/`.

[Thu87]   W. P. Thurston (1987). Hyperbolic structures on 3-manifolds, II: Surface groups and 3-manifolds which fiber over the circle. Preprint; electronic version 1998, `math.GT/9801045`.

Hideki Miyachi

Department of Mathematical Sciences
Tokyo Denky University
Ishizaka, Hatoyama, Hiki
Saitama, 329-0394, Japan

`miyachi@r.dendai.ac.jp`

AMS Classification: 30F40, 26A15, 30C20, 54F50

Keywords: Kleinian group, Cannon–Thurston map, modulus of continuity, limit set, Peano curve, Hausdorff dimension

Spaces of Kleinian Groups
Lond. Math. Soc. Lec. Notes **329**, 151–185

Cambridge University Press
Y. Minsky, M. Sakuma & C. Series (Eds.)

Variations of McShane's identity for punctured surface groups

Hirotaka Akiyoshi, Hideki Miyachi and Makoto Sakuma

Dedicated to Professor Yukio Matsumoto on the occasion of his sixtieth birthday

Abstract

Let M be an orientable complete hyperbolic 3-manifold of finite volume which fibers over the circle, with the fiber a punctured surface. Then each cusp torus of M has a Euclidean structure. We give a formula which expresses the modulus of the Euclidean torus in terms of the complex translation lengths of essential simple loops on the fiber. This generalizes Bowditch's formula on once-punctured torus bundles, which was obtained as a variation of McShane's identity. We also present a formula on the "width" of the limit set of a quasifuchsian punctured surface group. This generalizes the formula for quasifuchsian punctured torus groups, which had been obtained by the authors.

1. Introduction

In [McS91], G. McShane described the following remarkable identity concerning the lengths of simple closed geodesics on a once-punctured torus T with a complete hyperbolic structure of finite area (see also [Bow96]):

$$\sum_{\gamma \in \mathcal{S}} \frac{1}{1 + e^{l(\gamma)}} = \frac{1}{2}. \tag{1.1}$$

Here $\mathcal{S}$ denotes the set of all simple closed geodesics on T and $l(\gamma)$ denotes the length of a closed geodesic γ. McShane [McS98] generalized this identity to an identity for an arbitrary orientable complete hyperbolic surface of finite type with at least one puncture. (See also [McS04] for another variation of the identity for a hyperbolic once-punctured torus.) Recently, M. Mirzakhani [Mir03b, Mir03a] has generalized the identity for bordered hyperbolic surfaces and found beautiful applications of the identity. She used it to obtain a recursive formula for the Weil-Petersson volume of moduli spaces of such Riemann surfaces, and to obtain some counting estimates for simple closed geodesics on a surface. S. P. Tan informed us that he also has a variation of the identity for hyperbolic surfaces with cone singularities [Tan, TWZ04].

On the other hand, B. Bowditch has found 3-dimensional variations of the identity (1.1) for fuchsian punctured torus groups. In [Bow97], he proved that the same identity holds for quasifuchsian punctured torus groups, where the length is replaced with the complex length. He also described a formula which applies to hyperbolic once-punctured torus bundles in [Bow97]: namely, he proved a formula of the modulus of the cusp torus of a hyperbolic once-punctured torus bundle. Motivated by this work, the authors [AMS04] refined the identity in [Bow97] to a "width" formula of the limit set of a geometrically finite punctured torus group.

The purpose of this paper is to generalize the above 3-dimensional variations of the identity (1.1) to arbitrary punctured surface groups. To be precise, we prove the following formulae.

(i) A formula which expresses the width of the limit set of a quasifuchsian punctured surface group in terms of the complex translation lengths of closed geodesics (see Theorem 2.3).

(ii) A formula which expresses the modulus of the Euclidean torus of a hyperbolic punctured surface bundle in terms of the complex translation lengths of essential simple loops on the fiber (see Theorem 3.2).

This paper is organized as follows. In Sections 2 and 3, respectively, we present the explicit statements of the main Theorems 2.3 and 3.2. In order to state Theorem 3.2, we need a careful study of the dynamics of the action of a pseudo-Anosov homeomorphism of a punctured surface on the circle at infinity (see Lemma 3.5) and a lift of complex translation length function (see Lemma 3.8). In Section 4, we recall Mc-Shane's analysis of the simple complete geodesics in a hyperbolic punctured surface emanating from a puncture, which gives the starting point of the proof of the main theorems (see Proposition 4.1). In Section 5, we prove the locally uniform convergence of the infinite sum in Theorem 2.3 on the quasifuchsian space (see Proposition 5.1). The observation due to Birman and Series [BS85] (cf. [Mir03a]) that the number of simple closed geodesics in a hyperbolic surface grows at most polynomially with respect to the lengths holds the key to the proof of Proposition 5.1. In Section 6, we prove Theorem 2.3 by using McShane's result and Proposition 5.1. In Section 7, we prove Theorem 3.2 by using McShane's result and Proposition 7.6 concerning the convergence of the infinite sum in Theorem 3.2, where the proof of Proposition 7.6 is deferred to Section 8. A key ingredient of the proof of the proposition is the existence of a compact submanifold, in the infinite cyclic covering of the hyperbolic surface bundle, which contains all geodesics contributing to the infinite sum (see Lemma 8.2). Another key ingredient is Lemma 8.4, which follows from the observation that if a loop in a hyperbolic manifold is far away from its geodesic representative, then the "geodesic curvature" of the loop at the point where the distance attains the maximum should be large (see Figure 5 and Lemma 9.1). This idea was brought to us by Ken Bromberg. A rigorous proof to this observation is given in Section 9 by a differential geometric calculation.

We would like to express our deepest thanks to Caroline Series for stimulating conversations and encouragements. Actually this work grew out from conversations with her, while the authors were staying at the University of Warwick. We would also like to thank Brian Bowditch, Ken Bromberg, Young-Eun Choi, Greg McShane, Yair Minsky and Ser Peow Tan for their invaluable suggestions, conversations and comments. The last step of the proof was completed when the authors were attending the Newton Institute Programme, *The spaces of Kleinian groups and hyperbolic 3-manifolds*. We thank the Newton Institute and the University of Warwick for their wonderful hospitality.

Notation. We summarize notations used in this paper.

- $F = \mathbb{H}/\Gamma_0$: an orientable complete hyperbolic surface of finite area with a specified puncture, p.

- m: the meridian around p.

- $\rho_0 : \pi_1(F) \to \mathrm{PSL}(2, \mathbb{R})$: the holonomy representation of F with $\rho_0(\pi_1(F)) = \Gamma_0$ and $\rho_0(m) = \begin{pmatrix} 1 & 1 \\ 0 & 1 \end{pmatrix}$.

- $Q\mathcal{F}$ (resp. $\mathcal{F}$): the space of the equivalence classes of quasifuchsian (resp. fuchsian) representations of $\pi_1(F)$.

- $pl^{\pm}$: the bending laminations of $\rho \in Q\mathcal{F}$.

- $\lambda_\rho(\alpha)$: a lift to $\mathbb{C}$ of the complex translation length of $\rho(\alpha)$ (Definition 3.10, cf. Lemma 3.8).

- $\mathrm{width}_p(\Lambda)$: the width of limit set Λ at the puncture p.

- S (resp. $\mathcal{P}$): the set of isotopy classes of essential (resp. peripheral) simple loops in F.

- Δ (resp. $\vec{\Delta}$): the set of the isotopy classes of unoriented (resp. oriented) essential simple arcs in F with both ends in p.

- $\vec{\Delta}_L$ (resp. $\vec{\Delta}_R$): the left (resp. right) half of $\vec{\Delta}$ with respect to $pl^{\pm}$.

- For $\delta \in \Delta$ (or $\delta \in \vec{\Delta}$), $\alpha(\delta)$ and $\beta(\delta)$ denote the unordered pair of simple loops such that $\alpha(\delta) \cup \beta(\delta)$ bounds a punctured annulus containing δ.

- For $\delta \in \Delta$ (or $\delta \in \vec{\Delta}$),

$$h_\rho(\delta) = \frac{1}{1 + e^{\frac{1}{2}(\lambda_\rho(\alpha(\delta)) + \lambda_\rho(\beta(\delta)))}}.$$

- $\bar{F}$: the blow-up of F at the punctures.

- $\partial_p F$: the boundary component of $\bar{F}$ corresponding to p.

- $\varphi : F \to F$: a pseudo-Anosov homeomorphism which preserves the puncture p.

- B_φ: the F-bundle over S^1 with monodromy φ.

- $\partial_p B_\varphi$: the cusp torus of B_φ around the suspension of p.

- l: the longitude of $\partial_p B_\varphi$ (Definition 3.4).

- $\vec{\Delta}_\varphi$: a subset of $\vec{\Delta}$ associated with φ (Definition 3.6, cf. Remark 3.7).

- The homeomorphism φ naturally defines the following commutative diagrams:

$$
\begin{array}{ccc}
\bar{F} & \xleftarrow{\;\supset\;} & \partial_p F \\
\downarrow{\scriptstyle\varphi} & & \downarrow{\scriptstyle\varphi} \\
\bar{F} & \xleftarrow{\;\supset\;} & \partial_p F
\end{array}
$$

$$
\begin{array}{ccccccccc}
\mathbb{H} \cup \partial\mathbb{H} & \xleftarrow{\;\supset\;} & \partial\mathbb{H} - \{\infty\} & \longrightarrow & S_p^1 & \xleftarrow{\;\supset\;} & G & \xleftarrow{\;\supset\;} & \vec{\Delta} \\
\downarrow{\scriptstyle\tilde{\varphi}} & & \downarrow{\scriptstyle\tilde{\varphi}=\tilde{\varphi}_p} & & \downarrow{\scriptstyle\varphi_p} & & \downarrow{\scriptstyle\varphi_p} & & \downarrow{\scriptstyle\varphi_p} \\
\mathbb{H} \cup \partial\mathbb{H} & \xleftarrow{\;\supset\;} & \partial\mathbb{H} - \{\infty\} & \longrightarrow & S_p^1 & \xleftarrow{\;\supset\;} & G & \xleftarrow{\;\supset\;} & \vec{\Delta}
\end{array}
$$

2. Quasifuchsian groups and the widths of the limit sets

Quasifuchsian representations.   Let F be an orientable complete hyperbolic surface of finite type with at least one puncture. Let $\rho_0 : \pi_1(F) \to \mathrm{PSL}(2, \mathbb{R})$ be the holonomy representation and $\Gamma_0 := \rho_0(\pi_1(F))$ the holonomy group. A representation $\rho : \pi_1(F) \to \mathrm{PSL}(2, \mathbb{C})$ is said to be *type-preserving* if ρ sends the peripheral elements to parabolic transformations and ρ is irreducible. Two representations ρ and ρ' are said to be *equivalent* if ρ' is equal to the composition of ρ and an inner-automorphism of $\mathrm{PSL}(2, \mathbb{C})$. A type-preserving representation ρ is said to be *fuchsian* if it is equivalent to a discrete faithful representation into $\mathrm{PSL}(2, \mathbb{R})$. If ρ is fuchsian, the limit set of the image of ρ is a round circle. A type-preserving representation ρ is said to be *quasifuchsian* if it is quasiconformally equivalent to a fuchsian representation. This is equivalent to the condition that ρ is discrete faithful and the limit set of the image of ρ is homeomorphic to a circle. Let $Q\mathcal{F}$ (resp. $\mathcal{F}$) be the space of the equivalence classes of quasifuchsian (resp. fuchsian) representations of $\pi_1(F)$. Then the complex structure of $\mathrm{PSL}(2,\mathbb{C})$ descends to the complex structure on $Q\mathcal{F}$, and $\mathcal{F}$ is a totally real analytic submanifold of $Q\mathcal{F}$ with $\dim_{\mathbb{R}}\mathcal{F} = \dim_{\mathbb{C}}Q\mathcal{F}$. By Bers' simultaneous uniformization, the quasifuchsian space $Q\mathcal{F}$ is canonically identified with the product space $\mathrm{Teich}(F) \times \mathrm{Teich}(F)$ as complex manifold. In particular $Q\mathcal{F}$ is contractible.

Let p be a puncture and m be a *meridian* around p, i.e., a peripheral simple loop around p. We choose a base point for $\pi_1(F)$ on the circle m and denote the element of $\pi_1(F)$ represented by m by the same symbol. Pick an element $\gamma_0 \in \pi_1(F)$ represented by a non-peripheral loop. Then each element of $Q\mathcal{F}$ has a unique representative $\rho \in \mathrm{Hom}(\pi_1(F),\mathrm{PSL}(2,\mathbb{C}))$ which satisfies the following conditions:

$$\rho(m) = \begin{pmatrix} 1 & 1 \\ 0 & 1 \end{pmatrix}, \quad \mathrm{Fix}^+\rho(\gamma_0) = 0, \quad \mathrm{Fix}^+\rho(m\gamma_0 m^{-1}) = 1.$$

Here Fix^+ denotes the attractive fixed point of a loxodromic transformation. The correspondence

$$Q\mathcal{F} \ni [\rho] \mapsto \rho \in \mathrm{Hom}(\pi_1(F),\mathrm{PSL}(2,\mathbb{C}))$$

gives a holomorphic cross section of $Q\mathcal{F}$. Throughout this paper, we identify the space $Q\mathcal{F}$ with its image by the holomorphic section.

Complex translation length. Recall that the *complex translation length* $\lambda(A)$ of a loxodromic element $A \in \mathrm{PSL}(2,\mathbb{C})$ is defined to be the unique element of $\mathbb{C}/2\pi i\mathbb{Z}$ satisfying the following conditions:

(i) The real part $\Re(\lambda(A)) > 0$ is the translation length along the axis of A. Thus $\Re(\lambda(A)) = \min_{x\in\mathbb{H}^3} d(x,A(x))$, where d is the hyperbolic metric.

(ii) The imaginary part $\Im(\lambda(A))$ is the rotation angle of A around the axis of A.

If A is parabolic, then $\lambda(A)$ is defined to be $0 \in \mathbb{C}/2\pi i\mathbb{Z}$. Then $\lambda(A) \in \mathbb{C}/2\pi i\mathbb{Z}$ is characterized by

$$2\cosh\frac{\lambda(A)}{2} = \pm\mathrm{tr}(A), \quad \Re(\lambda(A)) \geq 0.$$

Let α be an *essential* simple loop in F, i.e., a simple loop in F which does not bound a disk nor a once-punctured disk in F. We abuse notation to denote an element of $\pi_1(F)$ represented by α by the same symbol. Then, for any quasifuchsian representation ρ of $\pi_1(F)$, $\rho(\alpha)$ is a loxodromic transformation. The correspondence $\rho \mapsto \lambda(\rho(\alpha))$ determines a holomorphic function $Q\mathcal{F} \to \mathbb{C}/2\pi i\mathbb{Z}$. Since $Q\mathcal{F}$ is contractible, this map lifts to a holomorphic function $Q\mathcal{F} \to \mathbb{C}$ which sends $\mathcal{F}$ into $\mathbb{R}$. We denote by $\lambda_\rho(\alpha)$ the complex number obtained as the image of $\rho \in Q\mathcal{F}$ by the holomorphic

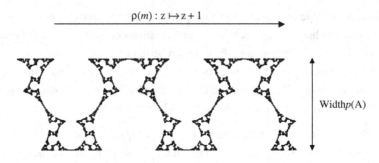

Figure 1: The width $\text{width}_p(\Lambda)$ of limit set Λ at p.

function, and continue to call it the *complex translation length* of $\rho(\alpha)$ (cf. [KS04, Section 6.1]). If α is a *peripheral* simple loop, i.e., α bounds a once-punctured disk in F, then we define $\lambda_\rho(\alpha) = 0 \in \mathbb{C}$.

Width of the limit set.   Let ρ be an element of $Q\mathcal{F}$, and let Γ be the quasifuchsian group obtained as the image of ρ. Then the quotient hyperbolic manifold $M = M_\rho := \mathbb{H}^3/\Gamma$ is homeomorphic to $F \times (-1, 1)$. Let M_0 be the convex core of M, that is, $M_0 = C/\Gamma$ where C is the convex hull of the limit set Λ of Γ in $\mathbb{H}^3$. Then we defined the *width* of Λ at the puncture p of F as follows (see [AMS04, Section 1]). Since $\rho(m) = \begin{pmatrix} 1 & 1 \\ 0 & 1 \end{pmatrix}$, the intersection $\Lambda \cap C$ is invariant by the translation $z \mapsto z+1$. Thus we have the following non-negative real number:

$$\text{width}_p(\Lambda) := m_p^+(\Lambda) - m_p^-(\Lambda),$$

where

$$m_p^+(\Lambda) := \max\{\Im z \mid z \in \Lambda \cap C\}, \quad m_p^-(\Lambda) := \min\{\Im z \mid z \in \Lambda \cap C\}.$$

We call $\text{width}_p(\Lambda)$ the *width* of the limit set Λ at the puncture p. A coordinate free definition of $\text{width}_p(\Lambda)$ is given as follows. Choose a cusp neighborhood C_p of the end of M corresponding to the puncture p. Then $A_p := \partial C_p \cap M_0$ is a Euclidean annulus, and $\text{width}_p(\Lambda)$ is equal to the modulus of A_p, that is,

$$\text{width}_p(\Lambda) = \frac{\text{(the distance between the components of } \partial A_p)}{\text{(the length of a component of } \partial A_p)}.$$

Bending laminations.   Note that (except when Γ is fuchsian) the convex core boundary ∂M_0 has two components, $\partial^\pm M_0$. Each of them has a structure of complete hyperbolic surface homeomorphic to F bent along a measured geodesic lamination, $pl^{\pm} \in \mathcal{ML}(F)$, called the *bending lamination* (see [EM87, Chapter 1]).

Punctured torus case. Suppose for a while that F is a once-punctured torus T. Then the projective measured lamination space $\mathcal{PML}(T)$ is identified with $S^1 = \mathbb{R} \cup \{\infty\}$ and the set $\mathcal{S}$ of the isotopy classes of essential simple loops in T can be thought of as the set of rational points in $\mathcal{PML}(T)$. So, the two projective measured laminations $[pl^{\pm}]$ divide S^1 into two closed intervals and hence they divide $\mathcal{S} - \{[pl^-], [pl^+]\}$ into two subsets, $\mathcal{S}_L$ and $\mathcal{S}_R$. Then the following formula was proved in [AMS04].

Theorem 2.1. *For any quasifuchsian punctured torus group* Γ*, the width* $\mathrm{width}_p(\Lambda)$ *of the limit set* Λ *is given by the following formula.*

$$\pm \mathrm{width}_p(\Lambda) = \Im \sum_{\alpha \in \mathcal{S}_L} \frac{1}{1 + e^{\lambda_p(\alpha)}} = -\Im \sum_{\alpha \in \mathcal{S}_R} \frac{1}{1 + e^{\lambda_p(\alpha)}}.$$

General case. To describe our theorem which generalizes the above formula, we need to introduce further notations. By $\mathcal{S}$ we denote the set of the isotopy classes of essential (unoriented) simple loops in a punctured surface F. A simple arc δ in F with both ends in the puncture p is said to be *essential* if it does not bound a monogon (i.e., a disk with one point removed from its boundary). By Δ (resp. $\vec{\Delta}$) we denote the set of the isotopy classes of unoriented (resp. oriented) essential simple arcs in F with both ends in p. We shall abuse notation to denote a simple loop or an arc and its isotopy class by the same symbol. For each essential arc $\delta \in \Delta$ (or $\delta \in \vec{\Delta}$) there is a unique (up to isotopy) unordered pair of simple loops $\alpha(\delta)$ and $\beta(\delta)$ such that $\alpha(\delta) \cup \beta(\delta)$ bounds a punctured annulus containing δ (cf. [McS98, Proposition 1]). These loops determine a pair of elements of $\mathcal{S} \cup \mathcal{P}$, where $\mathcal{P}$ is the set of the isotopy classes of peripheral simple loops in F. We note the following facts.

(i) If F is a punctured torus, then $\alpha(\delta) = \beta(\delta) \in \mathcal{S}$. Otherwise, $\alpha(\delta) \neq \beta(\delta)$.

(ii) One of $\alpha(\delta)$ and $\beta(\delta)$ belongs to $\mathcal{P}$ if and only if δ bounds a once-punctured monogon.

For each $\delta \in \Delta$ (or $\delta \in \vec{\Delta}$), set

$$h_p(\delta) := \frac{1}{1 + e^{\frac{1}{2}(\lambda_p(\alpha(\delta)) + \lambda_p(\beta(\delta)))}}.$$

Then the following theorem generalizes McShane's identity [McS98, Theorem 2] for fuchsian punctured surface groups and Bowditch's generalization [Bow97, Theorem 3] of (1.1) for quasifuchsian punctured torus groups.

Theorem 2.2. *For any* $\rho \in Q\mathcal{F}$, *we have*

$$\sum_{\delta \in \Delta} h_\rho(\delta) = \frac{1}{2}.$$

Let G be the set of the oriented complete simple geodesics in the hyperbolic sur-face $F = \mathbb{H}^2/\Gamma_0$ emanating from the puncture p. Then the set $\vec{\Delta}$ is regarded as a subset of G. Let $\tilde{G}$ be the set of oriented complete geodesics in $\mathbb{H}^2$ emanating from ∞ which projects to a simple geodesic in F. Then $\tilde{G}$ is identified with a subset of $\mathbb{R} = \partial\mathbb{H}^2 - \{\infty\}$ by associating each element $\tilde{\mu} \in \tilde{G}$ with its endpoint $z_{\rho_0}(\tilde{\mu}) \in \mathbb{R}$. This induces an identification of G with a subset of the circle $S_p^1 := \mathbb{R}/\langle\rho_0(m)\rangle = \mathbb{R}/\mathbb{Z}$.

Let $\rho \in Q\mathcal{F} - \mathcal{F}$ be a quasifuchsian representation, and let $|pl^\pm|$ be the underlying geodesic lamination of the bending laminations $pl^\pm$ of ρ. Then $|pl^\pm|$ is disjoint from a neighborhood of p, and hence we can find, for each $\varepsilon = \pm$, a complete simple geodesic $\mu^\varepsilon \in G$ which is disjoint from $|pl^\varepsilon|$. Then we have $\mu^- \neq \mu^+$, because $\rho \notin \mathcal{F}$. Since $\vec{\Delta}$ is identified with a subset of S_p^1, we obtain a partition of $\vec{\Delta} - \{\mu^-, \mu^+\}$ into two subsets, $\vec{\Delta}_L$ and $\vec{\Delta}_R$. We shall prove the following generalization of Theorem 2.1.

Theorem 2.3. *Let* ρ *be a quasifuchsian representation of* $\pi_1(F)$ *and* Γ *its image. Then the width of the limit set* Λ *of* Γ *at the puncture* p *is given by the following formula.*

$$\pm \text{width}_p(\Lambda) = \Im \sum_{\delta \in \vec{\Delta}_L} h_\rho(\delta) = -\Im \sum_{\delta \in \vec{\Delta}_R} h_\rho(\delta).$$

Remark 2.4. Since there are various choices of the pair $\{\mu^-, \mu^+\}$, the subsets $\vec{\Delta}_L$ and $\vec{\Delta}_R$ are not uniquely determined by ρ. However, by using the fact that $\rho(\gamma)$ is purely-hyperbolic for every essential simple loop γ which lies in $F - |pl^-|$ or $F - |pl^+|$ (cf. [KS93, Lemma 4.6]), we can see that the imaginary parts of the infinite sums do not depend on the choice of the subsets.

Comparing Theorems 2.1 and 2.3. In the remainder of this section, we describe the relationship between Theorems 2.1 and 2.3. Assume that the surface F is a once-punctured torus T. Then we consider the map $\psi : G \to \mathcal{PML}(T)$ defined as follows. Let μ be an element of G. Then there is a unique element $v \in \mathcal{PML}(T)$ whose under-lying geodesic lamination is disjoint from μ. (In fact, if $\mu \in \vec{\Delta}$, then $|v|$ is an essential simple loop; if $\mu \notin \vec{\Delta}$, then $|v|$ is obtained from the closure of μ in F by removing μ.) Then we define $\psi(\mu) = v$. Let τ be the hyper-elliptic involution on T. Then τ induces the half-rotation on S_p^1, which preserves the subsets G and $\vec{\Delta}$. We can easily see that the restriction of ψ to $\vec{\Delta}$ induces a τ-invariant two-to-one surjective map $\vec{\Delta} \to S$.

Let ρ be a quasifuchsian representation of $\pi_1(T)$ and μ^ε be elements of G such that $\mu^\varepsilon \cap |pl^\varepsilon| = \emptyset$ ($\varepsilon = \pm$). Then $\psi(\mu^\varepsilon) = [pl^\varepsilon]$ and the restriction $\psi|_{\vec{\Delta}_L}$ satisfies one of the

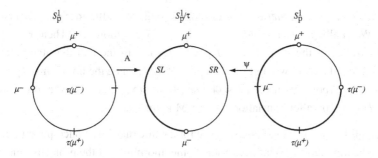

Figure 2: In each of the left and right circles, the bold circular arc corresponds to $\vec{\Delta}_L$

following conditions (see Figure 2):

(i) $\psi|_{\vec{\Delta}_L}$ is a bijection from $\vec{\Delta}_L$ onto $\mathcal{S}_L$ or $\mathcal{S}_R$.

(ii) $\psi|_{\vec{\Delta}_L}$ is a surjection from $\vec{\Delta}_L$ onto $\mathcal{S}$, and one of the following holds.

    (a) $(\psi|_{\vec{\Delta}_L})^{-1}(\alpha)$ consists of one or two elements according to whether α belongs to $\mathcal{S} - \mathcal{S}_R$.

    (b) $(\psi|_{\vec{\Delta}_L})^{-1}(\alpha)$ consists of one or two elements according to whether α belongs to $\mathcal{S} - \mathcal{S}_L$.

In the first case, the infinite sum $\sum_{\delta \in \vec{\Delta}_L} h_\rho(\delta)$ is equal to

$$\sum_{\alpha \in \mathcal{S}_L} \frac{1}{1 + e^{\lambda_\rho(\alpha)}} \quad \text{or} \quad \sum_{\alpha \in \mathcal{S}_R} \frac{1}{1 + e^{\lambda_\rho(\alpha)}}.$$

In the second case, the infinite sum is equal to

$$\sum_{\alpha \in \mathcal{S}_L} \frac{1}{1 + e^{\lambda_\rho(\alpha)}} + \sum_{\alpha \in \mathcal{S}} \frac{1}{1 + e^{\lambda_\rho(\alpha)}} \quad \text{or} \quad \sum_{\alpha \in \mathcal{S}_R} \frac{1}{1 + e^{\lambda_\rho(\alpha)}} + \sum_{\alpha \in \mathcal{S}} \frac{1}{1 + e^{\lambda_\rho(\alpha)}}.$$

Since $\sum_{\alpha \in \mathcal{S}} \frac{1}{1 + e^{\lambda_\rho(\alpha)}} = \frac{1}{2}$ by [Bow97, Theorem 3] (cf. Theorem 2.2), the imaginary part of the infinite sum is equal (modulo sign) to the infinite sum in Theorem 2.1.

3. Hyperbolic punctured surface bundles

In this section, we present a generalization of Bowditch's result [Bow97] on hyperbolic once-punctured torus bundles. Let $\varphi : F \to F$ be a pseudo-Anosov homeomorphism preserving the puncture p, and let B_φ be the F-bundle over S^1 with monodromy φ. Then B_φ admits a unique complete hyperbolic structure of finite volume, and each cusp torus carries a Euclidean structure. Let $\partial_p B_\varphi$ be the cusp torus around the suspension of p. A *meridian* m of $\partial_p B_\varphi$ is defined as the meridian around p of a fiber. We shall specify a *longitude* l of $\partial_p B_\varphi$ in Definition 3.4. Then the *modulus* Modulus($\partial_p B_\varphi$) of the cusp torus $\partial_p B_\varphi$, with respect to the meridian-longitude pair (m, l), is defined as follows: Let $\rho : \pi_1(B_\varphi) \to \mathrm{PSL}(2, \mathbb{C})$ be the holonomy representation of the hyperbolic manifold B_φ such that $\rho(m)$ is the parallel translation $z \mapsto z + 1$. Then $\rho(l)$ is the parallel translation $z \mapsto z + \mathrm{Modulus}(\partial_p B_\varphi)$.

To recall Bowditch's theorem, suppose for a while that F is a once-punctured torus T. Then the monodromy φ induces a self-homeomorphism of the projective measured lamination space $\mathcal{PML}(T) \cong S^1$ preserving the subset $\mathcal{S}$. This homeomorphism has two fixed points in $\mathcal{PML}(T)$, namely the stable and unstable laminations, ν^+ and ν^-, of the monodromy. Since ν^+ and ν^- are irrational, they determine a natural partition of $\mathcal{S}$ into two subsets $\mathcal{S}_L$ and $\mathcal{S}_R$. This in turn gives a partition of the quotient set $\mathcal{S}/\langle \varphi \rangle$ (which is identified with the set of essential simple loops on a fiber F modulo isotopy in the ambient 3-manifold B_φ) into two subsets $\mathcal{S}_L/\langle \varphi \rangle$ and $\mathcal{S}_R/\langle \varphi \rangle$. For two elements α and α' of $\mathcal{S}$ representing the same element in $\mathcal{S}/\langle \varphi \rangle$, the complex translation lengths of $\rho(\alpha)$ and $\rho(\alpha')$ coincide. So, the complex translation length $\lambda_\rho(\alpha) \in \mathbb{C}/2\pi i \mathbb{Z}$ is well-defined for $\alpha \in \mathcal{S}/\langle \varphi \rangle$. It should be noted that $e^{\lambda_\rho(\alpha)}$ is a well-defined complex number. Then the following theorem was proved by Bowditch [Bow97].

Theorem 3.1. *Let B_φ be a complete hyperbolic 3-manifold which fibers over the circle with fiber a once-punctured torus T with monodromy φ. Then the modulus Modulus($\partial_p B_\varphi$) of the cusp torus $\partial_p B_\varphi$, with respect to a suitable choice of a longitude l, is given by the following formula.*

$$\pm\mathrm{Modulus}(\partial_p B_\varphi) = \sum_{\alpha \in \mathcal{S}_L/\langle \varphi \rangle} \frac{1}{1 + e^{\lambda_\rho(\alpha)}} = -\sum_{\alpha \in \mathcal{S}_R/\langle \varphi \rangle} \frac{1}{1 + e^{\lambda_\rho(\alpha)}}.$$

In the general punctured surface bundle case, we study the action of the monodromy φ on the sets $\mathcal{G}$ and $\vec{\Delta}$, and specify a certain subset, $\vec{\Delta}_\varphi$, of $\vec{\Delta}$ in Definition 3.6 (cf. Remark 3.7). Then our generalization of Bowditch's result can be stated as follows.

Theorem 3.2. *Let B_φ be a complete hyperbolic 3-manifold which fibers over the circle with fiber F with monodromy φ that preserves the puncture p. Then the modulus Modulus($\partial_p B_\varphi$) of the cusp torus $\partial_p B_\varphi$, with respect to a suitable choice of a longitude l, is given by the following formula.*

$$\mathrm{Modulus}(\partial_p B_\varphi) = \pm \sum_{\delta \in \vec{\Delta}_\varphi} h_\rho(\delta).$$

In the above theorem, $h_\rho(\delta)$ is defined by

$$h_\rho(\delta) := \frac{1}{1 + e^{\frac{1}{2}(\lambda_\rho(\alpha(\delta)) + \lambda_\rho(\beta(\delta)))}},$$

where $\lambda_\rho(\alpha)$ with $\alpha \in \mathcal{S} \cup \mathcal{P}$ denotes a lift to $\mathbb{C}$ of the complex translation length of $\rho(\alpha)$ specified by Definition 3.10 below. In the remainder of this section, we give explicit definitions of the longitude l, the subset $\vec{\Delta}_\varphi \subset \vec{\Delta}$ and the complex translation length $\lambda_\rho(\alpha) \in \mathbb{C}$.

Behavior of φ on the boundary. Since φ is a pseudo-Anosov homeomorphism, there are measured foliations $\mathcal{F}^+$ and $\mathcal{F}^-$ satisfying the following conditions (cf. e.g. [Kap01, Section 11.4]).

(i) $\mathcal{F}^+$ and $\mathcal{F}^-$ are transversal, that is, their singular sets are equal, and $\mathcal{F}^+$ is transversal to $\mathcal{F}^-$ away from the singular set.

(ii) $\varphi(\mathcal{F}^+) = k\mathcal{F}^+$ and $\varphi(\mathcal{F}^-) = k^{-1}\mathcal{F}^-$ for some $k > 1$. Namely, φ preserves the singular foliations $\mathcal{F}^+$ and $\mathcal{F}^-$, and multiplies the measures by k and k^{-1} respectively.

For each puncture q of F, there is a neighborhood of q that is identified with a neighborhood of 0 of a complex plane, such that the $\mathcal{F}^+$ and $\mathcal{F}^-$ are given by $|\Im(z^{d/2}dz)|$ and $|\Re(z^{d/2}dz)|$, respectively, for some integer $d \geq -1$ (cf. e.g. [Gar87, Section 11.1], [Kap01, Section 11.3]). In particular, each of $\mathcal{F}^+$ and $\mathcal{F}^-$ has $d + 2$ (≥ 1) singular leaves landing at the puncture q. The number $d + 2$ is called the *degree* of $\mathcal{F}^\pm$ at q.

Let $\bar{F}$ be the compact surface with boundary obtained by adding the circle of rays from q for each puncture q. We denote this boundary circle by $\partial_q F$. Then the measured foliations $\mathcal{F}^\pm$ extend to measured foliations $\bar{\mathcal{F}}^\pm$ of $\bar{F}$. Each of $\bar{\mathcal{F}}^\pm$ has b (≥ 1) singular leaves landing in $\partial_q F$, where b is the degree of $\mathcal{F}^\pm$ at q. Moreover φ extends to a homeomorphism of $\bar{F}$, which we continue to denote by φ.

Since φ preserves the puncture p, $\varphi : \bar{F} \to \bar{F}$ induces a homeomorphism of the boundary circle $\partial_p F$. Let b be the degree of $\mathcal{F}^\pm$ at the puncture p, and let $\{x_1^\pm, x_2^\pm, \cdots, x_b^\pm\}$ be the endpoints of the singular leaves of $\mathcal{F}^\pm$ in $\partial_p F$. We assume that they are arranged on $\partial_p F$ in this cyclic order. Since φ preserves the singular leaves, there is a unique integer c with $0 \leq c < b$ such that φ acts on the sets $\{x_1^\pm, x_2^\pm, \cdots, x_b^\pm\}$ as the shift of indices by c. Set $n_0 = b/\gcd(b, c)$. Since φ is affine with respect to the singular Euclidean metric determined by the mutually transversal measured laminations $\mathcal{F}^+$ and $\mathcal{F}^-$, we have the following lemma.

Lemma 3.3. *The sets $\{x_1^+, x_2^+, \cdots, x_b^+\}$ and $\{x_1^-, x_2^-, \cdots, x_b^-\}$, respectively, are equal to the attractive and the repulsive fixed point sets of $\varphi^{n_0} : \partial_p F \to \partial_p F$, and they are arranged on $\partial_p F$ alternately.*

Specifying the longitude. Let $\bar{B}_\varphi = \bar{F} \times [0,1]/(x,0) \sim (\varphi(x),1)$ be the $\bar{F}$-bundle over S^1 with monodromy φ, and let $\partial_p \bar{B}_\varphi$ be the boundary component of $\bar{B}_\varphi$ corresponding to the puncture p of F. Namely, $\partial_p \bar{B}_\varphi = \partial_p F \times [0,1]/(x,0) \sim (\varphi(x),1)$. Then B_φ is identified with the interior of $\bar{B}_\varphi$, and the cusp torus $\partial_p B_\varphi$ is identified with $\partial_p \bar{B}_\varphi$.

Definition 3.4. By the *longitude* l of $\partial_p B_\varphi$, we mean the isotopy class of the simple loop in $\partial_p \bar{B}_\varphi$ obtained as the image of $\cup_{j=0}^{n_0-1} \varphi^j(x) \times [0,1]$, where n_0 is the natural number in Lemma 3.3 and x is a fixed point of φ^{n_0}.

Note that the meridian-longitude pair (m,l) defined in the above forms a basis of $H_1(\partial_p B_\varphi; \mathbb{Z})$ if and only if $n_0 = 1$. However, it always forms a basis of $H_1(\partial_p B_\varphi; \mathbb{Q})$ and hence the modulus of $\partial_p B_\varphi$ with respect to any basis of $H_1(\partial_p B_\varphi; \mathbb{Z})$ can be calculated from $\text{Modulus}(\partial_p B_\varphi)$. To be precise, let l_0 be an element of $H_1(\partial_p B_\varphi; \mathbb{Z})$ such that (m, l_0) is a basis of $H_1(\partial_p B_\varphi; \mathbb{Z})$. Then we see $l = n_0 l_0 + n_1 m$ for some integer n_1, and the modulus $\text{Modulus}_0(\partial_p B_\varphi)$ of $\partial_p B_\varphi$ with respect to this basis is given by

$$\text{Modulus}_0(\partial_p B_\varphi) = \frac{1}{n_0}\left(\text{Modulus}(\partial_p B_\varphi) - n_1\right).$$

In particular, the imaginary part is given by

$$\Im\left(\text{Modulus}_0(\partial_p B_\varphi)\right) = \frac{1}{n_0}\Im\left(\text{Modulus}(\partial_p B_\varphi)\right).$$

We also note that l can be regarded as the longitude of the cusp torus, corresponding to the puncture p, of the F-bundle $B_{\varphi^{n_0}}$ over S^1 with monodromy φ^{n_0}, which is a $\mathbb{Z}/n_0\mathbb{Z}$-covering of B_φ. Moreover (m,l) is a basis of the integral homology $H_1(\partial_p B_{\varphi^{n_0}})$ and we have

$$\text{Modulus}(\partial_p B_\varphi) = \text{Modulus}(\partial_p B_{\varphi^{n_0}}).$$

The action of φ on $\vec{\Delta}$ and $\mathcal{G}$. We may assume φ is the Teichmüller map and the conformal structure on $F = \mathbb{H}^2/\Gamma_0$ is absolutely φ-minimal in the sense of Bers, that is, it lies in the invariant axis of the action of φ on the Teichmüller space (see e.g. [IT92, Section 5.2]). Then $\varphi : F \to F$ is quasiconformal, and hence its lift to $\mathbb{H}^2$ extends to a homeomorphism of $\mathbb{H}^2 \cup \partial\mathbb{H}^2$. Let $\tilde{\varphi}$ be such a homeomorphism of $\mathbb{H}^2 \cup \partial\mathbb{H}^2$ which stabilizes $\infty = \text{Fix}(\rho_0(m))$. We denote by φ_p the homeomorphism of $S_p^1 = \mathbb{R}/\langle\rho_0(m)\rangle$ induced by the restriction of $\tilde{\varphi}$ to $\mathbb{R} = \partial\mathbb{H}^2 - \{\infty\}$.

For each $\tilde{\mu} \in \tilde{\mathcal{G}}$, consider the geodesic in $\mathbb{H}^2$ emanating from ∞ and ending at $\tilde{\varphi}(z_{\rho_0}(\tilde{\mu}))$, the image by $\tilde{\varphi}$ of the endpoint $z_{\rho_0}(\tilde{\mu})$ of $\tilde{\mu}$. Then it also belongs to $\tilde{\mathcal{G}}$. This determines a bijection $\tilde{\mathcal{G}} \to \tilde{\mathcal{G}}$, which in turn induces a bijection $\mathcal{G} \to \mathcal{G}$. After identifying G with a subset of S_p^1, the bijection is identified with the restriction of $\varphi_p : S_p^1 \rightarrow S_p^1$ to G. The following lemma describes the dynamics of φ_p.

Lemma 3.5. *Let n_0 be as in Lemma 3.3. Then $\varphi_p^{n_0}$ has finitely many attractive fixed points and repulsive fixed points, which are arranged on S_p^1 alternately. Moreover, for any component J of $S_p^1 - (\mathrm{Fix}^+(\varphi_p^{n_0}) \cup \mathrm{Fix}^-(\varphi_p^{n_0}))$ bounded by an attractive fixed point A^+ and a repulsive fixed point A^-, $\varphi_p^{n_0}$ maps every point $X \in J$ to a point strictly closer to A^+, and we have $\lim_j(\varphi_p^{n_0})^j(X) = A^+$ and $\lim_j(\varphi_p^{n_0})^{-j}(X) = A^-$.*

Proof. This lemma is an analogy of [CB88, Theorem 5.5] for pseudo-Anosov homeomorphisms of closed surfaces, and the proof is almost parallel to that in [CB88]. However, we could not find a reference which contains the proof in the punctured surface case. So, we include the proof for completeness. For simplicity, we prove the lemma by assuming $n_0 = 1$: the lemma for the general case can be proved similarly.

Let L be a singular leaf of $\mathcal{F}^\pm$ emanating from p. Since φ preserves the singular foliations $\mathcal{F}^\pm$ and multiplies the measures by $k^{\pm 1}$, L does not contain a singular point except p (cf. [Kap01, Lemma 11.42]). Thus every lift of L to $\mathbb{H}^2$ has a well-defined endpoint on $\partial\mathbb{H}^2$ and the endpoint is distinct from the origin by [MS85, Theorem]. Pick a lift $\widetilde{L}$ of L to $\mathbb{H}^2$ which emanates from ∞. Since φ preserves L (by the assumption $n_0 = 1$), we can assume $\widetilde{\varphi}$ preserves $\widetilde{L}$. Then $\widetilde{\varphi}$ fixes the endpoint of $\widetilde{L}$ in $\mathbb{R} = \partial\mathbb{H}^2 - \{\infty\}$, and hence $\mathrm{Fix}(\widetilde{\varphi}|_\mathbb{R}) \neq \emptyset$.

Now let v^+ and v^-, respectively, be the stable and unstable measured laminations of φ, and let $W^\pm$ be the component of $F - |v^\pm|$ containing the puncture p. We denote by $\overline{W}^\pm$ the metric completion of $W^\pm$ with the induced path metric. Then we see, by an argument parallel to [CB88, Lemma 5.3], that $\overline{W}^\pm$ is a "once-punctured finite sided ideal polygon"; namely, $\pi_1(\overline{W}^\pm)$ is the infinite cyclic group generated by a peripheral loop around p, and $\partial\overline{W}^\pm$ is a finite union of infinite geodesics, which are joined cyclically by ideal vertices (see Figure 3). In fact, if this is not the case then the boundary of the Nielsen core of $\overline{W}^\pm$ gives a φ-invariant family of mutually disjoint essential loops of F: this contradicts the assumption that φ is pseudo-Anosov.

Let $\widetilde{W}^\pm$ be the component of the inverse image of $W^\pm$ in $\mathbb{H}^2$ whose closure contains ∞. Then $\widetilde{W}^\pm$ is (the interior of) an infinite sided ideal polygon such that the set, $\widetilde{V}^\pm (\subset \mathbb{R})$, of the ideal vertices accumulates only at ∞ (see Figure 3). Since $\varphi(W^\pm) = W^\pm$ and $\widetilde{\varphi}(\infty) = \infty$, $\widetilde{\varphi}$ preserves $\widetilde{W}^\pm$ and hence $\widetilde{V}^\pm$. Hence every point of $\widetilde{V}^\pm$ must be fixed by $\widetilde{\varphi}$; otherwise, $\widetilde{\varphi}$ acts on $\widetilde{V}^\pm$ as a shift and hence $\widetilde{\varphi}|_\mathbb{R}$ cannot have a fixed point, a contradiction.

We follow the argument of [CB88, Proof of Theorem 5.5], and show that $\widetilde{V}^+$ and $\widetilde{V}^-$ are arranged alternately on $\mathbb{R}$ and that they are equal to the attractive and repulsive fixed point sets of $\widetilde{\varphi}|_\mathbb{R}$, respectively. To this end, let I be the closure of a component of

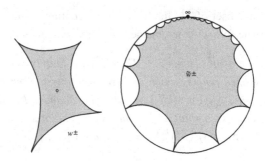

Figure 3: Once-punctured finite sided ideal polygon $W^\pm$ and the component $\widetilde{W}^\pm$ of its inverse image in $\mathbb{H}^2$ whose closure contains ∞.

$\mathbb{R} - V^+$. Then I is a *stable interval* for v^+ in the sense of [CB88, p.81, Definition], that is, for any two points P and Q in the interior int$I$ of I, there is a leaf $\widetilde{\delta}$ of the inverse image $\widetilde{\mathsf{v}}^+$ of v^+ in $\mathbb{H}^2$ whose endpoints separate P and Q from ∂I. This can be seen as follows as in [CB88, Lemma 5.4]. Since the leaf, $\widetilde{\delta}_I$, of $\widetilde{\mathsf{v}}^+$ with endpoints ∂I is not isolated, there is a sequence $\{\widetilde{\delta}_i\}$ of leaves of $\widetilde{\mathsf{v}}^+$ which tends to $\widetilde{\delta}_I$. Since $\widetilde{\mathsf{v}}^+$ has no closed leaves, there are only finitely many leaves of $\widetilde{\mathsf{v}}^+$ whose endpoints contain one of the points of ∂I (cf. [CB88, Lemma 4.5]). So we may assume that the endpoints of $\widetilde{\delta}_i$ are contained in int$I$. This shows that I is a stable interval of v^+.

Pick a simple closed geodesic γ in F. Then $\varphi^j(\gamma)$ converges to v^+ in $\mathcal{PML}(F)$, and hence $|\mathsf{v}^+|$ is contained in the limit $\lim \varphi^j(\gamma)$ in the Chabauty topology. Since $|\mathsf{v}^+|$ fills up F and since the image δ_I of $\widetilde{\delta}_I$ in $|\mathsf{v}^+|$ is dense in $|\mathsf{v}^+|$, γ intersects δ_I transversely. Therefore, some lift $\widetilde{\gamma}$ in $\mathbb{H}^2$ has endpoints $A \in$ int$I$ and $B \in \partial \mathbb{H}^2 - I$. Since $\widetilde{\varphi}$ preserves I (resp. $\partial \mathbb{H}^2 - intI$), $\widetilde{\varphi}^j(A)$ (resp. $\widetilde{\varphi}^j(B)$) converges to a point $A_\infty \in I$ (resp. $B_\infty \in \partial \mathbb{H}^2 - intI$). We show that $A_\infty \in \partial I$. Suppose to the contrary that $A_\infty \in$ int$I$. Then there is a leaf $\widetilde{\delta}$ of $\widetilde{\mathsf{v}}^+$ whose endpoints separate A_∞ from ∂I, because I is a stable interval for v^+. The geodesic δ_∞ with endpoints A_∞ and B_∞ projects to a leaf of $\lim \varphi^j(\gamma)$, and it intersects $\widetilde{\delta}$ transversely. This contradicts the fact $|\mathsf{v}^+| \subset \lim \varphi^j(\gamma)$. Hence $A_\infty \in \partial I$ and therefore A_∞ is an attractive fixed point of $\widetilde{\varphi}|_I$.

Let U be the open subinterval of I with endpoints A and A_∞. Then $\widetilde{\varphi}$ moves all points of U strictly closer to A_∞. Let V be a neighborhood of $\partial I - \{A_\infty\}$ in I such that V and $\widetilde{\varphi}(V)$ are disjoint from U. Since I is a stable interval, there is a leaf $\widetilde{\delta}$ of $\widetilde{\mathsf{v}}^+$ with endpoints $X \in U$ and $Y \in V$. It follows that $\widetilde{\varphi}(X)$ and $\widetilde{\varphi}(Y)$ separate X and Y from ∂I. Let X_∞ and Y_∞, respectively, be the limits of the sequences $\widetilde{\varphi}^{-j}(X)$ and $\widetilde{\varphi}^{-j}(Y)$. Suppose that $X_\infty \neq Y_\infty$. Then the closure I' of the component of $\partial \mathbb{H}^2 - \{X_\infty, Y_\infty\}$ that is not contained in I is a $\widetilde{\varphi}$-invariant stable interval, and both endpoints of I' are repulsive fixed points of $\widetilde{\varphi}|_{I'}$. However, the previous argument shows that one of the endpoints of I' is an attractive fixed point of $\widetilde{\varphi}|_{I'}$, a contradiction. Hence $X_\infty = Y_\infty$. This implies that $X_\infty = Y_\infty$ is the unique (repulsive) fixed point of $\widetilde{\varphi}$ contained in int$I$, and both points in ∂I are attractive fixed points of $\widetilde{\varphi}|_I$. Hence, the attractive fixed point set of $\widetilde{\varphi}|_\mathbb{R}$ is equal to V^+ and each component of $\mathbb{R} - V^+$ contains a unique repulsive fixed point of $\widetilde{\varphi}|_\mathbb{R}$. By applying the same argument to $\widetilde{\varphi}^{-1}$, we see that $\widetilde{V}^+$ and $\widetilde{V}^-$ are equal to the attractive and repulsive fixed point sets of $\widetilde{\varphi}|_\mathbb{R}$, respectively, and they are arranged alternately on $\mathbb{R}$. By taking the quotient by $\langle \rho_0(m) \rangle$, we obtain the first assertion of Lemma 3.5. The second assertion obviously follows from the first assertion. $\qquad\square$

Now the subset $\vec{\Delta}_\varphi$ of $\vec{\Delta}$ in Theorem 3.2 is defined as follows.

Definition 3.6. Let n_0 be the natural number in Lemma 3.5. Pick a connected component, J, of $S_p^1 - \mathrm{Fix}(\varphi_p^{n_0})$ and an element $\mu \in J \cap (G - \vec{\Delta})$, and let $[\mu, \varphi_p^{n_0}(\mu)]$ be the closed sub-interval of J bounded by μ and $\varphi_p^{n_0}(\mu)$. Then we define $\vec{\Delta}_\varphi := [\mu, \varphi_p^{n_0}(\mu)] \cap \vec{\Delta}$.

Remark 3.7. (1) There is a one-to-one correspondence between $\vec{\Delta}_\varphi$ and the quotient set $(J \cap \vec{\Delta})/\langle \varphi_p^{n_0} \rangle$, which in turn is a subset of $\vec{\Delta}/\langle \varphi_p^{n_0} \rangle$. Moreover $h_\rho(\varphi_p^{n_0}(\delta)) = h_\rho(\delta)$ for every $\delta \in \vec{\Delta}$. Thus we may identify $\vec{\Delta}_\varphi$ with the subset $(J \cap \vec{\Delta})/\langle \varphi_p^{n_0} \rangle$ of $\vec{\Delta}/\langle \varphi_p^{n_0} \rangle$. So the choice of μ in the definition of $\vec{\Delta}_\varphi$ is not essential.

(2) Throughout the remainder of this section and Section 7, we assume that μ and $\varphi_p^{n_0}(\mu)$ lie in this order with respect to the orientation of J induced by that of $\mathbb{R}$.

Complex translation length in the fiber group. Let ρ be the holonomy representation of the fiber group $\pi_1(F)$ in the hyperbolic manifold B_φ. Pick a point $\sigma \in \mathrm{Teich}(F)$, and let ρ_n be the element of $Q\mathcal{F}$ uniformizing $(\varphi_*^{-n}(\sigma), \varphi_*^n(\sigma))$ for each natural number n. Here φ_* denotes the automorphism of $\mathrm{Teich}(F)$ induced by φ. Then ρ_n converges to ρ strongly, because we know from the proof of Theorem 0.1 in [Thu87] (see [Thu87, §5]) that any subsequence of $\{\rho_n\}$ contains a subsequence converging to ρ strongly.

Lemma 3.8. *In the above situation, the sequence of the complex translation lengths $\lambda_{\rho_n}(\alpha)$ in $\mathbb{C}$ converges for each $\alpha \in S$. Moreover the limit $\lim \lambda_{\rho_n}(\alpha)$ does not depend on the choice of σ.*

Proof. We prepare some notations. Let d_T be the Teichmüller distance on $\mathrm{Teich}(F)$, and let d_{QF} be the metric on $Q\mathcal{F} \cong \mathrm{Teich}(F) \times \mathrm{Teich}(F)$ defined by

$$d_{QF}((\sigma, \tau), (\sigma', \tau')) = \max\{d_T(\sigma, \sigma'), d_T(\tau, \tau')\}.$$

Denote the automorphism of $Q\mathcal{F}$, induced by the automorphism $(\sigma, \tau) \mapsto (\varphi_*^{-1}(\sigma), \varphi_*(\tau))$ on $\mathrm{Teich}(F) \times \mathrm{Teich}(F)$, by the same symbol φ_*. Then it preserves the metric d_{QF}.

Now pick a point $\sigma \in \text{Teich}(T)$. Choose a path $\sigma_0 : [0,1] \to \text{Teich}(F)$ connecting σ and $\varphi_*(\sigma)$, and extend it to a path $\sigma(t)$ ($t \in (-\infty, \infty)$) in $\text{Teich}(F)$ by setting $\sigma(t) = \varphi_*^n(\sigma_0(t-n))$ when $0 \le t - n < 1$ for $n \in \mathbb{Z}$. Then it satisfies the condition $\varphi_*^n(\sigma(t)) = \sigma(t+n)$; in particular $\sigma(n) = \varphi_*^n(\sigma)$. Let ρ_t be the element of $Q\mathcal{F} \subset \text{Hom}(\pi_1(F), \text{PSL}(2,\mathbb{C}))$ uniformizing $(\sigma(-t), \sigma(t))$. Then we have $\varphi_*^n(\rho_t) = \rho_{t+n}$ by the definition of $\varphi_* : Q\mathcal{F} \to Q\mathcal{F}$.

Claim 3.9. *The path $[0, \infty) \ni t \mapsto \rho_t \in \text{Hom}(\pi_1(F), \text{PSL}(2,\mathbb{C}))$ extends to a continuous path from the closed interval $[0, \infty]$ by putting ρ_∞ to be the holonomy representation ρ of the fiber group in the hyperbolic manifold B_φ.*

Proof. Let $\{t_n\}$ be a sequence in $[0, \infty)$ such that $t_n \to \infty$. Take $m_n \in \mathbb{N}$ with $0 \le t_n - m_n < 1$. Then ρ_{m_n} and ρ_{t_n} are conjugate by a quasiconformal mapping f_n whose maximal dilatation is bounded above by a constant $d_0 := \max\{d_{QF}(\rho_0, \rho_t) \mid t \in [0,1]\}$, which is independent of n, because

$$d_{QF}(\rho_{m_n}, \rho_{t_n}) = d_{QF}(\varphi_*^{m_n}(\rho_0), \varphi_*^{m_n}(\rho_{t_n - m_n})) = d_{QF}(\rho_0, \rho_{t_n - m_n}) \le d_0.$$

By the choice of the holomorphic section $Q\mathcal{F} \to \text{Hom}(\pi_1(F), \text{PSL}(2,\mathbb{C}))$ (see Section 2), f_n is *normalized*, that is f_n fixes $\{0, 1, \infty\}$ pointwise. Applying the compactness of normalized e^{d_0}-quasiconformal mappings (cf. e.g. [IT92, Theorem 4.17]), we deduce that f_n converges to an e^{d_0}-quasiconformal mapping f_∞ (by taking a subsequence, if necessary). Thus, $\rho_{t_n}(\gamma) = f_n \circ \rho_{m_n}(\gamma) \circ f_n^{-1}$ converges to $f_\infty \circ \rho(\gamma) \circ f_\infty^{-1}$ for all $\gamma \in \pi_1(F)$. Since the limit set of $\rho(\pi_1(F))$ is the whole $\hat{\mathbb{C}}$, we conclude by the Sullivan's rigidity theorem (cf. e.g. [MT98, Theorem 5.20]) that f_∞ is conformal on the whole $\hat{\mathbb{C}}$. This implies that f_∞ coincides with the identity mapping on $\hat{\mathbb{C}}$ because of the normalization. Hence ρ_{t_n} converges to ρ.

It follows from the argument above that the path $t \mapsto \rho_t$ lands at ρ. Hence the path extends to a continuous path from $[0, \infty]$ by putting $\rho_\infty = \rho$.   □

Fix $\alpha \in \mathcal{S}$. Since $\rho_t \in Q\mathcal{F}$ for $t \in [0, \infty)$, we have a continuous map $[0, \infty) \ni t \mapsto \lambda_{\rho_t}(\alpha) \in \mathbb{C}$ (see Section 2). The composition of this map with the projection $\mathbb{C} \to \mathbb{C}/2\pi i\mathbb{Z}$ extends to a continuous map from the closed interval $[0, \infty]$, because every loxodromic transformation has a well-defined complex translation length in $\mathbb{C}/2\pi i\mathbb{Z}$. Since $\mathbb{C} \to \mathbb{C}/2\pi i\mathbb{Z}$ is a covering projection, we obtain a continuous extension of the map $\lambda_{\rho_t}(\alpha)$ to a map from the closed interval $[0, \infty]$ to $\mathbb{C}$. Hence $\lim \lambda_{\rho_t}(\alpha) = \lambda_{\rho_\infty}(\alpha) \in \mathbb{C}$ exists.

Finally, we show that $\lim \lambda_{\rho_t}(\alpha)$ does not depend on the choice of σ. Pick another point $\sigma' \in \text{Teich}(F)$, and consider the path $\sigma'(t)$ ($t \in (-\infty, \infty)$) in $\text{Teich}(F)$ as above. Let ρ_t' be the element of $Q\mathcal{F}$ uniformizing $(\sigma'(-t), \sigma'(t))$. Since $Q\mathcal{F}$ is contractible, there is a continuous map $D : [0,1]^2 \to Q\mathcal{F}$ such that $D(t,0) = \rho_t$, $D(t,1) = \rho_t'$, and

$\varphi_* \circ D(0,s) = D(1,s)$ for $t,s \in [0,1]$. The map D is extended to a continuous map on $\mathbb{R} \times [0,1]$ by putting $D(t,s) = \varphi_*^n \circ D(t-n,s)$, where n is the integer with $0 \le t-n < 1$. The relation $\varphi_*^n \circ D(t,s) = D(t+n,s)$ implies that for $(t,s) \in [0,1]^2$, $D(t,s)$ and ρ_t are conjugate by a quasiconformal mapping whose maximal dilatation is bounded by the constant $\max\{d_{QF}(D(t,0),D(t,s)) \mid (t,s) \in [0,1]^2\}$. By the same argument as above, we deduce that $D(t,s)$ extends to a continuous map on the closed disk $[0,\infty] \times [0,1]$ with $D(\infty,s) = \rho$ for $s \in [0,1]$. Therefore, we conclude that $\lim \lambda_{\rho_t}(\alpha) = \lim \lambda_{\rho_t'}(\alpha)$. This means that $\lim \lambda_{\rho_n}(\alpha)$ is independent of the choice of the reference point σ. $\quad\square$

Definition 3.10. Let ρ be the holonomy representation of the fiber group $\pi_1(F)$ in the hyperbolic manifold B_φ. Then for an essential simple loop $\alpha \in \mathcal{S}$, $\lambda_\rho(\alpha)$ denotes $\lim \lambda_{\rho_n}(\alpha) \in \mathbb{C}$ in Lemma 3.8.

4. McShane's analysis of $\mathcal{G}$

In this section we recall McShane's analysis of the set $\mathcal{G}$ of the oriented simple complete geodesics in a hyperbolic surface $F = \mathbb{H}^2/\Gamma_0$, where Γ_0 is the image of $\rho_0 \in \mathcal{F}$. As in the previous section, we identify $\mathcal{G}$ with the subspace of $S_p^1 = \mathbb{R}/\langle \rho_0(m) \rangle$. Here S_p^1 inherits the standard metric from that of $\mathbb{R}$. In particular, the total length of S_p^1 is 1. Then the following result has been proved by McShane [McS98, Theorem 4 and Proposition 3].

Proposition 4.1. *(1) $\vec{\Delta}$ consists of the isolated points of $\mathcal{G}$, and $\mathcal{G} - \vec{\Delta}$ is a Cantor set of measure 0.*

(2) For $\delta \in \vec{\Delta}$, let $J(\delta)$ be the maximal open interval in S_p^1 such that $J(\delta) \cap \mathcal{G} = \{\delta\}$. Then, generically, the two boundary points of $J(\delta)$ correspond to the elements of $\mathcal{G}$ which spiral to the oriented simple closed geodesics $\alpha(\delta)$ and $\beta(\delta)$, respectively. Here $\alpha(\delta)$ and $\beta(\delta)$ are oriented so that they are homologous to δ in the annulus obtained from the punctured annulus bounded by $\alpha(\delta) \cup \beta(\delta)$ through one point compactification. In the special case when $\alpha(\delta)$ or $\beta(\delta)$ is a peripheral circle around a puncture q, the corresponding boundary point of $J(\delta)$ is a simple oriented geodesic joining p to q.

(3) The length of $J(\delta)$ is equal to $h_{\rho_0}(\delta)$ for every $\delta \in \vec{\Delta}$.

McShane's original identity [McS98, Theorem 2] is obtained from the above proposition as follows. Since the measure of $\mathcal{G} - \vec{\Delta}$ is 0, the length of S_p^1 is equal to the infinite sum of the lengths of $J(\delta)$ where δ runs over all elements of $\vec{\Delta}$. Hence

$$1 = \sum_{\delta \in \vec{\Delta}} (\text{length of } J(\delta)) = \sum_{\delta \in \vec{\Delta}} h_{\rho_0}(\delta) = 2 \sum_{\delta \in \Delta} h_{\rho_0}(\delta).$$

By a similar argument, we obtain the following corollary.

Corollary 4.2. *Let μ_1 and μ_2 be elements of $G - \vec{\Delta} \subset S_p^1$. Let $[\mu_1, \mu_2]$ be the interval of S_p^1 such that $\partial[\mu_1, \mu_2] = \mu_2 - \mu_1$ with respect to the orientation induced from the natural orientation of S_p^1. Then the length of $[\mu_1, \mu_2]$ is equal to*

$$\sum_{\delta \in [\mu_1, \mu_2] \cap \vec{\Delta}} h_{\rho_0}(\delta).$$

5. Absolute convergence

In this section we prove the following proposition, which is used in the proof of Theorems 2.2 and 2.3.

Proposition 5.1. *For each $\rho \in Q\mathcal{F}$, the infinite sum $\sum_{\delta \in \vec{\Delta}} h_\rho(\delta)$ converges absolutely and uniformly on every compact subset of $Q\mathcal{F}$.*

To prove this proposition, let $l_\rho(\gamma) = \Re \lambda_\rho(\gamma)$ be the real translation length of $\rho(\gamma)$ for each $\gamma \in S$. Then we have the following lemma.

Lemma 5.2. *For any compact subset C of $Q\mathcal{F}$, there is a constant $k = k(C) > 1$ such that*

$$\frac{1}{k} l_{\rho_0}(\gamma) \leq l_\rho(\gamma) \leq k l_{\rho_0}(\gamma),$$

for any $\gamma \in S$ and $\rho \in C$. Here ρ_0 is a fixed element of $\mathcal{F}$.

Proof. Since C is compact, there is a constant $k > 1$ such that every $\rho \in C$ is k-quasiconformally equivalent to ρ_0. Hence we have the desired inequality by [JM79, Lemma 3] (cf. [Kap01, Theorem 8.57]). $\square$

We may assume F is not a thrice-punctured sphere. After interchanging $\alpha(\delta)$ and $\beta(\delta)$ if necessary, we may also assume $l_{\rho_0}(\alpha(\delta)) \geq l_{\rho_0}(\beta(\delta))$ for each $\delta \in \Delta$. Then we have $l_\rho(\alpha(\delta)) > 0$ for any $\rho \in Q\mathcal{F}$. Hence

$$
\begin{aligned}
|h_\rho(\delta)| &= \frac{1}{|1 + \exp\frac{1}{2}(\lambda_\rho(\alpha(\delta)) + \lambda_\rho(\beta(\delta)))|} \\
&\leq \frac{1}{|\exp\frac{1}{2}(\lambda_\rho(\alpha(\delta)) + \lambda_\rho(\beta(\delta)))| - 1} \\
&\leq \frac{1}{\exp\frac{1}{2}(l_\rho(\alpha(\delta)) + l_\rho(\beta(\delta))) - 1} \\
&\leq \frac{1}{\exp\frac{1}{2}(l_\rho(\alpha(\delta))) - 1} \times \frac{1}{\exp\frac{1}{2}(l_\rho(\beta(\delta)))}.
\end{aligned}
\tag{5.1}
$$

Set $U_\rho(\gamma) = \frac{1}{\exp\frac{1}{2}(l_\rho(\gamma))}$ for $\gamma \in S$. Then we have the following lemma.

Lemma 5.3. *For any compact subset C of $Q\mathcal{F}$, there is a finite subset S_0 of S, such that the following inequality holds for every $\gamma \in S - S_0$ and $\rho \in C$:*

$$\frac{1}{\exp\frac{1}{2}(l_\rho(\gamma)) - 1} \leq 2U_\rho(\gamma).$$

In particular, for every $\rho \in C$ and for every $\delta \in \Delta$ such that $\alpha(\delta) \notin S_0$, we have:

$$|h_\rho(\delta)| \leq 2U_\rho(\alpha(\delta))U_\rho(\beta(\delta)).$$

Proof. Note that the first inequality in the lemma is equivalent to

$$l_\rho(\gamma) \geq 2\log 2.$$

Hence an element $\gamma \in S$ does not satisfy the inequality only if

$$l_{\rho_0}(\gamma) < 2k\log 2,$$

where $k = k(C) > 1$ is the constant in Lemma 5.2. Let S_0 be the set of the elements of S satisfying this inequality. Then S_0 is a finite set and satisfies the first assertion. The second assertion follows from the first one and the inequality (5.1). $\square$

Lemma 5.4. *For any compact subset C of $Q\mathcal{F}$, $\sum_{\gamma \in S} U_\rho(\gamma)$ converges uniformly on C.*

Proof. We first recall a result of [BS85]. Let R be a fundamental domain for the action of $\Gamma_0 = \rho_0(\pi_1(F))$ on $\mathbb{H}^2$ such that R is a finite sided geodesic ideal polygon. Then the image, A, of ∂R in F consists of finitely many mutually disjoint simple complete geodesics each of which joins punctures of F. We identify S with the set of simple geodesics on the hyperbolic surface $F = \mathbb{H}^2/\Gamma_0$. Then each $\gamma \in S$ intersects A transversely: we denote the cardinality of the intersection $\gamma \cap A$ by $\|\gamma\|$.

Claim 5.5. *There is a positive constant c such that $l_{\rho_0}(\gamma) \geq c\|\gamma\|$ for any $\gamma \in S$.*

Proof. By [BS85, Lemma 3.1 and Corollary 5.3], this holds with finitely many exceptions. Since $\|\gamma\| \geq 1$ for any $\gamma \in S$, we have the desired assertion. $\square$

Set $S(n) = \{\gamma \in S \mid \|\gamma\| = n\}$. Then by [BS85, Lemma 2], there is a polynomial $P_0(n)$ such that $\#S(n) \leq P_0(n)$. On the other hand, for every $\gamma \in S(n)$ and $\rho \in C$, where C is a compact subset of $Q\mathcal{F}$, we have the following inequality by Lemma 5.2 and Claim 5.5:

$$l_\rho(\gamma) \geq \frac{1}{k}l_{\rho_0}(\gamma) \geq \frac{c}{k}\|\gamma\| = \frac{c}{k}n.$$

Here $k = k(C)$ is the constant in Lemma 5.2. Hence, for every $\rho \in C$,

$$\sum_{\gamma \in S} U_\rho(\gamma) = \sum_{n=1}^{\infty} \left(\sum_{\gamma \in S(n)} U_\rho(\gamma) \right) \leq \sum_{n=1}^{\infty} \frac{P_0(n)}{\exp(\frac{c}{2k}n)} < \infty.$$

So, $\sum_{\gamma \in S} U_\rho(\gamma)$ converges uniformly on the compact set C. □

Proof of Proposition 5.1. Let C be a compact subset of $Q\mathcal{F}$, and pick a real $\varepsilon > 0$. By Lemma 5.4, there is a finite subset S_ε of S containing the finite set S_0 in Lemma 5.3, such that, for every $\rho \in C$,

$$\sum_{\gamma \in S - S_\varepsilon} U_\rho(\gamma) < \varepsilon.$$

Let Δ_ε be the subset of Δ consisting of those elements δ such that $\{\alpha(\delta), \beta(\delta)\} \subset S_\varepsilon \cup \mathcal{P}$. Since $S_\varepsilon \cup \mathcal{P}$ is a finite set, Δ_ε is also a finite set. Moreover, by Lemma 5.3, we have:

$$\sum_{\delta \in \Delta - \Delta_\varepsilon} |h_\rho(\delta)| \leq 2 \sum_{\delta \in \Delta - \Delta_\varepsilon} U_\rho(\alpha(\delta)) U_\rho(\beta(\delta))$$

$$\leq 2 \left(\sum_{\alpha \in S - S_\varepsilon} U_\rho(\alpha) \right) \left(\sum_{\beta \in S \cup \mathcal{P}} U_\rho(\beta) \right)$$

$$\leq 2\varepsilon \left(\#\mathcal{P} + \sum_{\gamma \in S} U_\rho(\gamma) \right),$$

where $\#\mathcal{P}$ denotes the cardinality of $\mathcal{P}$. In the above, the second inequality follows from the following fact:

   If $\delta \in \Delta - \Delta_\varepsilon$ then after interchanging $\alpha(\delta)$ and $\beta(\delta)$ if necessary, we may assume $\alpha(\delta) \in S - S_\varepsilon$ and $\beta(\delta) \in S \cup \mathcal{P}$.

By Lemma 5.4, the last term is bounded above by a constant depending only on C. This completes the proof of Proposition 5.1. □

6. Proof of Theorems 2.2 and 2.3

Proof of Theorem 2.2. By Proposition 5.1, the correspondence $\rho \mapsto \sum_{\delta \in \Delta} h_\rho(\delta)$ defines a holomorphic function on the complex manifold $Q\mathcal{F}$. On the other hand, Mc-Shane's original identity (Corollary 4.2) shows that the holomorphic function takes the constant value 1 on the totally real submanifold $\mathcal{F}$. Hence the holomorphic function must be the constant function 1 by the theorem of identity. So, we have:

$$\sum_{\delta \in \Delta} h_\rho(\delta) = \frac{1}{2} \sum_{\delta \in \vec{\Delta}} h_\rho(\delta) = \frac{1}{2}.$$

$\square$

For each $\rho \in Q\mathcal{F}$, there is a quasiconformal mapping f_ρ of $\hat{\mathbb{C}}$ which is (ρ_0, ρ)-equivariant, i.e., $\rho(\gamma) = f_\rho \circ \rho_0(\gamma) \circ f_\rho^{-1}$, as homeomorphisms of $\hat{\mathbb{C}}$, for every $\gamma \in \pi_1(F)$. By the identification of $Q\mathcal{F}$ with a subspace of $\mathrm{Hom}(\pi_1(F), \mathrm{PSL}(2, \mathbb{C}))$ prescribed in Section 2, f_ρ fixes the three points 0, 1 and ∞. Though f_ρ is not unique, its restriction to the limit set $\hat{\mathbb{R}}$ of Γ_0 is uniquely determined by ρ. Thus we obtain a well-defined map

$$Q\mathcal{F} \times \hat{\mathbb{R}} \ni (\rho, z) \mapsto f_\rho(z) \in \hat{\mathbb{C}}. \tag{6.1}$$

This is holomorphic in the sense that the mapping $Q\mathcal{F} \ni \rho \mapsto f_\rho(z) \in \hat{\mathbb{C}}$ is holomorphic for any $z \in \hat{\mathbb{R}}$, because it is proved by Ahlfors and Bers [AB60] that the normalized quasiconformal mapping of $\hat{\mathbb{C}}$ depend holomorphically on the complex dilatation (cf. e.g. [Gar87, Chapter 1, Theorem 5], [IT92, Theorem 4.37]).

Recall that $\widetilde{\mathcal{G}}$ is the set of oriented complete geodesics in $\mathbb{H}^2$ emanating from ∞ which projects to a simple geodesic in $F = \mathbb{H}^2/\Gamma_0$. For $\widetilde{\mu} \in \widetilde{\mathcal{G}}$ and $\rho \in Q\mathcal{F}$, let $z_\rho(\widetilde{\mu}) = f_\rho(z_{\rho_0}(\widetilde{\mu}))$ be the image of the endpoint $z_{\rho_0}(\widetilde{\mu})$ of $\widetilde{\mu}$ by f_ρ. We denote by $\widetilde{\mu}_\rho$ the oriented geodesic emanating from ∞ and ending at $z_\rho(\widetilde{\mu})$. If $\mu \in \mathcal{G}$ is the image of $\widetilde{\mu}$ in F, then we denote by μ_ρ the image of $\widetilde{\mu}_\rho$ in the hyperbolic 3-manifold $\mathbb{H}^3/\rho(\pi_1(F))$, and call it the geodesic representative of μ in $\mathbb{H}^3/\rho(\pi_1(F))$.

Proof of Theorem 2.3. We fix a quasifuchsian representation $\rho_1 \in Q\mathcal{F}$. Let pl^ε be the bending lamination of ρ_1 and μ^ε an element of $\mathcal{G}$ disjoint from $|pl^\varepsilon|$ ($\varepsilon = \pm$). Then the geodesic representative $\mu_{\rho_1}^\varepsilon$ of μ in $M := \mathbb{H}^3/\rho_1(\pi_1(F))$ lies in the boundary $\partial^\varepsilon M_0$ of the convex core M_0 of M. Let $\widetilde{\mu}^\varepsilon$ be an element of $\widetilde{\mathcal{G}}$ which projects to μ^ε. Then $\widetilde{\mu}_{\rho_1}^\varepsilon$ lies in the boundary component, $\partial^\varepsilon C(\Lambda)$, of the convex hull of $C(\Lambda)$ of the limit set Λ of $\rho_1(\pi_1(F))$. By the arguments in the proof of [AMS04, Lemmas 3.1 and 3.2], we have

$$\mathrm{width}_p(\Lambda) = \Im \left(z_{\rho_1}(\widetilde{\mu}^+) - z_{\rho_1}(\widetilde{\mu}^-) \right). \tag{6.2}$$

Let $w : Q\mathcal{F} \to \mathbb{C}$ be the mapping defined by $w(\rho) = z_\rho(\widetilde{\mu}^+) - z_\rho(\widetilde{\mu}^-)$. Then w is holomorphic, because $w(\rho) = f_\rho(z_{\rho_0}(\widetilde{\mu}^+)) - f_\rho(z_{\rho_0}(\widetilde{\mu}^-))$ and the mapping (6.1) is holomorphic.

Case 1. $\mu^{\pm} \notin \vec{\Delta}$. Then, by Corollary 4.2, we have the following identity for every $\rho \in \mathcal{F}$:

$$w(\rho) = \sum_{\delta \in \vec{\Delta}_L} h_\rho(\delta) \quad \text{in } \mathbb{R}/\mathbb{Z},$$

where $\vec{\Delta}_L$ is the subset of Δ which lies between μ^- and μ^+ with respect to the cyclic order. By Proposition 5.1, we see that $\sum_{\delta \in \vec{\Delta}_L} h_\rho(\delta)$ is a holomorphic function of $\rho \in Q\mathcal{F}$. Hence, the above identity holds for every $\rho \in Q\mathcal{F}$ by the theorem of identity. By (6.2), we obtain the desired identity.

Case 2. $\mu^\varepsilon \in \vec{\Delta}$ for some ε. For simplicity we treat the case where $\mu^{\pm} \in \vec{\Delta}$. (The other cases can be treated in a similar way.) Let μ'^ε be an element of $G - \Delta$ which spirals to one of the oriented simple closed geodesics $\alpha(\mu^\varepsilon)$ or $\beta(\mu^\varepsilon)$. Then it is disjoint from pl^ε. Let $\vec{\Delta}'_L \sqcup \vec{\Delta}'_R$ be the partition of $\vec{\Delta} - \{\mu'^-, \mu'^+\} = \vec{\Delta}$ determined by $\{\mu'^-, \mu'^+\}$ by Proposition 4.1(2). Then the difference between the sets $\vec{\Delta}_L$ and $\vec{\Delta}'_L$ (resp. $\vec{\Delta}_R$ and $\vec{\Delta}'_R$) is at most $\{\mu^-, \mu^+\}$. Since Theorem 2.3 holds for Case 1, we have

$$\pm \text{width}_p(\Lambda) = \Im \sum_{\delta \in \vec{\Delta}'_L} h_\rho(\delta) = -\Im \sum_{\delta \in \vec{\Delta}'_R} h_\rho(\delta).$$

On the other hand, since μ^ε is disjoint from the bending locus $|pl^\varepsilon|$, the pair of loops $\{\alpha(\mu^\varepsilon), \beta(\mu^\varepsilon)\}$ lies in the flat piece of the convex core boundary. This implies that $h_\rho(\mu^\varepsilon)$ is real and hence

$$\Im \sum_{\delta \in \vec{\Delta}_L} h_\rho(\delta) = \Im \sum_{\delta \in \vec{\Delta}'_L} h_\rho(\delta), \qquad \Im \sum_{\delta \in \vec{\Delta}_R} h_\rho(\delta) = \Im \sum_{\delta \in \vec{\Delta}'_R} h_\rho(\delta).$$

Thus we obtain the desired identity. $\square$

7. Proof of Theorem 3.2

By taking a power of φ, we may assume the number n_0 in Definition 3.6 is equal to 1, because $\text{Modulus}(\partial_p B_\varphi) = \text{Modulus}(\partial_p B_{\varphi^{n_0}})$ by definition. Then the homeomorphism $\varphi : \partial_p F \to \partial_p F$ has finitely many attractive/repulsive fixed points, which are arranged on $\partial_p F$ alternately (see Lemma 3.3). Moreover, φ preserves each singular leaf of $\mathcal{F}^{\pm}$ emanating from p. Pick such a singular leaf L of $\mathcal{F}^+$. We may assume the meridian m intersects L transversely in a single point, x_0. Then after a bounded isotopy, we may assume that φ preserves L and fixes x_0. Then the longitude $l \subset \bar{B}_\varphi$ is homotopic to the loop which is obtained as the image of $\{x_0\} \times [0,1]$ in $B_\varphi = F \times [0,1]/(x,0) \sim (\varphi(x),1)$. We denote the image of $(x_0,0)$ in B_φ by the same symbol x_0, and identify F with the fiber $F \times 0$ in B_φ. Then the fundamental group $\pi_1(B_\varphi, x_0)$ is an HNN extension of $\pi_1(F, x_0)$ by the infinite cyclic group $\langle l \rangle$ with the

following relation:
$$l\gamma l^{-1} = \varphi_*(\gamma) \qquad (\gamma \in \pi_1(F, x_0)).$$

Let $\widetilde{L}$ be a lift of L to the universal cover $\mathbb{H}^2$ emanating from the parabolic fixed point ∞ of $\rho_0(m) = \begin{pmatrix} 1 & 1 \\ 0 & 1 \end{pmatrix}$. Let $\widetilde{\varphi}$ be the lift of φ which preserves $\widetilde{L}$.

Lemma 7.1. *For every $\gamma \in \pi_1(F, x_0)$, we have*

$$\rho_0(\varphi_*(\gamma)) = \widetilde{\varphi} \circ \rho_0(\gamma) \circ \widetilde{\varphi}^{-1}$$

as homeomorphisms of $\mathbb{H}^2$. In particular, for every non-peripheral element $\gamma \in \pi_1(F, x_0)$, we have

$$\mathrm{Fix}^+(\rho_0(\varphi_*(\gamma))) = \widetilde{\varphi}(\mathrm{Fix}^+(\rho_0(\gamma))).$$

Proof. Let $\widetilde{x}_0$ be the lift of x_0 contained in $\widetilde{L} \subset \mathbb{H}^2$. Then the representation $\rho : \pi_1(F, x_0) \to \mathrm{PSL}(2, \mathbb{R}) = \mathrm{Isom}^+(\mathbb{H}^2)$ gives *the identification of the fundamental group $\pi_1(F, x_0)$ with the covering transformation group via the point $\widetilde{x}_0$* in the following sense:

> Let γ be an element of $\pi_1(F, x_0)$ represented by a path $c : ([0, 1], \{0, 1\}) \to (F, x_0)$. Then the covering transformation $\rho_0(\gamma)$ maps $\widetilde{x}_0$ to the endpoint $\widetilde{c}(1)$, where $\widetilde{c}$ is the lift of c to $\mathbb{H}^2$ with $\widetilde{c}(0) = \widetilde{x}_0$.

In fact, we can check this for the meridian m by using the fact that $\widetilde{L}$ is a lift of L emanating from the parabolic fixed point of $\rho_0(m)$. This implies that the condition holds for every element of $\pi_1(F, x_0)$.

Now we identify the universal cover of B_φ with $\mathbb{H}^2 \times \mathbb{R}$ and identify the fundamental group $\pi_1(B_\varphi, x_0)$ with the covering transformation group via the point $(\widetilde{x}_0, 0)$. Then we see that each $\gamma \in \pi_1(F, x_0)$ is identified with the covering transformation $\rho_0(\gamma) \times id$ and that the longitude l is identified with the covering transformation $\widetilde{\varphi} \times (+1)$, where $(+1)$ denotes the translation of $\mathbb{R}$ by 1. Hence, the relation $l\gamma l^{-1} = \varphi_*(\gamma)$ implies the first assertion. The second assertion is an immediate consequence of the first one. $\square$

Recall that the set $\vec{\Delta}_\varphi$ is defined to be $[\mu, \varphi_p(\mu)] \cap \vec{\Delta}$, where $[\mu, \varphi_p(\mu)]$ is the closed sub-interval of a component, J, of $S_p^1 - \mathrm{Fix}(\varphi_p)$ bounded by μ and $\varphi_p(\mu)$ (see Definition 3.6). Here μ is an arbitrary element of $J \cap (G - \vec{\Delta})$. So we may assume that μ spirals to an oriented simple closed geodesic γ. Let $\widetilde{\mu}$ be a lift of the oriented geodesic $\mu \subset F = \mathbb{H}^2 / \Gamma_0$ to $\mathbb{H}^2$ emanating from ∞. Then there is an element of $\pi_1(F, x_0)$, denoted by the same symbol γ, which represents the closed geodesic γ such that $z_{\rho_0}(\widetilde{\mu}) = \mathrm{Fix}^+(\rho_0(\gamma))$, where $z_{\rho_0}(\widetilde{\mu})$ is the endpoint of $\widetilde{\mu}$ in $\mathbb{R} = \partial \mathbb{H}^2 - \{\infty\}$.

Lemma 7.2. *We have*

$$\mathrm{Fix}^+(\rho_0(\varphi_*(\gamma))) - \mathrm{Fix}^+(\rho_0(\gamma)) = \sum_{\delta \in \vec{\Delta}_\varphi} h_{\rho_0}(\delta).$$

Proof. By Lemma 7.1, $\mathrm{Fix}^+(\rho_0(\varphi_*(\gamma))) = \widetilde{\varphi}(\mathrm{Fix}^+(\rho_0(\gamma))) = \widetilde{\varphi}(z_{\rho_0}(\widetilde{\mu}))$. By the definition of $\varphi_p : \widetilde{G} \to \widetilde{G}$ (see Section 3), $z_{\rho_0}(\varphi_p(\widetilde{\mu})) = \widetilde{\varphi}(z_{\rho_0}(\widetilde{\mu}))$. Hence

$$\begin{aligned}
\mathrm{Fix}^+(\rho_0(\varphi_*(\gamma))) - \mathrm{Fix}^+(\rho_0(\gamma)) &= \widetilde{\varphi}(z_{\rho_0}(\widetilde{\mu})) - z_{\rho_0}(\widetilde{\mu}) \\
&= z_{\rho_0}(\varphi_p(\widetilde{\mu})) - z_{\rho_0}(\widetilde{\mu}).
\end{aligned}$$

Since $\widetilde{\varphi}(\widetilde{L}) = \widetilde{L}$, $\widetilde{\varphi}$ fixes the endpoint of $\widetilde{L}$ in $\mathbb{R} = \partial \mathbb{H}^2 - \{\infty\}$, and hence the fixed point set of $\widetilde{\varphi}|_\mathbb{R}$ is equal to the inverse image in $\mathbb{R}$ of the fixed point set of $\varphi_p : S_p^1 \to S_p^1$. So, $z_{\rho_0}(\widetilde{\mu})$ and $\widetilde{\varphi}(z_{\rho_0}(\widetilde{\mu}))$ lie in a single component, say $\widetilde{J}$, of $\mathbb{R} - \mathrm{Fix}(\widetilde{\varphi})$, which projects isometrically to the component J of $S_p^1 - \mathrm{Fix}(\varphi_p)$, with respect to the standard metric. Hence $z_{\rho_0}(\varphi_p(\widetilde{\mu})) - z_{\rho_0}(\widetilde{\mu})$ is equal to $\pm$ of the length of the interval $[\mu, \varphi_p(\mu)] \subset J \subset S_p^1$. The sign depends on whether μ and $\varphi_p(\mu)$ lies in J in this order with respect to the orientation of J induced by that of $\mathbb{R}$. Hence, by Corollary 4.2, we obtain the desired result, because of the assumption we made in Remark 3.7 (2). $\qquad\square$

Recall that we have chosen a holomorphic cross section $Q\mathcal{F} \to \mathrm{Hom}(\pi_1(F), \mathrm{PSL}(2,\mathbb{C}))$ so that $\rho(m) = \begin{pmatrix} 1 & 1 \\ 0 & 1 \end{pmatrix}$ for every $\rho \in Q\mathcal{F}$. Thus we can see that the proof of Lemma 7.2 works for every $\rho \in \mathcal{F}$, and we obtain the following lemma.

Lemma 7.3. *For any $\rho \in \mathcal{F} \subset \mathrm{Hom}(\pi_1(F), \mathrm{PSL}(2,\mathbb{C}))$, we have*

$$\mathrm{Fix}^+(\rho(\varphi_*(\gamma))) - \mathrm{Fix}^+(\rho(\gamma)) = \sum_{\delta \in \vec{\Delta}_\varphi} h_\rho(\delta).$$

Lemma 7.4. *For any $\rho \in Q\mathcal{F}$, we have*

$$\mathrm{Fix}^+(\rho(\varphi_*(\gamma))) - \mathrm{Fix}^+(\rho(\gamma)) = \sum_{\delta \in \vec{\Delta}_\varphi} h_\rho(\delta).$$

Proof. By Proposition 5.1, the infinite sum on the right hand side converges absolutely and locally uniformly on $Q\mathcal{F}$. Hence it is holomorphic on $Q\mathcal{F}$. On the other hand, the left hand side is obviously holomorphic on $Q\mathcal{F}$, and it coincides with the right hand side on the totally real analytic submanifold $\mathcal{F}$ by Lemma 7.3. Hence the identity holds for every $\rho \in Q\mathcal{F}$. $\qquad\square$

Let $\rho_\infty : \pi_1(B_\varphi, x_0) \to \mathrm{PSL}(2, \mathbb{C})$ be the holonomy representation of the complete hyperbolic structure of B_φ. Then we may assume that the restriction of ρ_∞ to $\pi_1(F, x_0)$, which we continue to denote by the same symbol, lies in the closure of the image of the holomorphic section $Q\mathcal{F} \to \mathrm{Hom}(\pi_1(F), \mathrm{PSL}(2,\mathbb{C}))$ (which we fixed in Section 2) and that the following identities hold.

$$\rho_\infty(m) = \begin{pmatrix} 1 & 1 \\ 0 & 1 \end{pmatrix}, \qquad \rho_\infty(l) = \begin{pmatrix} 1 & \mathrm{Modulus}(\partial_p B_\varphi) \\ 0 & 1 \end{pmatrix}.$$

Then we have the following lemma.

Lemma 7.5. *For any non-peripheral element* $\gamma \in \pi_1(F, x_0)$, *we have*

$$\mathrm{Fix}^+(\rho_\infty(\varphi_*(\gamma))) - \mathrm{Fix}^+(\rho_\infty(\gamma)) = \mathrm{Modulus}(\partial_p B_\varphi),$$

Proof.

$$\begin{aligned}
\mathrm{Fix}^+(\rho_\infty(\varphi_*(\gamma))) &= \mathrm{Fix}^+(\rho_\infty(l\gamma l^{-1})) \\
&= \rho_\infty(l)(\mathrm{Fix}^+(\rho_\infty(\gamma))) \\
&= \mathrm{Fix}^+(\rho_\infty(\gamma)) + \mathrm{Modulus}(\partial_p B_\varphi).
\end{aligned}$$

$\square$

We shall prove the following proposition in the next section.

Proposition 7.6. *Let* $\{\rho_n\}$ *be a sequence of quasifuchsian representations which converges strongly to* ρ_∞. *Then the infinite sum* $\sum_{\delta \in \vec{\Delta}_\varphi} h_{\rho_\infty}(\delta)$ *converges absolutely, and*

$$\sum_{\delta \in \vec{\Delta}_\varphi} h_{\rho_\infty}(\delta) = \lim \sum_{\delta \in \vec{\Delta}_\varphi} h_{\rho_n}(\delta).$$

By assuming this proposition, we can complete the proof of Theorem 3.2 as follows. Let $\{\rho_n\}$ be a sequence in $Q\mathcal{F}$ as in Lemma 3.8, which converges to ρ_∞ strongly. Then we have:

$$\begin{aligned}
\mathrm{Modulus}(\partial_p B_\varphi) &= \mathrm{Fix}^+(\rho_\infty(\varphi_*(\gamma))) - \mathrm{Fix}^+(\rho_\infty(\gamma)) && \text{by Lemma 7.5} \\
&= \lim \left(\mathrm{Fix}^+(\rho_n(\varphi_*(\gamma))) - \mathrm{Fix}^+(\rho_n(\gamma)) \right) \\
&= \lim \sum_{\delta \in \vec{\Delta}_\varphi} h_{\rho_n}(\delta) && \text{by Lemma 7.4} \\
&= \sum_{\delta \in \vec{\Delta}_\varphi} h_{\rho_\infty}(\delta) && \text{by Proposition 7.6.}
\end{aligned}$$

8. Proof of Proposition 7.6

For a type-preserving faithful discrete representation $\rho : \pi_1(F) \to \mathrm{PSL}(2, \mathbb{C})$ and $\gamma \in S$ such that $\rho(\gamma)$ is not parabolic, we denote by γ_ρ the geodesic representative of γ in the hyperbolic manifold $M_\rho = \mathbb{H}^3/\rho(\pi_1(F))$. Let ρ_∞ and ρ_n be the representations in Proposition 7.6, and set $M_\infty := M_{\rho_\infty}$ and $M_n := M_{\rho_n}$.

Lemma 8.1. *There is a horospherical neighborhood C_∞ of the cusps of M_∞ which is disjoint from the closed geodesics γ_{ρ_∞} for every $\gamma \in S$.*

Proof. It is well-known that there exists $\varepsilon_0 > 0$ such that, for any $\sigma \in \mathrm{Teich}(F)$, every simple closed geodesic on the hyperbolic surface (F, σ) is disjoint from the cuspidal components of the ε_0-thin part (cf. e.g. [CEG87, 2.2.4 Corollary]). By [Min00, Lemma 3.1], there exists $\varepsilon_1 > 0$ such that, if $g : (F, \sigma) \to M_\infty$ is a π_1-injective pleated surface then only the ε_0-thin part of the surface can be mapped into the ε_1-thin part of M_∞. Let C_∞ be the union of cuspidal parts of the ε_1-thin part of M_∞. Then C_∞ satisfies the desired property, because any element $\gamma \in S$ is realized by a π_1-injective pleated surface. $\qquad\square$

Let $S(\vec{\Delta}_\varphi)$ be the set of all $\gamma \in S$ such that γ is isotopic to $\alpha(\delta)$ or $\beta(\delta)$ for some $\delta \in \vec{\Delta}_\varphi$.

Lemma 8.2. *There is a compact submanifold K_∞ of M_∞ which contains the closed geodesic γ_{ρ_∞} for every $\gamma \in S(\vec{\Delta}_\varphi)$.*

Proof. Set $M_\infty^0 := M_\infty - \mathrm{int}C_\infty$, where C_∞ is as in Lemma 8.1. Note that M_∞^0 has precisely two ends, E^- and E^+. Pick a fundamental neighborhood system $\{U_i^+\}_{i=1}^\infty$ of E^+ such that $U_j^+ \supset \overline{U_{j+1}^+}$ ($j = 1, \cdots, \infty$). Suppose to the contrary that the assertion does not hold. Then we may assume that there is a sequence $\{\gamma_j\}$ in $S(\vec{\Delta}_\varphi)$ such that $(\gamma_j)_{\rho_\infty} \cap U_j^+ \neq \emptyset$.

Claim 8.3. *A subsequence of $\{(\gamma_j)_{\rho_\infty}\}$ exits the end E^+.*

Proof. Suppose the claim does not hold. Then there is a compact subset K of M_∞^0 such that $(\gamma_j)_{\rho_\infty} \cap K \neq \emptyset$ for every j. Let $g_j : (F, \sigma_j) \to M_\infty$ be a (marked) pleated surface in the correct homotopy class which realizes $(\gamma_j)_{\rho_\infty}$. By compactness of marked pleated surfaces meeting K (see [CEG87, Corollary 5.2.18]), the pleated surface g_j converges to a pleated surface $g : (F, \sigma) \to M_\infty$, that is, $g_j : F \to M_\infty$ (after precomposition with diffeomorphism of F isotopic to the identity) converges to $g : F \to M_\infty$ in the compact-open topology. Now pick a point $x_j \in (\gamma_j)_{\rho_\infty} \cap U_j^+ \subset g_j(F) \cap M_\infty^0$. Let F_0 be a compact neighborhood of $g^{-1}(M_\infty^0)$ in F. Then we may assume that $x_j = g_j(y_j)$ for some $y_j \in F_0$. Let y_∞ be the limit of a subsequence of $\{y_j\}$. Then the corresponding subsequence

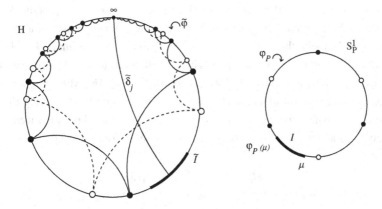

Figure 4: In each figure, the black/white dots represent attractive/repulsive fixed points.

of $\{x_j\} = \{g_j(y_j)\}$ converges to $g(y_\infty) \in g(F_0)$. Since M_∞^0 is closed, $g(y_\infty) \in M_\infty^0$. This contradicts the facts that $x_j \in U_j$ and that $\{U_j\}_{j=1}^\infty$ is a fundamental neighborhood system of the end E^+. $\qquad\square$

By the above claim and compactness of $\mathcal{PML}(F)$, we may assume that $t_j\gamma_j$, for suitable $t_j > 0$, converges to a measured lamination ν' in $\mathcal{ML}(F)$. By [Can93, Proposition 10.1], the support of ν' is equal to the ending lamination of E^+, which in turn coincides with the support $|\nu^+|$ of the stable lamination ν^+ of φ. Let γ_∞ be the limit of (a subsequence of) the geodesic representatives, in the hyperbolic surface F, of $\{\gamma_j\}$ in the Hausdorff topology. Then, it is known that $|\nu^+| \subset \gamma_\infty$ (cf. [Ota01, Remark A.3.2]).

On the other hand, by the definition of $S(\vec{\Delta}_\varphi)$, we may assume that $\gamma_j = \alpha(\delta_j)$ for some $\delta_j \in \vec{\Delta}_\varphi$. Let δ_∞ be the geodesic lamination obtained as the limit of (a subsequence of) δ_j in the Chabauty topology. Let $\widetilde{I}$ be a lift of the interval $I := [\mu, \varphi_p(\mu)] \subset S_p^1 - \mathrm{Fix}(\varphi_p)$ to $\partial\mathbb{H}^2 - \{\infty\}$. Then each δ_j has a unique lift $\widetilde{\delta}_j$ to $\mathbb{H}^2$ emanating from ∞ and ending at a point in $\widetilde{I}$ (see Figure 4). Hence δ_∞ contains a leaf L which is the image of a geodesic $\widetilde{L}$ joining ∞ to a point in I. Recall that the inverse image $\widetilde{|\nu^+|}$ of $|\nu^+|$ in $\mathbb{H}^2$ contains the boundary of the convex hull of $\mathrm{Fix}^+(\widetilde{\varphi}|_{\partial\mathbb{H}^2})$, where $\widetilde{\varphi}$ is the lift of φ such that $\mathrm{Fix}(\widetilde{\varphi}|_{\partial\mathbb{H}^2}) \supsetneqq \{\infty\}$ (see Proof of Lemma 3.5). Moreover, the interval $\widetilde{I}$ is contained in a component of $\partial\mathbb{H}^2 - \mathrm{Fix}(\widetilde{\varphi}|_{\partial\mathbb{H}^2})$. Hence, $\widetilde{L}$ must intersect $\widetilde{|\nu^+|}$ transversely, and therefore L intersects $|\nu^+|$ transversely (see Figure 4). However, this is impossible, because $\gamma_j \cap \delta_j = \emptyset$ and $|\nu^+| \subset \gamma_\infty$. Hence we obtain the desired result. $\qquad\square$

The standard frame at the origin of the Poincaré model of $\mathbb{H}^3$ determines a base frame ω_n of the hyperbolic manifold M_n ($n \in \mathbb{N} \cup \{\infty\}$). We may assume that (the origin of) ω_∞ lies in the compact submanifold K_∞. Since $\rho_n(\pi_1(F))$ converges to

$\rho_\infty(\pi_1(F))$ geometrically, there are smooth embeddings $f_n : K_\infty \to M_n$, defined for all n sufficiently large, such that f_n sends ω_∞ to ω_n and f_n tends to an isometry in the C^∞-topology. Namely, the lift $\tilde{f}_n : \tilde{K}_\infty \to \mathbb{H}^3$ of f_n to the inverse image $\tilde{K}_\infty$ of K_∞ in $\mathbb{H}^3$, sending the standard frame at the origin to itself, tends to the identity map in the compact-open C^∞-topology (see [BP92, Theorem E.1.13], [McM96, Section 2.2]). Since ρ_n converges to ρ_∞ algebraically, we may assume that the following diagram is commutative, where the vertical arrows represent homomorphisms induced by the inclusion maps and o_n denotes the origin of the frame ω_n for $n \in \mathbb{N} \cup \{\infty\}$:

$$
\begin{array}{ccc}
\pi_1(M_n, o_n) & \xleftarrow{\ \rho_n \circ \rho_\infty^{-1}\ } & \pi_1(M_\infty, o_\infty) \\[2mm]
\uparrow & & \uparrow \\[2mm]
(f_n)_*(\pi_1(K_\infty, o_\infty)) & \xleftarrow{\ (f_n)_*\ } & \pi_1(K_\infty, o_\infty)
\end{array}
$$

Fix a positive real number $r > 0$. We shall prove the following lemma in the next section.

Lemma 8.4. *There is a natural number N_1 which satisfies the following condition. For any $n \geq N_1$ and for any closed geodesic γ^* in M_∞ which lies in K_∞, the r-neighborhood of $f_n(\gamma^*)$ contains its geodesic representative in M_n.*

Recall that K_∞ contains the closed geodesic γ_{ρ_∞} for every $\gamma \in S(\vec{\Delta}_\varphi)$ (Lemma 8.2). We see from the commutative diagram that $f_n(\gamma_{\rho_\infty})$ has the geodesic γ_{ρ_n} as the geodesic representative. Hence the above lemma implies that γ_{ρ_n} is contained in the r-neighborhood of $f_n(\gamma_{\rho_\infty})$ in M_n for every $\gamma \in S(\vec{\Delta}_\varphi)$ and $n \geq N_1$.

Fix a positive real number $R > r$, and let K'_∞ be the closed R-neighborhood of K_∞ in M_∞. Then we may assume that $f_n : K_\infty \to M_n$ extends to an embedding of K'_∞ into M_n, which we continue to denote by f_n, such that the lift $\tilde{f}_n : \tilde{K}'_\infty \to \mathbb{H}^3$ tends to the identity map in the compact-open C^∞-topology. By the last condition, we can see that there is a natural number $N_2 \geq N_1$ such that the r-neighborhood of $f_n(K_\infty)$ in M_n is contained in $f_n(K'_\infty)$ for each $n \geq N_2$. Hence, the geodesic representative γ_{ρ_n} of $f_n(\gamma_{\rho_\infty})$ in M_n is contained in $f_n(K'_\infty)$ for every $\gamma \in S(\vec{\Delta}_\varphi)$ and $n \geq N_2$.

On the other hand, since $\tilde{f}_n$ tends to the identity map in the compact-open C^∞-topology and since K'_∞ is compact, there is a natural number $N_3 \geq N_2$ such that

$$
\frac{1}{2} \leq \frac{\|(f_n)_*(v)\|}{\|v\|} \leq 2
$$

for any $n \geq N_3$, $v \in T_x M_\infty - \{0\}$ and $x \in K'_\infty$. Hence we obtain the following inequalities for every $\gamma \in S(\vec{\Delta}_\varphi)$ and $n \geq N_3$.

$$l_{\rho_n}(\gamma) = l_{\rho_n}(\gamma_{\rho_n}) \leq l_{\rho_n}(f_n(\gamma_{\rho_\infty})) \leq 2l_{\rho_\infty}(\gamma_{\rho_\infty}) = 2l_{\rho_\infty}(\gamma),$$
$$l_{\rho_\infty}(\gamma) = l_{\rho_\infty}(\gamma_{\rho_\infty}) \leq l_{\rho_\infty}(f_n^{-1}(\gamma_{\rho_n})) \leq 2l_{\rho_n}(\gamma_{\rho_n}) = 2l_{\rho_n}(\gamma).$$

In the second inequality, we use the fact that $\gamma_{\rho_n} \subset f_n(K'_\infty)$. Since $\rho_n \in \mathcal{QF}$ for every $n \in \mathbb{N} \cup \{0\}$, the above inequality together with Lemma 5.2 implies the following.

Lemma 8.5. *There exists $k \geq 1$ such that*

$$\frac{1}{k}l_{\rho_0}(\gamma) \leq l_\rho(\gamma) \leq kl_{\rho_0}(\gamma)$$

for every $\rho \in \{\rho_n \mid n \geq 1\} \cup \{\rho_\infty\}$ and $\gamma \in S(\vec{\Delta}_\varphi)$.

Proof of Proposition 7.6. By using Lemma 8.5. we can show that $\sum_{\delta \in \vec{\Delta}_\varphi} h_\rho(\delta)$ converges absolutely and uniformly on $\{\rho_n \mid n \geq 1\} \cup \{\rho_\infty\}$ through an argument parallel to the proof of Proposition 5.1. Hence, for any $\varepsilon > 0$, there is a finite subset $\vec{\Delta}_{\varphi,\varepsilon}$ of $\vec{\Delta}_\varphi$ such that the following inequality holds for every $n \in \mathbb{N} \cup \{\infty\}$:

$$\left| \sum_{\delta \in \vec{\Delta}_\varphi - \vec{\Delta}_{\varphi,\varepsilon}} h_{\rho_n}(\delta) \right| < \varepsilon.$$

On the other hand, since ρ_n converges to ρ_∞ and since $\vec{\Delta}_{\varphi,\varepsilon}$ is a finite set, we have the following inequality for every sufficiently large n.

$$\left| \sum_{\delta \in \vec{\Delta}_{\varphi,\varepsilon}} h_{\rho_\infty}(\delta) - \sum_{\delta \in \vec{\Delta}_{\varphi,\varepsilon}} h_{\rho_n}(\delta) \right| < \varepsilon.$$

Hence

$$\left| \sum_{\delta \in \vec{\Delta}_\varphi} h_{\rho_\infty}(\delta) - \sum_{\delta \in \vec{\Delta}_\varphi} h_{\rho_n}(\delta) \right|$$

$$\leq \left| \sum_{\delta \in \vec{\Delta}_{\varphi,\varepsilon}} h_{\rho_\infty}(\delta) - \sum_{\delta \in \vec{\Delta}_{\varphi,\varepsilon}} h_{\rho_n}(\delta) \right| + \left| \sum_{\delta \in \vec{\Delta}_\varphi - \vec{\Delta}_{\varphi,\varepsilon}} h_{\rho_\infty}(\delta) \right| + \left| \sum_{\delta \in \vec{\Delta}_\varphi - \vec{\Delta}_{\varphi,\varepsilon}} h_{\rho_n}(\delta) \right|$$

$$\leq \varepsilon + \varepsilon + \varepsilon = 3\varepsilon.$$

This completes the proof of Proposition 7.6. □

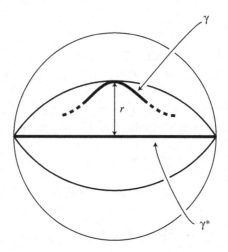

Figure 5: The Poincaré ball model for the hyperbolic space; the horizontal straight line segment represents the geodesic $\widetilde{\gamma}^*$, and the two circular arcs encircling $\widetilde{\gamma}^*$ represent the boundary of the r-neighborhood of $\widetilde{\gamma}^*$, to which $\widetilde{\gamma}$ tangents from the inside of the neighborhood.

9. Proof of Lemma 8.4

The key idea for the proof of Lemma 8.4 was brought to the authors by K. Bromberg: If a loop in a hyperbolic manifold is far away from its geodesic representative, then the "geodesic curvature" of the loop at the point where the distance attains the maximum should be large (see Figure 5).

The precise meaning of the idea is stated as follows.

Lemma 9.1. *Let $\gamma\colon \mathbb{R} \to \mathbb{H}^3$ be a smooth path and γ^* a complete geodesic in $\mathbb{H}^3$. Suppose the distance, $r(t)$, from the point $\gamma(t)$ to γ^* attains a local maximum at $t = t_0$. Then $\|(\nabla_{\dot\gamma}\dot\gamma)(t_0)\| \geq \|\dot\gamma(t_0)\|^2 \tanh r(t_0)$.*

Proof. We will use the cylindrical coordinates, $(r,\theta,z) \in \mathbb{R}_{\geq 0} \times S^1 \times \mathbb{R}$, around the geodesic γ^*; i.e., r is the distance from γ^*, θ represents the rotation around γ^*, and z represents the translation along γ^*. Then the hyperbolic metric at the point (r,θ,z) is of the form

$$dr^2 + \sinh^2 r d\theta^2 + \cosh^2 r dz^2.$$

Let $(r(t),\theta(t),z(t))$ be the expression of $\gamma(t)$ with respect to the cylindrical coordinates. Since $r(t)$ attains a local maximum at $t = t_0$, it follows that $\dot r(t_0) = 0$. Thus

$$\|\dot\gamma(t_0)\|^2 = \dot\theta(t_0)^2 \sinh^2 r(t_0) + \dot z(t_0)^2 \cosh^2 r(t_0)$$
$$\leq \{\dot\theta(t_0)^2 + \dot z(t_0)^2\} \cosh^2 r(t_0),$$

and hence

$$\dot{\theta}(t_0)^2 + \dot{z}(t_0)^2 \geq \frac{\|\dot{\gamma}(t_0)\|^2}{\cosh^2 r(t_0)}.$$

On the other hand, we obtain the following by a direct calculation:

$$\nabla_{\dot{\gamma}}\dot{\gamma} = \left(\ddot{r} - (\dot{\theta}^2 + \dot{z}^2)\sinh r \cosh r\right)\frac{\partial}{\partial r}$$
$$+ \left(\ddot{\theta} + 2\dot{r}\dot{\theta}\tanh r\right)\frac{\partial}{\partial \theta} + (\ddot{z} + 2\dot{r}\dot{z}\tanh r)\frac{\partial}{\partial z}.$$

Since $r(t)$ attains a local maximum at $t = t_0$, it follows that $\ddot{r}(t_0) \leq 0$. Thus

$$\|\nabla_{\dot{\gamma}}\dot{\gamma}(t_0)\| \geq \left|\ddot{r}(t_0) - (\dot{\theta}(t_0)^2 + \dot{z}(t_0)^2)\sinh r(t_0)\cosh r(t_0)\right|$$
$$= -\ddot{r}(t_0) + (\dot{\theta}(t_0)^2 + \dot{z}(t_0)^2)\sinh r(t_0)\cosh r(t_0)$$
$$\geq (\dot{\theta}(t_0)^2 + \dot{z}(t_0)^2)\sinh r(t_0)\cosh r(t_0)$$
$$\geq \frac{\|\dot{\gamma}(t_0)\|^2}{\cosh^2 r(t_0)}\sinh r(t_0)\cosh r(t_0) = \|\dot{\gamma}(t_0)\|^2\tanh r(t_0).$$

This completes the proof. □

Remark 9.2. The estimate in Lemma 9.1 is best possible. In fact, if $\gamma(t) = (r_0, \theta_0, t)$ in the cylindrical coordinate around the geodesic γ^* with constants $r_0 > 0$ and θ_0 (namely γ is equidistant from γ^* and lies in a hyperbolic plane containing γ^*), then the inequality is actually an equality.

Lemma 9.3. *Let $\gamma: \mathbb{R} \to M$ be a closed smooth curve in a hyperbolic 3-manifold M with geodesic representative γ^*. Suppose for some $0 < \varepsilon < 1$ and $r \in [0, \infty)$ that $\min\{\|\dot{\gamma}(t)\| \,|\, t \in \mathbb{R}\} > \varepsilon$ and that $\max\{\|(\nabla_{\dot{\gamma}}\dot{\gamma})(t)\| \,|\, t \in \mathbb{R}\} < \varepsilon^2\tanh r$. Then γ^* is contained in the r-neighborhood of γ.*

Proof. First, we will prove that γ is contained in the r-neighborhood of γ^*. Let $\tilde{\gamma}$ and $\tilde{\gamma}^*$, respectively, be lifts of γ and γ^* to $\mathbb{H}^3$, such that γ and γ^* share the same endpoints. Then they are invariant by a loxodromic transformation with axis $\tilde{\gamma}^*$, and hence $R := \max\{d(\tilde{\gamma}(t), \tilde{\gamma}^*) \,|\, t \in \mathbb{R}\}$ exists. Suppose that the maximum is attained at $t = t_0$. By Lemma 9.1, it follows that $\|(\nabla_{\dot{\tilde{\gamma}}}\dot{\tilde{\gamma}})(t_0)\| \geq \|\dot{\tilde{\gamma}}(t_0)\|^2\tanh R$. Then, by using the assumptions, we obtain:

$$\varepsilon^2\tanh r > \max\{\|(\nabla_{\dot{\gamma}}\dot{\gamma})(t)\| \,|\, t \in \mathbb{R}\} \geq \|(\nabla_{\dot{\tilde{\gamma}}}\dot{\tilde{\gamma}})(t_0)\|$$
$$\geq \|\dot{\tilde{\gamma}}(t_0)\|^2\tanh R\varepsilon^2\tanh R.$$

Hence $\tanh r > \tanh R$ and therefore $r > R$. So $\tilde{\gamma}$ is contained in the r-neighborhood of $\tilde{\gamma}^*$.

Next, we prove that γ^* is contained in the r-neighborhood of γ. Let x be any point of $\widetilde{\gamma}^*$ and Π the hyperplane of $\mathbb{H}^3$ which intersects $\widetilde{\gamma}^*$ perpendicularly at x. Since $\widetilde{\gamma}$ and $\widetilde{\gamma}^*$ have the same endpoints, some point, y, of $\widetilde{\gamma}$ is contained in Π. Notice that x is the nearest point in $\widetilde{\gamma}^*$ to y. Since $\widetilde{\gamma}$ is contained in the r-neighborhood of $\widetilde{\gamma}^*$, $d(x,y) < r$. Hence x is contained in the r-neighborhood of $\widetilde{\gamma}$. Thus $\widetilde{\gamma}^*$ is contained in the r-neighborhood of $\widetilde{\gamma}$, and therefore γ^* is contained in the r-neighborhood of γ. □

Proof of Lemma 8.4. Fix $r > 0$ and $0 < \varepsilon < 1$ arbitrarily. Since $\widetilde{f}_n$ converges to the identity map in the compact-open C^∞-topology, there is a natural number N_1 with the following property:

Let $\gamma^* : \mathbb{R} \to M_\infty$ be a closed geodesic in M_∞ which lies in the compact set K_∞ and is parametrized by the length. Let $\gamma_n : \mathbb{R} \to M_n$ be the smooth closed curve obtained as the composition $f_n \circ \gamma^*$. Then for any $n \geq N_1$, the following hold.

$$\min\{\|\dot{\gamma}_n(t)\| \,|\, t \in \mathbb{R}\} > \varepsilon, \quad \max\{\|(\nabla_{\dot{\gamma}_n}\dot{\gamma}_n)(t)\| \,|\, t \in \mathbb{R}\} < \varepsilon^2 \tanh r.$$

By Lemma 9.3, N_1 satisfies the desired property. This completes the proof. □

References

[AB60]   L. Ahlfors & L. Bers (1960). Riemann's mapping theorem for variable metrics. *Ann. of Math. (2)* **72**, 385–404.

[AMS04] H. Akiyoshi, H. Miyachi & M. Sakuma (2004). A refinement of mcshane's identity for quasifuchsian punctured torus groups. In *In the Tradition of Ahlfors and Bers, III*, number 355 in Contemp. Math., pp. 21–40.

[Bow96] B. H. Bowditch (1996). A proof of McShane's identity via Markoff triples. *Bull. London Math. Soc.* **28** (1), 73–78.

[Bow97] B. H. Bowditch (1997). A variation of McShane's identity for once-punctured torus bundles. *Topology* **36** (2), 325–334.

[BP92]   R. Benedetti & C. Petronio (1992). *Lectures on Hyperbolic Geometry.* Universitext. Springer-Verlag, Berlin.

[BS85]   J. S. Birman & C. Series (1985). Geodesics with bounded intersection number on surfaces are sparsely distributed. *Topology* **24** (2), 217–225.

[Can93] R. D. Canary (1993). Ends of hyperbolic 3-manifolds. *J. Amer. Math. Soc.* **6** (1), 1–35.

[CB88]   A. J. Casson & S. A. Bleiler (1988). *Automorphisms of Surfaces After Nielsen and Thurston.* Cambridge University Press, Cambridge.

[CEG87] R. D. Canary, D. B. A. Epstein & P. Green (1987). Notes on notes of Thurston. In *Analytical and Geometric Aspects of Hyperbolic Space (Coventry/Durham, 1984), London Math. Soc. Lecture Note Ser.,* volume 111, pp. 3–92. Cambridge Univ. Press, Cambridge.

[EM87]   D. B. A. Epstein & A. Marden (1987). Convex hulls in hyperbolic space, a theorem of Sullivan, and measured pleated surfaces. In *Analytical and Geometric Aspects of Hyperbolic Space (Coventry/Durham, 1984), London Math. Soc. Lecture Note Ser.,* volume 111, pp. 113–253. Cambridge Univ. Press, Cambridge.

[Gar87] F. P. Gardiner (1987). *Teichmüller Theory and Quadratic Differentials.* Pure and Applied Mathematics (New York). John Wiley & Sons Inc., New York.

[IT92]   Y. Imayoshi & M. Taniguchi (1992). *An Introduction to Teichmüller spaces.* Springer-Verlag, Tokyo.

[JM79]   T. Jørgensen & A. Marden (1979). Two doubly degenerate groups. *Quart. J. Math. Oxford Ser. (2)* **30** (118), 143–156.

[Kap01] M. Kapovich (2001). *Hyperbolic Manifolds and Discrete Groups, Progress in Mathematics,* volume 183. Birkhäuser Boston Inc., Boston, MA.

[KS93]   L. Keen & C. Series (1993). Pleating coordinates for the Maskit embedding of the Teichmüller space of punctured tori. *Topology* **32** (4), 719–749.

[KS04]   L. Keen & C. Series (2004). Pleating invariants for punctured torus groups. *Topology* **43** (2), 447–491.

[McM96] C. T. McMullen (1996). *Renormalization and 3-manifolds which Fiber over the Circle, Annals of Mathematics Studies,* volume 142. Princeton University Press, Princeton, NJ.

[McS91] G. McShane (1991). *A Remarkable Identity for Lengths of Curves.* Ph.D. thesis, University of Warwick.

[McS98] G. McShane (1998). Simple geodesics and a series constant over Teichmuller space. *Invent. Math.* **132** (3), 607–632.

[McS04] G. McShane (2004). Weierstrass points and simple geodesics. *Bull. London Math. Soc.* **36** (2), 181–187.

[Min00]   Y. Minsky (2000). Kleinian groups and the complex of curves. *Geometry and Topology* **4**, 117–148.

[Mir03a]   M. Mirzakhani (2003). Growth of the number of simple closed geodesics on hyperbolic surfaces. Preprint.

[Mir03b]   M. Mirzakhani (2003). Simple geodesics and Weil–Petersson volume of moduli spaces of bordered Riemann surfaces. Preprint.

[MS85]   A. Marden & K. Strebel (1985). On the ends of trajectories. In *Differential Geometry and Complex Analysis*, pp. 195–204. Springer, Berlin.

[MT98]   K. Matsuzaki & M. Taniguchi (1998). *Hyperbolic Manifolds and Kleinian Groups*. Oxford Mathematical Monographs. The Clarendon Press Oxford University Press, New York.

[Ota01]   J.-P. Otal (2001). *The Hyperbolization Theorem for Fibered 3-Manifolds*, *SMF/AMS Texts and Monographs*, volume 7. American Mathematical Society, Providence, RI.

[Tan]   S. P. Tan. Private communication, Newton Institute.

[Thu87]   W. P. Thurston (1987). Hyperbolic structures on 3-manifolds, II: Surface groups and 3-manifolds which fiber over the circle. Preprint; electronic version 1998, math.GT/9801045.

[TWZ04]   S. P. Tan, Y. L. Wong & Y. Zhang (2004). Generalizations of McShane's identity to hyperbolic cone-surfaces. Preprint.

Hirotaka Akiyoshi

Osaka City University Advanced
Mathematical Institute
Sugimoto, Sumiyoshi-ku
Osaka 558-8585
Japan

akiyoshi@sci.osaka-cu.ac.
jp

Hideki Miyachi

Department of Mathematical Sciences
Tokyo Denki University
Ishizaka, Hatoyama
Hiki, Saitama, 329-0349
Japan

miyachi@r.dendai.ac.jp

Makoto Sakuma

Department of Mathematics
Graduate School of Science
Osaka University
Machikaneyama-cho 1-16
Toyonaka, Osaka, 560-0043
Japan

sakuma@math.wani.osaka-u.ac.jp

AMS Classification: 58F17, 11J06, 30F40, 11D25, 32G15, 37F30, 57M50

Keywords: McShane's identity, complex translation length, quasifuchsian group,
surface bundle

Spaces of Kleinian Groups
Lond. Math. Soc. Lec. Notes **329**, 187–207

Cambridge University Press
Y. Minsky, M. Sakuma & C. Series (Eds.)

Train tracks and the Gromov boundary
of the complex of curves

Ursula Hamenstädt[1]

1. Introduction

Consider a compact oriented surface S of genus $g \geq 0$ from which $m \geq 0$ points, so-called *punctures*, have been deleted. We require that $3g - 3 + m \geq 2$; this rules out a sphere with at most 4 punctures and a torus with at most one puncture.

In [Har81], Harvey defined the *complex of curves* $C(S)$ for S. The vertices of this complex are free homotopy classes of essential simple closed curves on S, i.e. curves which are not freely homotopic into a puncture. The simplices in $C(S)$ are spanned by collections of such curves which can be realized disjointly. Thus the dimension of $C(S)$ equals $3g - 3 + m - 1$ (recall that $3g - 3 + m$ is the number of curves in a *pants decomposition* of S).

The *extended mapping class group* $\tilde{\mathcal{M}}_{g,m}$ of S is the group of all isotopy classes of homeomorphisms of S. It acts naturally on the complex of curves as a group of simplicial automorphisms. Even more is true: If S is not a torus with 2 punctures or a closed surface of genus 2, then the extended mapping class group is *precisely* the group of simplicial automorphisms of $C(S)$ (see [Iva02] for references and a sketch of the proof).

Providing each simplex in $C(S)$ with the standard euclidean metric of side-length 1 equips the complex of curves with the structure of a geodesic metric space whose isometry group is just $\tilde{\mathcal{M}}_{g,m}$ (except for the twice punctured torus and the closed surface of genus 2). However, this metric space is not locally compact. Masur and Minsky [MM99] showed that nevertheless the geometry of $C(S)$ can be understood quite explicitly. Namely, $C(S)$ is hyperbolic of infinite diameter. Recall that for some $\delta > 0$ a geodesic metric space is δ-*hyperbolic in the sense of Gromov* if it satisfies the δ-*thin triangle condition*: For every geodesic triangle with sides a, b, c the side c is contained in the δ-neighborhood of $a \cup b$. Later Bowditch [Bow02] gave a simplified proof of the result of Masur and Minsky which can also be used to compute explicit bounds for the hyperbolicity constant δ.

---

[1] Partially supported by Sonderforschungsbereich 611

A δ-hyperbolic geodesic metric space X admits a *Gromov boundary* which is defined as follows. Fix a point $p \in X$ and for two points $x, y \in X$ define the *Gromov product* $(x, y)_p = \frac{1}{2}(d(x, p) + d(y, p) - d(x, y))$. Call a sequence $(x_i) \subset X$ *admissible* if $(x_i, x_j)_p \to \infty$ $(i, j \to \infty)$. We define two admissible sequences $(x_i), (y_i) \subset X$ to be *equivalent* if $(x_i, y_i)_p \to \infty$. Since X is hyperbolic, this defines indeed an equivalence relation (see [BH99]). The Gromov boundary ∂X of X is then the set of equivalence classes of admissible sequences $(x_i) \subset X$. It carries a natural Hausdorff topology with the property that the isometry group of X acts on ∂X as a group of homeomorphisms. For the complex of curves, the Gromov boundary was determined by Klarreich [Kla99].

For the formulation of Klarreich's result, recall that a *geodesic lamination* for a complete hyperbolic structure of finite volume on S is a *compact* subset of S which is foliated into simple geodesics. A simple closed geodesic on S is a geodesic lamination with a single leaf. The space $\mathcal{L}$ of geodesic laminations on S can be equipped with the Hausdorff topology for compact subsets of S. With respect to this topology, $\mathcal{L}$ is compact and metrizable. A geodesic lamination is called *minimal* if each of its half-leaves is dense. A minimal geodesic lamination λ *fills up* S if every simple closed geodesic on S intersects λ transversely, i.e. if every complementary component of λ is an ideal polygon or a once punctured ideal polygon with geodesic boundary [CEG87].

A geodesic lamination is *maximal* if its complementary regions are all ideal triangles or once punctured monogons. Note that a geodesic lamination can be both minimal and maximal (this unfortunate terminology is by now standard in the literature). Each geodesic lamination λ is a *sublamination* of a maximal lamination, i.e. there is a maximal lamination which contains λ as a closed subset [CEG87]. For any minimal geodesic lamination λ which fills up S, the number of geodesic laminations which contain λ as a sublamination is bounded by a universal constant only depending on the topological type of the surface S. Namely, each such lamination μ can be obtained from λ by successively subdividing complementary components P of λ which are different from an ideal triangle or a once punctured monogon by adding a simple geodesic line which either connects two non-adjacent cusps of P or goes around a puncture in the interior of P. Note that every leaf of μ which is not contained in λ is necessarily isolated in μ.

We say that a sequence $(\lambda_i) \subset \mathcal{L}$ *converges in the coarse Hausdorff topology* to a minimal lamination μ which fills up S if every accumulation point of (λ_i) with respect to the Hausdorff topology contains μ as a sublamination. We equip the space $\mathcal{B}$ of minimal geodesic laminations which fill up S with the following topology. A set $A \subset \mathcal{B}$ is closed if and only if for every sequence $(\lambda_i) \subset A$ which converges in the coarse Hausdorff topology to a lamination $\lambda \in \mathcal{B}$ we have $\lambda \in A$. We call this topology on $\mathcal{B}$ the *coarse Hausdorff topology*. Using this terminology, Klarreich's result [Kla99] can be formulated as follows.

Theorem 1.1.

(i) *There is a natural homeomorphism Λ of $\mathcal{B}$ equipped with the coarse Hausdorff topology onto the Gromov boundary $\partial C(S)$ of the complex of curves $C(S)$ for S.*

(ii) *For $\mu \in \mathcal{B}$ a sequence $(c_i) \subset C(S)$ is admissible and defines the point $\Lambda(\mu) \in \partial C(S)$ if and only if (c_i) converges in the coarse Hausdorff topology to μ.*

In the paper [Kla99], Klarreich formulates her result using *measured foliations* on the surface S, i.e. topological foliations F on S equipped with a *transverse translation invariant measure*. The space $\mathcal{MF}$ of measured foliations can be equipped with the weak*-topology which is metrizable and hence Hausdorff. This topology projects to a metrizable topology on the space $\mathcal{PMF}$ of *projective measured foliations* which is the quotient of $\mathcal{MF}$ under the natural action of the positive half-line $(0, \infty)$. A topological foliation on S is called *minimal* if it does not contain a trajectory which is a simple closed curve. For every minimal topological foliation F, the set of projective measured foliations whose support equals F is a closed subset of $\mathcal{PMF}$. It follows that the quotient Q of the space of minimal projective measured foliations under the measure forgetting equivalence relation is a Hausdorff space as well. Note that the extended mapping class group of S acts on Q as a group of homeomorphisms. Klarreich shows that Q can naturally be identified with the Gromov boundary of the complex of curves.

There is a natural map ι which assigns to a measured foliation F on S a *measured geodesic lamination* $\iota(F)$, i.e. a geodesic lamination λ together with a transverse translation invariant measure supported in λ. The geodesic lamination λ is the closure of the set of geodesics which are obtained by straightening the non-singular trajectories of the foliation (see [Lev83] for details), together with the natural image of the transverse measure. A measured geodesic lamination can be viewed as a locally finite Borel measure on the space of unoriented geodesics in the hyperbolic plane which is invariant under the action of the fundamental group of S. Thus the space $\mathcal{ML}$ of measured geodesic laminations on S can be equipped with the restriction of the weak*-topology on the space of all such measures. With respect to this topology, the map ι is a homeomorphism of $\mathcal{MF}$ onto $\mathcal{ML}$ which factors to a homeomorphism of the space $\mathcal{PMF}$ of projective measured foliations onto the space $\mathcal{PML}$ of *projective measured laminations*, i.e. the quotient of $\mathcal{ML}$ under the natural action of $(0, \infty)$. This homeomorphism maps the space of minimal projective measured foliations onto the space $\mathcal{MPML}$ of projective measured geodesic laminations whose support is a minimal geodesic lamination which fills up S. Since every minimal geodesic lamination is the support of a transverse translation invariant measure (compare the expository article [Bon01] for a discussion of this fact and related results), the image of $\mathcal{MPML}$ under the natural forgetful map Π which assigns to a projective measured geodesic lamination its support equals the set $\mathcal{B}$. As a consequence, our above theorem is just a reformulation of the result of Klarreich provided that the coarse Hausdorff topology on $\mathcal{B}$ is induced from the weak*-topology on $\mathcal{MPML}$ via the surjective map Π.

For this it suffices to show that the map Π is continuous and closed. To show continuity, let $(\mu_i) \subset \mathcal{MPML}$ be a sequence of projective measured geodesic laminations. Assume that $\mu_i \to \mu \in \mathcal{MPML}$ in the weak*-topology, so that the support $\Pi(\mu)$ of μ is contained in $\mathcal{B}$. Since the space of geodesic laminations equipped with the Hausdorff topology is compact, up to passing to a subsequence we may assume that the laminations $\Pi(\mu_i) \in \mathcal{B}$ converge as $i \to \infty$ in the Hausdorff topology to a geodesic lamination $\tilde{\lambda}$. Then $\tilde{\lambda}$ necessarily contains the support $\Pi(\mu) \in \mathcal{B}$ of μ as a sublamination and therefore $\Pi(\mu_i) \to \Pi(\mu)$ in the coarse Hausdorff topology. Note however that $\tilde{\lambda}$ may contain isolated leaves which are not contained in the support of μ [CEG87]. Since $\mathcal{MPML}$ and $\mathcal{B}$ are Hausdorff spaces, this shows that the map Π is indeed continuous.

To show that the map Π is closed, let $A \subset \mathcal{MPML}$ be a closed set and let $(\mu_i) \subset A$ be a sequence with the property that $(\Pi(\mu_i)) \subset \mathcal{B}$ converges in the coarse Hausdorff topology to a lamination $\lambda \in \mathcal{B}$. Up to passing to a subsequence we may assume that the geodesic laminations $\Pi(\mu_i)$ converge in the usual Hausdorff topology to a lamination $\tilde{\lambda}$ containing λ as a sublamination. Since the space of projective measured laminations is compact, after passing to another subsequence we may assume that the projective measures μ_i converge in the weak*-topology to a projective measure μ. Then μ is necessarily supported in $\tilde{\lambda}$. Now λ fills up S by assumption and therefore every transverse measure on $\tilde{\lambda}$ is supported in λ. Thus we have $\mu \in \mathcal{MPML}$, $\lambda = \Pi(\mu)$ and, moreover, $\mu \in A$ since $A \subset \mathcal{MPML}$ is closed. This shows that Π is closed and consequently our theorem is just the main result of [Kla99].

Klarreich's proof of the above theorem relies on Teichmüller theory and the results of Masur and Minsky in [MM99]. In this note we give a more combinatorial proof which uses train tracks and a result of Bowditch [Bow02]. We discuss the relation between the complex of train tracks and the complex of curves in Section 2. The proof of the theorem is completed in Section 3.

2. The train track complex

A *train track* on S is an embedded 1-complex $\tau \subset S$ whose edges (called *branches*) are smooth arcs with well-defined tangent vectors at the endpoints. At any vertex (called a *switch*) the incident edges are mutually tangent. Through each switch there is a path of class C^1 which is embedded in τ and contains the switch in its interior. In particular, the branches which are incident on a fixed switch are divided into "incoming" and "outgoing" branches according to their inward pointing tangent at the switch. Each closed curve component of τ has a unique bivalent switch, and all other switches are at least trivalent. The complementary regions of the train track have negative Euler characteristic, which means that they are different from discs with $0, 1$ or 2 cusps at the boundary and different from annuli and once-punctured discs with no cusps at the boundary. We always identify train tracks which are isotopic. Train tracks were invented by Thurston to study the structure of the mapping class group. A detailed account on train tracks can be found in [PH92] and [Mos].

A train track is called *generic* if all switches are at most trivalent. The train track τ is called *transversely recurrent* if every branch b of τ is intersected by an embedded simple closed curve $c = c(b) \subset S$ which intersects τ transversely and is such that $S - \tau - c$ does not contain an embedded *bigon*, i.e. a disc with two corners at the boundary.

Recall that a geodesic lamination for a complete hyperbolic structure of finite volume on S is a compact subset of S which is foliated into simple geodesics. Particular geodesic laminations are simple closed geodesics, i.e. laminations which consist of a single leaf. A geodesic lamination λ is called minimal if each of its half-leaves is dense in λ. Thus a simple closed geodesic is a minimal geodesic lamination. A minimal geodesic lamination with more than one leaf has uncountably many leaves. Every geodesic lamination λ is a disjoint union of finitely many minimal components and a finite number of non-compact isolated leaves. Each of the isolated leaves of λ either is an isolated closed geodesic and hence a minimal component, or it *spirals* about one or two minimal components ([CEG87], [Ota96]).

A geodesic lamination λ is *maximal* if all its complementary components are ideal triangles or once punctured monogons. A geodesic lamination is called *complete* if it is maximal and can be approximated in the Hausdorff topology by simple closed geodesics. Every minimal geodesic lamination is a sublamination of a complete geodesic lamination [Ham04]. The space $\mathcal{CL}$ of complete geodesic laminations on S equipped with the Hausdorff topology is compact.

A geodesic lamination or a train track λ is *carried* by a transversely recurrent train track τ if there is a map $F : S \to S$ of class C^1 which is homotopic to the identity and maps λ to τ in such a way that the restriction of its differential dF to every tangent line of λ is non-singular. Note that this makes sense since a train track has a tangent line everywhere. A train track τ is called *complete* if it is generic and transversely recurrent and if it carries a complete geodesic lamination [Ham04].

A half-branch $\tilde{b}$ in a generic train track τ incident on a switch v is called *large* if every arc $\rho : (-\varepsilon, \varepsilon) \to \tau$ of class C^1 which passes through v meets the interior of $\tilde{b}$. A branch b in τ is called *large* if each of its two half-branches is large; in this case b is necessarily incident on two distinct switches, and it is large at both of them (for all this, see [PH92]).

There is a simple way to modify a complete train track τ to another complete train track. Namely, if e is a large branch of τ then we can perform a right or left *split* of τ at e as shown in Figure 1 below. The split τ' of a train track τ is carried by τ. If τ is complete and if $\lambda \in C\mathcal{L}$ is carried by τ, then for every large branch e of τ there is a unique choice of a right or left split of τ at e with the property that the split track τ' carries λ, and τ' is complete. In particular, a complete train track τ can always be split at any large branch e to a complete train track τ'; however there may be a choice of a right or left split at e such that the resulting track is not complete any more (compare p.120 in [PH92]).

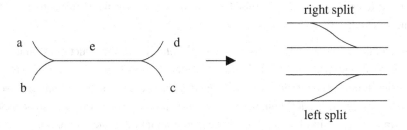

Figure 1

Let $\mathcal{TT}$ be the set of all isotopy classes of complete train tracks on S. We connect two train tracks τ, τ' with a directed edge if τ' can be obtained from τ by a single split at a large branch e. This provides $\mathcal{TT}$ with the structure of a locally finite directed metric graph. The *mapping class group* $\mathcal{M}_{g,m}$ of all isotopy classes of orientation preserving homeomorphisms of S acts naturally on $\mathcal{TT}$ as a group of simplicial isometries. The following result is shown in [Ham04].

Theorem 2.1. *The train track complex $\mathcal{TT}$ is connected, and the action of the mapping class group on $\mathcal{TT}$ is proper and cocompact.*

A *transverse measure* on a train track τ is a nonnegative weight function μ on the branches of τ satisfying the *switch condition*: For every switch s of τ, the sum of the weights over all incoming branches at s is required to coincide with the sum of the weights over all outgoing branches at s. The set $V(\tau)$ of all transverse measures on τ is a closed convex cone in a linear space and hence topologically it is a closed cell. The train track is called *recurrent* if it admits a transverse measure which is positive on every branch. We call such a transverse measure μ *positive*, and we write $\mu > 0$. A complete train track τ is recurrent [Ham04].

A transverse measure μ on τ is called a *vertex cycle* [MM99] if μ spans an extreme ray in $V(\tau)$. Up to scaling, every vertex cycle μ is a *counting measure* of a simple closed curve c which is carried by τ. This means that for a carrying map $F : c \rightarrow \tau$ and every open branch b of τ the μ-weight of τ equals the number of connected components of $F^{-1}(b)$. More generally, every integral transverse measure μ for τ defines uniquely a simple *weighted geodesic multicurve*, i.e there are simple closed pairwise disjoint geodesics $c_1, \ldots, c_\ell$ and a carrying map $F : \cup_i c_i \to \tau$ such that $\mu = \sum a_i \nu_i$ where $a_i > 0$ is a positive integer and where ν_i is the counting measure for c_i. We have.

Lemma 2.2. *Let c be a simple closed curve which is carried by τ, with carrying map $F : c \to \tau$. Then c defines a vertex cycle on τ only if $F(c)$ passes through every branch of τ at most twice, with different orientation.*

Proof. Let $F : c \to \tau$ be a carrying map for a simple closed curve $c : S^1 \to S$ which defines a vertex cycle μ for τ. Assume to the contrary that there is a branch b of τ with the property that Fc passes through b twice in the same direction. Then there is a closed nontrivial subarc $[p, q] \subset S^1$ with nontrivial complement such that $F \circ c[p, q]$ and $F \circ c[q, p]$ are closed (not necessarily simple) curves on τ. For a branch e of τ define $\nu(e)$ to be the number of components of $(F \circ c[p, q])^{-1}(e)$. Then ν is a nontrivial nonnegative integral weight function on the branches of τ which clearly satisfies the switch condition, and the same is true for $\mu - \nu$. As a consequence, the transverse measure μ can be decomposed into a nontrivial sum of integral transverse measures which contradicts our assumption that μ is a vertex cycle for τ. This proves the lemma. $\square$

In the sequel we mean by a vertex cycle of a complete train track τ an *integral* transverse measure on τ which is the counting measure of a simple closed curve c on S carried by τ and which spans an extreme ray of $V(\tau)$; we also use the notion vertex cycle for the simple closed curve c. As a consequence of Lemma 2.2 and the fact that the number of branches of a complete train track on S only depends on the topological type of S, the number of vertex cycles for a complete train track on S is bounded by a universal constant (see [MM99]).

Recall that the *intersection number* $i(\gamma, \delta)$ between two simple closed geodesics γ, δ equals the minimal number of intersection points between representatives of the free homotopy classes of γ, δ. This intersection number extends bilinearly to a pairing for weighted simple geodesic multicurves on S. The following corollary is immediate from Lemma 2.2. For its formulation, for a transverse measure μ on a train track τ denote by $\mu(\tau)$ the *total mass of* μ, i.e. $\mu(\tau) = \sum_b \mu(b)$ where b runs through the branches of τ. We have.

Corollary 2.3. *Let $\mu \in V(\tau)$ be an integral transverse measure on τ which defines the weighted simple geodesic multicurve c. Let ξ be any vertex cycle of τ; then $i(c, \xi) \leq 2\mu(\tau)$.*

Proof. Let c be any simple closed curve which is carried by the complete train track τ and denote by μ the counting measure on τ defined by c. Write $n = \mu(\tau)$; then there is a *trainpath* of length n, i.e. a C^1-immersion $\rho : [0,n] \to \tau$ which maps each interval $[i, i+1]$ onto a branch of τ and which parametrizes the image of c under a carrying map $c \to \tau$. We then can deform ρ with a smooth homotopy to a closed curve $\rho' : [0,n] \to S$ which is mapped to ρ by a carrying map and is such that for each $i \leq n$, $\rho'[i, i+1]$ intersects τ in at most one point contained in the interior of the branch $\rho[i, i+1]$.

Now let ξ be any vertex cycle of τ. By Lemma 2.2, ξ can be parametrized as a trainpath $\sigma : [0,s] \to \tau$ which passes through every branch of τ at most twice. Then the number of intersection points between σ and ρ' is not bigger than $2n = 2\mu(\tau)$. This shows the corollary for simple closed curves c which are carried by τ. The case of a general weighted simple geodesic multicurve carried by τ then follows from linearity of counting measures and the intersection form. $\qquad\square$

Since the distance in $C(S)$ between two simple closed curves a, c is bounded from above by $2i(a,c) + 1$ [MM99], we obtain from Lemma 2.2 and Corollary 2.3 the existence of a number $D > 0$ with the property that for every train track $\tau \in \mathcal{TT}$, the distance in $C(S)$ between any two vertex cycles of τ is at most D.

Define a map $\Phi : \mathcal{TT} \to C(S)$ by assigning to a train track $\tau \in \mathcal{TT}$ a vertex cycle $\Phi(\tau)$ for τ. Every such map is roughly $\mathcal{M}_{g,m}$-equivariant. Namely, for $\psi \in \mathcal{M}_{g,m}$ and $\tau \in \mathcal{TT}$, the distance between $\Phi(\psi(\tau))$ and $\psi(\Phi(\tau))$ is at most D. Denote by d both the distance on $\mathcal{TT}$ and on $C(S)$. We have.

Corollary 2.4. *There is a number $C > 0$ such that $d(\Phi(\tau), \Phi(\eta)) \leq Cd(\tau, \eta)$ for all $\tau, \eta \in \mathcal{TT}$.*

Proof. Let $\alpha : [0,m] \to \mathcal{TT}$ be any (simplicial) geodesic. Then for each i, either the train track $\alpha(i+1)$ is obtained from $\alpha(i)$ by a single split or $\alpha(i)$ is obtained from $\alpha(i+1)$ by a single split. Assume that $\alpha(i+1)$ is obtained from $\alpha(i)$ by a single split. Then there is a natural carrying map $F : \alpha(i+1) \to \alpha(i)$. By Lemma 2.2 and the definition of a split, via this carrying map the counting measure of a vertex cycle c on $\alpha(i+1)$ defines an integral transverse measure on $\alpha(i)$ whose total mass is bounded from above by a universal constant. Thus by Corollary 2.3, the intersection number between c and any vertex cycle of $\alpha(i)$ is bounded from above by a universal constant. Then the distance in $C(S)$ between c and any vertex cycle on $\alpha(i)$ is uniformly bounded as well [MM99]. This shows the corollary. $\qquad\square$

Define a *splitting sequence* in $\mathcal{TT}$ to be a (simplicial) map $\alpha : [0,m] \to \mathcal{TT}$ with the property that for each i the train track $\alpha(i+1)$ can be obtained from $\alpha(i)$ by a single split.

We use now a construction of Bowditch [Bow02]. Recall the definition of the intersection form i on simple geodesic multicurves. For simple geodesic multicurves

α, β on S with $i(\alpha, \beta) > 0$ and $a > 0, r > 0$ define

$$L_a(\alpha, \beta, r) = \{\gamma \in C(S) \mid \max\{ai(\gamma, \alpha), i(\gamma, \beta)/ai(\alpha, \beta)\} \leq r\}.$$

Our next goal is to link the sets $L_a(\alpha, \beta, r)$ to splitting sequences. For this recall that a *pants decomposition* of S is a collection of $3g - 3 + m$ pairwise disjoint mutually not freely homotopic simple closed essential curves on S, i.e. these curves are not contractible and not freely homotopic into a puncture. Let $P = \{\gamma_1, \ldots, \gamma_{3g-3+m}\}$ be a pants decomposition for S. Then there is a special family of complete train tracks with the property that for a train track τ in this family, each pants curve γ_i admits a closed neighborhood A diffeomorphic to an annulus and such that $\tau \cap A$ is diffeomorphic to a *standard twist connector* depicted in Figure 2. Such a train track clearly carries each pants curve from the pants decomposition P; we call it *adapted* to P (see [PH92]). The set of train tracks adapted to a pants decomposition os S is invariant under the action of $\mathcal{M}_{g,m}$. We show.

 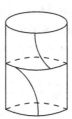

Figure 2

Lemma 2.5. *There is a number $k \geq 1$ with the following property. Let $\tau_0 \in \mathcal{TT}$ be adapted to a pants decomposition P of S, let $(\tau_i)_{0 \leq i \leq m} \subset \mathcal{TT}$ be a splitting sequence issuing from τ_0 and let α be a simple multicurve consisting of vertex cycles for τ_m. Then there is a monotonous surjective function $\kappa : (0, \infty) \to \{0, \ldots, m\}$ such that $\kappa(s) = 0$ for all sufficiently small $s > 0, \kappa(s) = m$ for all sufficiently large $s > 0$ and that for all $s \in (0, \infty)$ there is a vertex cycle of $\tau_{\kappa(s)}$ which is contained in $L_s(\alpha, P, k)$.*

Proof. Let P be a pants decomposition for S and let β be an arbitrary simple multicurve on S. Let $k > 1$ and assume that there is a curve $\gamma \in C(S)$ with the property that $0 < c = i(P, \gamma)i(\gamma, \beta) \leq ki(P, \beta)$. Write $b = i(P, \gamma)/i(P, \beta), a = i(\gamma, \beta)/c$; then $abi(P, \beta) = 1$ and $\max\{ai(P, \gamma), bi(\beta, \gamma)\} \leq k$. As a consequence, we have $\gamma \in L_a(P, \beta, k)$. Thus for the proof of our lemma we only have to show the existence of a number $k > 0$ with the following property. Let ζ be a train track which is adapted to a pants decomposition $P = \{\gamma_1, \ldots, \gamma_{3g-3+m}\}$ and let $\zeta : [0, m] \to \mathcal{TT}$ be a splitting sequence issuing from $\zeta(0) = \zeta$. Let $j > 0$ be such that the distance in $C(S)$ between every vertex cycle of $\zeta(j)$ and every vertex cycle of $\zeta(0)$ is at least 3. Let ρ be a vertex cycle for $\zeta(m)$; then there is a vertex cycle $\alpha(j)$ for $\zeta(j)$ such that

$$i(\rho,\alpha(j))\Big(\sum_{i=1}^{3g-3+m} i(\alpha(j),\gamma_i)\Big) \leq k \sum_{i=1}^{3g-3+m} i(\rho,\gamma_i). \qquad (2.1)$$

Since $\zeta(0)$ is adapted to the pants decomposition P, every pants curve of P is a vertex cycle for $\zeta(0)$. Moreover, for each $i \leq 3g-3+m$ there is a branch b_i of $\zeta(0)$ contained in an annulus A_i about γ_i and such that the counting measure v_i for γ_i is the *unique* vertex cycle of $\zeta(0)$ which gives positive mass to b_i. Thus the counting measure μ of any simple closed curve c which is carried by $\zeta(0)$ can be decomposed in a unique way as $\mu = \mu_0 + \sum_{i=1}^{3g-3+m} n_i v_i$ where $n_i \geq 0$ and where μ_0 is an integral transverse measure for $\zeta(0)$ with $\mu_0(b_i) = 0$ for all i. The intersection number of the curve c with a pants curve γ_i equals the μ_0-weight of the *large* branch e_i of $\zeta(0)$ contained in the annulus A_i. In particular, the intersection number of c and γ_i coincides with the intersection number of γ_i and the simple weighted multicurve c_0 defined by the transverse measure μ_0. Moreover, since the complement of P in S does not contain any essential closed curve which is not homotopic into a boundary component or a cusp, there is a constant k_0 only depending on the topological type of S with the property that

$$\mu_0(\zeta(0)) \geq \sum_{i=1}^{3g-3+m} i(c,\gamma_i) = \sum_{i=1}^{3g-3+m} i(c_0,\gamma_i) \geq \mu_0(\zeta(0))/k_0. \qquad (2.2)$$

Consider again the splitting sequence $\zeta : [0,m] \to \mathcal{T}\mathcal{T}$ and let $j \leq m$ be such that the distance in $C(S)$ between every vertex cycle of $\zeta(j)$ and every vertex cycle of $\zeta(0)$ is at least 3. Let ρ be a vertex cycle for the train track $\zeta(m)$. Since $\zeta(m)$ is carried by $\zeta(j)$, the curve ρ defines a counting measure η on $\zeta(j)$. This counting measure can (perhaps non-uniquely) be written in the form $\eta = \sum_{i=1}^{d} a_i \xi_i$ where ξ_i $(i=1,\ldots,d)$ are the vertex cycles of $\zeta(j)$ and $a_i \geq 0$ are nonnegative integers. The number d of these vertex cycles is bounded from above by a universal constant and by Lemma 2.2, the total mass of each of these vertex cycles ξ_i is bounded from above by a universal constant as well. Therefore, there is a universal number $q > 0$ and there is some $i \leq d$ such that $a_i \geq \eta(\zeta(j))/q$. After reordering we may assume that $i = 1$. Write $\xi = \xi_1$; Corollary 2.3 shows that $i(\rho,\xi) \leq 2\eta(\zeta(j)) \leq 2qa_1$.

On the other hand, by our assumption on $\zeta(j)$ the distance in $C(S)$ between ξ and each of the curves γ_i is at least 3. Thus ξ is mapped via the carrying map $\zeta(j) \to \zeta(0)$ to a curve in $\zeta(0)$ which together with each of the pants curves of P fills up S. Then ξ defines a counting measure μ on $\zeta(0)$, and we have $\mu = \mu_0 + \sum_{i=1}^{3g-3+m} p_i v_i$ for some $p_i \geq 0$ with $\mu_0 \neq 0$. By inequality (2.2), the sum of the intersection numbers between ξ and the curves γ_i is contained in the interval $[\mu_0(\zeta(0))/k_0, \mu_0(\zeta(0))]$. On the other hand, by the choice of ξ and the fact that the carrying map $\zeta(j) \to \zeta(0)$ maps the convex cone $V(\zeta(j))$ of transverse measures on $\zeta(j)$ *linearly* into the convex cone of transverse measures on $\zeta(0)$, the counting measure for our curve ρ viewed as a curve which is carried by $\zeta(0)$ is of the form $a_1\mu + \mu'$; in particular, we have

$$\sum_i i(\rho, \gamma_i) \geq a_1\mu_0(\zeta(0))/k_0 \geq \eta(\zeta(j))\mu_0(\zeta(0))/qk_0 \geq i(\rho, \xi)\mu_0(\zeta(0))/2qk_0. \quad (2.3)$$

As a consequence of inequalities (2.2), (2.3) we have

$$i(\rho, \xi)\sum_i i(\xi, \gamma_i) \leq 2qk_0\sum_i i(\rho, \gamma_i). \quad (2.4)$$

This completes the proof of the lemma. □

For any metric space (X, d) and any $L \geq 1$, a curve $\gamma : (a, b) \to X$ is called an *L-quasigeodesic* if for all $a < s < t < b$ we have

$$d(\gamma(s), \gamma(t))/L - L \leq t - s \leq Ld(\gamma(s), \gamma(t)) + L.$$

Since $C(S)$ is a δ-hyperbolic geodesic metric space for some $\delta > 0$, every L-quasigeodesic of finite length is contained in a uniformly bounded neighborhood of a geodesic in $C(S)$. Call a path $\gamma : [0, m] \to C(S)$ an *unparametrized L-quasigeodesic* if there is some $s > 0$ and a homeomorphism $\sigma : [0, s] \to [0, m]$ such that the path $\gamma \circ \sigma : [0, s] \to C(S)$ is an L-quasigeodesic. The image of every unparametrized L-quasigeodesic in $C(S)$ of finite length is contained in a uniformly bounded neighborhood of a geodesic.

The following corollary is the key step toward the investigation of the Gromov boundary of $C(S)$. It was first shown by Masur and Minsky [MM04], with a different proof.

Corollary 2.6. *There is a number $Q > 0$ such that the image under Φ of every splitting sequence in $\mathcal{TT}$ is an unparametrized Q-quasigeodesic.*

Proof. Recall the definition of the sets $L_a(\alpha, \beta, r)$ for $\alpha, \beta \in C(S)$. Bowditch [Bow02] showed that there is a number $r_0 > 0$ with the following property. Assume that $\alpha, \beta \in C(S)$ *fill up* S, i.e. the distance $d(\alpha, \beta)$ between α and β in $C(S)$ is at least 3; then we have.

(i) $L_a(\alpha, \beta, r_0) \neq \emptyset$ for all $a > 0$.

(ii) For every $r > 0$, $a > 0$ the diameter of $L_a(\alpha, \beta, r)$ is bounded from above by a universal constant only depending on r.

(iii) For $r > r_0$ there is a constant $q(r) > 0$ with the following property. For $a > 0$ choose some $\gamma(a) \in L_a(\alpha, \beta, r)$; then $\gamma : (0, \infty) \to C(S)$ is an unparametrized $q(r)$-quasigeodesic with $d(\gamma(s), \alpha) \leq q(r)$ for all sufficiently large $s > 0$ and $d(\gamma(s), \beta) \leq q(r)$ for all sufficiently small $s > 0$.

Let again $\alpha, \beta \in C(S)$ be such that α, β fill up S. For $r > r_0$ define

$$L(\alpha, \beta, r) = \cup_a L_a(\alpha, \beta, r).$$

By property (iii) above and hyperbolicity of the complex of curves, there is a number $D(r) > 0$ only depending on r such that $L(\alpha, \beta, r)$ is contained in a tubular neighborhood of radius $D(r)$ about a geodesic connecting α to β.

Now let P be any pants decomposition for S containing the curve α and assume that $\gamma \in L_a(P, \beta, r)$ for some $r > 0$. Let α' be a pants curve of P so that $i(\alpha', \beta) = \max\{i(v, \beta) \mid v \in P\}$; then we have $\gamma \in L_a(\alpha', \beta, (3g - 3 + m)r)$. As a consequence of this, hyperbolicity of $C(S)$ and Lemma 2.5, the image under Φ of the splitting sequence ζ, connecting a train track $\zeta(0)$ adapted to P to a train track $\zeta(m)$ which carries β as a vertex cycle, is contained in a uniformly bounded neighborhood of any geodesic in $C(S)$ connecting α to β. Since this consideration applies to *every* splitting sequence, "backtracking" of the assignment $j \to \Phi(\zeta(j))$ is excluded. From this the lemma is immediate.   □

Remark 2.7. More generally, the proof of Corollary 2.6 also shows the following. Let $\zeta, \eta \in \mathcal{TT}$ and assume that η is carried by ζ. Let c be any simple closed curve which is carried by η; then $\Phi(\eta)$ is contained in a uniformly bounded neighborhood of a geodesic arc in $C(S)$ connecting $\Phi(\zeta)$ to c.

3. Proof of the theorem

Fix again a complete hyperbolic metric on S of finite volume. Recall that a *measured geodesic lamination* on S is a geodesic lamination equipped with a transverse translation invariant measure. As in the introduction we equip the space $\mathcal{ML}$ of measured geodesic laminations with the restriction of the weak*-topology. The Dirac mass on any simple closed geodesic c on S defines a measured geodesic lamination. The intersection of weighted simple geodesic multicurves extends to a continuous symmetric bilinear form i on $\mathcal{ML}$ which is called the *intersection form*. The support of a measured geodesic lamination μ for S is minimal and fills up S if and only if $i(\mu, v) > 0$ for every measured geodesic lamination v on S whose support does not coincide with the support of μ. The space $\mathcal{PML}$ of projective measured laminations on S is the quotient of $\mathcal{ML}$ under the natural action of the multiplicative group $(0, \infty)$; it is homeomorphic to a sphere of dimension $6g - 6 + 2m - 1$ [FLP79], in particular, it is compact.

The complex of curves naturally embeds into $\mathcal{PML}$ by assigning to a simple closed geodesic its projectivized transverse Dirac mass.

Projective measured geodesic laminations can be used to study infinite sequences in the complex of curves. Denote again by d the distance on $C(S)$. We have the following.

Lemma 3.1. *Let $(c_i) \subset C(S)$ be a sequence which converges in $\mathcal{PML}$ to a projective measured lamination whose support λ_0 is minimal and fills up S. Let $k > 0$ and assume that $a_i \in C(S)$ is such that $d(a_i, c_i) \leq k$; then up to passing to a subsequence, the sequence (a_i) converges in $\mathcal{PML}$ to a projective measured geodesic lamination supported in λ_0.*

Proof. We use an argument of Luo as explained in the proof of Proposition 4.6 of [MM99]. Namely, choose a continuous section $\iota : \mathcal{PML} \to \mathcal{ML} - \{0\}$ of the projection $\mathcal{ML} - \{0\} \to \mathcal{PML}$. Then every simple closed geodesic c on S defines a measured geodesic lamination $\hat{c} \in \iota(\mathcal{PML})$. Let $(c_i) \subset C(S)$ be a sequence of simple closed geodesics. Assume that the sequence $(\hat{c}_i)$ converges in $\iota(\mathcal{PML})$ to a measured geodesic lamination ν_0 whose support λ_0 is minimal and fills up S.

Let $(a_i) \subset C(S)$ be a sequence with $d(a_i, c_i) \leq k$ for a fixed number $k > 0$. By passing to a subsequence we may assume that $d(c_i, a_i)$ is independent of i, i.e. we may assume that $d(c_i, a_i) = k$ for all i. Then for each i there is a curve $c_i^1 \in C(S)$ which is disjoint from c_i and such that $d(c_i^1, a_i) = k - 1$. Up to passing to a subsequence, the sequence $(\hat{c}_i^1) \subset \iota(\mathcal{PML})$ converges weakly to a measured geodesic lamination $\nu_1 \in \iota(\mathcal{PML})$. Since $i(\hat{c}_i^1, \hat{c}_i) = 0$ for all i, by continuity of the intersection form we have $i(\nu_0, \nu_1) = 0$ and therefore ν_1 is supported in λ_0. Proceeding inductively we conclude that up to passing to a subsequence, the measured laminations $\hat{a}_i$ defined by the curves a_i converge in $\iota(\mathcal{PML})$ to a measured lamination which is supported in λ_0. This shows the lemma. $\qquad\square$

Consider again the train track complex $\mathcal{TT}$. For $\tau \in \mathcal{TT}$ denote by $A(\tau) \subset \mathcal{CL}$ the set of all complete geodesic laminations carried by τ. Then $A(\tau)$ is open and closed in $\mathcal{CL}$. Following [Ham04], define a *full splitting sequence* in $\mathcal{TT}$ to be a sequence $\alpha : [0, \infty) \to \mathcal{TT}$ with the property that for every $i \geq 0$, the train track $\alpha(i + 1)$ is obtained by splitting $\alpha(i)$ at each of the large branches precisely once. If $\tau \in \mathcal{TT}$ is arbitrary and if $\lambda \in \mathcal{CL}$ is a complete geodesic lamination which is carried by τ, then λ determines uniquely a full splitting sequence $\alpha_{\tau, \lambda}$ issuing from τ by requiring that each of the train tracks $\alpha_{\tau, \lambda}(i)$ carries λ, and $\cap_i A(\alpha_{\tau, \lambda}(i)) = \{\lambda\}$ [Ham04]. Recall the definition of the map $\Phi : \mathcal{TT} \to C(S)$. By Corollary 2.6, there is a universal number $Q > 0$ such that the curve $i \to \Phi(\alpha_{\tau, \lambda}(i))$ is an unparametrized Q-quasigeodesic in $C(S)$. This means that this curve defines a quasiisometric embedding of the half-line $[0, \infty)$ into $C(S)$ if and only if the diameter in $C(S)$ of the set $\Phi(\alpha_{\tau, \lambda}[0, \infty))$ is infinite.

Let $\mathcal{B}$ be the set of all minimal geodesic laminations on S which fill up S, equipped with the coarse Hausdorff topology. Recall that $\mathcal{B}$ is a Hausdorff space. The next statement is immediate from Lemma 3.1.

Corollary 3.2. *Let $\lambda \in CL$ be a complete geodesic lamination which contains a sublamination $\lambda_0 \in \mathcal{B}$. Let $\tau \in TT$ be a train track which carries λ; then the diameter of the set $\Phi(\alpha_{\tau,\lambda}[0,\infty)) \subset C(S)$ is infinite.*

Proof. Let $\lambda \in CL$ be a complete geodesic lamination which contains a sublamination $\lambda_0 \in \mathcal{B}$. Assume that λ is carried by a train track $\tau \in TT$. Denote by $\alpha_\lambda = \alpha_{\tau,\lambda}$ the full splitting sequence issuing from τ which is determined by λ. We have to show that the diameter of the set $\Phi(\alpha_\lambda[0,\infty))$ is infinite. For this recall that $\cap_i A(\alpha_\lambda(i)) = \{\lambda\}$. Since for each i the curve $\Phi(\alpha_\lambda(i))$ is carried by $\alpha_\lambda(i)$, the curves $\Phi(\alpha_\lambda(i))$ viewed as projective measured laminations converge up to passing to a subsequence as $i \to \infty$ in PML to a projective measured geodesic lamination which is supported in λ_0. Thus by Lemma 3.1, there is no curve $a \in C(S)$ with $d(\Phi(\alpha_\lambda(i)),a) \leq k$ for a fixed number $k > 0$ and all i and hence the diameter in $C(S)$ of the set $\Phi(\alpha_\lambda[0,\infty))$ is indeed infinite. $\qquad\square$

As in the introduction, we call a sequence $(c_i) \subset C(S)$ admissible if for a fixed $p \in C(S)$ we have $(c_i,c_j)_p \to \infty\ (i,j \to \infty)$. Two admissible sequences $(a_i),(c_i) \subset C(S)$ are equivalent if $(a_i,c_i)_p \to \infty(i \to \infty)$. The Gromov boundary $\partial C(S)$ of $C(S)$ is the set of equivalence classes of admissible sequences in $C(S)$. Note that any quasigeodesic ray in $C(S)$ defines an admissible sequence. We use Corollary 3.2 to show the following.

Lemma 3.3. *There is an injective map $\Lambda : \mathcal{B} \to \partial C(S)$.*

Proof. Fix a pants decomposition P of S. Then there is a finite collection $\tau_1,\ldots,\tau_\ell \subset TT$ of train tracks adapted to P with the property that every complete geodesic lamination $\lambda \in CL$ is carried by one of the tracks τ_i (see [PH92], [Ham04]). Let $\mathcal{A} \subset CL$ be the set of all complete geodesic laminations which contain a sublamination $\lambda_0 \in \mathcal{B}$. For $\lambda \in \mathcal{A}$ let τ_j be a train track from our collection $\tau_1,\ldots,\tau_\ell$ which carries λ and let $\alpha_\lambda : [0,\infty) \to TT$ be the full splitting sequence issuing from τ_j which is determined by λ. By Corollary 2.6 and Corollary 3.2, there is a universal number $Q > 0$ with the property that the curve $i \to \Phi(\alpha_\lambda(i))$ is an unparametrized Q-quasigeodesic of infinite diameter. Hence this curve defines a point $\tilde{\Lambda}(\lambda) \in \partial C(S)$.

There is a natural continuous projection $\pi : \mathcal{A} \to \mathcal{B}$ which maps a lamination $\lambda \in \mathcal{A}$ to its unique minimal sublamination $\pi(\lambda) \in \mathcal{B}$. We claim that $\tilde{\Lambda}(\lambda) = \tilde{\Lambda}(\mu)$ for $\lambda,\mu \in \mathcal{A}$ if $\pi(\lambda) = \pi(\mu) = \lambda_0$. For this extend the map Φ to the collection of all recurrent train tracks on S by assigning to such a train track σ a vertex cycle $\Phi(\sigma)$ of σ. Since the minimal sublamination $\lambda_0 = \pi(\lambda)$ of λ fills up S and is carried by each of the train tracks $\alpha_\lambda(i)$, the image of λ_0 under a carrying map $\lambda \to \alpha_\lambda(i)$ is a *recurrent* subtrack $\hat{\alpha}_\lambda(i)$ of $\alpha_\lambda(i)$ which is *large*. This means that $\hat{\alpha}_\lambda(i)$ is a train track on S which is a subset of $\alpha_\lambda(i)$ and whose complementary components do not contain an essential simple closed curve which is not homotopic into a puncture. By Lemma 2.2 and Corollary 2.3, the distance in $C(S)$ between $\Phi(\alpha_\lambda(i))$ and $\Phi(\hat{\alpha}_\lambda(i))$ is bounded by a universal constant.

Up to isotopy, the train tracks $\hat{\alpha}_\lambda(i)$ converge as $i \to \infty$ in the Hausdorff topology to the lamination λ_0 (see [Mos],[Ham04]). Since λ_0 is a sublamination of μ, for every $i > 0$ there is a number $j(i) > 0$ such that the train track $\hat{\alpha}_\lambda(j(i))$ is carried by $\alpha_\mu(i)$ (see [Ham04]). By the remark following the proof of Corollary 2.6, this implies that $\Phi(\alpha_\mu(i))$ is contained in a uniformly bounded neighborhood of $\Phi(\alpha_\lambda[0,\infty))$. Since $i \geq 0$ was arbitrary, the Hausdorff distance between the Q-quasigeodesics in $C(S)$ defined by λ, μ is bounded and hence we have $\tilde{\Lambda}(\lambda) = \tilde{\Lambda}(\mu)$ as claimed. Thus there is a map $\Lambda : \mathcal{B} \to \partial C(S)$ such that $\tilde{\Lambda} = \Lambda \circ \pi$.

We claim that the map Λ is injective. For this let $\lambda_0 \neq \mu_0 \in \mathcal{B}$ and let $\lambda \in \pi^{-1}(\lambda_0) \subset \mathcal{A}, \mu \in \pi^{-1}(\mu_0) \subset \mathcal{A}$. By Corollary 2.6 and Corollary 3.2, the image under Φ of full splitting sequences $\alpha_\lambda, \alpha_\mu \in \mathcal{TT}$ determined by λ, μ are unparametrized Q-quasigeodesics in $C(S)$ of infinite diameter. Thus by the definition of Λ, we have $\Lambda(\lambda_0) = \Lambda(\mu_0)$ only if the Hausdorff distance between $\Phi(\alpha_\lambda[0,\infty))$ and $\Phi(\alpha_\mu[0,\infty))$ is finite.

Assume to the contrary that this is the case. Then there is a number $D > 0$ and for every $i > 0$ there is a number $j(i) > 0$ such that $d(\Phi(\alpha_\lambda(i)), \Phi(\alpha_\mu(j(i)))) \leq D$. Since $d(\Phi(\alpha_\lambda(0)), \Phi(\alpha_\lambda(i))) \to \infty$ we have $j(i) \to \infty$ $(i \to \infty)$ by Corollary 2.4. Therefore, up to passing to a subsequence, the curves $\Phi(\alpha_\lambda(i))$, $\Phi(\alpha_\mu(j(i)))$ viewed as projective measured geodesic laminations converge as $i \to \infty$ to projective measured geodesic laminations ν_0, ν_1 supported in λ_0, μ_0. But λ_0, μ_0 fill up S and do not coincide and hence this contradicts Lemma 3.1. $\qquad\square$

The Gromov boundary $\partial C(S)$ of $C(S)$ admits a natural Hausdorff topology which can be described as follows. Extend the Gromov product $(,)_p$ to a product on $\partial C(S)$ by defining $(\xi, \zeta)_p = \sup_{(x_i),(y_j)} \liminf_{i,j \to \infty} (x_i, y_j)_p$ where the supremum is taken over all admissible sequences representing the points ξ, ζ. We have $(\xi, \zeta)_p = \infty$ if and only if $\xi = \zeta \in \partial C(S)$. A subset U of $\partial C(S)$ is a neighborhood of a point $\xi \in \partial C(S)$ if and only if there is a number $\varepsilon > 0$ such that $\{\zeta \in \partial C(S) \mid e^{-(\xi,\zeta)_p} < \varepsilon\} \subset U$ (compare [BH99]).

We say that a sequence $(c_i) \subset C(S)$ *converges in the coarse Hausdorff topology* to a lamination $\mu \in \mathcal{B}$ if every accumulation point of (c_i) with respect to the Hausdorff topology contains μ as a sublamination. The next lemma completes the proof of our theorem from the introduction.

Lemma 3.4.

(i) *The map* $\Lambda : \mathcal{B} \to \partial C(S)$ *is a homeomorphism.*

(ii) *For* $\mu \in \mathcal{B}$, *a sequence* $(c_i) \subset C(S)$ *is admissible and defines the point* $\Lambda(\mu) \in \partial C(S)$ *if and only if* $c_i \to \mu$ *in the coarse Hausdorff topology.*

Proof. We show first the following. Let $(c_i) \subset C(S)$ be an admissible sequence, i.e. a sequence with the property that $(c_i, c_j)_p \to \infty$ $(i, j \to \infty)$. Then there is some $\lambda_0 \in \mathcal{B}$ such that (c_i) converges in the coarse Hausdorff topology to λ_0.

For this we first claim that there is a number $b > 0$ and an admissible sequence $(a_j) \subset C(S)$ which is equivalent to (c_i) (i.e. which satisfies $(a_i, c_i)_p \to \infty$) and such that the assignment $j \to a_j$ is a b-quasigeodesic in $C(S)$.

Namely, let $j > 0$ and choose a number $n(j) > j$ such that $(c_\ell, c_n)_p \geq j$ for all $\ell, n \geq n(j)$. By hyperbolicity, this means that there is a point $a_j \in C(S)$ with $d(p, a_j) \geq j$ and the property that for $n \geq n(j)$, every geodesic connecting c_n to p passes through a neighborhood of the point a_j of uniformly bounded diameter not depending on j. By construction, the sequence $(a_j) \subset C(S)$ is contained in a b-quasigeodesic for a number $b > 0$ only depending on the hyperbolicity constant, and this quasigeodesic defines the same equivalence class as the sequence (c_j). As a consequence, we may assume without loss of generality that (c_i) is a uniform quasigeodesic. By the considerations in Section 2 we may moreover assume that there is a splitting sequence $(\tau_j)_{j \geq 0} \subset \mathcal{TT}$ and a strictly increasing function $\sigma : \mathbb{N} \to \mathbb{N}$ such that $c_i = \Phi(\tau_{\sigma(i)})$ where $\Phi : \mathcal{TT} \to C(S)$ assigns to a train track τ a vertex cycle for τ.

By Lemma 2.5 there is a number $k > 0$ with the property that for all $0 < i < j$ there is a vertex cycle $a_{i,j} \in C(S)$ for $\tau_{\sigma(i)}$ such that

$$i(c_0, a_{i,j}) i(a_{i,j}, c_j) \leq k i(c_0, c_j) \quad \text{for } 0 < i < j. \tag{3.1}$$

Note that this inequality is invariant under multiplication of the simple closed curve c_j with an arbitrary positive weight. Let again $\iota : \mathcal{PML} \to \mathcal{ML} - \{0\}$ be a continuous section and for $j > 0$ let $\hat{c}_j \in \iota(\mathcal{PML})$ be a multiple of c_j. By passing to a subsequence we may assume that the sequence $(\hat{c}_j)$ converges in the space of measured geodesic laminations to a measured geodesic lamination μ.

We claim that the support of μ is a minimal geodesic lamination which fills up S. For this we argue by contradiction and we assume otherwise. Then there is a simple closed curve c on S with $i(c, \mu) = 0$ (it is possible that the curve c is a minimal component of the support of μ). Replace the quasigeodesic (c_i) by an equivalent quasigeodesic, again denoted by (c_i), which issues from $c = c_0$ and which eventually coincides with the original quasigeodesic. Such a quasigeodesic exists by hyperbolicity of $C(S)$. Since the number of vertex cycles for a fixed train track is bounded from above by a universal constant, after passing to a subsequence and using a standard diagonal argument we may assume that the curve $a_{i,j}$ is independent of $j > i$; we denote this curve by a_i. Inequality (3.1) and continuity of the intersection form then implies that $i(c, a_i)i(a_i, \hat{c}_j) \leq k_0 i(c, \mu) = 0$ for all $i > 0$. Since $d(c, a_i) \geq d(c, c_i) - k_0$ for all i, for $i > k_0 + 2$ the intersection numbers $i(c, a_i)$ are bounded from below by a universal constant and therefore $i(a_i, \mu) = 0$ for all $i > 0$. If the support of μ contains a simple closed curve component a, then this just means that the set $\{a_i \mid i > 0\} \subset C(S)$ is contained in the $k_0 + 1$-neighborhood of a which is impossible. Otherwise μ has a minimal component μ_0 which fills a nontrivial bordered subsurface S_0 of S, and $i(\mu_0, a) > 0$ for every simple closed curve a in S which is contained in S_0 and which is not freely homotopic into a boundary component or a cusp. Since $i(a_i, \mu) = 0$ by assumption, the curves a_i do not have an essential intersection with S_0 which means that $i(a_i, a) = 0$ for every simple closed essential curve a in S_0. Again we deduce that the set $\{a_i \mid i > 0\} \subset C(S)$ is bounded. Together we obtain a contradiction which implies that indeed the support of μ is a minimal geodesic lamination $\lambda_0 \in \mathcal{B}$ which fills up S.

Let λ_i be a complete geodesic lamination which contains c_i as a minimal component. By passing to a subsequence we may assume that the laminations λ_i converge in the Hausdorff topology to a complete geodesic lamination λ. Since the measured laminations $\hat{c}_j$ converge in the weak*-topology to μ, the lamination λ necessarily contains λ_0 as a sublamination.

Let α_λ be a full splitting sequence determined by λ. For every $i > 0$ the set of complete geodesic laminations which are carried by $\alpha_\lambda(i)$ is an open neighborhood of λ in $\mathcal{CL}$. Thus for every $i > 0$ there is a number $j(i) > 0$ with the property that for every $j \geq j(i)$ the geodesic c_j is carried by $\alpha_\lambda(i)$. From the remark following Corollary 2.6 we conclude that $\Phi(\alpha_\lambda(i))$ is contained in a uniformly bounded neighborhood of any geodesic connecting c_j to $\Phi(\alpha_\lambda(0))$. As a consequence, the image under Φ of the full splitting sequence α_λ defines the same point in the Gromov boundary of $C(S)$ as (c_j). In other words, the point in $\partial C(S)$ defined by (c_j) equals $\Lambda(\lambda_0)$ and the map Λ is surjective. Hence by Lemma 3.3, the map Λ is a bijection. Moreover, if $(c_i) \subset C(S)$ is any admissible sequence and if (c_{i_j}) is any subsequence with the property that the curves c_{i_j} converge in the Hausdorff topology to a geodesic lamination λ, then λ contains a lamination $\lambda_0 \in \mathcal{B}$ as a minimal component, and (c_i) defines the point $\Lambda(\lambda_0) \in \partial C(S)$. In particular, for every admissible sequence $(c_i) \subset C(S)$ the curves (c_i) converge in the coarse Hausdorff topology to the lamination $\lambda_0 = \Lambda^{-1}((c_i)) \in \mathcal{B}$. This shows our above claim.

Let again $\mathcal{L}$ be the space of all geodesic laminations on S equipped with the Hausdorff topology. Let $\mathcal{A} \subset \mathcal{L}$ be the set of all laminations containing a minimal sublamination which fills up S. Let again $\pi : \mathcal{A} \to \mathcal{B}$ be the projection which associates to a lamination $\lambda \in \mathcal{A}$ its unique minimal component. Let $\lambda_0 \in \mathcal{B}$ and let $L = \pi^{-1}(\lambda_0) \subset \mathcal{A}$ be the set of all geodesic laminations which contain λ_0 as a sublamination. Since λ_0 fills up S, the set L is finite. We call a subset V of $C(S) \cup \mathcal{B}$ a *neighborhood* of λ_0 *in the coarse Hausdorff topology of* $C(S) \cup \mathcal{B}$ if there is a neighborhood W of L in $\mathcal{L}$ such that $V \supset (W \cap C(S)) \cup \pi(W \cap \mathcal{A})$.

For $\xi \in \partial C(S)$ and $c \in C(S)$ write $(c,\xi)_p = \sup_{(x_i)} \liminf_{i \to \infty} (c, x_i)_p$ where the supremum is taken over all admissible sequences (x_i) defining ξ. A subset U of $C(S) \cup \partial C(S)$ is called a neighborhood of $\xi \in \partial C(S)$ if there is some $\varepsilon > 0$ such that U contains the set $\{\zeta \in C(S) \cup \partial C(S) \mid e^{-(\xi,\zeta)_p} < \varepsilon\}$. In the sequel we identify $\mathcal{B}$ and $\partial C(S)$ with the bijection Λ. In other words, we view a point in $\partial C(S)$ as a minimal geodesic lamination which fills up S, i.e. we suppress the map Λ in our notation. To complete the proof of our lemma it is now enough to show the following. A subset U of $C(S) \cup \partial C(S)$ is a neighborhood of $\lambda_0 \in \mathcal{B} = \partial C(S)$ if and only if U is a neighborhood of λ_0 in the coarse Hausdorff topology.

For this let $\lambda_0 \in \mathcal{B}$, let $L = \pi^{-1}(\lambda_0) \subset \mathcal{A}$ be the collection of all geodesic laminations containing λ_0 as a sublamination and let $p = \Phi(\tau)$ for a train track $\tau \in \mathcal{TT}$ which carries each of the laminations $\lambda \in L$ (see [Ham04] for the existence of such a train track τ). Let $\varepsilon > 0$ and let $U = \{\zeta \in C(S) \cup \partial C(S) \mid e^{-(\lambda_0,\zeta)_p} < \varepsilon\}$. Let $\lambda_1, \ldots, \lambda_s \subset L$ be the collection of all complete geodesic laminations contained in L and for $i \leq s$ let α_i be the full splitting sequence issuing from $\alpha_i(0) = \tau$ which is determined by λ_i. By hyperbolicity and the remark after Corollary 2.6, there is a universal constant $\chi > 0$ with the property that for each $i \leq s, j \geq 0$ every geodesic connecting p to a curve $c \in C(S)$ which is carried by $\alpha_i(j)$ passes through the χ-neighborhood of $\Phi(\alpha_i(j))$. Since $\Phi(\alpha_i)$ is an unparametrized quasigeodesic which represents the point $\lambda_0 \in \partial C(S)$, this implies that for every $\varepsilon > 0$ there is a number $j > 0$ such that $e^{-(c,\lambda_0)_p} < \varepsilon$ and $e^{-(\mu,\lambda_0)_p} < \varepsilon$ for all simple closed curves $c \in C(S)$ and all laminations $\mu \in \mathcal{B}$ which are carried by one of the train tracks $\alpha_i(j)$ $(i = 1, \ldots, s)$. Since the set of all geodesic laminations which are carried by the train tracks $\alpha_i(j)$ $(i = 1, \ldots, s)$ is a neighborhood of L in $\mathcal{L}$ with respect to the Hausdorff topology (see [Ham04]), we conclude that a neighborhood of λ_0 in $\partial C(S) \cup C(S)$ is a neighborhood of λ_0 in the coarse Hausdorff topology as well.

To show that a neighborhood of $\lambda_0 \in \mathcal{B}$ in the coarse Hausdorff topology contains a set of the form $\{\zeta \in C(S) \cup \partial C(S) \mid e^{-(\lambda_0,\zeta)_p} < \varepsilon\}$ we argue by contradiction. Let again $L = \pi^{-1}(\lambda_0) \subset \mathcal{A}$ be the collection of all geodesic laminations containing λ_0 as a sublamination. Assume that there is an open neighborhood $W \subset L$ of L in the Hausdorff topology with the property that $\pi(W \cap \mathcal{A}) \cup (W \cap C(S))$ does not contain a neighborhood of λ_0 in $C(S) \cup \partial C(S)$. Let $(c_i) \subset C(S)$ be a sequence which represents $\lambda_0 \in \mathcal{B} = \partial C(S)$. By our above consideration, every accumulation point of $(c_i) \subset L$

with respect to the Hausdorff topology is contained in L. By our assumption, there is a sequence $i_j \to \infty$, a sequence $(a_j) \subset C(S)$ and a sequence $R_j \to \infty$ such that $(c_{i_j}, a_j)_p \geq R_j$ and that $a_j \notin W$. By passing to a subsequence we may assume that the curves a_j converge in the Hausdorff topology to a lamination $\zeta \notin W$. However, since $(c_{i_j}, a_j)_p \to \infty$, the sequence (a_j) is admissible by assumption and equivalent to (c_j) and therefore by our above consideration, (a_j) converges in the coarse Hausdorff topology to λ_0. Then $a_j \in W$ for all sufficiently large j which is a contradiction. This shows our above claim and completes the proof of our lemma. $\qquad\square$

Acknowledgement. I am grateful to the referee for pointing out the reference [MM04] to me and for many other helpful suggestions.

References

[BH99] M. R. Bridson & A. Haefliger (1999). *Metric Spaces of Non-positive Curvature*. Springer-Verlag, Berlin.

[Bon01] F. Bonahon (2001). Geodesic laminations on surfaces. In *Laminations and Foliations in Dynamics, Geometry and Topology (Stony Brook, NY, 1998)*, *Contemp. Math.*, volume 269, pp. 1–37. Amer. Math. Soc., Providence, RI.

[Bow02] B. Bowditch (2002). Intersection numbers and the hyperbolicity of the curve complex. Preprint.

[CEG87] R. D. Canary, D. B. A. Epstein & P. Green (1987). Notes on notes of Thurston. In *Analytical and Geometric Aspects of Hyperbolic Space (Coventry/Durham, 1984)*, *London Math. Soc. Lecture Note Ser.*, volume 111, pp. 3–92. Cambridge Univ. Press, Cambridge.

[FLP79] A. Fathi, F. Laudenbach & V. Poenaru (1979). Travaux de Thurston sur les surfaces. *Astérisque* **66–67**.

[Ham04] U. Hamenstädt (2004). Train tracks and mapping class groups I. Preprint.

[Har81] W. J. Harvey (1981). Boundary structure of the modular group. In *Riemann Surfaces and Related Topics: Proceedings of the 1978 Stony Brook Conference (State Univ. New York, Stony Brook, N.Y., 1978)*, *Ann. of Math. Stud.*, volume 97, pp. 245–251. Princeton Univ. Press, Princeton, N.J.

[Iva02] N. V. Ivanov (2002). Mapping class groups. In *Handbook of Geometric Topology*, pp. 523–633. North-Holland, Amsterdam.

[Kla99] E. Klarreich (1999). The boundary at infinity of the curve complex and the relative Teichmüller space. Preprint.

[Lev83]   G. Levitt (1983). Foliations and laminations on hyperbolic surfaces. *Topology* **22** (2), 119–135.

[MM99]   H. A. Masur & Y. N. Minsky (1999). Geometry of the complex of curves. I. Hyperbolicity. *Invent. Math.* **138** (1), 103–149.

[MM04]   H. Masur & Y. Minsky (2004). Quasiconvexity in the curve complex. In W. Abikoff & A. Haas (eds.), *In the Tradition of Ahlfors and Bers, III*, *Contemp. Math.*, volume 355, pp. 309–320. Amer. Math. Soc.

[Mos]   L. Mosher. Train track expansions of measured foliations. Unpublished manuscript.

[Ota96]   J.-P. Otal (1996). Le théorème d'hyperbolisation pour les variétés fibrées de dimension 3. *Astérisque* **235**.

[PH92]   R. C. Penner & J. L. Harer (1992). *Combinatorics of Train Tracks*. Princeton University Press, Princeton, NJ.

Ursula Hamenstädt

Mathematisches Institut
Universität Bonn
Beringstraße 1
D-53115 Bonn
Germany

ursula@math.uni-bonn.de

AMS Classification: 30F60

Keywords: Complex of curves, Gromov boundary, train tracks, geodesic laminations

Spaces of Kleinian Groups
Lond. Math. Soc. Lec. Notes **329**, 209–218

Cambridge University Press
Y. Minsky, M. Sakuma & C. Series (Eds.)

This article is in the public domain, 2005

The pants complex has only one end

Howard Masur and Saul Schleimer[1]

1. Definitions and statement of the main theorem

The purpose of this note is to prove the following theorem:

Theorem 4.1. *Let S be a closed, connected, orientable surface with genus $g(S) \geq 3$. Then the pants complex of S has only one end. In fact, there are constants $K = K(S)$ and $M = M(S)$ so that: if $R > M$, and P and Q are pants decompositions at distance greater than KR from a basepoint, then P and Q may be connected by a path which remains at least distance R from the basepoint.*

A *pants decomposition* of S consists of $3g(S) - 3$ disjoint essential non-parallel simple closed curves on S. Each component of the complement of the curves is a three-holed sphere; a *pants*. Then the *pants complex* $\mathcal{P}(S)$ is the metric graph whose vertices are pants decompositions of S, up to isotopy. Two vertices P, P' are connected by an edge of length one if P, P' differ by an *elementary move*. In an elementary move all curves of the pants are fixed except for one curve α. Remove α and let V be the component of the complement of the remaining curves which is not a pants. Then V contains α and is either a once-holed torus or a four-holed sphere. Now α is replaced by any curve β contained in V that intersects α minimally; in the torus case once, and in the sphere case twice. All edges of $\mathcal{P}(S)$ are assigned length 1. We let $d(\cdot, \cdot)$ be the distance function in $\mathcal{P}(S)$. The pants complex $\mathcal{P}(S)$ is known to be connected [HT80].

A path metric space (X, d) has *one end* if for any basepoint $O \in X$ and any radius R the complement of $B_R = B_R(O)$, the ball of radius R centered at O, has only one unbounded component. It is easy to see that the definition does not depend on the choice of the point O. Clearly having one end is a quasi-isometry invariant of path metric spaces. So, following Brock [Bro03] (see also [Bro02]) and Wolpert [Wol87], our theorem implies:

Corollary 1.1. *Fix S a closed, connected, orientable surface with genus three or higher. The Teichmüller space of S, equipped with the Weil-Peterson metric, has only one end.*

---

[1]This research is partially supported by NSF grant DMS0244472 (H. Masur) and was previously partially supported by NSF grant DMS0102069 (S. Schleimer).

Finally, recall that the *curve complex* $C(S)$ is the complex whose k-simplices consist of $k+1$ distinct isotopy classes of essential simple closed curves on S that have disjoint representatives on S. Or, in the case of a once-holed torus and four-holed sphere, $C(S)$ is the Farey graph. From the metric point of view we will only be interested in the 1-skeleton of $C(S)$. Each edge is assigned length 1. We let $d_S(\cdot, \cdot)$ denote the distance function in $C(S)$.

Acknowledgement   We thank Yair Minsky for critiquing an earlier version of this paper.

2. The set of handle curves is connected

In this section we prove two combinatorial facts. First, the set of handle curves in the curve complex is connected and second, any pants decomposition is a bounded distance (in the pants complex) from a decomposition containing a handle curve.

Again assume S is a closed, connected, orientable surface with genus three or greater. We will call α a *handle curve* in S if α separates S into two surfaces: the once-holed torus $S(\alpha)$ and the rest of the surface.

We will need the following result. It was first proved by Farb and Ivanov [FI03] by different methods. Another proof has been given by McCarthy and Vautaw [MV03] by methods similar to ours. We include a proof for completeness.

Proposition 2.1. *If $g(S) \geq 3$, the subcomplex $\mathcal{H}(S) \subset C(S)$ of handle curves is connected.*

Remark 2.2. Note that the hypothesis $g(S) \geq 3$ cannot be removed; it is easy to check that $\mathcal{H}(S)$, when S has genus 2, is an infinite collection of points.

Remark 2.3. Note that Proposition 2.1 immediately implies that the set of separating curves in $C(S)$ is also connected.

Remark 2.4. Our proof of Proposition 2.1 generalizes to the case $\partial S \neq \emptyset$. An interesting open question is the higher connectivity of $\mathcal{H}(S)$.

Before we begin the proof we will require a bit of terminology. Let $i(\cdot, \cdot)$ denote the geometric intersection number of two essential simple closed curves in S. Also, if δ is a separating curve in S we say that an arc β' is a *wave* for δ if $\beta' \cap \delta = \partial \beta'$ and β' is essential as a properly embedded arc in $S \smallsetminus \delta$. We say that two waves β' and β'' for δ *link* if $\beta' \cap \beta'' = \emptyset$, both β' and β'' meet the same side of δ, and $\partial \beta'$ separates $\partial \beta''$ inside δ. Figure 1 shows a pair of linking waves.

Finally we define *double surgery* as follows. Suppose we are given a linking pair of waves β' and β'' for an essential separating curve δ. Form the closed regular neighborhood $U = \text{neigh}(\delta \cup \beta' \cup \beta'')$. Let δ' be the component of ∂U which is not homotopic

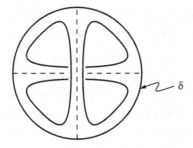

Figure 1: The dotted lines are β' and β''.

to δ. We say that δ' is obtained from δ via double surgery along β' and β''. Again, see Figure 1 for a picture of δ, δ', and U. Note that δ' is necessarily a separating curve and is disjoint from δ. Furthermore, the curves δ and δ' cobound a two-holed torus. We deduce that δ' is also essential as long as the component of $S \setminus \delta$ containing $\beta' \cup \beta''$ is not a handle.

We are now equipped to prove the proposition:

Proof of Proposition 2.1. Let $\alpha, \beta \in \mathcal{H}(S)$ be handle curves and S a closed orientable surface of genus at least three. Suppose that α and β are *tight*: α has been isotoped to make $|\alpha \cap \beta| = i(\alpha, \beta)$. If $i(\alpha, \beta) = 0$ then there is nothing to prove. If $i(\alpha, \beta) > 0$ we will find a curve $\gamma \in \mathcal{H}(S)$ with $i(\gamma, \alpha) = 0$ and $i(\gamma, \beta) < i(\alpha, \beta)$. By induction, γ will be connected to β in $\mathcal{H}(S)$, proving the proposition.

We find γ via the following inductive procedure. Recall that $S(\alpha)$ is the handle which α bounds. To begin, we define $\delta_0 \subset S \setminus S(\alpha)$ to be a parallel copy of α, still intersecting β tightly. At stage k by induction we will be given an essential separating curve δ_k where

- $i(\alpha, \delta_k) = 0$,

- δ_k is tight with respect to β, and

- $i(\delta_k, \beta) < i(\delta_{k-1}, \beta)$, if $k > 0$.

Let T_k be the component of $S \setminus \delta_k$ which does not contain α. If T_k is a handle, then we take $\gamma = \delta_k$ and we are done with the inductive procedure. If $i(\delta_k, \beta) = 0$ then we may take γ to be any handle curve inside T_k. As this γ satisfies $i(\alpha, \gamma) = i(\beta, \gamma) = 0$ finding γ would finish the proposition. From now on we assume that T_k is not a handle and that $i(\delta_k, \beta) > 0$.

We now attempt to do a double surgery of δ_k into T_k. Either we will find γ directly or the curve resulting from double surgery, δ_{k+1}, will satisfy the induction hypothesis.

As the geometric intersection with β is always decreasing, this procedure will stop after finitely many steps yielding the desired handle curve.

So all that remains is to do the double surgery. Suppose for the moment that $\beta', \beta'' \subset \beta \cap T_k$ are linking waves for δ_k. As described above we may form δ_{k+1} via a double surgery along β' and β''. Isotope δ_{k+1}, in the complement of δ_k, to be tight with respect to β. As noted in the definition of double surgery, δ_{k+1} is an essential separating curve which is disjoint from α. Finally note that $i(\delta_{k+1}, \beta) \leq i(\delta_k, \beta) - 4$. Thus all of the induction hypotheses are satisfied.

Suppose now that we cannot find linking waves among the arcs $\beta \cap T_k$. Choose instead an *outermost* wave $\beta' \subset \beta \cap T_k$: that is, there exists an arc $\delta'_k \subset \delta_k$ such that $\delta'_k \cap \beta = \partial \delta'_k = \partial \beta'$. See Figure 2.

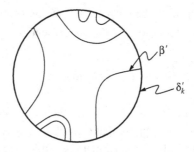

Figure 2: The arc β' is an outermost wave.

Here there are two remaining cases. If $\delta'_k \cup \beta'$ is a separating curve take $\delta_{k+1} = \delta'_k \cup \beta'$ and note that the induction hypotheses are easily verified. The final possibility is that $\delta'_k \cup \beta'$ is not separating. See Figure 3.

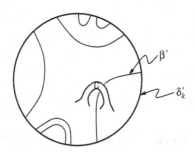

Figure 3: The curve $\beta' \cup \delta'_k$ is nonseparating.

Since $\delta'_k \cup \beta'$ is not separating there exists a properly embedded essential arc $\beta'' \subset T_k$ such that $\beta'' \cap \beta = \emptyset$ and $|\beta'' \cap \delta'_k| = 1$. Then β' and β'' link. Do a double surgery along these waves to obtain δ_{k+1}. Isotope δ_{k+1}, in the complement of δ_k, to be tight with respect to β. Again, all of the induction hypotheses are easily verified, as we have $i(\delta_{k+1},\beta) \leq i(\delta_k,\beta) - 2$. This completes the second induction step and hence completes the proof of Proposition 2.1. $\qquad\square$

We also require

Lemma 2.5. *There is a constant $M = M(S)$ such that the pants decompositions containing a handle curve are M-dense in the space of all pants decompositions*

Proof. The mapping class group acts co-compactly on the space of pants decompositions. $\qquad\square$

3. Subsurface projections and distances

Here we set forth three lemmas studying the pants complex. The first is a slight refinement of an idea of Masur and Minsky [MM00], the second gives a condition for a pants decomposition to lie outside of a large ball about the origin in $\mathcal{P}(S)$, and the third provides us with useful paths lying outside of such a ball.

Fix attention on a subsurface $W \subset S$ which has ∂W essential in S and which is not an annulus or a pants (a three-holed sphere). Suppose that γ is an essential simple closed curve in S which is not isotopic to a boundary component of ∂W. Suppose further that γ either lies in the interior of W or has non-zero geometric intersection with ∂W. Isotope γ to be tight with respect to ∂W.

We briefly define the *subsurface projection* $\pi_W(\gamma)$ (see [MM00] for a more through discussion). If $\gamma \subset W$ then set $\pi_W(\gamma) = \gamma$. If not, then for every arc $\alpha \subset \gamma \cap W$ take every curve of $\partial(\mathrm{neigh}(\alpha \cup \partial W)))$ which is not isotopic into ∂W. Let $\pi_W(\gamma)$ be this set of curves and note that $\pi_W(\gamma) \subset C(W)$ has diameter at most 2 ([MM00] Lemma 2.3).

Similarly, given a pants decomposition P we may project each curve of P into W. We denote the resulting image $\pi_W(P) \subset C(W)$. This again has diameter at most 2. By $d_W(P,P')$ we mean the distance in the curve complex of W between the sets $\pi_W(P)$ and $\pi_W(P')$.

Let $[x]_C$ be the function on $\mathbb{N}$ giving zero if $x < C$ and giving x if $x \geq C$. We will need the following result from [MM00] (see Theorem 6.12 and Section 8 of that paper):

Lemma 3.1. *There is a constant $C_0 = C_0(S) \geq 1$ such that for any $D \geq C_0$ there are constants $\lambda_1 = \lambda_1(D) \geq D$ and $\varepsilon_1 = \varepsilon_1(D) \geq 0$ with the following property: for any pants decompositions P and P' we have*

$$\frac{1}{\lambda_1} \sum_V [d_V(P,P')]_D - \varepsilon_1 \leq d(P,P') \leq \lambda_1 \sum_V [d_V(P,P')]_D + \varepsilon_1, \qquad (3.1)$$

where the sums range over subsurfaces $V \subset S$ with essential boundary and where V is neither an annulus nor a pants.   □

We have a final definition: Choose a basepoint $O \in \mathcal{P}(S)$. Fix $R > 0$. A curve $\alpha'' \subset S$ is *R-distant* from the basepoint O if, for any pants decomposition P containing α'', we have $d(P, O) > R$.

Fix now $C > \max\{2, C_0\}$. Let $\lambda_1 = \lambda_1(C) \geq C$ and $\varepsilon_1 = \varepsilon_1(C) \geq 0$ as in Lemma 3.1.

Lemma 3.2. *Fix $R > 1$. Fix a handle curve α and some curve $\alpha'' \subset S(\alpha)$ satisfying $d_{S(\alpha)}(O, \alpha'') > \lambda_1(R + \varepsilon_1) + 2$. Then α'' is R-distant from O.*

Proof. Since $\lambda_1 > C$ and $R > 1$ we have $d_{S(\alpha)}(O, \alpha'') \geq C$. Fix any pants decomposition P containing α''. As $\pi_{S(\alpha)}(P)$ has diameter at most two we have

$$d_{S(\alpha)}(P, O) \geq d_{S(\alpha)}(\alpha'', O) - 2.$$

So, by the left inequality of Equation 3.1 we have

$$d(P, O) \geq \frac{1}{\lambda_1}[d_{S(\alpha)}(P, O)]_C - \varepsilon_1 = \frac{1}{\lambda_1}d_{S(\alpha)}(P, O) - \varepsilon_1 \geq \frac{1}{\lambda_1}(d_{S(\alpha)}(\alpha'', O) - 2) - \varepsilon_1 > R.$$

□

As the Farey graph for $S(\alpha)$ has infinite diameter, and as the diameter of $\pi_{S(\alpha)}(O)$ is bounded, such curves α'' exist in abundance. We now turn to the existence of paths lying outside of the R-ball about the basepoint. As a bit of notation let $B_R = B_R(O)$ be the ball of radius R centered at the basepoint O.

Note that it follows from Equation 3.1 that projections of size exactly C or $C + 1$ cannot account for the entire pants distance between P and P'. Namely there are constants $\lambda_2 = \lambda_2(C) > 1$ and $\varepsilon_2 = \varepsilon_2(C) > 0$ such that

$$\sum_V [d_V(P, P')]_{C+2} \leq \sum_V [d_V(P, P')]_C \leq \lambda_2 \sum_V [d_V(P, P')]_{C+2} + \varepsilon_2. \tag{3.2}$$

Choose $K = K(C) > 0$ so that for all $R \geq 1$,

$$\frac{1}{2\lambda_2\lambda_1}\left((K-1)R - \varepsilon_1 - \lambda_1\varepsilon_2 - 2\lambda_2\lambda_1 - \lambda_2\lambda_1^2(R + \varepsilon_1)\right) > \lambda_1(R + \varepsilon_1) \tag{3.3}$$

Lemma 3.3. *Suppose P_0 is a pants decomposition of S such that $P_0 \notin B_{(K-1)R}$ and P_0 contains a curve α which bounds a handle $S(\alpha)$. Then there is a pants decomposition P_1 and a path $\{P_{i/n}\}_{i=0}^n$ so that*

- $P_{0/n} = P_0$, $P_{n/n} = P_1$,

- $P_{i/n}$ *differs from* $P_{(i+1)/n}$ *by a single elementary move,*

- *for all i, $P_0|(S \setminus S(\alpha)) = P_{i/n}|(S \setminus S(\alpha))$,*

- *for all i, $P_{i/n} \notin B_R$*

- *The curve $\alpha'' = P_1 \cap S(\alpha)$ is R-distant from O.*

Proof. Let $\alpha' \in P_{0/n} = P_0$ be the curve contained in $S(\alpha)$. Consider a geodesic segment in the Farey graph connecting α' to $\beta \in \pi_{S(\alpha)}(O)$, where β is chosen as close as possible to α'. Extend this segment through α' to a geodesic ray L in the direction opposite β. The ray L meets the segment only at α'. Move along L distance more than $\lambda_1(R + \varepsilon_1) + 2$ from α' to a point α''. Let $P_{i/n}$ be the path obtained by making elementary moves along the curves in L and fixing the pants in $S \setminus S(\alpha)$. This path has all of the desired properties except perhaps the fourth. (The fifth follows from Lemma 3.2). It remains to show that $P_{i/n} \notin B(R)$.

There are two cases. Suppose first that $d_{S(\alpha)}(\beta, \alpha') > \lambda_1(R + \varepsilon_1) + 2$. Then by Lemma 3.2 for any i the curve $P_{i/n} \cap S(\alpha)$ is R-distant. So $P_{i/n} \notin B(R)$ and Lemma 3.3 holds in this case.

Next suppose that $d_{S(\alpha)}(\beta, \alpha') \leq \lambda_1(R + \varepsilon_1) + 2$. Then by Equation 3.1 and the right hand side of Equation 3.2

$$(K - 1)R \leq d(P_0, O) \leq \lambda_1 \sum_V [d_V(P_0, O)]_C + \varepsilon_1 \leq$$

$$\leq \lambda_2 \lambda_1 \sum_V [d_V(P_0, 0)]_{C+2} + \lambda_1 \varepsilon_2 + \varepsilon_1 \leq$$

$$\leq \lambda_2 \lambda_1 \sum_{V \neq S(\alpha)} [d_V(P_0, O)]_{C+2} + 2\lambda_2 \lambda_1 + \lambda_2 \lambda_1^2(R + \varepsilon_1) + \lambda_1 \varepsilon_2 + \varepsilon_1.$$

Let V be any subsurface disjoint from $S(\alpha)$. Since $P_{i/n}$ is constant in V, the projection $\pi_V(P_{i/n})$ is constant. Now let V be a subsurface that intersects $S(\alpha)$ or strictly contains $S(\alpha)$. Since $\alpha \in P_{i/n}$, it follows that $\pi_V(P_{i/n})$ contains $\pi_V(\alpha)$. Since each $\pi_V(P_{i/n})$ has diameter at most 2, $d_V(P_{i/n}, O) \geq d_V(P_0, O) - 2$. Thus for any subsurface V not isotopic to $S(\alpha)$, as $C > 2$, we have $[d_V(P_{i/n}, O)]_C \geq \frac{1}{2}[d_V(P_0, O)]_{C+2}$. Thus for all i,

$$\sum_{V \neq S(\alpha)} [d_V(P_{i/n}, O)]_C \geq \frac{1}{2} \sum_{V \neq S(\alpha)} [d_V(P_0, 0)]_{C+2} \geq$$

$$\geq \frac{1}{2\lambda_2\lambda_1}\left((K-1)R-\varepsilon_1-\lambda_1\varepsilon_2-2\lambda_2\lambda_1-\lambda_2\lambda_1^2(R+\varepsilon_1)\right) > \lambda_1(R+\varepsilon_1),$$

the last inequality following from Equation 3.3. So, by the left-hand side of Equation 3.1, for all i we have $d(P_{i/n}, O) > R$.   □

4. Proof of the theorem

Recall the statement:

Theorem 4.1. *Let S be a closed, connected, orientable surface with genus $g(S) \geq 3$. Then the pants complex of S has only one end. In fact, there are constants $K = K(S)$ and $M = M(S)$ so that, if $R > M$, any pants decompositions P and Q, at distance greater than KR from a basepoint, can be connected by a path which remains at least distance R from the basepoint.*

Proof. We take M as defined in Section 2 and K as defined in Section 3.

Using Lemma 2.5 move P and Q a distance at most $M < R$ to obtain pants decompositions P_0 and Q_0. The lemma gives handle curves $\alpha_P \in P_0$, $\alpha_Q \in Q_0$. Also, $P_0, Q_0 \notin B_{(K-1)R}$.

Apply Lemma 3.3 twice in order to connect P_0 and Q_0 to pants decompositions P_1 and Q_1 satisfying all of the conclusions of the lemma. Let $\alpha_P'' \in P_1$ and $\alpha_Q'' \in Q_1$ be the R-distant curves lying in the handles $S(\alpha_P)$ and $S(\alpha_Q)$ respectively.

We must now construct a path from P_1 to Q_1. Consider first the case where $\alpha_P \neq \alpha_Q$.

Applying Proposition 2.1 we connect α_P to α_Q by a path of handle curves in $\mathcal{H}(S)$. Label these $\{\alpha_i\}_{i=1}^n$ where $\alpha_1 = \alpha_P$, $\alpha_n = \alpha_Q$, and $n > 1$. Note that in this step the hypothesis $g(S) > 2$ is used. Choose, for $i \in \{2, 3, \ldots, n-1\}$, any R-distant curve $\alpha_i'' \subset S(\alpha_i)$. This requires Lemma 3.2. Again, for $i \in \{2, 3, \ldots, n-1\}$, extend the pair α_i, α_i'' to a pants decomposition P_i. Finally set $P_n = Q_1$.

We connect P_i to P_{i+1} by a path where every pants decomposition in the first part of the path contains α_i and α_i'' and every pants decomposition in the rest of the path contains α_{i+1} and α_{i+1}''. (This is possible because $\mathcal{P}(S \smallsetminus S(\alpha_i)) \cong \mathcal{P}(S \smallsetminus S(\alpha_{i+1}))$ are connected and because $S(\alpha_i)$ is disjoint from $S(\alpha_{i+1})$.) By Lemma 3.2, this path lies outside of the ball of radius R and we are done.

In the case which remains $\alpha_P = \alpha_Q$. Here there is no need for Proposition 2.1. Instead we choose any handle curve β which is disjoint from α_P. Note that β exists as $g(S) > 2$. Using Lemma 3.2 choose a R-distant $\beta'' \subset S(\beta)$ and extend this to a pants decomposition P_2. Set $P_3 = Q_1$. Connect P_1 to P_2 to P_3 as in the previous paragraph. This completes the proof.   □

References

[Bro02] J. F. Brock (2002). Pants decompositions and the Weil-Petersson metric. In *Complex Manifolds and Hyperbolic Geometry (Guanajuato, 2001)*, *Contemp. Math.*, volume 311, pp. 27–40. Amer. Math. Soc., Providence, RI.

[Bro03] J. F. Brock (2003). The Weil–Petersson metric and volumes of 3-dimensional hyperbolic convex cores. *J. Amer. Math. Soc.* **16** (3), 495–535 (electronic).

[FI03] B. Farb & N. V. Ivanov (2003). The Torelli geometry and its applications. To appear in *Math. Res. Lett.*, `math.GT/0311123`.

[HT80] A. Hatcher & W. Thurston (1980). A presentation for the mapping class group of a closed orientable surface. *Topology* **19** (3), 221–237.

[MM00] H. A. Masur & Y. N. Minsky (2000). Geometry of the complex of curves. II. Hierarchical structure. *Geom. Funct. Anal.* **10** (4), 902–974.

[MV03] J. D. McCarthy & W. R. Vautaw (2003). Automorphisms of Torelli groups. Preprint, `math.GT/0311250`.

[Wol87] S. A. Wolpert (1987). Geodesic length functions and the Nielsen problem. *J. Differential Geom.* **25** (2), 275–296.

Howard Masur

Department of Mathematics
University of Illinois at Chicago
851 South Morgan Street
Chicago, Illinois 60607
USA

masur@math.uic.edu

Saul Schleimer

Department of Mathematics
Rutgers
110 Frelinghuysen Road
Piscataway, New Jersey 08854
USA

saulsch@math.rutgers.edu

AMS Classification: 30F99, 57M60

Keywords: Pants decompositions, pants complex, metric ends

Spaces of Kleinian Groups
Lond. Math. Soc. Lec. Notes **329**, 219–231

Cambridge University Press
Y. Minsky, M. Sakuma & C. Series (Eds.)

The Weil–Petersson geometry of the five-times punctured sphere

Javier Aramayona

Abstract

We give a new proof that the completion of the Weil–Petersson metric on Teichmüller space is Gromov-hyperbolic if the surface is a five-times punctured sphere or a twice-punctured torus. Our methods make use of the synthetic geometry of the Weil–Petersson metric.

1. Introduction

The large scale geometry of Teichmüller space has been a very important tool in different aspects of the theory of hyperbolic 3-manifolds. Within this context, a natural question to ask is whether Teichmüller space, with a given metric, is hyperbolic in the sense of Gromov (or Gromov hyperbolic, for short). In general, the answer is negative. In their paper [BF01], the authors prove the following: if Σ is a surface of genus g and with p punctures, with $3g - 3 + p > 2$, then the Teichmüller space of Σ, endowed with the Weil–Petersson metric, is not Gromov hyperbolic. In the case when $3g - 3 + p = 2$ (that is, when the surface is a sphere with five punctures or a twice-punctured torus) they show that the Weil–Petersson Teichmüller space is Gromov hyperbolic. The proof makes reference to very deep results by Masur and Minsky [MM99] on the Gromov hyperbolicity of the curve complex. We remark that Behrstock [Beh05] has also used the geometric structure of the curve complex to give a new proof of the hyperbolicity of the Weil–Petersson metric for these "low-complexity" cases. The aim of this paper if to give a direct proof of the Gromov hyperbolicity of the Weil–Petersson Teichmüller space in the case $3g - 3 + p = 2$. More specifically, we show the following result.

Theorem 1.1. *If Σ is the five-times punctured sphere or the twice punctured torus, then $\overline{T}_{WP}(\Sigma)$ is Gromov hyperbolic, where $\overline{T}_{WP}(\Sigma)$ denotes the metric completion of the Weil–Petersson metric on the Teichmüller space $T(\Sigma)$ of Σ.*

Brock [Bro03] showed that, for every hyperbolic surface Σ, $\overline{T}_{WP}(\Sigma)$ is quasiisometric to the pants complex (see [Bro03] for definitions) $C_P(\Sigma)$ of the surface Σ. Since Gromov hyperbolicity is a quasiisometry invariant, we get the following result:

Corollary 1.2. *If* Σ *is the five-times punctured sphere or the twice punctured torus,* *then the pants complex* $C_P(\Sigma)$ *of* Σ *is Gromov hyperbolic.* □

Let Σ be a hyperbolic surface of genus g and with p punctures and let $\mathcal{T}(\Sigma)$ be the Teichmüller space of Σ. The Weil–Petersson metric on $\mathcal{T}(\Sigma)$ is a non-complete metric of negative sectional curvatures (see [Wol03]). However, we note that Theorem 1.1 cannot be obtained as a consequence of the negative sectional curvatures of the Weil–Petersson metric, since these curvatures have been shown not to be bounded away from zero (see [Hua03]).

Let $\mathcal{T}_{WP}(\Sigma)$ denote the Teichmüller space of Σ endowed with the Weil–Petersson metric. The metric completion $\overline{\mathcal{T}}_{WP}(\Sigma)$ of $\mathcal{T}_{WP}(\Sigma)$ is the *augmented Teichmüller space* (see [Mas76]), i.e. the set of marked metric structures on Σ with nodes on a (possibly empty) collection of different homotopy classes of essential simple closed curves on Σ. Here, a curve is *essential* if it is not null-homotopic nor homotopic to a puncture. From now on we will refer to a homotopy class of essential simple closed curves on Σ simply as a *curve*, unless otherwise stated.

Let $C = C(\Sigma)$ be the *curve complex* of Σ, as defined in [Har81]. Recall that this is a finite-dimensional simplicial complex whose vertices correspond to *curves* on Σ, and that a subset $A \subseteq V(C)$ spans a simplex in C if the elements of A can be realised disjointly on Σ. Note that inclusion determines a partial order on the set of simplices of C. Following [Wol03], we can define a map $\Lambda : \overline{\mathcal{T}}_{WP}(\Sigma) \to C \cup \{\emptyset\}$ that assigns, to a point in $u \in \overline{\mathcal{T}}_{WP}(\Sigma)$, the (possibly empty) simplex of C spanned by the curves of Σ on which u has nodes. The space $\overline{\mathcal{T}}_{WP}(\Sigma)$ is the union of the level sets of Λ. Then, $\overline{\mathcal{T}}_{WP}(\Sigma)$ has the structure of a *stratified* space, where the level sets of Λ are the *strata*; observe that $\Lambda^{-1}(\{\emptyset\}) = \mathcal{T}_{WP}(\Sigma)$ and that two strata always intersect over a stratum. We will refer to $\Lambda^{-1}(\Lambda(u))$ as the *stratum containing u* and we will say that it has *label* $\Lambda(u)$. The strata of $\overline{\mathcal{T}}_{WP}(\Sigma)$ are isometric embeddings of products of lower dimensional Teichmüller spaces (which come from subsurfaces of Σ) with their corresponding Weil–Petersson metrics (see [Wol03]). It is clear that the stratum with label a collection of curves that determine a pants decomposition on Σ consists of only one point in $\overline{\mathcal{T}}_{WP}(\Sigma)$. Let $\mathrm{Mod}(\Sigma)$ be the mapping class group of Σ, i.e. the group of self-homeomorphisms of Σ up to homotopy. It is known (see [Abi80]) that $\mathrm{Mod}(\Sigma)$ acts cocompactly on $\overline{\mathcal{T}}_{WP}(\Sigma)$. The space $\overline{\mathcal{T}}_{WP}(\Sigma)$ is not locally compact: a point in $\overline{\mathcal{T}}_{WP}(\Sigma) \setminus \mathcal{T}_{WP}(\Sigma)$ does not admit a relatively compact neighbourhood. Indeed, let u be a point in $\overline{\mathcal{T}}_{WP}(\Sigma) \setminus \mathcal{T}_{WP}(\Sigma)$, which corresponds to a surface with a node on the simple closed curve α (and possibly more), and consider the Dehn twist T_α along α. Then the T_α-orbit of any point lies in every neighbourhood of u (see [Wol03]). Nevertheless, individual strata are locally compact, since they are (products of) lower dimensional Teichmüller spaces.

The following result summarises some deep and remarkable facts about the geometry of the Weil–Petersson metric on Teichmüller space (see Section 2 for the relevant definitions). Part (i) is due to S. Yamada [Yam01]; (ii) is due to Wolpert [Wol03] and (iii) is due to Daskalopoulos and Wentworth [DW03]. Let us note that all these results rely on previous work by Wolpert on the Weil–Petersson metric.

Theorem 1.3 ([DW03], [Wol03], [Yam01]). *Let Σ be a surface of hyperbolic type and let $\overline{\mathcal{T}}_{WP}(\Sigma)$ be the completion of the Weil–Petersson metric on $\mathcal{T}_{WP}(\Sigma)$. Then,*

(i) *The space $\overline{\mathcal{T}}_{WP}(\Sigma)$ is a* CAT(0) *space.*

(ii) *The closure of a stratum in $\overline{\mathcal{T}}_{WP}(\Sigma)$ is convex and complete in the induced metric.*

(iii) *The open geodesic segment $[u,v] \setminus \{u,v\}$ from u to v lies in the stratum with label $\Lambda(u) \cap \Lambda(v)$.* □

Let X_0 be the Teichmüller space of the five-times punctured sphere $\Sigma_{0,5}$ endowed with the Weil–Petersson metric and let X be its metric completion. We will write $X_F = X \setminus X_0$. Since a pants decomposition of $\Sigma_{0,5}$ corresponds to two disjoint curves on $\Sigma_{0,5}$, we get that a stratum in X has label either α or $\alpha\beta$, where α and β are disjoint curves on $\Sigma_{0,5}$. The closure of a stratum of the form S_α is given by $\overline{S}_\alpha = S_\alpha \cup (\bigcup_{\beta \in \mathcal{B}} S_{\alpha\beta})$, where $\mathcal{B}$ is the set of curves that are disjoint from α. We observe that any curve α separates $\Sigma_{0,5}$ into two subsurfaces, namely a four-times punctured sphere and a three-times punctured sphere. Then S_α corresponds to the Teichmüller space of the four-times punctured sphere component of $\Sigma_{0,5} \setminus \alpha$, since the Teichmüller space of the other subsurface is trivial. Also, recall that a stratum of the form $S_{\alpha\beta}$ is a single point in X.

We prove Theorem 1.1 in the case where the surface is a sphere with five punctures. Using the techniques from Sections 2 and 3 we immediately get the result for the twice-punctured torus. The only difference between the two cases is the nature of the strata in $\mathcal{T}_{WP}(\Sigma)$ which arise from pinching a single curve on the surface. In the first case, these strata correspond to the Teichmüller space of a four-times punctured sphere; in the second, they correspond either to the Teichmüller space of a four-times punctured sphere or to the Teichmüller space of a once-punctured torus, depending on whether or not the curve giving rise to such a stratum separates the surface. In both cases, $\mathcal{T}_{WP}(\Sigma)$ has the same structure as a stratified space.

Acknowledgements. I would like to thank Brian Bowditch for introducing me to the problem and for the interesting conversations we had about the topic. I would also like to thank the referee for his/her very useful comments.

2. Preliminaries

In order to give a proof of Theorem 1.1 we will have to make use of some geometric properties of CAT(0) spaces. Let us begin by recalling the definition of a CAT(0) space. Let Y be a *geodesic metric space*, that is, a space in which every two points can be connected by a path which realises their distance; such a path is called a *geodesic* between the two points. By a *triangle* in Y we will mean three points $x_1, x_2, x_3 \in Y$, the *vertices* of T, and three geodesics connecting them pairwise. We will write $[x_i, x_j]$ for the geodesic side of T with endpoints x_i and x_j. Throughout this article we will denote the euclidean plane by $\mathbb{E}^2$ and the euclidean distance in $\mathbb{E}^2$ by d_e.

Definition 2.1. Let Y be a geodesic metric space and let T be a triangle in Y with vertices x_1, x_2, x_3. A *comparison triangle* $\overline{T}$ for T in $\mathbb{E}^2$ is a geodesic triangle in $\mathbb{E}^2$ with vertices $\overline{x}_1, \overline{x}_2$ and $\overline{x}_3$ such that $d(x_i, x_j) = d_e(\overline{x}_i, \overline{x}_j)$ for all $i, j = 1, 2, 3$. Given a point $p \in [x_i, x_j]$, for some $i, j = 1, 2, 3$ distinct, a *comparison point* $\overline{p}$ for p is a point $\overline{p} \in [\overline{x}_i, \overline{x}_j]$ such that $d(x_i, p) = d_e(\overline{x}_i, \overline{p})$ and $d(x_j, p) = d_e(\overline{x}_j, \overline{p})$.

Definition 2.2. Let T be a triangle in the geodesic metric space Y. We say that the triangle T satisfies the CAT(0) *inequality* if for any points $p \in [x_i, x_j]$ and $q \in [x_j, x_k]$, for $i, j, k = 1, 2, 3$ distinct, we have that $d(p, q) \le d_e(\overline{p}, \overline{q})$, where $\overline{p}$ and $\overline{q}$ are comparison points for p and q, respectively, in the comparison triangle $\overline{T} \subseteq \mathbb{E}^2$. We will say that the space X is a CAT(0) *space* if every triangle in X satisfies a CAT(0) inequality.

The next three results about CAT(0) spaces are well-known; they will be crucial in our main argument. For a proof see, for instance, [BH99].

Theorem 2.3. *If Y is a CAT(0) space, then the distance function on Y is convex along geodesics, that is, if $\sigma, \sigma' : [0, 1] \to Y$ are geodesics in Y parametrised proportional to arc-length, then*

$$d(\sigma(t), \sigma'(t)) \le t d(\sigma(0), \sigma'(0)) + (1 - t) d(\sigma(1), \sigma'(1)),$$

for all $t \in [0, 1]$. □

Corollary 2.4. *Every CAT(0) space is uniquely geodesic.* □

Theorem 2.5. *Let Y be a CAT(0) space and let C be a complete convex subset of Y. Given $x \in Y$ there exists a point $\pi(x) \in C$ such that $d(x, \pi(x)) = \inf_{c \in C} d(x, c)$. Moreover, the map $\pi : Y \to C$ is distance non-increasing, that is, for all $x, y \in Y$ we have that $d(x, y) \ge d_C(\pi(x), \pi(y))$, where d_C denotes the subspace metric.* □

When interested in the large-scale geometry of a CAT(0) space, a natural question to ask is what are the obstructions for such a space to be Gromov hyperbolic. An answer to this question was given by Bowditch [Bow95] and Bridson [Bri95] separately, generalising a result announced by Gromov. They showed the following:

Theorem 2.6 ([Bow95], [Bri95]). *Let Y be a complete, locally compact* CAT(0) *space which admits a cocompact isometric group action. Then, either Y is Gromov hyperbolic or else it contains a totally geodesic embedding of a euclidean plane.*

Let X be the completion of the Weil–Petersson metric on the Teichmüller space of the five-times punctured sphere. We are going to show, using some of the techniques in [Bow95], that if X is not Gromov hyperbolic then there exists an isometrically embedded euclidean disc in one of the strata of X, which is impossible since all the sectional curvatures of a stratum of X are strictly negative. More specifically, we are going to construct, for each $n \in \mathbb{N}$, a map $\phi_n : \mathbb{D} \to X$, where $\mathbb{D}$ is the unit disc in $\mathbb{E}^2$, such that

$$\lambda_n d_e(x,y) \leq d(\phi_n(x), \phi_n(y)) \leq d_e(x,y),$$

for all $x, y \in \mathbb{D}$, and where $\lambda_n \in (0,1)$ with $\lambda_n \to 1$ as $n \to \infty$. If the space X were locally compact, one could take the limit of ϕ_n as n tends to infinity, obtaining in this way an isometric embedding of the euclidean unit disc in X. Even though X is not locally compact, we will be able to use the structure of X as a stratified space and the fact that individual strata are locally compact to obtain such an isometrically embedded disc. We note that we will not use the cocompact isometric action of the mapping class group on X to construct these maps; this action will only be used in the arguments in the next section.

The construction of the maps ϕ_n is totally analogous to the one in [Bow95], where Bowditch shows the following result (we remark that the results in [Bow95] are more general than the ones we present here).

Lemma 2.7 ([Bow95]). *Let Y be a* CAT(0) *space. Given $n \in \mathbb{N}$ and $\varepsilon > 0$, there exists a map $\phi_{n,\varepsilon} : ([-n,n] \cap \mathbb{Z})^2 \to Y$ such that*

$$|i - i'| - \varepsilon \leq d(\phi_{n,\varepsilon}(i,j), \phi_{n,\varepsilon}(i',j)) \leq |i - i'| \tag{1}$$

and

$$|j - j'| - \varepsilon \leq d(\phi_{n,\varepsilon}(i,j), \phi_{n,\varepsilon}(i,j')) \leq |j - j'| \tag{2}$$

$\square$

Remark 2.8. For the sake of completeness, we now give a brief account on Bowditch's construction. Let $n \in \mathbb{N}$ and $\varepsilon > 0$ and let q be a natural number bigger than $\frac{2n^2}{\varepsilon}$. Let $\sigma : [0, qn] \to Y$ be a geodesic segment in Y and let y be a point in Y at distance at least n from σ. Let $\tau_i : [0, d(y, \sigma(i))] \to Y$ be the unique geodesic from $\sigma(i)$ to y, for all $i \in [0, qn] \cap \mathbb{N}$. Bowditch then sets, for all $i, j = 0, \ldots, n$ and for all $p = 0, \ldots, q-1$, $\phi_p(i,j) = \tau_{pn+i}(j)$ and shows that there exists a number $p = 0, \ldots, q-1$ such that ϕ_p

satisfies (1) and (2).

Bowditch then shows that if, in addition, the space Y is not Gromov hyperbolic then the maps ϕ_n satisfy the following non-degeneracy condition.

Lemma 2.9 ([Bow95]). *Let Y be a* CAT(0) *space and suppose Y is not Gromov hyperbolic. Then the maps ϕ_n in the result above satisfy, in addition, that*

$$d(\phi_{n,\varepsilon}(i,j), \phi_{n,\varepsilon}(i+1,j+1)) \geq 1/2 \tag{3}$$

and

$$d(\phi_{n,\varepsilon}(i,j+1), \phi_{n,\varepsilon}(i+1,j)) \geq 1/2 \tag{4}$$

for all $n \in \mathbb{N}$. $\square$

We now give an extension, using standard arguments about CAT(0) spaces, to Bowditch's construction. It will play a central role in the proof of Theorem 1.1.

Lemma 2.10. *Let Y be a* CAT(0) *space and suppose that Y is not Gromov hyperbolic. Fix a number $N \in \mathbb{N}$ and consider $K = [-N,N] \times [-N,N] \subseteq \mathbb{R}^2$, endowed with the euclidean metric. Then there exists a sequence $(\phi_n)_{n \in \mathbb{N}}$ of maps $\phi_n : K \to Y$ such that*

$$\lambda_n d_e(x,y) \leq d(\phi_n(x), \phi_n(y)) \leq d_e(x,y),$$

where $\lambda_n \in (0,1)$ for all $n \in \mathbb{N}$ and $\lambda_n \to 1$ as $n \to \infty$.

Proof. Let $(\varepsilon_n)_{n \in \mathbb{N}}$ be a sequence of positive real numbers tending to zero and let $\phi_n = \phi_{n,\varepsilon_n}$ be the map described in the previous two Lemmas.

First, we are going to extend the map ϕ_n to K as follows: let $\sigma_n : [0,1] \to Y$ be the unique geodesic segment in Y from $\phi_n(0,0)$ to $\phi_n(1,0)$. Similarly, let $\sigma'_n : [0,1] \to Y$ be the geodesic segment in Y from $\phi_n(0,1)$ to $\phi_n(1,1)$. Here, and from now on, we assume that all geodesics are parametrised proportional to arc-length. For $t \in [0,1]$, let $\tau_t^n : [0,1] \to Y$ be the unique geodesic in Y connecting $\sigma_n(t)$ and $\sigma'_n(t)$. We set $\phi_n(t,s) := \tau_t^n(s)$, for all $t,s \in [0,1]$. We can extend ϕ_n to K by performing this construction on each square $[i,i+1] \times [j,j+1]$ for all $i,j \in [-N, N-1] \cap \mathbb{Z}$, provided n is large enough so that ϕ_n is defined on the whole of K. Note that, up to extracting a subsequence, we can assume that this is always the case.

We now proceed to show that the maps ϕ_n are continuous on K. Let $0 \leq t,t',s,s' \leq 1$. From the convexity of the distance function we obtain that

$$
\begin{aligned}
d(\tau_t^n(s),\tau_{t'}^n(s)) &\le sd(\tau_t^n(0),\tau_{t'}^n(0))+(1-s)d(\tau_t^n(1),\tau_{t'}^n(1))\\
&= sd(\sigma_n(t),\sigma_n(t'))+(1-s)d(\sigma_n'(t),\sigma_n'(t'))\\
&= |t-t'|.
\end{aligned}
$$

From this inequality and the fact that $\tau_t^n : [0,1] \to Y$ is a geodesic paremetrised protional to arc length for every $t \in [0,1]$, we deduce that

$$
\begin{aligned}
d(\phi_n(t,s),\phi_n(t',s')) &= d(\tau_t^n(s),\tau_{t'}^n(s'))\\
&\le d(\tau_t^n(s),\tau_t^n(s'))+d(\tau_t^n(s'),\tau_{t'}^n(s'))\\
&= |s-s'|+|t-t'|,
\end{aligned}
$$

and thus ϕ_n is continuous on $[0,1] \times [0,1]$ for all $n \in \mathbb{N}$. A totally analogous argument gives the result for K.

Notation 2.11. Let L be a straight line in K. We will say that the images of L under ϕ_n *get arbitrarily close to being geodesic* in Y if the Hausdorff distance between $\phi_n(L)$ and the geodesic in Y with the same endpoints tends to zero as n grows.

Next, we show that the images of horizontal and vertical lines in K under ϕ_n get arbitrarily close to being parallel geodesics in Y. Let L be a horizontal line in K of the form $[-N,N] \times \{j\}$, for some $j \in [-N,N] \cap \mathbb{Z}$. The fact that $\phi_n(L)$ gets arbitrarily close to being geodesic in Y follows directly from inequality (1) and the definition of the maps ϕ_n. Now, since the distance function on Y is convex along geodesics (Theorem 2.3) we get, from inequality (2), that

$$
d(\phi_n(t,j),\phi_n(t,j+1)) \le 1,
$$

for all $t \in [-N,N]$, $j \in [-N,N-1] \cap \mathbb{Z}$ and $n \in \mathbb{N}$. The convexity of the distance function on Y yields that the real function $[t \to d(\phi_n(t,j),\phi_n(t,j+1))]$ tends, as n grows, to a real function which is convex (and also bounded, from (2)). But a real function which is convex and bounded must be constant and therefore $d(\phi_n(t,j),\phi_n(t,j+1)) \to 1$ as $n \to \infty$, from (1).

Consider now an arbitrary horizontal line $L = [-N,N] \times \{s_0\}$ in K, for some $s_0 \in [-N,N]$. For the simplicity of the notation we will assume that $0 \le s_0 \le 1$. Let $\rho_n : [0,1] \to Y$ be the geodesic in Y between $\phi_n(-N,s_0)$ and $\phi_n(N,s_0)$. Let $c_n = \phi_n([-N,N] \times \{0\})$ and $c_n' = \phi_n([-N,N] \times \{1\})$ for all $n \in \mathbb{N}$. We know, from the discussion above, that $d(c_n(t),c_n'(t)) \to 1$ as $n \to \infty$ for all $t \in [-N,N]$. Note that c_n (resp.

c'_n) is geodesic when restricted to $[i, i+1] \times \{0\}$ (resp. $[i, i+1] \times \{1\}$). From the convexity of the distance function we get that $d(c_n(t), \rho_n(t)) \to s_0$ and $d(c'_n(t), \rho_n(t)) \to 1 - s_0$ as $n \to \infty$. But $d(\phi_n(t, s_0), c_n(t)) \to s_0$ and $d(\phi_n(t, s_0), c'_n(t)) \to 1 - s_0$ as $n \to \infty$, since $[s \to \phi_n(t, s)]$ is geodesic for a fixed t. Thus the images of horizontal lines in K under ϕ_n are paths that get arbitrarily close to being parallel geodesics $n \to \infty$. Note that this implies that, up to taking a subsequence, the maps ϕ_n are injective on K.

By a completely analogous convexity argument, we can deduce that the images of a vertical segment of the form $\{i\} \times [-N, N]$ under ϕ_n, where $i \in [-N, N] \cap \mathbb{Z}$, get arbitrarily close to being geodesic in Y as n grows. Similarly we obtain that the images of any two vertical lines in K under ϕ_n get arbitrarily close to being parallel geodesics as $n \to \infty$.

So the maps ϕ_n are continuous, injective and map horizontal (and vertical) lines in K to paths in Y which get arbitrarily close to being parallel geodesics. Let d_n be the pull-back metric on K determined by ϕ_n. Since $d_n((-N, -N), (N, N))$ is bounded above and below (note that, in particular, it is bounded below away from 0, from (3) and (4)) we can assume that $d_n((-N, -N), (N, N)) \to a > 0$, up to extracting a subsequence. Also recall, from inequalities (1) and (2), that $d_n((-N, -N), (-N, N)) \to 2N$ and $d_n((-N, -N), (N, -N)) \to 2N$ as $n \to \infty$.

Let $f : \mathbb{R}^2 \to \mathbb{R}^2$ be an affine map such that $f(t, 0) = (t, 0)$, for all $t \in \mathbb{R}$, and that $d_e(f(-N, -N), f(N, N)) = a$. By considering $\phi_n \circ f^{-1}$, which we denote again by ϕ_n abusing notation, we deduce that the maps ϕ_n, restricted to K, satisfy that

$$d_e(x, y) - \varepsilon_n \le d(\phi_n(x), \phi_n(y)) \le d_e(x, y),$$

for all $x, y \in K$. Consider the restriction of ϕ_n to the unit disc $\mathbb{D}$ in $\mathbb{R}^2$. It follows immediately from the last inequality that there exists a sequence $(\lambda_n)_{n \in \mathbb{N}}$ of real numbers, with $\lambda_n = \lambda_n(\varepsilon_n)$, $\lambda_n \in (0, 1)$ and $\lambda_n \to 1$ as $n \to \infty$, such that

$$\lambda_n d_e(x, y) \le d(\phi_n(x), \phi_n(y)) \le d_e(x, y),$$

as desired. $\square$

3. Proof of the Theorem

Let us begin with some technical results that will be important in the proof of Theorem 1.1.

Remark 3.1. Let Σ be a hyperbolic surface. In [Wol03], Corollary 21, Wolpert shows that for a stratum S defined by the vanishing of the length sum $l = l_1 + \ldots l_n$, where l_i

corresponds to the length of some curve α_i on the surface for $i = 1, \ldots, n$, the distance to the stratum is given locally as $d(p,S) = (2\pi l)^{1/2} + O(l^2)$. In particular, if $(u_m)_{m \in \mathbb{N}}$ is a sequence in X such that $d(u_m, S) \to 0$ as $m \to \infty$ we get that $l_{u_m}(\alpha_i) \to 0$ as $m \to \infty$, for all $i = 1, \ldots, n$, where $l_{u_m}(\alpha)$ denotes the length of α in u_m.

Lemma 3.2. *Let S_α and S_β be two different strata in X_F such that $\overline{S}_\alpha \cap \overline{S}_\beta \neq \emptyset$. Let $(u_n)_{n \in \mathbb{N}}$ and $(v_n)_{n \in \mathbb{N}}$ be sequences in X such that $d(u_n, \overline{S}_\alpha) \to 0$, $d(v_n, \overline{S}_\beta) \to 0$, and $d(u_n, v_n) \to 0$ as $n \to \infty$. Then $d(u_n, S_{\alpha\beta}) \to 0$ as $n \to \infty$ (and thus $d(v_n, S_{\alpha\beta}) \to 0$ as well).*

Proof. Suppose, for contradiction, that the result is not true, that is, there exists $R > 0$ such that $d(u_n, S_{\alpha\beta}) > R$ for infinitely many n. Let Γ be the mapping class group of $\Sigma_{0,5}$. Since the action of Γ on X is cocompact there is a compact subset Z of X and a sequence $(\gamma_n)_{n \in \mathbb{N}}$ in Γ such that $\gamma_n(u_n) \in Z$ for all $n \in \mathbb{N}$. Note that $d(u_n, v_n) \to 0$ as $n \to \infty$ if and only if $d(\gamma_n(u_n), \gamma_n(v_n)) \to 0$ as $n \to \infty$, since Γ acts isometrically on X and therefore $d(\gamma_n(u_n), \gamma_n(v_n)) = d(u_n, v_n)$ for all $n \in \mathbb{N}$. We replace u_n and v_n by $\gamma_n(u_n)$ and $\gamma_n(v_n)$, respectively; abusing notation we denote the new points by u_n and v_n again. Then, up to extracting a subsequence, we get that the sequence $(u_n)_{n \in \mathbb{N}}$ converges to a point $u_0 \in X$, since Z is compact.

Since $d(u_n, \overline{S}_\alpha) \to 0$ as $n \to \infty$, then $l_{u_n}(\alpha) \to 0$ as $n \to \infty$. Similarly, we have that $l_{v_n}(\beta) \to 0$ as $n \to \infty$. From the remark above and since $d(u_n, S_{\alpha\beta}) > R$ for infinitely many n, it follows that there must exist a constant $C = C(R) > 0$ such that $l_{u_n}(\beta) > C$, for infinitely many $n \in \mathbb{N}$. Therefore, for the simple closed curve β we have that $l_{u_n}(\beta) > C > 0$, for infinitely many n, and $l_{v_n}(\beta) \to 0$ as $n \to \infty$, which is impossible since we know from the hypotheses that $d(u_n, v_n) \to 0$ as $n \to \infty$. $\square$

As a trivial consequence, we get the following corollary:

Corollary 3.3. *Given $r > 0$ and a stratum of the form $S_{\alpha\beta}$, let $B(r) = B(S_{\alpha\beta}, r)$ be the ball of radius r around $S_{\alpha\beta}$. Then, there exists $D = D(r) > 0$ such that the following holds:*

Let $(u_n)_{n \in \mathbb{N}}$ and $(v_n)_{n \in \mathbb{N}}$ be two sequences in X such that $d(u_n, \overline{S}_\alpha \setminus (B(r) \cap \overline{S}_\alpha)) \to 0$ and $d(v_n, \overline{S}_\beta \setminus (B(r) \cap \overline{S}_\beta)) \to 0$ as $n \to \infty$. Then $d(u_n, v_n) > D$ for infinitely many $n \in \mathbb{N}$. $\square$

Proof of Theorem 1.1. Suppose, for contradiction, that X is not Gromov hyperbolic. Since X is a CAT(0) space, we can construct a sequence of maps $(\phi_n : \mathbb{D} \to X)_{n \in \mathbb{N}}$ as described in the previous section. Using the structure of X as a stratified space, there is a stratum $S \subseteq X$ such that $\phi_n(\mathbb{D}) \subseteq \overline{S}$ (note that S may have label $\emptyset$, so that $\overline{S} = X$). Also, S cannot have a pants decomposition as label, since recall that such a stratum consists of only one point). The stratum S corresponds to the Weil–Petersson

Teichmüller space of a properly embedded hyperbolic subsurface $\Sigma_S \subseteq \Sigma_{0,5}$ (possibly with $\Sigma_S = \Sigma_{0,5}$). Let Γ_0 be the mapping class group of Σ_S, and recall that Γ_0 acts cocompactly on $\overline{S}$. We will denote $\overline{S} \setminus S$ by S_F. We now have two possibilities:

(a) There exists a point $x_0 \in \mathbb{D}$ and a constant $\delta = \delta(x_0)$ such that $d(\phi_n(x_0), S_F) > \delta$, for infinitely many $n \in \mathbb{N}$. We can assume, up to passing to X/Γ_0 and lifting back, that the sequence $(\phi_n(x_0))_{n \in \mathbb{N}}$ converges to a point $w \in S$, since the action of Γ_0 on $\overline{S}$ is cocompact.

Notation 3.4. We will write $D_e(a, r)$ to denote the euclidean disc in $\mathbb{E}^2$ with centre a and radius r.

Consider $D_e(x_0, \eta)$, where $\eta = \min(\delta/4, 1 - d_e(0, x_0))$. Then,

$$d(\phi_n(x), S_F) \geq d(\phi_n(x_0), S_F) - d(\phi_n(x), \phi_n(x_0)) \geq \delta - 2\eta \geq \delta/2 > 0,$$

for all $x \in D_e(x_0, \eta)$. So, $d(\phi_n(D_e(x_0, \eta)), S_F) > \delta/2$, for all $n \in \mathbb{N}$. Since S is locally compact, the maps ϕ_n converge (maybe after passing to a subsequence) on $D_e(x_0, \eta)$ to a map ϕ. Therefore, $\phi(D_e(x_0, \eta))$ is a copy of a euclidean disc in S, which is impossible since all its sectional curvatures are strictly negative.

(b) Otherwise, for all $x \in \mathbb{D}$, $d(\phi_n(x), S_F) \to 0$ as $n \to \infty$. Again we can assume (up to the action of Γ) that $\phi_n(0) \to w'$ as $n \to \infty$, for some $w' \in S_F$.

Remark 3.5. One of the consequences of the *Collar Lemma* is that if α and β are intersecting curves on a hyperbolic surface Σ, then their lengths cannot be very small simultaneously. In the light of this result, it is possible to show (see [Wol03], Corollary 22) that there exists a constant $k_0 = k_0(\Sigma)$ such that two strata $S_1, S_2 \subseteq \overline{\mathcal{T}}_{WP}(\Sigma) \setminus \mathcal{T}_{WP}(\Sigma)$ either have intersecting closures or they satisfy $d(S_1, S_2) \geq k_0$.

Let $k_0 = k_0(\Sigma_S)$ be the constant given in the remark above and consider the maps ϕ_n restricted to $D_e(0, k_0/3)$. We may as well assume that $k_0 \leq 1$. From Remark 3.1, we deduce that S must have label $\emptyset$. Otherwise, if S had label α, for some simple closed curve α, the fact that $d(\phi_n(x), S_F) \to 0$ as $n \to \infty$, for all $x \in D_e(0, k_0/3)$ would imply that there exists a curve β in $\Sigma_{0,5}$, disjoint from α, such that $d(\phi_n(x), S_{\alpha\beta}) \to 0$ as $n \to \infty$, for all $x \in D_e(0, k_0/3)$. But we know that a stratum of the form $S_{\alpha\beta}$ consists of only one point, which contradicts the construction of the maps ϕ_n since we know that the maps ϕ_n contract distances by at most a factor λ_n. Thus we assume, from now on, that S has label $\emptyset$ and so $\overline{S} = X$ and $S_F = X_F$.

We observe that, given $u \in X_F$, there are at most two non-trivial strata, say S_α and S_β, in X_F such that $u \in \overline{S}_\alpha \cap \overline{S}_\beta$. These strata correspond to two (possibly equal) simple closed curves on the surface on which u has nodes. Note that in the case when there are exactly two strata, we get that $\{u\} = \bar{S}_\alpha \cap \bar{S}_\beta$. So we deduce that $d(\phi_n(x), \bar{S}_\alpha \cup \bar{S}_\beta) \to 0$ as $n \to \infty$, for all $x \in D_e(0, k_0/3)$.

Our next aim is to show that there exists $x_1 \in D_e(0, k_0/3)$ and $k_1 = k_1(x_1) > 0$, with $k_1 \leq k_0/3$, so that the images under ϕ_n of the points in $D_e(x_1, k_1)$ get uniformly arbitrarily close to the same stratum as $n \to \infty$. Using this result, we will be able to define a distance non-decreasing projection from $D_e(x_1, k_1)$ to the closure of that particular stratum.

Suppose that $\phi_n(0) \to w' \in \bar{S}_\alpha \setminus S_{\alpha\beta}$ as $n \to \infty$. In particular, we can choose $k_0' < k_0$ small enough so that there exists $r > 0$ with $d(\phi_n(x), S_{\alpha\beta}) > r$, for all $x \in D_e(0, k_0')$ and for all $n \in \mathbb{N}$. Let $D = D(r)$ be the constant given in Corollary 3.3 and consider $D_e(0, k)$, where $k = \min(D/3, k_0'/3)$. We claim that we can take (x_1, k_1) to be $(0, k)$. Suppose, for contradiction, that the images of $D_e(0, k)$ do not get uniformly arbitrarily close to $\bar{S}_\alpha$; that is, there exists $K_0 > 0$, a subsequence $(\phi_m)_{m \in \mathbb{N}} \subseteq (\phi_n)_{n \in \mathbb{N}}$ and points $u_m \in \phi_m(D_e(0, k))$ such that $d(u_m, \bar{S}_\alpha) \geq K_0$ for all $m \in \mathbb{N}$. We know that, for all $m \in \mathbb{N}$, $u_m = \phi_m(x_m)$ for some $x_m \in D_e(0, k)$ and thus, up to a subsequence, $x_m \to y \in D_e(0, k)$. From the properties of the maps ϕ_n we have that $d(u_m, \phi_m(y)) \to 0$ as $n \to \infty$. Therefore, $d(u_m, \bar{S}_\alpha \cup \bar{S}_\beta) \to 0$ as $m \to \infty$ and thus $d(u_m, \bar{S}_\beta) \to 0$, since we know that $d(u_m, \bar{S}_\alpha) \geq K_0$ for all m. This contradicts Corollary 3.3, since the points u_m and $\phi_m(0)$ lie in $X \setminus B(r)$, $d(\phi_m(0), \bar{S}_\alpha) \to 0$, $d(u_m, \bar{S}_\beta) \to 0$ but $d(u_m, \phi_m(0)) \leq D/3$.

The case $\phi_n(0) \to w' \in S_{\alpha\beta}$ as $n \to \infty$ is dealt with in complete analogy, considering x_1 to be any point in $D_e(0, k_0/3) \setminus \{0\}$ and and defining k_1 in a similar way as we did above.

Since X is a CAT(0) space and $\bar{S}_\alpha$ is complete and convex we can consider the projection $\pi: X \to \bar{S}_\alpha$ as defined in Theorem 2.5. Recall that this projection is distance non-increasing; in particular, for all $x, y \in D_e(x_1, k_1)$ we have that

$$d(\pi(\phi_n(x)), \pi(\phi_n(y))) \leq d(\phi_n(x), \phi_n(y)) \leq d_e(x, y).$$

Choose a sequence $(\delta_m)_{m \in \mathbb{N}}$ of positive reals such that $\delta_m \to 0$ as $m \to \infty$. Up to a subsequence we can assume that, given $m \in \mathbb{N}$, $d(\phi_n(x), \bar{S}_\alpha) < \delta_m$, for $n \geq m$ and for all $x \in D_e(x_1, k_1)$. We have that

$$d(\phi_n(x), \phi_n(y)) \leq d(\phi_n(x), \pi(\phi_n(x))) + d(\phi_n(y), \pi(\phi_n(y))) + d(\pi(\phi_n(x)), \pi(\phi_n(y))),$$

and thus, for all $m \in \mathbb{N}$,

$$d(\pi(\phi_m(x)), \pi(\phi_m(y))) \geq d(\phi_m(x), \phi_m(y)) - 2\delta_m \geq \lambda_m d_e(x, y) - 2\delta_m,$$

Let $\psi_n = \pi \circ \phi_n : D_e(x_1, k_1) \to \overline{S}_\alpha$. Then there exists a sequence $(\lambda_n')_{n \in \mathbb{N}}$ of positive real numbers tending to 1 so that

$$\lambda_n' d_e(x,y) \leq d(\psi_n(x), \psi_n(y)) \leq d_e(x,y).$$

(we could simply take $\lambda_n' = \lambda_n - 2\delta_n$, since $k_1 < 1$ and therefore $2\delta_n < 2\delta_n d_e(x,y)$ for all $x, y \in D_e(x_1, k_1)$). We are now back to the situation described at the beginning of Section 3, this time in a stratum of the form $\overline{S}_\alpha$. Reasoning in a totally analogous way to cases (a) and (b) we get that either we can find an isometrically embedded copy of a euclidean disc in S_α or else we can find $x_2 \in D_e(x_1, k_1)$ and $k_2 > 0$ such that $D_e(x_2, k_2) \subseteq D_e(x_1, k_1)$ and $d(\phi_n(x), S_{\alpha\beta}) \to 0$ as $n \to \infty$, for all $x \in D_e(x_2, k_2)$, which is impossible since we know that ϕ_n decreases distances at most by a factor λ_n. In any case, we get a contradiction.

Therefore X is Gromov hyperbolic, as desired. □

References

[Abi80]   W. Abikoff (1980).   *The Real Analytic Theory of Teichmüller Space.* Springer, Berlin.

[Beh05]   J. Behrstock (2005). Asymptotic geometry of the mapping class group and Teichmüller space. Preprint, `math.GT/0502367`.

[BF01]   J. Brock & B. Farb (2001).   Curvature and rank of Teichmüller space. Preprint, `math.GT/0109045`.

[BH99]   M. R. Bridson & A. Haefliger (1999). *Metric Spaces of Non-positive Curvature.* Springer-Verlag, Berlin.

[Bow95]   B. H. Bowditch (1995).   Minkowskian subspaces of non-positively curved metric spaces. *Bull. London Math. Soc.* **27** (6), 575–584.

[Bri95]   M. R. Bridson (1995). On the existence of flat planes in spaces of nonpositive curvature. *Proc. Amer. Math. Soc.* **123** (1), 223–235.

[Bro03]   J. F. Brock (2003).   The Weil–Petersson metric and volumes of 3-dimensional hyperbolic convex cores. *J. Amer. Math. Soc.* **16** (3), 495–535 (electronic).

[DW03]   G. Daskalopoulos & R. Wentworth (2003). Classification of Weil–Petersson isometries. *Amer. J. Math.* **125** (4), 941–975.

[Har81]   W. J. Harvey (1981). Boundary structure of the modular group. In *Riemann Surfaces and Related Topics: Proceedings of the 1978 Stony Brook Conference (State Univ. New York, Stony Brook, N.Y., 1978), Ann. of Math. Stud.,* volume 97, pp. 245–251. Princeton Univ. Press, Princeton, N.J.

[Hua03]   Z. Huang (2003). Asymptotic flatness of the Weil–Petersson metric on Teichmüller space. Preprint, math.DG/0312419.

[Mas76]   H. Masur (1976). Extension of the Weil–Petersson metric to the boundary of Teichmuller space. *Duke Math. J.* **43** (3), 623–635.

[MM99]   H. A. Masur & Y. N. Minsky (1999). Geometry of the complex of curves. I. Hyperbolicity. *Invent. Math.* **138** (1), 103–149.

[Wol03]   S. A. Wolpert (2003). Geometry of the Weil–Petersson completion of Teichmüller space. In *Surveys in Differential Geometry, Vol. VIII (Boston, MA, 2002),* pp. 357–393. Int. Press, Somerville, MA.

[Yam01]   S. Yamada (2001). Weil–Petersson completion of Teichmüller spaces and mapping class group actions. Preprint, math.DG/0112001.

Javier Aramayona

Mathematics Institute
University of Warwick
Coventry, CV4 7AL
U.K.

jaram@maths.warwick.ac.uk

AMS Classification: 57M50, 51K10

Keywords: Weil–Petersson metric, Gromov-hyperbolic, CAT(0) space

Spaces of Kleinian Groups   Cambridge University Press

Lond. Math. Soc. Lec. Notes **329**, 233–245   Y. Minsky, M. Sakuma & C. Series (Eds.)

Convexity of geodesic-length functions: a reprise

Scott A. Wolpert

Abstract

New results on the convexity of geodesic-length functions on Teichmüller space are presented. A formula for the Hessian of geodesic-length is presented. New bounds for the gradient and Hessian of geodesic-length are described. A relationship of geodesic-length functions to Weil-Petersson distance is described. Applications to the behavior of Weil-Petersson geodesics are discussed.

1. Introduction

In this research brief we describe a new approach to the work [Wol87] (esp. Secs. 3 and 4), as well as new results and applications of the convexity of geodesic-length functions on the Teichmüller space $\mathcal{T}$. Our overall goal is to obtain an improved understanding of the convexity behavior of geodesic-length functions along Weil-Petersson (WP) geodesics. Applications are presented in detail for the $CAT(0)$ geometry of the augmented Teichmüller space. A complete treatment of results is in preparation [Wol04]. Convexity of geodesic-length functions has found application for the convexity of Teichmüller space [Bro02, Bro03, DS03, Ker83, Ker92, McM00, SS01, SS99, Wol87, Yeu03], for the convexity of the WP metric completion [DW03, MW02, Wol03, Yam01], for the study of harmonic maps into Teichmüller space [DKW00, Yam99, Yam01], and for the action of the mapping class group [DW03, MW02]. We consider marked Riemann surfaces R with complete hyperbolic metrics possibly with cusps and consider the lengths of closed geodesics. The length of the unique geodesic in a prescribed free homotopy class provides a function on the Teichmüller space. Specifically for σ a closed curve on R, let $\ell_\sigma(R)$ denote the length of the geodesic homotopic to σ; more generally for μ a geodesic current [Bon88], let $\ell_\mu(R)$ denote the total-length of the geodesic current for R.

A closed geodesic σ on R determines a cyclic cover of R by a geometric cylinder C. For ℓ the length of σ the geometric cylinder is represented as $\mathbb{H}/ < t \to e^\ell t >$ for $\mathbb{H}$ the upper half-plane with coordinate t; for $w = \exp(2\pi i \frac{\log t}{\ell})$ the cylinder is further represented as the concentric annulus $\{e^{-\frac{2\pi^2}{\ell}} < |w| < 1\}$ in the plane. We discovered in [Wol87](Sec. 4) that the potential operator for the Beltrami equation on C is diagonalized by the $\mathbf{S}^1$ rotation action of the cylinder and that the potential equation can be solved term-by-term for the corresponding Fourier expansions. The special properties for the potential theory generalize the properties for the function theory of the cylinder. For instance holomorphic differentials on R, lifted to C, admit Laurent (Fourier) expansions. The WP dual of the Hessian of ℓ_σ, a quadratic form for holomorphic quadratic differentials on R, has Hermitian and complex-bilinear components *diagonalized* by the terms of the corresponding Laurent expansions [Wol87](see Lemmas 4.2 and 4.4.) We further found that the contribution for a single Laurent term is a positive definite form. At this time, we have simplified the considerations of the Hessian and are now able to effect a straightforward comparison to the Petersson pairing for holomorphic quadratic differentials [Wol04]. The simplified considerations provide the basis for an improved understanding of the Hessian and of convexity. In the following paragraphs we outline the approach and results. We close the discussion by providing several applications complete with proofs.

2.   The Hessian of geodesic-length

We introduce for μ a geodesic current a natural function $\mathbb{P}_\mu$ on R. We begin with the geometry of the space of complete geodesics on the hyperbolic plane. For $\mathbb{H}$ the upper half plane with boundary $\check{\mathbb{R}} = \mathbb{R} \cup \{\infty\}$, the space of complete geodesics on $\mathbb{H}$ is given as $G = \check{\mathbb{R}} \times \check{\mathbb{R}} \setminus \{diagonal\}/\{interchange\}$. A point p of $\mathbb{H}$ is at finite distance $d(p,\sigma)$ to a complete geodesic σ and so $e^{-2d(p,\sigma)}$ defines a *Gaussian* on G. The natural area measure on G is $\omega = (a-b)^{-2}da\,db$ in terms of the *endpoint coordinates* $(a,b)/\sim$. The measure $e^{-2d(p,\sigma)}\omega$ is finite for G. Finiteness is noted as follows. A point z of $\mathbb{H}$, its conjugate $\bar{z}$, and the boundary points (a,b) have cross ratio $cr(z,a,b) = \frac{(a-b)\Im z}{|z-a||z-b|}$. The simple inequality $cr^2(z,a,b) \geq e^{-2d(z,ab)}$ is established by considering the point triple $(i,a,-a)$. Finiteness of the measure now follows from the inequality $e^{-2d(i,\widehat{ab})}\omega \leq (1+a^2)^{-1}(1+b^2)^{-1}da\,db$ for the point triple (i,a,b). A geodesic current μ for R naturally lifts to the upper half plane; the lift is a positive measure $d\mu$ on the space G of complete geodesics. For R represented as the quotient $\mathbb{H}/\Gamma$ the integral

$$\mathbb{P}_\mu(p) = \int_G e^{-2d(p,\sigma)}d\mu(\sigma) \tag{2.1}$$

defines a Γ-invariant function on $\mathbb{H}$, *the mean-squared inverse exponential-distance of p to μ*. Finiteness of the integral is established by comparing $d\mu$ to ω. The construction for $\mathbb{P}_\mu$ is motivated by the construction for the classical Petersson series representing the differential $d\ell_\sigma$ of the geodesic-length on $\mathcal{T}$ [Gar75, Gar86]. The reader can check that $\mu \to \mathbb{P}_\mu$ is a continuous mapping from the space of geodesic currents to the space of continuous functions on R. The central role of $\mathbb{P}_\mu$ in studying geodesic-length functions and the total-length of geodesic currents is discussed and demonstrated below.

From Kodaira–Spencer deformation theory the infinitesimal deformations of R are represented by the Beltrami differentials $\mathcal{H}(R)$ harmonic with respect to the hyperbolic metric [Ahl61]. A harmonic Beltrami differential is the symmetric tensor given as $\overline{\varphi}(ds^2)^{-1}$ for φ a holomorphic quadratic differential with at most simple poles at the cusps and ds^2 the hyperbolic metric tensor. At R the differential on $\mathcal{T}$ of the geodesic-length of ℓ_σ is bounded for $v \in \mathcal{H}(R)$ as

$$|d\ell_\sigma(v)| \leq \frac{8}{\pi} \int_R |v| \, \mathbb{P}_\sigma \, dA$$

for dA the hyperbolic area element and from applying the inequality $|(\frac{\Im z}{\overline{z}})^2| \leq 4e^{-2\widehat{d(z,0\infty)}}$ and the formula of F. Gardiner [Gar75]. By taking limits the integral bound is generalized to the total-length of laminations. Ahlfors noted [Ahl61] for second-order deformations defined by harmonic Beltrami differentials that the WP Levi-Civita connection is Euclidean to zeroth order in the following sense. A Γ-invariant Beltrami differential v on $\mathbb{H}$ determines a one-parameter family as follows. For the complex parameter ε small there is a suitable self-homeomorphism f^ε of $\mathbb{H}$ satisfying $f^\varepsilon_{\overline{z}} = \varepsilon v f^\varepsilon_z$. The homeomorphism f^ε serves to compare the quotients $\mathbb{H}/\Gamma$ and $\mathbb{H}/f^\varepsilon \circ \Gamma \circ (f^\varepsilon)^{-1}$. For a basis of harmonic Beltrami differentials $v_1, \ldots, v_n$ and small complex parameters ε_* and $v(\varepsilon) = \sum_j \varepsilon_j v_j$ the association $(\varepsilon_1, \ldots, \varepsilon_n)$ to $\mathbb{H}/f^{v(\varepsilon)} \circ \Gamma \circ (f^{v(\varepsilon)})^{-1}$ in effect provides a local coordinate for $\mathcal{T}$. Ahlfors found for a basis of harmonic Beltrami differentials that the local coordinates for $\mathcal{T}$ are normal: the first derivatives of the WP metric tensor vanish at the origin [Ahl61]. The observation is used in the calculation of the WP Riemannian Hessian $\ddot{\ell}_\mu(v,v)$.

Our analysis of the Hessian consists of three considerations. We consider the metric cover of the cylinder C by an infinite horizontal strip S in $\mathbb{C}$ with the $\mathbf{S}^1$ rotation action of the cylinder lifting to an $\mathbb{R}$ action by Euclidean translations of the strip. We purposefully normalize the covering S so that a Euclidean horizontal translation by δ is a hyperbolic isometry with translation length δ. First, we consider the formula for the variation of the translation length ℓ of the covering of C. For z the complex coordinate for the strip and f^ε the suitable self-homeomorphism of S the translation equivariance provides that $\ell^\varepsilon = f^\varepsilon(z+\ell) - f^\varepsilon(z)$. We find for v a harmonic Beltrami differential defining a deformation and $\mathcal{F}$ a fundamental domain for the metric covering of S to C the first variation

$$\dot{\ell} = \frac{1}{\pi} Re \int_{\mathcal{F}} \frac{d}{d\varepsilon} f^\varepsilon_{\overline{z}} \, i dz d\overline{z} = \frac{1}{\pi} Re \int_{\mathcal{F}} v \, i dz d\overline{z}$$

and the second variation

$$\ddot{\ell} = \frac{1}{\pi} Re \int_{\mathcal{F}} \frac{d^2}{d\varepsilon^2} f^\varepsilon_{\overline{z}} \, i dz d\overline{z} = \frac{2}{\pi} Re \int_{\mathcal{F}} v f_z \, i dz d\overline{z}$$

for f a suitable solution of the potential equation $f_{\bar{z}} = v$. The second variation formula should be compared to the considerably more involved formula of Theorem 3.2 of [Wol87]. Second, we consider the Fourier expansion of v on S relative to the translation group of the covering to C. From Corollary 2.5 and formulas (4.1) of [Wol87] the potential equation $f_{\bar{z}} = v$ admits a term-by-term solution relative to the Fourier expansion of v. In particular for the Beltrami differential with series expansion

$$v = -4\sin^2 \Im z \, \overline{\sum a_n e^{\varepsilon n z}}, \quad \varepsilon = \frac{2\pi i}{\ell},$$

we find that

$$f_z = 2\left(e^z \Re \sum a_n \frac{e^{\varepsilon n z}-1}{\varepsilon n - 1} - e^{-z} \Re \sum a_n \frac{e^{\varepsilon n z + 1}}{\varepsilon n + 1}\right).$$

The quantity f_z is a linear form in the Fourier expansion of v. The expansion enables calculation of the above variation integral term-by-term and the calculation is a special feature for harmonic Beltrami differentials. Third, we simplify the resulting term-by-term expressions to obtain an exact formula in terms of the operator

$$A[\varphi] = \zeta^{-1} \int^{\zeta} t^2 \varphi \, dt$$

for quadratic differentials φ invariant by $t \to e^{\ell} t$ on $\mathbb{H}$ with coordinate t, and the Hermitian form

$$Q(\beta, \delta) = \int_{1<|t|<e^{\ell}} \beta \bar{\delta} \, (Imt)^2 \frac{i}{2} dt d\bar{t}.$$

In [Wol87](Theorem 2.4) we found that $A[\varphi]$ is associated to the Eichler integral of φ. The overall resulting final formula

$$\ddot{\ell} = \frac{32}{\pi} Q(A,A) - \frac{16}{\pi} Q(A, \bar{A}) \tag{2.2}$$

is the replacement for the intricate formulas of Lemmas 4.2 and 4.4 of [Wol87]. The formula can be compared to the formula of Gardiner for the first variation [Gar75]. Bounds for the Hessian of ℓ_σ in terms of $\mathbb{P}_\sigma$ and the Petersson product can be derived by comparing the two Hermitian forms

$$Q(A,A) \quad \text{and} \quad \int_{1<|t|<e^{\ell}} |\varphi|^2 (Imt)^4 \frac{i}{2} \frac{dt d\bar{t}}{|t|^2}$$

where $(Imt)^{-1}|t|$ is comparable to the exponential-distance of t to the imaginary axis. The bounds are straightforward since the Hermitian forms are diagonalized by the Fourier expansion of φ.

3. Convexity results

We find that for the total-length ℓ_μ of a geodesic current its complex Hessian on $\mathcal{T}$, a Hermitian form on $\mathcal{H}(R)$, is bounded in terms of the integral pairing with factor $\mathbb{P}_\mu$ and hyperbolic area element

$$\int_R v\overline{\rho}\,\mathbb{P}_\mu dA \leq \frac{3\pi}{16}\partial\overline{\partial}\ell_\mu(v,\rho) \leq 16\int_R v\overline{\rho}\,\mathbb{P}_\mu dA$$

for $v,\rho \in \mathcal{H}(R)$. Since $\int_R v\overline{\rho}\,dA$ is the WP pairing $\langle v,\rho\rangle_{WP}$, we have the following comparison of Hermitian forms

$$\langle\,,\mathbb{P}_\mu\rangle_{WP} \leq \frac{3\pi}{16}\partial\overline{\partial}\ell_\mu \leq 16\langle\,,\mathbb{P}_\mu\rangle_{WP}.$$

The strict convexity of geodesic-length functions and of the total-length of geodesic currents is an immediate consequence of the positivity of $\mathbb{P}_\mu$. We find further consequences of our calculations and considerations of $\mathbb{P}_\mu$. The first and second derivatives of total-lengths ℓ_λ, ℓ_μ actually satisfy general comparisons

$$|d\ell_\lambda(v)d\ell_\mu(v)| < \ell_\lambda\ddot{\ell}_\mu(v,v) + \ell_\mu\ddot{\ell}_\lambda(v,v) \tag{3.1}$$

and

$$4|\partial\ell_\lambda(v)\overline{\partial}\ell_\mu(v)| < \ell_\lambda\partial\overline{\partial}\ell_\mu(v,v) + \ell_\mu\partial\overline{\partial}\ell_\lambda(v,v). \tag{3.2}$$

The complex Hessian and WP Riemannian Hessian of a total-length ℓ_μ also satisfy a general comparison

$$\partial\overline{\partial}\ell_\mu \leq \ddot{\ell}_\mu \leq 3\partial\overline{\partial}\ell_\mu.$$

A basic consequence of the formulas is the observation that the first and second derivatives of a geodesic-length ℓ_σ are bounded in terms of the supremum norm of $\mathbb{P}_\sigma$ on R. The magnitude of $\mathbb{P}_\sigma$ can in turn be analyzed in terms of the *thick-thin* decomposition of the surface [Wol92, II, Sec. 2]. For σ a simple closed geodesic a suitable decomposition of R has three regions: *i) thick* ; *ii)* cusps and *thin* collars not intersecting σ; and *iii) thin* collars which σ crosses. For the first region since $e^{-2d(p,\sigma)}$ satisfies a mean value estimate and the injectivity radius is uniformly bounded below the supremum of $\mathbb{P}_\sigma$ is bounded by the L^1-norm $\|\mathbb{P}_\sigma\| = \frac{4}{3}\ell_\sigma$. For the second region the distance to σ is at least the distance δ to the region boundary and the supremum can be bounded using the general inequality $e^\delta\rho > c$ bounding the exponential-distance and the injectivity radius for a collar or cusp [Wol92, II, Lem. 2.1]. For the third region the supremum of $\mathbb{P}_\sigma$ is bounded in terms of the reciprocal injectivity radius, which from the general inequality is bounded by $e^{\ell_\sigma/2}$ since σ crosses the collar.

We accordingly find in complete generality that there exists constants c_*, c_{**} independent of R such that the gradient of the geodesic-length of a simple curve is bounded in terms of the geodesic-length itself

$$\langle \operatorname{grad} \ell_\sigma, \operatorname{grad} \ell_\sigma \rangle_{WP} \leq c_*(\ell_\sigma + \ell_\sigma^2 e^{\ell_\sigma/2}) \tag{3.3}$$

and for the relative systole $sys_{rel}(R)$, the least (closed) geodesic-length for R, that

$$c_{**}(sys_{rel}(R))^{4\dim_{\mathbb{C}}\mathcal{T}} \langle \ , \ \rangle_{WP} \leq \partial\bar{\partial}\ell_\sigma \leq c_*(1 + \ell_\sigma e^{\ell_\sigma/2}) \langle \ , \ \rangle_{WP}. \tag{3.4}$$

In brief the first and second derivatives of a simple geodesic-length relative to the WP metric are universally bounded in terms of the geodesic-length and the relative systole. The bound (3.3) can be compared to the familiar universal bound $\|d\ell_\sigma\|_T \leq 2\ell_\sigma$ for the differential relative to the Teichmüller metric [Gar75]. The *degeneration* of $\mathbb{P}_\sigma$ can be further analyzed in terms of the *thick-thin* decomposition [Wol92, II, Sec. 2].

For the study of geodesic currents and applications of geodesic-lengths it is desirable to have bounds (dependent on R) proportional to the geodesic-length. We find for compact subsets of the moduli space of Riemann surfaces that there are general uniform bounds. In particular we have the following.

Theorem 3.1. *Given $\mathcal{T}$, there are functions c_1 and c_2 such that for a curve σ*

$$c_1(sys_{rel}(R))\ell_\sigma \leq \mathbb{P}_\sigma \leq c_2(sys_{rel}(R))\ell_\sigma$$

with $c_1(s)$ an increasing function vanishing at the origin and $c_2(s)$ a decreasing function tending to infinity at the origin. For the total-length of a geodesic current μ

$$c_1(sys_{rel}(R))\ell_\mu \langle \ , \ \rangle_{WP} \leq \partial\bar{\partial}\ell_\mu \leq c_2(sys_{rel}(R))\ell_\mu \langle \ , \ \rangle_{WP}.$$

In summary for compact subsets of the moduli space of Riemann surfaces the Hessian of geodesic-length is proportional to the product of the geodesic-length and the WP pairing.

The first-derivative second-derivative comparison inequalities (3.1), (3.2) provide for new convexity results.

Theorem 3.2. *For the closed curves $\alpha_1, \ldots, \alpha_n$ their geodesic-length sum $\ell_{\alpha_1} + \cdots + \ell_{\alpha_n}$ satisfies: $(\ell_{\alpha_1} + \cdots + \ell_{\alpha_n})^{1/2}$ is strictly convex along WP geodesics, $\log(\ell_{\alpha_1} + \cdots + \ell_{\alpha_n})$ is strictly plurisubharmonic, and $(\ell_{\alpha_1} + \cdots + \ell_{\alpha_n})^{-1}$ is strictly plurisuperharmonic.*

S. K. Yeung showed with a detailed analysis of the first-derivative and second-derivative that $(\ell_{\alpha_1} + \cdots + \ell_{\alpha_n})^{-a}$, $0 < a < 1$, is strictly plurisuperharmonic [Yeu03]. He applied the result to study the behavior of line bundles and L^2-sections over $\mathcal{T}$.

C. McMullen found and used that an ℓ_α^{-1} has complex Hessian uniformly bounded relative to the Teichmüller metric [McM00, Thm. 3.1]. The present result offers an elaboration: a sum $(\ell_{\alpha_1} + \cdots + \ell_{\alpha_n})^{-1}$ is plurisuperharmonic with complex Hessian bounded as

$$-\partial\bar\partial((\ell_{\alpha_1} + \cdots + \ell_{\alpha_n})^{-1}) < (2\partial\bar\partial(\ell_{\alpha_1} + \cdots + \ell_{\alpha_n}))(\ell_{\alpha_1} + \cdots + \ell_{\alpha_n})^{-2}. \tag{3.5}$$

4. The $CAT(0)$ geometry of the augmented Teichmüller space

Applications of geodesic-length convexity are provided by considering the augmented Teichmüller space $\overline{\mathcal{T}}$ with the completion of the WP metric [Abi77, Ber74, Mas76]. $\overline{\mathcal{T}}$ is the space of marked possibly noded Riemann surfaces; $\overline{\mathcal{T}}$ is a non locally compact space [Abi77, Ber74]. $\overline{\mathcal{T}}$ is a $CAT(0)$ metric space [DW03, MW02, Wol03, Yam01]. The geometry of $CAT(0)$ spaces is developed in detail in Bridson-Haefliger [BH99]. For a metric space a *geodesic triangle* is prescribed by a triple of points and a triple of joining length-minimizing curves. A characterization of curvature for metric spaces is provided in terms of distance-comparisons to geodesic triangles in constant curvature spaces. For a $CAT(0)$ space the distance and angle measurements for a triangle are bounded by the corresponding measurements for a Euclidean triangle with the corresponding edge-lengths [BH99, Chap. II.1, Prop. 1.7].

$\overline{\mathcal{T}}$ with the completion of the WP metric is a *stratified* unique geodesic space with the strata intrinsically characterized by the metric geometry [Wol03, Thm. 13]. The stratum containing a given point is the union of all open length-minimizing segments containing the point. To characterize the strata structure in-the-large consider a reference topological surface F for the marking and $C(F)$, the partially ordered set *the complex of curves*. A k-simplex of $C(F)$ consists of $k+1$ distinct nontrivial free homotopy classes of nonperipheral mutually disjoint simple closed curves. Consider Λ the natural labeling-function from $\overline{\mathcal{T}}$ to $C(F) \cup \{0\}$. For a marked noded Riemann surface (R, f) with $f : F \to R$, the labeling $\Lambda((R, f))$ is the simplex of free homotopy classes on F mapped to the nodes on R. The level sets of Λ are the strata of $\overline{\mathcal{T}}$ [Abi77, Ber74]. The strata of $\overline{\mathcal{T}}$ are lower-dimensional Teichmüller spaces; each stratum with its natural WP metric isometrically embeds into the completion $\overline{\mathcal{T}}$ [Mas76]. The unique WP geodesic $\widehat{pq}$ connecting $p, q \in \overline{\mathcal{T}}$ is contained in the closure of the stratum with label $\Lambda(p) \cap \Lambda(q)$ (see [Wol03, Thm. 13]). The open segment $\widehat{pq} - \{p, q\}$ is a solution of the WP geodesic differential equation on the stratum with label $\Lambda(p) \cap \Lambda(q)$. It follows from Theorem 3.1 that a geodesic-length function finite on $\widehat{pq}$ is necessarily strictly convex and on the open segment differentiable.

A complete, convex subset C of a $CAT(0)$ space is the base for an *orthogonal projection*, [BH99, Chap. II.2, Prop. 2.4]. For a general point p there is a unique point, *the projection of p*, on C such that the connecting geodesic realizes the distance to C. The projection is a retraction that does not increase distance. The distance d_C to C is a convex function satisfying $|d_C(p) - d_C(q)| \leq d(p,q)$, [BH99, Chap. II.2, Prop. 2.5]. Examples of complete, convex sets C are: points, complete geodesics, and fixed-point sets of isometry groups. In the case of $\overline{\mathcal{T}}$ since geodesics coincide at most at endpoints, the fibers of a projection are filled out by the geodesics realizing distance between their points and the base. In the case of $\overline{\mathcal{T}}$ the closure of each individual stratum is complete and convex, thus the base of a projection [Wol03, Thm. 13]. For simple disjoint closed curves the relation of the quantity $\ell^{1/2} = (\ell_{\alpha_1} + \cdots + \ell_{\alpha_n})^{1/2}$ to a stratum of $\overline{\mathcal{T}}$ was considered in [Wol03, Cor. 21]. The expansion of the WP metric about a stratum [Wol03, Cor. 4] enabled us to give an expansion for the distance to a stratum. The expansion combines with the comparison inequality (3.1) to provide an inequality for distance in-the-large. In the following the quantity $\ell^{1/2}$ serves as a *Busemann function* for the stratum of vanishing.

Theorem 4.1. *For closed curves $\alpha_1, \ldots, \alpha_n$ represented by simple disjoint distinct free homotopy classes, let S be the closed stratum of $\overline{\mathcal{T}}$ defined by the vanishing of $\ell = \ell_{\alpha_1} + \cdots + \ell_{\alpha_n}$. The WP distance of a point p to S satisfies in terms of $\ell(p)$: in general $d_{WP}(p, S) \leq (2\pi\ell)^{1/2}$ and locally for ℓ small, $d_{WP}(p, S) = (2\pi\ell)^{1/2} + O(\ell^2)$.*

Corollary 4.2. *For β represented by a simple free homotopy class the WP gradient of ℓ_β satisfies $\langle \operatorname{grad} \ell_\beta, \operatorname{grad} \ell_\beta \rangle_{WP} \geq \frac{2}{\pi} \ell_\beta$. As above, for $\alpha_1, \ldots, \alpha_n$ and β represented by disjoint distinct free homotopy classes: for $\gamma(s), 0 \leq s \leq s_0$ the unit-speed distance-realizing WP geodesic connecting S to p the derivatives of $(2\pi\ell)^{1/2}$ and ℓ_β along γ satisfy $\frac{d}{ds}(2\pi\ell)^{1/2}(\gamma(s)) \geq 1$ and $\frac{d}{ds} \ell_\beta(\gamma(s)) \geq 0$.*

G. Riera has recently obtained an exact formula for $\langle \operatorname{grad} \ell_\alpha, \operatorname{grad} \ell_\beta \rangle_{WP}$ as an infinite sum for the lengths of the minimal geodesics connecting α to β [Rie03]. The above lower bound for $\langle \operatorname{grad} \ell_\beta, \operatorname{grad} \ell_\beta \rangle_{WP}$ also follows from his formula. The lower bound and the bound (3.3) can be combined to show that the *injectivity radius* inj_{WP} (the minimal distance to a proper sub stratum in $\overline{\mathcal{T}}$) of $\mathcal{T}$ is comparable to the square root of the least geodesic-length. In particular the bounds provide for positive constants c_*, c_{**} and c_{***} such that for $\ell = \ell_{\alpha_1} + \cdots + \ell_{\alpha_n}$, $\ell \leq c_*$ then $c_* \ell \leq \langle \operatorname{grad} \ell, \operatorname{grad} \ell \rangle_{WP} \leq c_{**} \ell$ and for $\ell(R) \geq c_*$, $d_{WP}(R, R') < c_{***}$ then $\ell(R') \geq c_*/2$. The overall bound $c' \, inj_{WP} \leq (sys_{rel})^{1/2} \leq c'' \, inj_{WP}$ for positive constants is a consequence of the fact that inj_{WP} and $(sys_{rel})^{1/2}$ are comparable for small values and are bounded in general.

We now present in detail two further applications for the behavior of WP geodesics. The first is *Brock's approximation by rays to maximally noded surfaces* [Bro02]. J. Brock noted that the $CAT(0)$ geometry and the observation of Bers on bounded partitions [Ber74] provide for an approximation of infinite WP geodesics. First note that the (incomplete) finite length WP geodesics from a point of $\mathcal{T}$ to the marked noded Riemann surfaces can be extended to include their endpoints in $\overline{\mathcal{T}}$. As a consequence of the $CAT(0)$ geometry the initial unit tangents for such geodesics from a point to a stratum provide for a Lipschitz map from the stratum to the unit tangent sphere of the point. Accordingly the image of $\overline{\mathcal{T}} - \mathcal{T}$ in each unit tangent sphere has measure zero and consequently the infinite length geodesic rays have tangents dense in each tangent sphere. In particular to approximate rays it suffices to approximate the infinite length rays.

From the result of Bers there is a positive constant $L_{g,n}$ depending only on the genus and number of punctures such that each surface has a maximal collection of simple closed curves $\alpha_1, \ldots, \alpha_{3g-3+n}$ (a partition) with total geodesic-length bounded by $L_{g,n}$. By Corollary 4.2 each point of $\mathcal{T}$ is at most distance $(2\pi L_{g,n})^{1/2}$ to a maximally noded Riemann surface. To approximate an infinite ray γ in $\mathcal{T}$ with initial point p, consider a point q on the ray with $d_{WP}(p,q)$ large. The point q is at distance at most $(2\pi L_{g,n})^{1/2}$ to a maximally noded Riemann surface $q^{\#}$. Since $\overline{\mathcal{T}}$ is a unique geodesic space the triple $(p, q, q^{\#})$ determines a geodesic triangle. The comparison to a Euclidean triangle provides that the initial angle between $\widehat{pq}$ and $\widehat{pq^{\#}}$ is $O(L_{g,n}^{1/2} d_{WP}(p,q)^{-1})$ [BH99, Chap. II.1, Prop. 1.7]. Further since γ has infinite length the geodesic differential equation on $\mathcal{T}$ ensures that *close initial tangents provides for close initial segments*. It further follows that initial segments of $\widehat{pq}$ and $\widehat{pq^{\#}}$ are close. The considerations are summarized with the following.

Theorem 4.3. *In $\mathcal{T}$ the infinite length geodesic rays and the rays to maximally noded Riemann surfaces each have initial tangents dense in each tangent space.*

Brock discovered that the situation for finite rays is different: convergence of initial ray segments to finite rays does not provide for convergence of entire rays [Bro02], [Wol03, Sec. 7]. Rays approximating a finite ray can behave in a special way. At this time an additional question is to understand infinite rays asymptotic to a stratum.

As our second application we present a construction for asymptotic rays. Begin with the Teichmüller space $\mathcal{T}'$ for a surface with $n > 0$ punctures and $\mathcal{A}$ the axis for a pseudo Anosov mapping class. The existence and uniqueness of a pseudo Anosov axis was first established in the work of G. Daskalopoulos and R. Wentworth [DW03, Thm. 1.1]. In [Wol03, Thm. 25] the result was also obtained as an application of the classification of limits of geodesics and the general study of *translation length* [BH99, Chap. II.6]. Let $\{R'\}$ be the family of marked Riemann surfaces forming the axis $\mathcal{A}$ and R'' a particular Riemann surface with n punctures. We view $\{R'\}$ as a surface bundle over $\mathcal{A}$ and R'' as a bundle over a point. We introduce a formal bijective pairing of the punctures of $\{R'\}$ with the punctures of R'' and consider the sum of surface bundles along fibers $\{R'\} + R''$ as a family of marked noded Riemann surfaces. The nodes are the paired punctures. For g the formal genus of the family let $\mathcal{T}$ be the Teichmüller space of genus g surfaces with the length function $\ell = \ell_{\alpha_1} + \cdots + \ell_{\alpha_n}$

defining the stratum S containing $\{R'\} + R''$ (the nodes have free homotopy classes $\alpha_1, \ldots, \alpha_n$). Further let γ (reducible and partially pseudo Anosov) be an element of the mapping class group for $\mathcal{T}$ given as a sum of the pseudo Anosov (for $\{R'\}$) and the identity (for R''). The mapping class γ fixes $\alpha_1, \ldots, \alpha_n$ and the action of γ extends to S with the extension acting as the product of the pseudo Anosov on $\mathcal{T}'$ and the identity on $\mathcal{T}(R'')$.

We proceed and describe the construction of a geodesic ray in $\mathcal{T}$ asymptotic to S. First observe that the relative systole is periodic along $\mathcal{A}$ and consequently that sys_{rel} is bounded below along $\mathcal{A}$ by a positive constant c. It now follows from the gradient bound (3.3) that there exists a positive constant δ such that any surface R of $\mathcal{T}$ closer in $\overline{\mathcal{T}}$ to $\{R'\} + R''$ than δ satisfies $0 \leq \ell_{\alpha_j} < c/3$ and for $\beta \neq \alpha_1, \ldots, \alpha_n$ (or a power of an α_j) then $\ell_\beta(R) \geq 2c/3$. In particular the only *short* primitive geodesics on such an R are $\alpha_1, \ldots, \alpha_n$. We are ready to form the candidate ray asymptotic to S by a limiting process. For a sequence of points along $\{R'\} + R''$ tending to forward infinity, connect the reference point R by a WP geodesic to each point of the sequence. The point R in $\mathcal{T}$ has relatively compact neighborhoods and consequently we can select a convergent subsequence of the connecting geodesics. Denote the resulting limit as G. We will verify that the limit is an infinite ray. We are interested in the behavior of three functions on G: ℓ_{α_j}, $d_{WP}(\ ,\{R'\} + R'')$ and $d_{WP}(\ ,S)$. On each geodesic connecting R to a point of $\{R'\} + R''$ each of the functions is convex (see the above on orthogonal projections). Further each function vanishes at the far endpoint of each connecting geodesic. It follows that each function is strictly decreasing on each connecting geodesic and consequently that each function is non increasing on the limit G. Now the classification of geodesic (with lengths tending to infinity) limits provides that either: a limit is an infinite ray or at a fixed distance from the basepoint: the limiting rays successively approach and then strictly recede from a stratum [Wol03, Prop. 23]. As already noted along G the only possible *small* geodesic-lengths have non increasing length functions: the second limiting behavior is precluded and consequently the limit is an infinite ray.

We will now show that ℓ_{α_j} and $d_{WP}(\ ,S)$ tend to zero along G. For this sake consider $\mathcal{N}_\delta$: the points in $\overline{\mathcal{T}}$ at distance at most δ from $\{R'\} + R''$. The closed set $\mathcal{N}_\delta$ is stabilized by the action of the mapping class γ, as well as by the Dehn twists τ_j about α_j. Since the only possible short primitive geodesicsf for a surface in $\mathcal{N}_\delta$ are $\alpha_1, \ldots, \alpha_n$ it can be shown that the quotient of $\mathcal{N}_\delta$ by the action of the group generated by γ and the τ_j is compact. Now for a sequence of points along G tending to infinity consider the associated sequence of forward direction rays. Since the quotient $\mathcal{N}_\delta$ is compact we can select a convergent subsequence of rays translated by appropriate compositions with powers of γ and the τ_j [Wol03, Prop. 23]. The resulting limit is a geodesic G_0 in $\mathcal{N}_\delta$. Since ℓ_{α_j}, $d_{WP}(\ ,\{R'\} + R'')$ and $d_{WP}(\ ,S)$ are non increasing along G, each function has a limit along G and consequently each function is actually constant on $\mathcal{G}_0$. In particular each ℓ_{α_j} is constant on $\mathcal{G}_0$; Theorem 3.1 provides that each ℓ_{α_j} vanishes on $\mathcal{G}_0$. It further follows from Theorem 4.1 that $d_{WP}(\ ,\mathcal{S})$ vanishes on $\mathcal{G}_0$ and thus that $\mathcal{G}$ is asymptotic to $\mathcal{S}$, as proposed. Finally in closing we note that if the Teichmüller space $\mathcal{T}(R'')$ is not a singleton then the product $\mathcal{T}(R') \times \mathcal{T}(R'')$ contains *Euclidean flats* and γ stabilizes parallel lines $\{R'\}+R''$, $\{R'\}+R'''$. We expect families of asymptotic rays in this case.

References

[Abi77]   W. Abikoff (1977). Degenerating families of Riemann surfaces. *Ann. of Math. (2)* **105** (1), 29–44.

[Ahl61]   L. V. Ahlfors (1961). Some remarks on Teichmüller's space of Riemann surfaces. *Ann. of Math. (2)* **74**, 171–191.

[Ber74]   L. Bers (1974). Spaces of degenerating Riemann surfaces. In *Discontinuous Groups and Riemann Surfaces (Proc. Conf., Univ. Maryland, College Park, Md., 1973)*, pp. 43–55. Ann. of Math. Studies, No. 79. Princeton Univ. Press, Princeton, N.J.

[BH99]   M. R. Bridson & A. Haefliger (1999). *Metric Spaces of Non-positive Curvature*. Springer-Verlag, Berlin.

[Bon88]   F. Bonahon (1988). The geometry of Teichmüller space via geodesic currents. *Invent. Math.* **92** (1), 139–162.

[Bro02]   J. F. Brock (2002). The Weil–Petersson visual sphere. Preprint.

[Bro03]   J. F. Brock (2003). The Weil–Petersson metric and volumes of 3-dimensional hyperbolic convex cores. *J. Amer. Math. Soc.* **16** (3), 495–535 (electronic).

[DKW00]   G. Daskalopoulos, L. Katzarkov & R. Wentworth (2000). Harmonic maps to Teichmüller space. *Math. Res. Lett.* **7** (1), 133–146.

[DS03]   R. Diaz & C. Series (2003). Limit points of lines of minima in Thurston's boundary of Teichmueller space. *Algebr. Geom. Topol.* **3**, 207–234.

[DW03]   G. Daskalopoulos & R. Wentworth (2003). Classification of Weil–Petersson isometries. *Amer. J. Math.* **125** (4), 941–975.

[Gar75]   F. P. Gardiner (1975). Schiffer's interior variation and quasiconformal mapping. *Duke Math. J.* **42**, 371–380.

[Gar86]   F. P. Gardiner (1986).   A correspondence between laminations and quadratic differentials. *Complex Variables Theory Appl.* **6** (2-4), 363–375.

[Ker83]   S. P. Kerckhoff (1983). The Nielsen realization problem. *Ann. of Math. (2)* **117** (2), 235–265.

[Ker92]   S. P. Kerckhoff (1992). Lines of minima in Teichmüller space. *Duke Math. J.* **65** (2), 187–213.

[Mas76]   H. Masur (1976). Extension of the Weil–Petersson metric to the boundary of Teichmuller space. *Duke Math. J.* **43** (3), 623–635.

[McM00]   C. T. McMullen (2000). The moduli space of Riemann surfaces is Kähler hyperbolic. *Ann. of Math. (2)* **151** (1), 327–357.

[MW02]   H. Masur & M. Wolf (2002). The Weil–Petersson isometry group. *Geom. Dedicata* **93**, 177–190.

[Rie03]   G. Riera (2003). A formula for the Weil–Petersson product of quadratic differentials. Preprint.

[SS99]   P. Schmutz Schaller (1999). Systoles and topological Morse functions for Riemann surfaces. *J. Differential Geom.* **52** (3), 407–452.

[SS01]   P. Schmutz Schaller (2001). A cell decomposition of Teichmüller space based on geodesic length functions. *Geom. Funct. Anal.* **11** (1), 142–174.

[Wol87]   S. A. Wolpert (1987). Geodesic length functions and the Nielsen problem. *J. Differential Geom.* **25** (2), 275–296.

[Wol92]   S. A. Wolpert (1992). Spectral limits for hyperbolic surfaces. I, II. *Invent. Math.* **108** (1), 67–89, 91–129.

[Wol03]   S. A. Wolpert (2003). Geometry of the Weil–Petersson completion of Teichmüller space. In *Surveys in Differential Geometry VIII: Papers in Honor of Calabi, Lawson, Siu and Uhlenbeck*, pp. 357–393. Intl. Press, Cambridge, MA.

[Wol04]   S. A. Wolpert (2004). Convexity of geodesic-length functions. In preparation.

[Yam99]   S. Yamada (1999). Weil-Peterson convexity of the energy functional on classical and universal Teichmüller spaces. *J. Differential Geom.* **51** (1), 35–96.

[Yam01]   S. Yamada (2001). Weil-Petersson Completion of Teichmüller Spaces and Mapping Class Group Actions. Preprint.

[Yeu03]   S.-K. Yeung (2003). Bounded smooth strictly plurisubharmonic exhaustion functions on Teichmüller spaces. *Math. Res. Lett.* **10** (2-3), 391–400.

Scott A. Wolpert

Mathematics Department
University of Maryland
College Park, MD 20742
U.S.A. saw@math.umd.edu

AMS Classification: 32G15, 30F60, 53C22

Keywords: Teichmüller spaces, geodesic-length functions, convexity

Spaces of Kleinian Groups   Cambridge University Press
Lond. Math. Soc. Lec. Notes **329**, 247–257   Y. Minsky, M. Sakuma & C. Series (Eds.)

© A. Marden, 2005

A proof of the Ahlfors finiteness theorem

Albert Marden

1. Introduction

The modern theory of hyperbolic 3-manifolds began with the Ahlfors Finiteness Theorem. It states that the quotient $\Omega(G)/G$ of the ordinary set of a finitely generated Kleinian group G is a finite union of surfaces, each of which is a closed surface with at most a finite number of punctures and cone points. We will formally state Ahlfors' theorem in §4. His 1964 proof [Ahl66] involved delicate analytic estimates for automorphic forms. In his enthusiasm, he forgot to rule out the possibility of infinitely many triply punctured spheres (or triangle groups more generally). The omission was soon rectified both by Bers [Ber67] with further analysis of the forms, and more elegantly by Greenberg [Gre67] using a lemma akin to Selberg's lemma. See [Gre77] and [Kra72] for expositions of these proofs. As the theory of hyperbolic manifolds developed around 1970 it was recognized that much of the proof can be carried out more simply and naturally using topological considerations [Mar74]. The topological approach was completed when the theory of the compact core and relative core was brought in, first by Kulkarni–Shalen [KS89], and then in its full implementation by Feighn–McCullough [FM87]. What then remains to do for the proof of Ahlfors' theorem is to rule out boundary components of the 3-manifold being topological disks or more generally bordered Riemann surfaces.

In the wake of Sullivan's proof of the no wandering domain theorem [Sul85] the whole shebang, especially the analytic part, became greatly simplified. The process began with Sullivan himself [Sul85], and continued with Bers [Ber87], and recently Kapovich [Kap01, pp. 110, 206-210].

The proof[1] set out below was developed at the 1986 MSRI conference on complex analysis. Motivated by Kapovich's recent proof, we have decided to put ours out for public view as well. The overall pattern is the same as the other proofs, starting with Sullivan; the difference is in the simplicity of the details.

The proof we will give is based on I.N. Baker's proof [Bak84, Thm. 6.1] of a "no wandering domain theorem" for iterations of certain classes of entire functions. It depends on some basic Teichmüller theory, and some knowledge of fuchsian and Kleinian groups. Beyond this, it is fairly elementary.

---

[1] Thanks to V. Markovic and the referee for useful comments.

An alternate path to Ahlfors' theorem has become available through the recently announced confirmations of the Tameness Conjecture for hyperbolic 3-manifolds by Agol [Ago04] and then by Calegari–Gabai [CG04]. Tameness implies Ahlfors' theorem, although a proof via this route is surely overkill and obscures the natural deformation explanation. What it does clearly expose is that the Finiteness Theorem in its entirety is mandated by the inner topological/geometric properties of hyperbolic manifolds.

2. A basic lemma

We will use a simple but useful fact inspired by a lemma of Baker.

Lemma 2.1. *There exists $K > 1$ such that if F is any K-quasiconformal mapping of $\mathbb{S}^2$ that fixes $0, 1, \infty$, then*

$$|F(z) - z| \leq \frac{1}{2}, \quad \text{for all } z \text{ with } |z| \leq 1.$$

Corollary 2.2. *If F fixes instead ζ, ζ', ∞, then*

$$|F(z) - z| \leq \frac{1}{2}|\zeta - \zeta'|, \quad \text{for all } z \text{ with } |z - \zeta| \leq |\zeta' - \zeta|.$$

Proof. The Corollary is a renormalized form of the Lemma. The Lemma is proved by a normal family argument. The family of K-quasiconformal mappings of $\mathbb{S}^2$ fixing three points is a normal family. Therefore every infinite sequence has a uniformly convergent subsequence. Any limit function is again a K-quasiconformal map of $\mathbb{S}^2$. By taking K sufficiently close to one, we can ensure that the inequality is satisfied. For a sequence of the K_n-quasiconformal maps with $K_n \to 1$ would converge to the identity.   □

3. Mappings homotopic to the identity

The following result is central to our proof (compare with [EM88]). It is what ensures that lifts of certain maps of a region (typically with fractal boundary) to its universal cover $\mathbb{D}$ extend to pointwise fix $\partial\mathbb{D}$.

Proposition 3.1. *There exists $K > 1$ with the following property. Suppose that*

- *$U \subset \mathbb{C}$ is either simply connected $\neq \mathbb{C}$, or is infinitely connected without isolated boundary components.*

- *$F : \mathbb{C} \to \mathbb{C}$ is a K-quasiconformal mapping with the properties:*

(i) *F fixes every point of ∂U.*

(ii) *The restriction F_U of F maps $U \to U$, and*

(iii) *is homotopic to the identity in U.*

Let $\pi : \mathbb{D} = \{z : |z| < 1\} \to U$ denote either a Riemann map to U or projection from the universal cover of U depending on whether U is simply connected or not. There exists a lift $F^* = \pi^{-1} \circ F_U \circ \pi$ of F_U such that the extension to $\partial \mathbb{D}$ of $F^* : \mathbb{D} \to \mathbb{D}$ fixes every point on $\partial \mathbb{D}$.

Proof. Recall that a quasiconformal mapping $\mathbb{D} \to \mathbb{D}$ extends to $\partial \mathbb{D}$ so as to become a homeomorphism of the closed disk.

Step 1. Let Γ denote the fuchsian group of cover transformations where we allow $\Gamma = \{\mathrm{id}\}$ if U is simply connected. A map $F : U \to U$ is homotopic to the identity if and only if[2] it has a lift F^* that satisfies

$$F^* \circ T(z) = T \circ F^*(z), \quad \text{for all } z \in \mathbb{D}, \ T \in \Gamma.$$

Since F and hence F^* are quasiconformal, F^* has an extension to the closed disk $\overline{\mathbb{D}}$ which is a homeomorphism. It necessarily fixes every fixed point of Γ and by continuity fixes every limit point of Γ.

Therefore if the limit set $\Lambda(\Gamma) = \partial \mathbb{D}$, we are finished. So assume this is not the case.

Step 2. Let I be a component of $\partial \mathbb{D} \setminus \Lambda(\Gamma)$. It is an open interval with the property that $T(I) \cap I = \emptyset$ for all $T \in \Gamma, T \neq \mathrm{id}$. For suppose this were not true so that $T(I) \cap I \neq \emptyset$ for some $T \in \Gamma$. Then T would have to map I onto itself, in which case T would fix its endpoints. If this were the case, I would project to an isolated boundary component of U, contrary to our assumption in the non-simply connected case.

Choose a closed subinterval $I_0 \subset I$.

Let $\sigma \subset \mathbb{D}$ be a circular arc between the endpoints of I_0. We can choose σ so close to I_0 that the region $X \subset \mathbb{D}$, bounded by σ and I_0, has the following properties:

- $T(X) \cap X = \emptyset$ for all $T \in \Gamma, T \neq \mathrm{id}$.

- The axis of no element of Γ penetrates X.

The assumption that no boundary components of U are isolated implies that all elements $\neq \mathrm{id}$ of Γ are hyperbolic.

---

[2]The necessity is a consequence of lifting the homotopy. The sufficiency comes from taking $F_t^*(z)$ to be the point of distance tL_z along the geodesic arc of length L_z from z to $F^*(z)$, $0 \leq t \leq 1$.

Before moving down from $\mathbb{D}$ to U, recall that an ideal boundary component of U is defined in terms of an equivalence class of nested sequences of connected open sets that nest down on the component. Nested sequences for different ideal boundary components are eventually disjoint from each other. In fact, given two ideal boundary components, there is a simple geodesic in U that separates them.

Now $\pi : X \to \pi(X) \subset U$ is an embedding. We claim that there is an ideal boundary component J of U such that any sequence $\{z_n\} \subset X$ that accumulates to I_0 projects to a sequence $\{\pi(z_n)\} \subset \pi(X)$ that accumulates to J. For suppose to the contrary that the sequence accumulates to two ideal boundary components of U. There is a geodesic $\alpha \subset U$ that separates the components. A lift of α would necessarily penetrate into X, contrary to our choice of X.

Furthermore, the ideal boundary component J cannot reduce to a single point in $\mathbb{S}^2$, since the conformal embedding π of X cannot send the interval I_0 to a single point. One way of confirming this is in terms of the bounded harmonic function on X with boundary values 0 on I_0 and 1 on σ. It projects to a bounded harmonic function in $\pi(X)$. If $\pi(I_0)$ were a single point, the projected bounded harmonic function would have boundary values 1 a.e., and would necessarily be a constant.

Therefore the euclidean diameter of J is not zero. We may thus make X smaller if necessary so that the following holds for the euclidean distance d_e:

$$d_e(\pi(z), \partial U) = d_e(\pi(z), J) < \frac{1}{2} \operatorname{diam}(J), \quad \text{for all } z \in X.$$

Given $z \in X$, let $\zeta \in J$ be a nearest point to $\pi(z)$. We can find another point $\zeta' \in J$ such that

$$|\pi(z) - \zeta| = |\zeta' - \zeta|.$$

There is such a point, for otherwise the diameter of J would have to be less than $2 d_e(\pi(z), J)$.

We are finally in a position to apply Corollary 2.2 to a K-quasiconformal $F(z)$ with fixed points ζ, ζ', ∞. Here K must be chosen to satisfy Lemma 2.1. We then find that for our choice of $z \in X$,

$$|F(\pi(z)) - \pi(z)| \le \frac{1}{2} d_e(\pi(z), \partial U), \tag{3.1}$$

since $|\pi(z) - \zeta| = d_e(\pi(z), \partial U)$.

Step 3. Given two points $a, b \in U$, let $[a, b]$ denote the shortest hyperbolic distance between them. Let $d_h(\cdot, \cdot)$ denote hyperbolic distance in $\mathbb{D}$.

Corresponding to our point $z \in X$, find $T_z \in \Gamma$ with the following property. Among all points in the Γ-orbit of $F^*(z)$ in $\mathbb{D}$, $T_z(F^*(z))$ is a closest point to z in the hyperbolic metric. Then,

$$d_h(z, T_z F^*(z)) = [\pi(z), F\pi(z)]. \tag{3.2}$$

Consider the disk

$$\Delta = \{w \in U : |w - \pi(z)| < d_e(\pi(z), \partial U)\}.$$

From equation (3.1), we know $F(\pi(z)) \subset \Delta$. Moreover,

$$[\pi(z), F\pi(z)] \le d_\Delta(\pi(z), F(\pi(z))) \le \log 3. \tag{3.3}$$

Here $d_\Delta(\cdot, \cdot)$ is hyperbolic distance in Δ and $\log 3$ is the hyperbolic distance in Δ from the origin to a point halfway to the boundary. We are thus using (3.1).

We now want to apply (3.3) to any choice of $z \in X$. We can do this, because once K is prescribed, each allowable F fixes all points of ∂U. Hence (3.3) does not depend on the z-dependent choices of ζ, ζ' on the boundary component J of U.

We can therefore conclude that for *any* $z \in X$, there exists $T_z \in \Gamma$ for which,

$$d_h(z, T_z F^*(z)) < \log 3. \tag{3.4}$$

Step 4. Let now $\{z_n\} \subset X$ be a sequence with limit point $p \in I_0$. Denote the corresponding sequence of elements of Γ by $\{T_n = T_{z_n}\}$. The sequence of values $\{F^*(z_n)\}$ converges to $F^*(p)$ since F^* extends continuously to $\partial \mathbb{D}$. In view of (3.4),

$$d_h(z_n, T_n F^*(z_n)) < \log 3. \tag{3.5}$$

Suppose first the sequence $\{T_n\}$ has infinitely many distinct elements. Since Γ is discrete there is a subsequence $\{T_m\}$ which converges to a constant $c \in \Lambda(\Gamma)$, uniformly on compact subsets of $\mathbb{D} \cup (\partial \mathbb{D} \setminus \Lambda(\Gamma))$. In particular $\lim T_m F^*(z_m) = c$. But $\lim z_m = p$, while $c \notin I_0$ since it is a limit point. We have a contradiction to (3.5).

We conclude that the sequence $\{T_m\}$ contains only a finite number of distinct elements. We may then assume all the elements are the same transformation $T \in \Gamma$. Once again we bring in (3.5) and see that we must have $p = \lim z_m = T F^*(p)$, or

$$F^*(p) = T^{-1}(p). \tag{3.6}$$

So far in our analysis, T depends on $p \in I_0$. But F^* is a homeomorphism of $\partial \mathbb{D}$ while Γ is discrete. Therefore Equation (3.6) holds in a maximal subinterval of I_0 which is both open and closed. The conclusion is that T satisfying (3.6) holds for all points on the interior of I_0, then for all points on the interior of the original interval

I, and by continuity the endpoints of *I* as well. As we found in Step 1, the endpoints q_1, q_2 of *I* are fixed points of F^*. This makes them fixed points of *T* as well. The only possibility left is that $T =$ id.

We conclude that $F^*(p) = p$ for all points $p \in I$ and hence for all points $p \in \partial \mathbb{D}$. □

4. Deformations

Suppose *R* is an arbitrary hyperbolic Riemann surface with possible border ∂R. Represent *R* in its universal cover $\mathbb{D}$, $R = \mathbb{D}/\Gamma$. The border is then represented as $\partial R = (\partial \mathbb{D} \setminus \Lambda(\Gamma))/\Gamma$. In addition to a possible border, *R* may have isolated ideal boundary points which are called punctures.

The Teichmüller space $\mathfrak{T}(R)$ is the space of pairs $\{(S, f)\}$ where $f : R \to S$ is quasiconformal, subject to the equivalence $(S, f) \equiv (S_1, f_1)$ if and only if $f_1 \circ f^{-1} : S \to S_1$ is homotopic to a conformal map in such a way that the border remains pointwise fixed. (The maps *f* automatically extend to punctures.) The Teichmüller space is a complex manifold. We take as the origin the equivalence class of (R, id).

The Teichmüller space is infinite dimensional if and only if the vector space $Q(R)$ of quadratic differentials φdz^2 with bounded norm on R has infinite dimension. Here

$$\|\varphi\| = \int \int_R |\varphi| dx dy < \infty.$$

A *maximal* Riemann surface is one that cannot be properly embedded in a larger surface in such a way that the inclusion of fundamental groups is an isomorphism. In particular this means that the surface does not have a border. For example, the complement in $\mathbb{C}$ of the Cantor set is a maximal surface. Equivalently, a maximal Riemann surface is one such that the limit set of its fuchsian covering group in $\mathbb{D}$ is $\partial \mathbb{D}$.

Thus if *R* is not maximal and/or has infinite topological type, $\mathfrak{T}(R)$ is infinite dimensional. The following statement is well known in the field.

Lemma 4.1. *Suppose R is a Riemann surface which is not maximal and/or is of infinite topological type. Given any integer $N > 0$, there is a complex dimension N submanifold V of a neighborhood containing the origin in $\mathfrak{T}(R)$.*

A quasiconformal deformation of *R* is a quasiconformal map $R \to S$. Two maps f, f_1 are equivalent if there is a conformal map *g* such that $f_1 = g \circ f$. Thus under the hypothesis of Lemma 4.1, each $\sigma \in V$ corresponds to a quasiconformal map $f_\sigma : R \to S$. Different choices of $\sigma \in V$ result in inequivalent quasiconformal maps. Furthermore the maps f_σ depend holomorphically on σ. Here $f_0 = \text{id}$.

5. Ahlfors' finiteness theorem

Suppose now Γ is a Kleinian group with limit set $\Lambda(\Gamma)$ and complementary set of discontinuity $\Omega(\Gamma)$. The corresponding 3-orbifold and its boundary, which may be empty, is

$$\mathcal{M}(\Gamma) = \mathbb{H}^3 \cup \Omega(\Gamma)/\Gamma, \quad \partial\mathcal{M}(\Gamma) = \Omega(\Gamma)/\Gamma.$$

Ahlfors' Finiteness Theorem. If Γ is finitely generated, then $\partial\mathcal{M}(\Gamma)$ has at most a finite number of components. Each component is a closed Riemann surface[3] with at most a finite number of punctures and a finite number of cone (branch) points.

Not included in the statement of the theorem are the facts (i) there are at most a finite number of parabolic conjugacy classes in G and (ii) there are at most a finite number of conjugacy classes of finite subgroups. For parabolic and elliptic fixed points are not always detectable in $\partial\mathcal{M}(\Gamma)$. Proofs of (i) are contained in [Sul81] and [FM87], and a proof of (ii) is in [FM91].

The topological method is also well suited to give bounds on the Euler characteristic, etc., of the boundary in terms of the number of generators of Γ.

According to Selberg's Lemma, there is a torsion free normal subgroup $\Gamma_0 \subset \Gamma$ of finite index. This means that $\mathcal{M}(\Gamma_0)$ is a finite sheeted, regular cover of $\mathcal{M}(\Gamma)$. Therefore it suffices to prove the theorem for Γ_0. Consequently we can assume that Γ itself contains no elliptic elements.

If the conclusion of Ahlfors' theorem were false, then some boundary components would have an infinite dimensional space of deformations. On the other hand, since Γ is finitely generated it can have only a finitely dimensional space of deformations: think in terms of varying the entries in the generating matrices. The strategy of this and all proofs is to derive a contradiction from this state of affairs.

However any proof by counting parameters, including Ahlfors' argument, is bound to be unable to deal with the following eventuality. Namely the possibility of $\partial\mathcal{M}$ having infinitely many triply punctured spheres as components. For all triply punctured spheres are conformally equivalent and hence have no deformations at all. Instead of a deformation argument, a separate argument is required to rule out this possibility. As mentioned earlier, the most beautiful and elementary of these is the argument of Leon Greenberg [Gre67]. In the proof below, we will assume that this case does not arise.

Proof. Step 1. Suppose the group $\Gamma = \langle T_1, \ldots, T_N \rangle$ has N generators. The matrix in $SL(2, \mathbb{C})$ corresponding to each generator depends on 3 complex entries. Therefore

---

[3] A closed surface is one which is compact, without boundary.

Γ depends on at most $3N$ complex parameters; it will actually be fewer in particular because there are likely various relations in Γ, including one that arises for each parabolic conjugacy class.

We want to prove that each boundary component R of $\mathcal{M}(\Gamma)$ is a closed surface with at most a finite number of punctures.

Suppose to the contrary that this is not true of a component R. Apply Lemma 4.1 to find an $3N+1$ dimensional submanifold V of a neighborhood of the origin of $\mathfrak{T}(R)$. Each point $\sigma \in V$ corresponds to a quasiconformal deformation f_σ of R.

Step 2. Before proceeding we will move up to $\mathbb{S}^2$. Fix a component $\Omega \subset \Omega(\Gamma)$ lying over R. Thus $R = \Omega/\mathrm{Stab}(\Omega)$ with $\mathrm{Stab}(\Omega) = \{T \in \Gamma : T(\Omega) = \Omega\}$.

Consider the Beltrami differential of f_σ on R, $\sigma \in V$,

$$\frac{\bar{\partial} f_\sigma}{\partial f_\sigma}.$$

Denote its lift to Ω by μ_σ. According to deformation theory, μ_σ is holomorpic in σ, $\mu_0 = 0$. It automatically has the invariance

$$\mu_\sigma(T(z))\frac{\overline{T'(z)}}{T'(z)} = \mu_\sigma(z), \quad \text{for all } T \in \mathrm{Stab}(\Omega), z \in \Omega. \tag{5.1}$$

We can extend each μ_σ from Ω to the full Γ-orbit of Ω in such a way that (5.1) is satisfied for all elements of Γ and points $\{z\}$ of the orbit. Then extend the resulting μ_σ to all $\mathbb{S}^2$ by setting $\mu_\sigma(z) = 0$ whenever z is not in the Ω-orbit. The invariance (5.1) then holds for all $z \in \mathbb{S}^2$.

Now solve the Beltrami equation $\bar{\partial} G_\sigma = \mu_\sigma \partial G_\sigma$ on $\mathbb{S}^2$. The solution G_σ is a quasiconformal map of $\mathbb{S}^2$. It is uniquely determined if we impose the normalization that $0, 1, \infty$ are to be fixed. In particular $G_0 = \mathrm{id}$. The invariance condition (5.1) implies that for each σ, there is an isomorphism $\theta_\sigma : \Gamma \to \theta_\sigma(\Gamma)$ with the property

$$G_\sigma \circ T(z) = \theta_\sigma(T) \circ G_\sigma(z), \quad \text{for all } T \in \Gamma, z \in \mathbb{S}^2. \tag{5.2}$$

The image group $\theta_\sigma(\Gamma)$ is another Kleinian group. In particular $\theta_0 = \mathrm{id}$.

The quasiconformal mapping G_σ is conformal on the interior of the complement of the Γ-orbit of Ω. The restriction of the family $\{G_\sigma\}$ to Ω projects to a family $\{g_\sigma\}$ of deformations of R with $g_0 = \mathrm{id}$. Each g_σ solves on R the same Beltrami equation as our original f_σ. Therefore $g_\sigma = h_\sigma \circ f_\sigma$ where h_σ is a conformal mapping of the surface $f_\sigma(R) \to g_\sigma(R)$.

Step 3. Consider the map

$$\sigma \in V \mapsto \langle \theta_\sigma(T_1), \ldots, \theta_\sigma(T_N) \rangle \subset \mathrm{PSL}(2, \mathbb{C})^N.$$

The entries of the normalized matrices of the generators $\theta_\sigma(T_i)$ are holomorphic functions of $\sigma \in V$. That is, we have an analytic map of V into a neighborhood of the identity.

By the implicit function theorem, there is a complex 1-dimensional section $\sigma = \sigma(s)$ of V, with $\sigma(0) = 0$, such that

$$\theta_\sigma(T_i) = T_i, \ \sigma = \sigma(s), \ s \in D = \{s : |s| < \delta\}, \ 1 \leq i \leq N.$$

Because the $\{T_i\}$ are generators of Γ, $\theta_\sigma(T) = T$ for all $T \in \Gamma$, where $\sigma = \sigma(s)$ with $s \in D$. The relation (5.2) then implies that G_σ fixes each fixed point of T, for all $T \in \Gamma$. Since the fixed points are dense in the limit set $\Lambda(\Gamma)$, by continuity G_σ fixes each point of $\Lambda(\Gamma)$. This is true for all $\sigma = \sigma(s)$, $s \in D$.

In particular, G_σ fixes each boundary point of the component $\Omega \subset \Omega(\Gamma)$.

Now we bring in Proposition 3.1. After further restricting δ if necessary, we may assume that $G_\sigma(\Omega) = \Omega$ for all $\sigma = \sigma(s)$, $s \in D$. Upon restricting δ even more if necessary, we may assume that each G_σ is K-quasiconformal with K given by Proposition 3.1.

Take the universal cover (or Riemann map image if Ω is simply connected) of Ω to be $\mathbb{D}$. This is also the universal covering of $R = \Omega/\mathrm{Stab}(\Omega)$. According to Proposition 3.1, each G_σ, $\sigma = \sigma(s)$, has a lift $G_\sigma^* : \mathbb{D} \to \mathbb{D}$ which extends to the identity on $\partial \mathbb{D}$. But this in turn implies that each G_σ^* induces the identity deformation of the fuchsian covering group of R. This is a contradiction.

We can now conclude that each component of $\partial \mathcal{M}(\Gamma)$ is a closed surface with at most a finite number of punctures.

Almost the same argument shows that $\partial \mathcal{M}(\Gamma)$ cannot have more than $3N$ components which are not triply punctured spheres. For each such component has at least a 1-complex parameter family of deformations, and deformations of separate components are independent of each other. $\square$

References

[Ago04] I. Agol (2004). Tameness of hyperbolic 3-manifolds. Preprint, math.GT/0405568v1.

[Ahl66] L. V. Ahlfors (1966). *Lectures on Quasiconformal Mappings*. Manuscript prepared with the assistance of Clifford J. Earle, Jr. Van Nostrand Mathematical Studies, No. 10. D. Van Nostrand Co., Inc., Toronto, Ont.-New York-London.

[Bak84] I. N. Baker (1984). Wandering domains in the iteration of entire functions. *Proc. London Math. Soc. (3)* **49** (3), 563–576.

[Ber67] L. Bers (1967). On Ahlfors' finiteness theorem. *Amer. J. Math.* **89**, 1078–1082.

[Ber87] L. Bers (1987). On Sullivan's proof of the finiteness theorem and the eventual periodicity theorem. *Amer. J. Math.* **109** (5), 833–852.

[CG04] D. Calegari & D. Gabai (2004). Shrinkwrapping and the taming of hyperbolic 3-manifolds. Preprint, math.GT/0407161.

[EM88] C. J. Earle & C. McMullen (1988). Quasiconformal isotopies. In *Holomorphic Functions and Moduli, Vol. I (Berkeley, CA, 1986), Math. Sci. Res. Inst. Publ.*, volume 10, pp. 143–154. Springer, New York.

[FM87] M. Feighn & D. McCullough (1987). Finiteness conditions for 3-manifolds with boundary. *Amer. J. Math.* **109** (6), 1155–1169.

[FM91] M. Feighn & G. Mess (1991). Conjugacy classes of finite subgroups of Kleinian groups. *Amer. J. Math.* **113** (1), 179–188.

[Gre67] L. Greenberg (1967). On a theorem of Ahlfors and conjugate subgroups of Kleinian groups. *Amer. J. Math.* **89**, 56–68.

[Gre77] L. Greenberg (1977). Finiteness theorems for Fuchsian and Kleinian groups. In *Discrete Groups and Automorphic Functions (Proc. Conf., Cambridge, 1975)*, pp. 199–257. Academic Press, London.

[Kap01] M. Kapovich (2001). *Hyperbolic Manifolds and Discrete Groups, Progress in Mathematics*, volume 183. Birkhäuser Boston Inc., Boston, MA.

[Kra72] I. Kra (1972). *Automorphic Forms and Kleinian Groups*. W. A. Benjamin, Inc., Reading, Mass.

[KS89] R. S. Kulkarni & P. B. Shalen (1989). On Ahlfors' finiteness theorem. *Adv. Math.* **76** (2), 155–169.

[Mar74] A. Marden (1974). The geometry of finitely generated Kleinian groups. *Ann. of Math. (2)* **99**, 383–462.

[Sul81]   D. Sullivan (1981). A finiteness theorem for cusps. *Acta Math.* **147** (3-4), 289–299.

[Sul85]   D. Sullivan (1985). Quasiconformal homeomorphisms and dynamics. I. Solution of the Fatou-Julia problem on wandering domains. *Ann. of Math. (2)* **122** (3), 401–418.

Albert Marden

School of Mathematics
University of Minnesota
Minneapolis, MN 55455
U.S.A.

am@math.umn.edu

AMS Classification: 30F40, 32Q45

Keywords: Ahlfors Finiteness Theorem, Kleinian groups, hyperbolic 3-manifolds

Spaces of Kleinian Groups   Cambridge University Press
Lond. Math. Soc. Lec. Notes **329**, 259–282   Y. Minsky, M. Sakuma & C. Series (Eds.)

On the automorphic functions for Fuchsian groups of genus two

Yohei Komori

Dedicated to Professor Kyoji Saito
on the occasion of his sixtieth birthday

Abstract

In terms of the presentation of a Fuchsian group Γ of genus two, we give an expression of the Riemann surface $\mathbb{H}^2/\Gamma$ as a hyperelliptic curve. We describe branched points as functions of the presentation of Γ i.e. functions on the Teichmüller space of genus two. In this procedure, we construct a Γ-automorphic function on the upper half plane $\mathbb{H}^2$ which is analogous to the Weierstraß $\wp$-function. I.Kra's algorithm of the construction of non-vanishing Poincaré series is the key tool for the construction of this automorphic function. Making use of our results, we can construct Teichmüller modular functions for the case of genus two as the case of the elliptic modular function.

1. Introduction

Let Γ be a marked Fuchsian group of genus two, a presentation of a Fuchsian group of genus two

$$\Gamma := \langle a_1, b_1, a_2, b_2 \mid \prod_{i=1}^{2}[a_i, b_i] \rangle \subset \mathrm{PSL}(2, \mathbb{R})$$

where $[a_i, b_i] := a_i b_i a_i^{-1} b_i^{-1}$ is the commutator of a_i and b_i. Then the quotient space $\mathbb{H}^2/\Gamma$ of the upper half plane $\mathbb{H}^2$ by Γ is a compact Riemann surface of genus two. Because every compact Riemann surface of genus two is a hyperelliptic curve i.e. a branched double covering over the Riemann sphere $P^1(\mathbb{C})$, there exist five complex numbers $q_1, \cdots, q_5 \in \mathbb{C}$ such that $\mathbb{H}^2/\Gamma$ can be written as a hyperelliptic curve

$$y^2 = \prod_{i=1}^{5}(x - q_i)$$

where we note that the point at infinity $\infty \in P^1(\mathbb{C})$ is also a branched point. The main purpose of this paper is to describe the set of branched points $\{q_i\}$ in terms of the presentation of a Fuchsian group Γ.

To explain how to attack this problem, we first review the same problem for the case of genus one, an elliptic curve. Let $L = \mathbb{Z} \cdot \omega_1 + \mathbb{Z} \cdot \omega_2$ where $\omega_1, \omega_2 \in \mathbb{C}$ with $Im(\omega_1/\omega_2) > 0$ be a lattice acting on the complex plane $\mathbb{C}$ by the parallel transform

$$\mathbb{C} \times L \quad \to \quad \mathbb{C}$$
$$(z, \, m \cdot \omega_1 + n \cdot \omega_2) \quad \mapsto \quad z + m \cdot \omega_1 + n \cdot \omega_2$$

where $m, n \in \mathbb{Z}$. Then the quotient space $\mathbb{C}/L$ is a compact Riemann surface of genus one and has an expression as an elliptic curve

$$y^2 = \prod_{i=1}^{3} (x - q_i)$$

where $\infty \in P^1(\mathbb{C})$ is also a branched point. We can write down the set of branched points $\{q_i\}$ in terms of the generators ω_1 and ω_2 of L as follows.

First of all we construct a discrete subgroup G of the $\mathbb{C}$-analytic automorphism group $\mathrm{Aut}(\mathbb{C})$ of $\mathbb{C}$ containing L as a subgroup of index two and $\mathbb{C}/G$ is isomorphic to $P^1(\mathbb{C})$. In practice G can be constructed from L by adding the involutive element ι of $\mathrm{Aut}(\mathbb{C})$ defined by $\iota(z) = -z$. In this case the set of half periods $\left\{0, \frac{\omega_1}{2}, \frac{\omega_2}{2}, \frac{\omega_1 + \omega_2}{2}\right\}$ is the complete set of representatives of G-fixed points of $\mathbb{C}$, where $z_0 \in \mathbb{C}$ is called a G-fixed point if there exists a non-trivial element $g \in G$ such that $g(z_0) = z_0$. Because $\mathbb{C}/G$ is isomorphic to $P^1(\mathbb{C})$, the field of G-automorphic functions on $\mathbb{C}$ i.e. the field of meromorphic functions on $\mathbb{C}$ invariant under G, is isomorphic to the field of rational functions of one complex variable. Moreover it is naturally isomorphic to the meromorphic function field of $\mathbb{C}/G$. Hence next we find a generator h(z) of the field of G-automorphic functions on $\mathbb{C}$ satisfying $h(0) = \infty$. In this case we can take the Weierstraß $\wp$ function

$$\wp(z) := \frac{1}{z^2} + \sum_{\omega \in L - \{0\}} \left\{ \frac{1}{(z - \omega)^2} - \frac{1}{\omega^2} \right\}$$

as a generator of the field of G-automorphic functions on $\mathbb{C}$. Consequently we get a presentation of the Riemann surface $\mathbb{C}/L$ as an elliptic curve

$$y^2 = (x - \wp(\frac{\omega_1}{2}))(x - \wp(\frac{\omega_2}{2}))(x - \wp(\frac{\omega_1 + \omega_2}{2})).$$

The next diagram shows the process of the above argument

$$
\begin{array}{l}
\mathbb{C} \\
\downarrow \\
\mathbb{C}/L \\
\downarrow \qquad \searrow \text{ double covering} \\
\mathbb{C}/G \qquad \xrightarrow{\;\wp\;} \qquad P^1(\mathbb{C}).
\end{array}
$$

In this paper we will follow the same procedure for the case of genus two

$$
\begin{array}{l}
\mathbb{H}^2 \\
\downarrow \\
\mathbb{H}^2/\Gamma \\
\downarrow \qquad \searrow \text{ double covering} \\
\mathbb{H}^2/G \qquad \xrightarrow{\;h\;} \qquad P^1(\mathbb{C}).
\end{array}
$$

More precisely, in section two we will construct the Fuchsian group G of genus zero which contains Γ as a subgroup of index two. In practice we will get a presentation of G as a Fuchsian group of genus zero

$$
G = \langle \gamma_1, \cdots, \gamma_6 \mid \gamma_1^2 = \cdots = \gamma_6^2 = \gamma_1\gamma_2\cdots\gamma_6 \rangle.
$$

Let $p_i \in \mathbb{H}^2$ be the fixed point of $\gamma_i \in G$ in $\mathbb{H}^2$, then the set $\{p_i\}$ is the complete set of representatives of G-fixed points of $\mathbb{H}^2$. In section three, we will show that a generator h of the field of G-automorphic functions on $\mathbb{H}^2$ satisfying $h(p_6) = \infty$ can be written by means of Poincaré series for G of weight 6 as follows

$$
h(z) = \frac{\sum_{\gamma \in G} \frac{1}{\gamma z - p_6} P(\gamma z)\gamma'(z)^3}{\sum_{\gamma \in G} P(\gamma z)\gamma'(z)^3}
$$

where P is a rational function holomorphic on $\mathbb{H}^2$ and $P(p_6) \neq 0$. Here the order formula of G-automorphic functions on $\mathbb{H}^2$ is the key idea and to explain it, expository parts of this section became a little bit long. This function h is analogous to the Weierstraß $\wp$ function for the case of genus one. In this case, the main problem is to show the non-vanishing of the denominator of the expression of h i.e. to find a rational function P(z) which gives a non-trivial Poincaré series $\sum_{\gamma \in G} P(\gamma z)\gamma'(z)^3$. We will solve this problem by the work of I.Kra [Kra84] about the vanishing of Poincaré series and will review his argument in section four. We call his algorithm of the construction of non-vanishing Poincaré series *Kra's algorithm*. And finally we will find such a rational function P(z) explicitly in section five and consequently get the expression of

$\mathbb{H}^2/\Gamma$ as a hyperelliptic curve

$$y^2 = \prod_{i=1}^{5}(x - h(p_i)).$$

This is the outline of this paper.

Let's review the case of genus one once again. After the normalization of the lattice L with $L = \mathbb{Z} \cdot \tau + \mathbb{Z} \cdot 1$ where $\tau \in \mathbb{H}^2$, we can consider $\{q_i(\tau)\}$ as functions on the Teichmüller space $\mathbb{H}^2$ of elliptic curves. Moreover from these functions $\{q_i(\tau)\}$, we can get the λ-function $\lambda(\tau)$ and finally construct the elliptic modular function $J(\tau)$ whose theory has been one of the most fascinating part of the complex function theory and still now on. My primary motivation is to construct the theory of Teichmüller modular functions for the case of genus two and the result of this paper should be a starting point of this approach. In practice jointing with the work of Igusa ([Igu60]), we can construct Teichmüller modular functions on the Fricke space i.e. the Teichmüller space of genus two.

Finally I would like to thank my supervisor Kyoji Saito for providing me with many ideas and remarks during the course of this work.

2. Construction of the Fuchsian group G of genus zero

Let Γ be a marked Fuchsian group of genus two

$$\Gamma = \langle a_1, b_1, a_2, b_2 \mid \prod_{i=1}^{2}[a_i, b_i]\rangle \subset \mathrm{PSL}(2, \mathbb{R}).$$

The hyperbolic axis of a_1 and that of b_1 intersect in the positive sense with respect to the orientation of $\mathbb{H}^2$, where a hyperbolic axis is oriented from the repelling fixed point to the attractive one. Then the configuration of the axes of a_i, b_i and $c_1 := [a_1, b_1]$ on $\mathbb{H}^2$ (i=1,2) is shown in Figure 1 where we consider these axes in the unit disk $\mathbb{D}$ ([Kee71]).

For i=1,2, let $z_i \in \mathbb{H}^2$ be the intersection point of the axis of a_i and b_i, and $\delta_i \in \mathrm{PSL}(2, \mathbb{R})$ be the elliptic element of order 2 whose fixed point in $\mathbb{H}^2$ is z_i.

Proposition 2.1. *The group G generated by Γ and δ_1 is a Fuchsian group of genus zero containing Γ as a subgroup of index two.*

Proof. First we show that G has the coset decompostion

$$G = \Gamma \cup \Gamma\delta_1$$

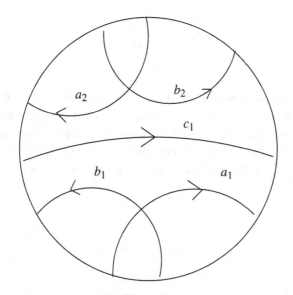

Figure 1: The configuration of the axes of generators of Γ.

which implies that Γ is a subgroup of index two in G. For $i = 1, 2$, define $\alpha_i, \beta_i \in$ PSL$(2, \mathbb{R})$ by

$$a_i = \alpha_i \cdot \delta_i$$
$$b_i = \beta_i \cdot \delta_i.$$

Then we can check that α_i, β_i are also elliptic elements of order two.

From the definition of α_i, β_i and δ_i,

$$a_i \delta_i = \delta_i a_i^{-1}, \ a_i^{-1} \delta_i = \delta_i a_i,$$

$$b_i \delta_i = \delta_i b_i^{-1}, \ b_i^{-1} \delta_i = \delta_i b_i.$$

Moreover the commutator relation $\prod_{i=1}^{2} [a_i, b_i]$ reduces

$$\alpha_1 \delta_1 \beta_1 = \beta_2 \delta_2 \alpha_2$$

which implies

$$\delta_2 = a_2^{-1} b_2^{-1} a_1 b_1 \delta_1.$$

Hence any element of G which is a finite word of elements of Γ and δ_1 can be written as either γ or $\gamma \delta_1$ where $\gamma \in \Gamma$.

Next we show that G has a presentation as a Fuchsian group of genus zero. It is easy to check that $\alpha_i, \beta_i, \delta_i \ (i = 1, 2)$ are generator system of G and satisfying the following relations

$$\alpha_i^2 = \delta_i^2 = \beta_i^2 = id. \quad (i = 1, 2)$$

$$\alpha_1 \cdot \delta_1 \cdot \beta_1 \cdot \alpha_2 \cdot \delta_2 \cdot \beta_2 = id. \tag{2.1}$$

Hence it is enough to show that any relation can be reduced from these relations. But we already knew that any word of G can be written as either γ or $\gamma\delta_1$ by means of these relations. Suppose that γ is trivial. Then it must be deduced from the commutator relation $\prod_{i=1}^2 [a_i, b_i]$ which is equivalent to the relation 2.1. Supose that $\gamma\delta_1$ is trivial. Then γ must be equal to δ_1 which contradicts to the fact that Γ is torsion-free. $\qquad \square$

Corollary 2.2.

(i) *Define*

$$\gamma_1 := \alpha_1 \quad \gamma_2 := \delta_1 \quad \gamma_3 := \beta_1$$
$$\gamma_4 := \alpha_2 \quad \gamma_5 := \delta_2 \quad \gamma_6 := \beta_2.$$

Then G has the presentation of a Fuchsian group of genus zero

$$G = \langle \gamma_1, \cdots, \gamma_6 \mid \gamma_1^2 = \cdots = \gamma_6^2 = \gamma_1\gamma_2 \cdots \gamma_6 \rangle.$$

(ii) *Let p_i be the fixed point of γ_i in $\mathbb{H}^2$, then the set $\{p_1, \cdots, p_6\}$ is the complete set of representatives of G-fixed points in $\mathbb{H}^2$.*

Remark 2.3. In practice $\{p_1, \cdots, p_6\}$ are half-periods in hyperbolic sense; the fixed points of α_i and β_i in $\mathbb{H}^2$ are $a_i^{\frac{1}{2}}(z_i)$ and $b_i^{\frac{1}{2}}(z_i)$ respectively, where $a_i^{\frac{1}{2}} \in \mathrm{PSL}(2, \mathbb{R})$ is the square root of a_i i.e. the unique element of $\mathrm{PSL}(2, \mathbb{R})$ with $(a_i^{\frac{1}{2}})^2 = a_i$.

3. Construction of a generator of the field of G-automorphic functions on $\mathbb{H}^2$

In this section, we will show that a generator of the field of meromorphic G-automorphic functions on $\mathbb{H}^2$ can be written by means of meromorphic G-automorphic forms of weight 6. For this purpose, we first review the relation between the G-automorphic forms of weight 2k on $\mathbb{H}^2$ and the k-th differentials on the Riemann surface $\mathbb{H}^2/G$ when G is a cocompact Fuchsian group in general.

Definition *A G-automorphic form of weight 2k $(k \geq 0)$ is a meromorphic function* f(z) *on $\mathbb{H}^2$ which satisfies the following identity*

$$f(gz)g'(z)^k = f(z) \quad \text{where} \quad g'(z) = \frac{\partial g}{\partial z}(z)$$

for $g \in G$ and $z \in \mathbb{H}^2$.

Definition For a local chart $(U_\alpha, \varphi_\alpha)_{\alpha \in A}$ of the Riemann surface $S = \mathbb{H}^2/G$, the collection $\{f_\alpha\}_{\alpha \in A}$ of the meromorphic functions f_α on $\varphi_\alpha(U_\alpha) \subset \mathbb{C}$ is a *k-th differential* on S ($k \geq 0$) provided that it satisfies the transformation law defined by

$$f_\alpha = f_\beta \circ \varphi_\beta \circ \varphi_\alpha^{-1} \cdot ((\varphi_\beta \circ \varphi_\alpha^{-1})')^k$$

on $\varphi_\alpha(U_\alpha \cap U_\beta)$ if $U_\alpha \cap U_\beta \neq \phi$. Let t_α be the coordinate on $\varphi_\alpha(U_\alpha)$. Then the above equation can be written as

$$f_\alpha(t_\alpha) = f_\beta(t_\beta) \cdot \left(\frac{dt_\beta}{dt_\alpha} \right)^k.$$

From the definition, 0-th differential on S is nothing but a meromorphic function on S and the product of m-th, and n-th differentials is a m+n-th differential. Moreover the set $\Omega_k(S)$ of k-th differentials on S is a 1-dimensional vector space over $K(S) := \Omega_0(S)$ the field of meromorphic functions on S. For $\omega \in \Omega_k(S)$, the order of ω at $p \in S$ is independent of the choice of a chart U_α, and we denote it by $v_p(\omega)$. Moreover since S is compact, $(\omega) := \sum_{p \in S} v_p(\omega) \cdot p$ is in fact a finite sum and put $deg(\omega) := \sum_{p \in S} v_p(\omega)$. Then the theorem of Riemann-Roch tells that $deg(\omega) = k(2g - 2)$ where g is the genus of S.

Now we will see the relation between the G-automorphic forms of weight 2k and the k-th differentials on S. First, let f(z) be a G-automorphic form of weight 2k. For $z \in \mathbb{H}^2$, we assume $\pi(z) \in U_\alpha (\subset S)$ where $\pi : \mathbb{H}^2 \to S = \mathbb{H}^2/G$ is the projection. Then the meromorphic function f_α on $\varphi_\alpha(U_\alpha)$ defined by

$$f_\alpha(\varphi_\alpha \circ \pi(z)) := \frac{f(z)}{(\varphi_\alpha \circ \pi)'(z)^k}$$

is well defined and the collection $\{f_\alpha\}_{\alpha \in A}$ is a k-th differential on S. We write this differential by $\pi_* f$. Conversely let $\omega = \{f_\alpha\}_{\alpha \in A}$ be a k-th differential on S, then the function f(z) on $\mathbb{H}^2$ defined by

$$f(z) := f_\alpha(\varphi_\alpha \circ \pi(z)) \cdot (\varphi_\alpha \circ \pi)'(z)^k \quad (z \in \mathbb{H}^2)$$

is well defined and a G-automorphic form of weight 2k. We write this by $\pi^* \omega$. Then π_* and π^* give the isomorphism between the set of G-automorphic forms of weight 2k and the set of k-th differentials on S, i.e. $\pi^* \circ \pi_* = \pi_* \circ \pi^* = id$.

Hence if we write a G-automorphic form f(z) like $f(z) = \pi^* \omega$ and denote the order of the isotropy subgroup $G_p := \{g \in G | g(p) = p\}$ of G at $p \in \mathbb{H}^2$ by n_p, the order

$v_p(f)$ of f(z) at $p \in \mathbb{H}^2$ can be written as

$$
\begin{aligned}
v_p(f) &= n_p \cdot v_{\pi(p)}(f_\alpha \circ \varphi_\alpha) + k \cdot v_p((\varphi_\alpha \circ \pi)') \\
&= n_p \cdot v_{\pi(p)}(\omega) + k(n_p - 1)
\end{aligned}
$$

if $\pi(p) \in U_\alpha \subset S$.

Next we apply the above formula for the case that k=3 and G defined in Section 2

$$
G = \langle \gamma_1, \cdots, \gamma_6 : \gamma_1^2 = \cdots = \gamma_6^2 = \gamma_1 \cdots \gamma_6 = id. \rangle.
$$

Let f(z) be a G-automorphic form of weight $2 \times 3 = 6$. Because the order of the isotropy subgroup G_{p_k} of G at $p_k \in \mathbb{H}^2$ where p_k is the fixed point of γ_k in $\mathbb{H}^2$, is 2 and the genus of $\mathbb{H}^2/G$ is 0, we have

$$
v_{p_k}(f) = 2v_{\pi(p_k)}(\pi_* f) + 3 \quad (k = 1, \cdots, 6) \tag{3.1}
$$

$$
deg(\pi_* f) = \sum_{p \in \mathbb{H}^2/G} v_p(\pi_* f) = -6 \tag{3.2}
$$

From 3.1, $v_{p_k}(f)$ is odd. In particular, if g(z) is a non-trivial holomorphic G-automorphic form of weight 6, then because of 3.2, the zeros of g(z) is precisely equal to the G-orbits of $\{p_k\}$ $(k = 1, \cdots, 6)$ and all of them are simple zeros. Hence if such g(z) exists (and in fact we can check the existence of g(z) by using the dimension formula of the space of holomorphic G-automorphic forms of weight 6 ([Shi71] Chapter 2, 2.6.), it is unique up to constant multiple.

On the other hand, let f(z) be a meromorphic G-automorphic form of weight 6 whose poles are only simple poles at the G-orbit of p_6. Then $v_{\pi(p_6)}(\pi_* f) = \frac{-1-3}{2} = -2$, hence by using 3.2, the type of the configuration of the zeros of f(z) is one of the following two types.

(i) f(z) has simple zeros at the G-orbits of $p_1, \cdots, p_5$ and of $q \in \mathbb{H}^2$ which is not a fixed point of G and f(z) has no more zeros. In this case

$$
\begin{aligned}
v_{\pi(p_k)}(\pi_* f) &= -1 \; (k = 1, \cdots, 5) \\
v_{\pi(p_6)}(\pi_* f) &= -2 \\
v_{\pi(q)}(\pi_* f) &= 1.
\end{aligned}
$$

(ii) f(z) has simple zeros at the G-orbits of $p_{i_1}, \cdots, p_{i_4}$ and triple zeros at the G-orbit of p_{i_5} where $\{i_1, \cdots, i_5\}$ is a permutation of $\{1, \cdots, 5\}$ and f(z) has no more zeros. In this case

$$v_{\pi(p_{i_j})}(\pi_* f) = -1 \ (j = 1, \cdots, 4)$$
$$v_{\pi(p_{i_5})}(\pi_* f) = 0$$
$$v_{\pi(p_6)}(\pi_* f) = -2.$$

If the configuration of the zeros of f(z) is of type 1, the function $h(z) := \frac{f(z)}{g(z)}$ where g(z) is a non-trivial holomorphic G-automorphic form of weight 6, has simple zeros at the G-orbit of q , double poles at the G-orbit of p_6 and h(z) has no more zeros and poles. Therefore since h(z) is a G-automorphic function on $\mathbb{H}^2$, if we consider h(z) as the function on $\mathbb{H}^2/G$, h(z) has only a simple zero at $\pi(q)$ and a simple pole at $\pi(p_6)$. Similarly, if the configuration of the zeros of f(z) is of type 2, h(z) has double zeros at the G-orbit of p_{i_5}, double poles at the G-orbit of p_6 and h(z) has no more zeros and poles. Therefore as the function on $\mathbb{H}^2/G$, h(z) has only a simple zero at $\pi(p_{i_5})$ and a simple pole at $\pi(p_6)$. Because $\mathbb{H}^2/G$ is isomorphic to $P^1(\mathbb{C})$, the meromorphic function field on $\mathbb{H}^2/G$ is isomorphic to the rational function field of one compex variable, hence its element h(z) is a generator if and only if h(z) has only one simple zero and pole. Moreover the meromorphic function field on $\mathbb{H}^2/G$ can be naturally identified with the field of G-automorphic functions on $\mathbb{H}^2$, we have just proved that a generator h(z) of the field of G-automorphic functions on $\mathbb{H}^2$ satisfying $h(p_6) = \infty$ can be constructed from a non-trivial holomorphic G-automorphic form g(z) of weight 6 and a meromorphic G-automorphic form f(z) of weight 6 whose poles are only simple poles at the G-orbit of p_6.

Moreover we can construct the above f(z) from g(z) by the following well known method ([Leh64] Chapter 5, section 2).

Lemma 3.1. *If a non-trivial holomorphic G-automorphic form g(z) of weight 6 is given by the Poincaré series*

$$g(z) = \sum_{\gamma \in G} P(\gamma z)\gamma'(z)^3$$

where P(z) is a holomorphic function on $\mathbb{H}^2$ with $P(p_6) \neq 0$, then the following function f(z)

$$f(z) = \sum_{\gamma \in G} \frac{1}{\gamma z - p_6} P(\gamma z)\gamma'(z)^3$$

is well defined and a G-automorphic form of weight 6 whose poles are only simple poles at the G-orbit of p_6.

Summarizing the above arguments, we get the following result of this section.

Proposition 3.2. *If we have a non-trivial holomorphic G-automorphic form g(z) of weight 6 as a Poincaré series*

$$g(z) = \sum_{\gamma \in G} P(\gamma z)\gamma'(z)^3$$

where P(z) is a holomorphic function on $\mathbb{H}^2$ with $P(p_6) \neq 0$, then the function h(z) defined by

$$h(z) := \frac{\sum_{\gamma \in G} \frac{1}{\gamma z - p_6} P(\gamma z)\gamma'(z)^3}{\sum_{\gamma \in G} P(\gamma z)\gamma'(z)^3}$$

is a generator of the field of G-automorphic functions on $\mathbb{H}^2$ satisfying $h(p_6) = \infty$.

In the remaining sections, we will find a rational function P(z) holomorphic on $\mathbb{H}^2$ which gives a non-trivial holomorphic G-automorphic form g(z) of weight 6 stated in Proposition 3.2.

4. Kra's algorithm for the construction of non-trivial Poincaré series

In this section following [Kra84] and [Kra72] mainly, we will review Kra's algorithm for the construction of cusp forms of finitely generated Fuchsian groups of the first kind by means of Poincaré series. We will see in the next section that this algorithm works well to find a rational function P(z) which gives a non-trivial Poincaré series described in the previous section.

4.1. Notations

We will use the following notations in this section.
$G :=$ a finitely generated Fuchsian group of the first kind
$\Lambda := \mathbb{R} \cup \{\infty\}$ the limit set of G
$\Omega := P^1(\mathbb{C}) - \Lambda$ the regions of discontinuity of G.
For a function $f : \Omega \to \mathbb{C}$ and $A \in \mathrm{PSL}(2,\mathbb{R})$,

$$(A_{q,r}^* f)(z) \quad := \quad f(Az)A'(z)^q \overline{A'(z)}^r \text{ where } A'(z) := \frac{\partial A}{\partial z}(z)$$

$$A_q^* \quad := \quad A_{q,0}^*.$$

4.2. The space of cusp forms on Ω of weight 2q $(q \geq 2)$

We define the space $A_q(\Omega, G)$ of cusp forms on Ω of weight 2q $(q \geq 2)$ by

$$A_q(\Omega, G) := \{\varphi : \Omega \to \mathbb{C} : \quad 1) \quad \varphi \text{ is holomorphic on } \Omega.$$
$$2) \quad \gamma_q^* \varphi = \varphi, \text{ for } \gamma \in G$$
$$3) \quad \sup_{z \in \Omega} |Im(z)|^q |\varphi(z)| < \infty\}$$

The space $A_q(\Omega, G)$ has the Petersson scalar product defined by

$$\langle \varphi, \psi \rangle_G := i \int \int_{G \backslash \Omega} |Im(z)|^{2q} \varphi(z) \overline{\psi(z)} \frac{dz \wedge d\bar{z}}{|Im(z)|^2}$$

for $\varphi, \psi \in A_q(\Omega, G)$.

For $A \in PSL(2, \mathbb{R})$, AGA^{-1} is also a Fuchsian group and A_q^* is a $\mathbb{C}$-linear isomorphism from $A_q(\Omega, AGA^{-1})$ to $A_q(\Omega, G)$

$$A_q^* : A_q(\Omega, AGA^{-1}) \quad \to \quad A_q(\Omega, G)$$
$$\varphi \quad \mapsto \quad A_q^* \varphi \ .$$

In the following, we will construct a basis of $A_q(\Omega, G)$ by applying the Poincaré series operator to the space of rational functions $R_q(\Lambda)$ defined in the next subsection.

4.3. The space of rational functions $R_q(\Lambda)$ $(q \geq 2)$

For $q \geq 2$, we define the family of rational functions $R_q(\Lambda)$ by

$$R_q(\Lambda) := \{f(z) : \text{rational function on } P^1(\mathbb{C}) : \quad (1) \quad f(z) \text{ is holomorphic on } \Omega;$$
$$(2) \quad \text{all poles of } f(z) \text{ are simple poles and located on } \Lambda;$$
$$(3) \quad f(z) = O(|z|^{-2q+1}) \ (z \to \infty)\}.$$

In the condition (2) above, we define the value of f(z) at $z = \infty$ by

$$f(\infty) := (-1)^q \lim_{z \to \infty} z^{2q} f(z).$$

$R_q(\Lambda)$ is a complex vector space and for $A \in \mathrm{PSL}(2, \mathbb{R})$, A_q^* defines a $\mathbb{C}$-linear automorphism of $R_q(\Lambda)$

$$A_q^* : R_q(\Lambda) \quad \rightarrow \quad R_q(\Lambda)$$
$$f \quad \mapsto \quad A_q^* f \ .$$

In particular, $f \in R_q(\Lambda)$ has (simple)poles at $Av_1, \cdots, Av_n \in \Lambda$ if and only if $A_q^* f \in R_q(\Lambda)$ has (simple)poles at $v_1, \cdots, v_n \in \Lambda$.

4.4. The Poicaré series operator Θ_q $(q \geq 2)$

For $q \geq 2$, we define the $\mathbb{C}$-linear map $\Theta_q = \Theta_{q,G}$ from $R_q(\Lambda)$ to $A_q(\Omega, G)$ by

$$\Theta_q : R_q(\Lambda) \quad \rightarrow \quad A_q(\Omega, G)$$
$$f \quad \mapsto \quad \Theta_q f := \sum_{\gamma \in G} \gamma_q^* f$$

which is called *the Poincaré series operator*.

Next theorem due to Bers shows that the operator Θ_q is an important notion for the study of cusp forms.

Theorem 4.1 ([Ber73]). *The map Θ_q is well defined, i.e. for $f \in R_q(\Lambda)$, $\Theta_q f$ converges uniformly on compact subsets of Ω. Moreover this map is surjective hence any cusp form of $A_q(\Omega, G)$ can be written as a Poincaré series $\Theta_q f$ $(f \in R_q(\Lambda))$.*

Because $\dim R_q(\Lambda)$ the dimension of $R_q(\Lambda)$ over $\mathbb{C}$ is infinite, this result is not effective for the construction of a basis of $A_q(\Omega, G)$. Quantitative improvement of this theorem is the following result due to Kra.

Theorem 4.2 ([Kra84]). *Fix a system of generators of G, $\gamma_0 = id, \gamma_1, \cdots, \gamma_N \in G$, and $(2q\text{-}1)$ distinct points $\{v_k\}$ of Λ which are G-fixed points. Put*

$$S(\vec{v}) \quad := \quad \{\gamma_j(v_k) \in \Lambda : j = 0, \cdots, N, \ k = 1, \cdots, 2q - 1\}$$
$$R_q(S(\vec{v})) \quad := \quad \{f(z) \in R_q(\Lambda) : f(z) \text{ is holomorphic on } \Lambda - S(\vec{v})\}$$

(i.e. $\dim R_q(S(\vec{v})) \leq (2q - 1)N$). Then the restriction of Θ_q to the subspace $R_q(S(\vec{v}))$ of $R_q(\Lambda)$ is also surjective. Hence we can choose a basis of $A_q(\Omega, G)$ from $\Theta_q(R_q(S(\vec{v})))$.

Remark 4.3. In his paper [Kra84], Kra shows this theorem for a finitely generated non elementary Kleinian group under modified conditions of $\{v_k\}$. Because we will use his result for our group G defined in section 2, we restrict his theorem to the statement of Theorem 4.2.

In the following we review the idea of the proof of Theorem 4.2 which leads us to Kra's algorithm for the construction of non-trivial Poincaré series. For this purpose we need to review the notions of the Eichler cohomology and the Eichler integral.

4.5. The Eichler cohomology

For $q \geq 2$, let Π_{2q-2} be the complex vector space consisting of polynomials in one complex variable of degree at most 2q-2 (i.e. $\dim \Pi_{2q-2} = 2q - 1$). We define the (right) action of G on Π_{2q-2} by

$$
\begin{aligned}
\Pi_{2q-2} \times G &\rightarrow \Pi_{2q-2} \\
(P, \gamma) &\mapsto P \cdot \gamma := \gamma_{1-q}^* P.
\end{aligned}
$$

A mapping $\chi : G \rightarrow \Pi_{2q-2}$ is a *cocycle* provided

$$
\chi(\gamma_1 \cdot \gamma_2) = \chi(\gamma_1) \cdot \gamma_2 + \chi(\gamma_2) \ (\gamma_1, \gamma_2 \in G).
$$

Such a cocycle is a *coboundary* if there exists some fixed $P \in \Pi_{2q-2}$ such that

$$
\chi(\gamma) = P \cdot \gamma - P \ (\gamma \in G).
$$

A cocycle χ is called *parabolic* if for any parabolic element η of G, there exists $P \in \Pi_{2q-2}$ such that

$$
\chi(\eta) = P \cdot \eta - P.
$$

We use the following notations.
$Z^1(G, \Pi_{2q-2}) :=$ the complex vector space of cocycles for G,
$B^1(G, \Pi_{2q-2}) :=$ the complex vector space of coboundaries for G,
$PZ^1(G, \Pi_{2q-2}) :=$ the complex vector space of parabolic cocycles for G.
Then *the Eichler cohomology* $H^1(G, \Pi_{2q-2})$ and *the parabolic cohomology* $PH^1(G, \Pi_{2q-2})$ are defined by

$$
\begin{aligned}
H^1(G, \Pi_{2q-2}) &:= Z^1(G, \Pi_{2q-2}) / B^1(G, \Pi_{2q-2}) \\
PH^1(G, \Pi_{2q-2}) &:= PZ^1(G, \Pi_{2q-2}) / B^1(G, \Pi_{2q-2}).
\end{aligned}
$$

For $A \in \mathrm{PSL}(2, \mathbb{R})$, A_{1-q}^* defines a $\mathbb{C}$-linear isomorphism

$$
\begin{aligned}
A_{1-q}^* : H^1(AGA^{-1}, \Pi_{2q-2}) &\rightarrow H^1(G, \Pi_{2q-2}) \\
\chi &\mapsto A_{1-q}^* \chi .
\end{aligned}
$$

Remark 4.4. The next mapping is isomorphic

$$\Pi_{2q-2} \quad \to \quad B^1(G, \Pi_{2q-2})$$
$$P \quad \mapsto \quad \chi(\gamma) = P \cdot \gamma - P, \ \gamma \in G.$$

Hence

$$\dim B^1(G, \Pi_{2q-2}) = 2q - 1.$$

Moreover

$$\dim Z^1(G, \Pi_{2q-2}) \leq (2q-1)N$$
$$\dim H^1(G, \Pi_{2q-2}) \leq (2q-1)(N-1)$$

where N is the number of the generators of G.

4.6. The Eichler integral

Let $D \subset P^1(\mathbb{C})$ be a non-void G-invariant set. A $\mathbb{C}$-valued function F(z) on D is *an Eichler integral for G of order 1-q* $(q \geq 2)$ if for $\gamma \in G$ there exists a polynomial $\chi(\gamma) \in \Pi_{2q-2}$ such that

$$\gamma_{1-q}^* F - F = \chi(\gamma)|_D.$$

Then for $\gamma \in G$, $\chi(\gamma) \in \Pi_{2q-2}$ is uniquely determined by F and the mapping χ defined by

$$\chi : G \quad \to \quad \Pi_{2q-2}$$
$$\gamma \quad \mapsto \quad \chi(\gamma)$$

satisfies the cocycle condition. We call this cocycle *the period of the Eichler integral* F and denote it by *pdF*.

The value of an Eichler integral at ∞ is defined by

$$F(\infty) := (-1)^{1-q} \lim_{z \to \infty} z^{2-2q} F(z)$$

if this limit exists.

If F is an Eichler integral for G, $A_{1-q}^* F$ is an Eichler integral for AGA^{-1} ($A \in \text{PSL}(2, \mathbb{R})$) and

$$F(Av) = 0 \Longleftrightarrow A_{1-q}^* F(v) = 0.$$

In the next section by using an Eichler integral, we can define the $\mathbb{C}$-anti-linear mapping β^* (which is called *the Bers map*) from $A_q(\Omega, G)$ to $H^1(G, \Pi_{2q-2})$, and we transform the problem of the cusp forms to the problem of the linear algebra on

$H^1(G,\Pi_{2q-2})$.

4.7. The Bers map β^*

For $\varphi \in A_q(\Omega, G)$, a canonical generalized Beltrami differential for G, $\mu(z)$ is defined by

$$\mu(z) := |Im(z)|^{2q-2}\overline{\varphi(z)}.$$

For $A \in PSL(2, \mathbb{R})$

$$A^*_{1-q,1}\mu = |Im(z)|^{2q-2}\overline{A^*_q\varphi}.$$

Hence in particular for $\gamma \in G$,

$$\gamma^*_{1-q,1}\mu = \mu.$$

A continuous function $F(z)$ on $\mathbb{C}$ is *a potential for μ* if it satisfies

$$
\begin{aligned}
F(z) &= O(|z|^{2q-2}) \ (z \to \infty) \\
\frac{\partial F(z)}{\partial \bar{z}} &= \mu(z)
\end{aligned}
$$

in the sense of generalized derivatives. We define the value of a potential $F(z)$ at $z = \infty$ by

$$F(\infty) := (-1)^{1-q} \lim_{z \to \infty} z^{2-2q} f(z)$$

if this limit exists. Then

$$F(\infty) = 0 \Longleftrightarrow F(z) = o(|z|^{2q-2}) \ (z \to \infty).$$

For $A \in PSL(2, \mathbb{R})$, $A^*_{1-q}F$ is a potential for $A^*_{1-q}\mu$ and for $v \in P^1(\mathbb{C})$,

$$F(Av) = 0 \Longleftrightarrow A^*_{1-q}F(v) = 0.$$

Next lemma is an easy consequence from the definition of a potential.

Lemma 4.5. *If F_1 and F_2 are potentials for a Beltrami differential μ, then $F_1 - F_2 \in \Pi_{2q-2}$*

Therefore a potential F for μ which vanishes at $(2q-1)$ distinct points $\{v_k\}$ is unique if it exists. Next theorem shows the existence of such F explicitly.

Theorem 4.6. *(Bers [Kra72]) For $\varphi \in A_q(\Omega, G)$, the unique potential F_φ for $\mu := |Im(z)|^{2q-2}\overline{\varphi}$ which vanishes at $(2q-1)$ distinct points $\{v_k\}$ of Λ can be written as follows*

$$F_\varphi(z) = \frac{(z-v_1)\cdots(z-v_{2q-1})}{2\pi i} \int\int_\Omega \frac{\mu(\zeta)d\zeta \wedge d\bar{\zeta}}{(\zeta-z)(\zeta-v_1)\cdots(\zeta-v_{2q-1})}$$

where we introduce the convention that if $v_k = \infty$ for some k, then the terms $(z - v_k)$ and $(\zeta - v_k)$ are dropped from this formula.

Because $\gamma_{1-q}^* F_\varphi$ is also a potential for $\gamma_{q-1,1}^* \mu = \mu$, Lemma 4.5 tells that $\gamma_{1-q}^* F_\varphi - F_\varphi \in \Pi_{2q-2}$. Hence F_φ is an Eichler integral and by using its period, we obtain the $\mathbb{C}$-anti-linear mapping β^* which is called *the Bers map*

$$\beta^* : A_q(\Omega, G) \quad \to \quad H^1(G, \Pi_{2q-2})$$
$$\varphi \quad \mapsto \quad [pd\, F_\varphi] \ .$$

Theorem 4.7 (Bers [Kra72]).

$$\beta^* : A_q(\Omega, G) = A_q(\mathbb{H}^2, G) \oplus A_q(\mathbb{H}_-^2, G) \to PH^1(G, \Pi_{2q-2})$$

where $\mathbb{H}_-^2$ is the lower half plane, is a $\mathbb{C}$-anti-linear isomorphism.

Fix (2q-1) distinct points $\{v_k\}$ of Λ. For $A \in \mathrm{PSL}(2, \mathbb{R})$, let $F_{1-q}(\Omega, AGA^{-1})$ be the set of the unique potentials $A_{1-q}^* F_\varphi$ for $\mu = |Im(z)|^{2q-2} \overline{A_q^* \varphi}$ where $\varphi \in A_q(\Omega, G)$, vanishing at $\{A(v_k)\}$. Then we have the following diagram

$$
\begin{array}{ccc}
R_q(\Lambda) & \xrightarrow{A_q^*} & R_q(\Lambda) \\
\Theta_{q, AGA^{-1}} \downarrow & & \downarrow \Theta_{q, G} \\
A_q(\Omega, AGA^{-1}) & \xrightarrow{A_q^*} & A_q(\Omega, G) \\
potential \downarrow & & \downarrow potential \\
F_{1-q}(\Omega, AGA^{-1}) & \xrightarrow{A_{1-q}^*} & F_{1-q}(\Omega, G) \\
period \downarrow & & \downarrow period \\
H^1(AGA^{-1}, \Pi_{2q-2}) & \xrightarrow{A_{1-q}^*} & H^1(G, \Pi_{2q-2}).
\end{array}
$$

4.8. The proof of Theorem 4.2

Next result is a corollary of Theorem 4.7.

Proposition 4.8. *Let $\{v_k\}$ be (2q-1) distinct points of Λ which are fixed points of G, F_φ be the potential for $\mu = |Im(z)|^{2q-2}\overline{\varphi}$ vanishing at $\{v_k\}$ and $S(\vec{v}) := \{\gamma_j(v_k)\}$ where $\gamma_0 = id, \gamma_1, \cdots, \gamma_N \in G$ are generators of G. Then for $\varphi \in A_q(\Omega, G)$,*

$$\varphi \equiv 0 \Longleftrightarrow F_\varphi|_{S(\vec{v})} \equiv 0.$$

where $F_\varphi|_{S(\vec{v})}$ is the restriction of F_φ to $S(\vec{v})$. Hence the mapping

$$A_q(\Omega, G) \to F_{1-q}(\Omega, G) \to F_{1-q}(\Omega, G)|_{S(\vec{v})}$$

is a bijection. Therefore if we put $d = \dim A_q(\Omega, G)$, *then there exist* $F_1, \cdots, F_d \in F_{1-q}(\Omega, G)$ *and* $w_1, \cdots, w_d \in S(\vec{v}) - \{v_1, \cdots, v_{2q-1}\}$ *such that*

$$F_j(w_k) = \delta_{jk} \ (1 \le j, k \le d).$$

Next proposition gives the proof of Theorem 4.2.

Proposition 4.9. *For* $w \in S(\vec{v}) - \{v_1, \cdots, v_{2q-1}\}$, *we define a rational function* $f(w, z) \in R_q(S(\vec{v}))$ *by*

$$f(w, z) \quad := \quad -\frac{1}{2\pi} \frac{1}{z - w} \prod_{j=1}^{2q-1} \frac{w - v_j}{z - v_j} \quad \text{if } w < \infty$$

$$f(w, z) \quad := \quad -\frac{1}{2\pi} (-1)^q \prod_{j=1}^{2q-1} \frac{1}{z - v_j} \quad \text{if } w = \infty$$

with the usual convention that if $v_j = \infty$ *for some* j *then the terms* $z - v_j$ *and* $w - v_j$ *are dropped from this formula. Then for* $\varphi \in A_q(\Omega, G)$

$$F_\varphi(w) = \langle \Theta_q f(w, z), \varphi \rangle_G.$$

Hence because of the non-degenerateness of the Petersson scalar product $\langle , \rangle_G$ *on* $A_q(\Omega, G)$, $\{\Theta_q f(w_j, z)\}_{j=1, \cdots, d}$ *is a basis of* $A_q(\Omega, G)$ *for* $w_1, \cdots, w_d \in S(\vec{v}) - \{v_1, \cdots, v_{2q-1}\}$ *stated in the Proposition 4.8.*

Corollary 4.10. *For* $w_1, \cdots, w_d \in S(\vec{v}) - \{v_1, \cdots, v_{2q-1}\}$ *where* $d = \dim A_q(\Omega, G)$, $\{\Theta_q f(w_j, z)\}_{j=1, \cdots, d}$ *is a basis of* $A_q(\Omega, G)$ *if and only if the* $\mathbb{C}$-*linear mapping*

$$\begin{aligned} F_{1-q}(\Omega, G) \quad &\to \quad \mathbb{C}^d \\ F \quad &\mapsto \quad (F(w_1), \cdots, F(w_d)) \end{aligned}$$

is an isomorphism. In particular for $w \in S(\vec{v}) - \{v_1, \cdots, v_{2q-1}\}$

$$\Theta_q f(w, z) \ne 0 \iff F(w) \ne 0 \ \text{for some } F \in F_{1-q}(\Omega, G).$$

4.9. Kra's algorithm for the construction of a non-trivial Poincaré series

For the construction of non-trivial Poincaré series, we have seen in the previous sub-section that we need only to know whether $F(w) = 0$ or not for some $w \in S(\vec{v}) - \{v_1, \cdots, v_{2q-1}\}$ and some $F \in F_{1-q}(\Omega, G)$. If $h_k \in G$ has $v_k \in \Lambda$ as a fixed point and a hyperbolic element $h \in G$ has $w \in S(\vec{v}) - \{v_1, \cdots, v_{2q-1}\}$ with $w < \infty$ as a fixed point , then we can compute $F(w)$ as follows. First we define the subspace $\overline{Z}^1(G, \Pi_{2q-2})$ of

$Z^1(G,\Pi_{2q-2})$ by

$$\overline{Z}^1(G,\Pi_{2q-2}) := \left\{\chi \in Z^1(G,\Pi_{2q-2})|\chi(h_k)(v_k) = 0 \ (k = 1,\cdots,2q-1)\right\}.$$

Then $\overline{Z}^1(G,\Pi_{2q-2})$ is naturally isomorphic to $H^1(G,\Pi_{2q-2})$. Moreover $pdF \in \overline{Z}^1(G,\Pi_{2q-2})$ for $F \in F_{1-q}(\Omega,G)$ from the definition of $F_{1-q}(\Omega,G)$. Hence if $w \in S(\vec{v}) - \{v_1,\cdots,v_{2q-1}\}$ with $w < \infty$ and a hyperbolic element $h \in G$ with $h(w) = w$,

$$
\begin{aligned}
pdF(h)(w) &= h^*_{1-q}F(w) - F(w) = F(w)(h'(w)^{1-q} - 1) \\
F(w) &= \frac{pdF(h)(w)}{h'(w)^{1-q} - 1}.
\end{aligned}
$$

This argument shows that if there exists $\chi \in \overline{Z}^1(G,\Pi_{2q-2}) \cap PZ^1(G,\Pi_{2q-2})$ with $\chi(w) \neq 0$, because such χ can be written as pdF for some $F \in F_{1-q}(\Omega,G)$ by Theorem 4.7, $F(w) \neq 0$ hence $\Theta_q f(w,z) \neq 0$ by Corollary 4.10.

We call these procedure *the Kra's algorithm* for the construction of a non-trivial Poincaré series which consists of the following steps:

(i) Fix a system of generators of G, $\gamma_0 = id, \gamma_1,\cdots,\gamma_N \in G$.

(ii) Fix $(2q-1)$ distinct points $\{v_k\}$ of Λ and $h_k \in G$ with $h_k(v_k) = v_k$.

(iii) Find a parabolic cocycle $\chi \in PZ^1(G,\Pi_{2q-2})$ with $\chi(h_k)(v_k) = 0 \ (k = 1,\cdots,2q-1)$.

(iv) Find $w \in S(\vec{v}) := \{\gamma_j(v_k) \in \Lambda \mid j = 0,\cdots,N, k = 1,\cdots,2q-1\} - \{v_1,\cdots,v_{2q-1}\}$ with $w < \infty$ and a hyperbolic element $h \in G$ with $h(w) = w$ satisfying $\chi(h)(w) \neq 0$.

(v) Then $\Theta_q f(w,z)$ is a non-trivial cusp form of $A_q(\Omega,G)$.

In the next section, we will see that this algorithm works nicely for the case of q=3 and the Fuchsian group G of genus zero defined in section two.

5. Construction of a non-trivial Poincaré series. Main result

In this section by means of Kra's algorithm discussed in the previous section, we will find a rational function P(z) holomorphic on $\mathbb{H}^2$ with $P(p_6) \neq 0$ which gives a non-trivial Poincaré series

$$\sum_{\gamma \in G} P(\gamma z)\gamma'(z)^3.$$

Let G be a Fuchsian group of genus zero defined in section 2. For $j = 1, \cdots, 5$, define the hyperbolic element $h_j := \gamma_j \gamma_6$ where $\gamma_1, \cdots, \gamma_6$ are generators of G stated in section two , and let v_j be the attractive fixed point of h_j. Then the repelling fixed point of h_j is $\gamma_6 v_j = \gamma_j v_j$. Put $h := h_1 = \gamma_1 \gamma_6$ and $w := \gamma_6 v_1 = \gamma_1 v_1$. Then $h(w) = w$ and $w \in S(\bar{v}) - \{v_1, \cdots, v_5\}$.

If we can find a cocycle $\chi \in Z^1(G, \Pi_4)$ with

$$\chi(h_j)(v_j) \;=\; 0 \;\; (j = 1, \cdots, 5)$$
$$\chi(h)(w) \;\neq\; 0$$

then the Kra's algorithm tells that

$$\Theta_3 \Big(\frac{1}{z - w} \prod_{j=1}^{5} \frac{1}{z - v_j} \Big) \neq 0 \;\; \text{in } A_3(\Omega, G).$$

For this purpose, we first assume that $z = \infty$ is a fixed point of $c^{\frac{1}{2}} = \gamma_1 \gamma_2 \gamma_3$. Then because h_j and $c^{\frac{1}{2}}$ don't commute, they don't have a common fixed point. Hence

$$v_j < \infty, \;\; \gamma_6 v_j = \gamma_j v_j < \infty \;\; (j = 1, \cdots, 5).$$

In particular $w = \gamma_6 v_1 < \infty$.

Lemma 5.1. *There exists a cocycle $\chi \in Z^1(G, \Pi_4)$ such that*

$$\chi(h_j)(v_j) = 0 \;\; (j = 1, \cdots, 5)$$

and the zeros of $\chi(\gamma_6) \in \Pi_4$ are $p_6, \bar{p}_6, v_2$ and $\gamma_6 v_2 = \gamma_2 v_2$ where $p_6 \in \mathbb{H}^2$ is the fixed point of γ_6 and $\bar{p}_6$ is the complex conjugate of p_6.

Proof. The equalities of the cocycle condition of $\chi \in Z^1(G, \Pi_4)$

$$0 = \chi(\gamma_k^2) = \chi(\gamma_k) \cdot \gamma_k + \chi(\gamma_k) \;\; (k = 1, \cdots, 6)$$

shows that there exist $s_k, t_k \in \mathbb{C} \;\; (k = 1, \cdots, 6)$ such that

$$\chi(\gamma_k) = s_k (z - p_k)^3 (z - \bar{p}_k) + t_k (z - p_k)(z - \bar{p}_k)^3.$$

Hence the above equations determine 12 dimensional vector space. Because the identities

$$\chi(\gamma_1 \cdots \gamma_6) \;=\; 0$$
$$\chi(h_j)(v_j) \;=\; 0 \;\; (j = 1, \cdots, 5)$$

are 10 linear equations of s_k, t_k $(k = 1, \cdots, 6)$, we can find a solution $\{s_k, t_k\}$ i.e. a cocycle $\chi \in Z^1(G, \Pi_4)$ which satisfies

$$\chi(\gamma_6)(v_2) = \chi(\gamma_6)(\gamma_6 v_2) = 0.$$

$\square$

Let $\chi \in Z^1(G, \Pi_4)$ be a cocycle defined in Lemma 5.1. We will show that $\chi(h)(w) \neq 0$. From the equality

$$
\begin{aligned}
\chi(h_1)(v_1) &= \chi(\gamma_1\gamma_6)(v_1) \\
&= (\chi(\gamma_1) \cdot \gamma_6 + \chi(\gamma_6))(v_1) \\
&= \chi(\gamma_1)(\gamma_6 v_1)\gamma_6'(v_1)^{-2} + \chi(\gamma_6)(v_1) \\
&= \chi(\gamma_1)(\gamma_1 v_1)\gamma_6'(v_1)^{-2} + \chi(\gamma_6)(v_1) \\
&= -\chi(\gamma_1)(v_1)(\frac{\gamma_1'(v_1)}{\gamma_6'(v_1)})^2 + \chi(\gamma_6)(v_1) = 0
\end{aligned}
$$

we obtain that

$$\chi(\gamma_1)(v_1) = (\frac{\gamma_6'(v_1)}{\gamma_1'(v_1)})^2 \cdot \chi(\gamma_6)(v_1).$$

Because v_1 is the attractive fixed point of $h_1 = \gamma_1\gamma_6$,

$$
\begin{aligned}
\left|\frac{\gamma_6'(v_1)}{\gamma_1'(v_1)}\right| &= |\gamma_1'(\gamma_1 v_1)\gamma_6'(v_1)| \\
&= |\gamma_1'(\gamma_6 v_1)\gamma_6'(v_1)| \\
&= |(\gamma_1\gamma_6)'(v_1)| < 1.
\end{aligned}
$$

On the other hand

$$
\begin{aligned}
\chi(h)(w) &= \chi(\gamma_1\gamma_6)(\gamma_6 v_1) \\
&= (\chi(\gamma_1) \cdot \gamma_6 + \chi(\gamma_6))(\gamma_6 v_1) \\
&= \chi(\gamma_1)(v_1)\gamma_6'(\gamma_6 v_1)^{-2} + \chi(\gamma_6)(\gamma_6 v_1) \\
&= \chi(\gamma_1)(v_1)\gamma_6'(v_1)^2 - \chi(\gamma_6)(v_1)\gamma_6'(v_1)^2 \\
&= \{\chi(\gamma_1)(v_1) - \chi(\gamma_6)(v_1)\}\gamma_6'(v_1)^2 \\
&= \left\{(\frac{\gamma_6'(v_1)}{\gamma_1'(v_1)})^2 - 1\right\}\chi(\gamma_6)(v_1)\gamma_6'(v_1)^2.
\end{aligned}
$$

Because of the choice of a cocycle χ, the zeros of $\chi(\gamma_6)$ are $p_6, \bar{p}_6 \in \mathbb{H}^2$ and $v_2, \gamma_6 v_2 = \gamma_2 v_2 \in \mathbb{R}$. On the other hand $h = h_1$ and h_2 are not commutative,

$$v_1 \neq v_2 \text{ and } v_1 \neq \gamma_2 v_2$$

hence $\chi(\gamma_6)(v_1) \neq 0$. This shows $\chi(h)(w) \neq 0$ and we obtain

$$\Theta_3 \left(\frac{1}{z - \gamma_1 v_1} \prod_{j=1}^{5} \frac{1}{z - v_j} \right) \neq 0$$

by Kra's algorithm.

Next we consider the case $c_1^{\frac{1}{2}}(\infty) \neq \infty$. There exists $A \in PSL(2, \mathbb{R})$ such that $z = \infty$ is a fixed point of $Ac_1^{\frac{1}{2}}A^{-1}$. Then the previous argument shows that if we put

$$P(z) = \frac{1}{z - A\gamma_1 v_1} \prod_{j=1}^{5} \frac{1}{z - Av_j}$$

then

$$\Theta_{3, AGA^{-1}} P \neq 0 \text{ in } A_3(\Omega, AGA^{-1}).$$

Therefore the diagram

$$
\begin{array}{ccc}
R_3(\Lambda) & \xrightarrow{A_3^*} & R_3(\Lambda) \\
\Theta_{3, AGA^{-1}} \downarrow & & \downarrow \Theta_{3, G} \\
A_3(\Omega, AGA^{-1}) & \xrightarrow{A_3^*} & A_3(\Omega, G)
\end{array}
$$

shows that if

$$Az = \frac{pz + q}{rz + s}$$

then

$$A_3^* P(z) = \frac{r(\gamma_1 v_1) + s}{z - \gamma_1 v_1} \prod_{j=1}^{5} \frac{r v_j + s}{z - v_j}$$

gives

$$\Theta_{3, G} A_3^* P \neq 0 \text{ in } A_3(\Omega, G).$$

Hence

$$\Theta_{3, G} \left(\frac{1}{z - \gamma_1 v_1} \prod_{j=1}^{5} \frac{1}{z - v_j} \right) \neq 0$$

where we use the convention that if $\gamma_1 v_1 = \infty$ or $v_j = \infty$, then the terms $z - \gamma_1 v_1$ or $z - v_j$ are dropped.

Remark 5.2. Because the poles of $P(z)$ are located on $\Lambda = \mathbb{R} \cup \{\infty\}$,

$$\overline{\Theta_3 P(z)} = \Theta_3 P(\bar{z}).$$

Hence

$$\Theta_3 P \neq 0 \ \text{in} \ A_3(\Omega, G)$$

implies

$$\Theta_3 P \neq 0 \ \text{in} \ A_3(\mathbb{H}^2, G).$$

Summarizing the previous arguments, we conclude the main theorem.

Theorem 5.3. *For a presentation of a Fuchsian group Γ of genus two*

$$\Gamma = \langle a_1, b_1, a_2, b_2 \mid \prod_{i=1}^{2} [a_i, b_i] \rangle \subset \mathrm{PSL}(2, \mathbb{R})$$

let $\delta_i \in \mathrm{PSL}(2, \mathbb{R})$ $(i = 1, 2)$ be the elliptic element of order 2 whose fixed point in $\mathbb{H}^2$ is the intersection point z_i of the axes of a_i and b_i. And let $\alpha_i (resp. \ \beta_i) \in \mathrm{PSL}(2, \mathbb{R})$ $(i = 1, 2)$ be the elliptic element of order 2 whose fixed point in $\mathbb{H}^2$ is $a_i^{\frac{1}{2}}(z_i) (resp. \ b_i^{\frac{1}{2}}(z_i))$. Put

$$\gamma_1 = \alpha_1, \quad \gamma_2 = \delta_1, \quad \gamma_3 = \beta_1$$
$$\gamma_4 = \alpha_2, \quad \gamma_5 = \delta_2, \quad \gamma_6 = \beta_2$$

and let $p_k \in \mathbb{H}^2$ $(k = 1, \cdots, 6)$ be the fixed point of γ_k. Then the Riemann surface $\mathbb{H}^2/\Gamma$ of genus two has the following expression as a hyperelliptic curve

$$y^2 = \prod_{k=1}^{5} (x - h(p_k))$$

where

$$h(z) = \frac{\sum_{\gamma \in G} \frac{1}{\gamma z - p_6} P(\gamma z) \gamma'(z)^3}{\sum_{\gamma \in G} P(\gamma z) \gamma'(z)^3}$$

$$P(z) = \frac{1}{z - \gamma_1 v_1} \prod_{j=1}^{5} \frac{1}{z - v_j}$$

where v_j is the attractive fixed point of the hyperbolic element $\gamma_j \gamma_6$ $(j = 1, \cdots, 5)$ and we use the convention that if $\gamma_1 v_1 = \infty$ or $v_j = \infty$, then the terms $z - \gamma_1 v_1$ or $z - v_j$ are dropped from this formula.

References

[Ber73] L. Bers (1973). Poincaré series for Kleinian groups. *Comm. Pure Appl. Math.* **26**, 667–672.

[Igu60] J.-i. Igusa (1960). Arithmetic variety of moduli for genus two. *Ann. of Math.* (2) **72**, 612–649.

[Kee71] L. Keen (1971). On Fricke moduli. In *Advances in the Theory of Riemann Surfaces (Proc. Conf., Stony Brook, N.Y., 1969)*, pp. 205–224. Ann. of Math. Studies, No. 66. Princeton Univ. Press, Princeton, N.J.

[Kra72] I. Kra (1972). *Automorphic Forms and Kleinian Groups*. W. A. Benjamin, Inc., Reading, Mass.

[Kra84] I. Kra (1984). On the vanishing of and spanning sets for Poincaré series for cusp forms. *Acta Math.* **153** (1-2), 47–116.

[Leh64] J. Lehner (1964). *Discontinuous Groups and Automorphic Functions*. Mathematical Surveys, No. VIII. American Mathematical Society, Providence, R.I.

[Shi71] G. Shimura (1971). *Introduction to the Arithmetic Theory of Automorphic functions*. Publications of the Mathematical Society of Japan, No. 11. Iwanami Shoten, Publishers, Tokyo.

Yohei Komori

Department of Mathematics
Osaka City University
Sugimoto, Sumiyoshi-ku
Osaka, 558-8585
Japan

komori@sci.osaka-cu.ac.jp

AMS Classification: 30F40, 32G05

Keywords: automorphic functions, Kleinian groups, Poincaré series, Teichmüller spaces

Spaces of Kleinian Groups
Lond. Math. Soc. Lec. Notes **329**, 283–299

Cambridge University Press
Y. Minsky, M. Sakuma & C. Series (Eds.)

Boundaries for two-parabolic Schottky groups

Jane Gilman[1]

1. Introduction

In this paper we survey various results about the Schottky parameter space for a two parabolic generator group and the smooth boundary for the classical Schottky parameter space lying inside what is known as the Riley slice (of Schottky space).

The problem can be formulated in a number of equivalent different settings: in terms of the topology and the geometry of hyperbolic three-manifolds, in purely algebraic terms, or in a combination of these. Since each of these use terminology whose exact meaning has evolved over time, we survey some of the basic terminology for Schottky groups, non-separating disjoint circle groups, noded surfaces and their various representation spaces. This is done in sections 2, 5, 3, and 4 respectively. In the introduction (sections 1.1, 1.2, and 1.3) we state the main theorems taking the liberty of using some terms whose precise definitions are deferred to the later sections. We begin with the algebraic formulation which may be the quickest way to approach the problem.

1.1. Two Parabolics after Lyndon–Ullman

In the Lyndon–Ullman formulation, we consider two by two matrices, S and T where

$$S = \begin{pmatrix} 1 & 0 \\ 1 & 1 \end{pmatrix} \text{ and } T = \begin{pmatrix} 1 & 2\lambda \\ 0 & 1 \end{pmatrix}.$$

We write $T = T_\lambda$ to emphasize that T depends upon the complex number λ. We let G_λ be the group generated by S and T_λ so that $G_\lambda = \langle S, T_\lambda \rangle$.

In 1969, Lyndon and Ullman asked the question, "for what values of λ is G_λ a free group?". They found certain regions in the complex λ-plane that assured that G_λ would be free for λ in one of these regions. The regions are symmetric about the real and imaginary axes. The portions in the first quadrant of three of the five regions they found, F_1, F_2, F_3, are shown in the diagram of figure 1.

---

[1] Supported in part by NSA grant #0G2-186 and the Rutgers Research Council.

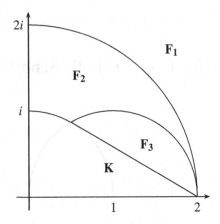

Figure 1: Some of the Lyndon–Ullman Free Region(s) in $\mathbb{C}$.

Around 1980 David Wright wrote a computer program to plot all the points λ for which G_λ is free and discrete. His picture (shown as figure 2) has come to be known as the *Riley slice of Schottky Space* although under the original definitions of Schottky space it is not technically a slice of Schottky space or even a subset of Schottky space, but rather the quotient of a slice of Schottky space, which we will term *Schottky Parameter Space.*

In Wright's plot the boundary appears to be fractal in nature, but this is not known for a fact [MSW02]. Keen and Series [KS94] have studied moving towards the boundary along the lines that are roughly orthogonal to the boundary and are faintly visible in figure 2. These are called *pleating rays* and geometrically these are points where the pleating locus of the convex hull in $\mathbb{H}^3$ has a particular form (see [KS94]). The geometric condition implies the algebraic condition that the trace of certain words in the generators of G_λ take on real values, but this algebraic condition alone is not sufficient for a point to be on a pleating ray.

1.2. Classical and Non-classical groups

In [GW02] Gilman and Waterman described the parameter space of classical Schottky groups affording two parabolic generators within the larger parameter space of all Schottky groups with two parabolic generators. Two parabolic Schottky groups belong to the boundary of purely loxodromic Schottky space of rank two. The Gilman–Waterman result uses the most general definition of Schottky group (see section 2). This gave a smooth boundary (except for two points) and the boundary equations are given by portions of two intersecting parabolas. These are depicted superimposed upon the Wright plot in figure 3 and also in figure 4.

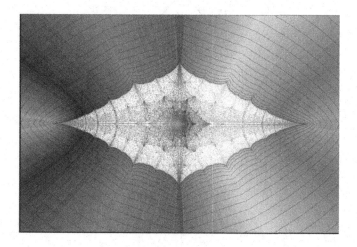

Figure 2: The Riley Slice

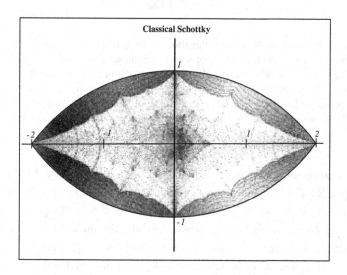

Figure 3: Classical Schottky boundary superimposed on the Riley Slice

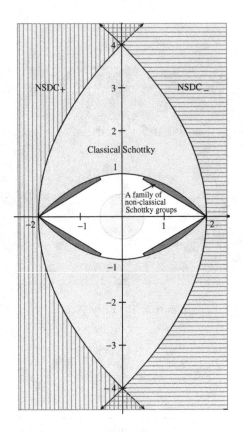

Figure 4: Superimposed Boundary Parabolas. Each point $\lambda \in \mathbb{C}$ corresponds to a two-generator group. The darkest region shows a one parameter family of non-classical Schottky groups. The line-shaded subset of the classical Schottky groups comprises the non-separating disjoint circle groups (NSDC groups). The unshaded region consists of additional non-classical Schottky groups together with degenerate groups, isolated discrete groups and non-discrete groups. Points inside the Shimizu-Leutchbecher-Jørgensen circle ($|\lambda| < \frac{1}{2}$) are non-discrete groups.

The existence of non-classical Schottky groups was proved by A. Marden in his 1974 paper [Mar74]. However, his proof was not constructive. In 1975 Zarrow ([Zar75]) claimed to give an example of a non-classical Schottky group, but in 1988 Sato showed that Zarrow's example was erroneous. Zarrow's construction only gave a group that was not marked classical (on the given set of generators) but might be classical on a different set of generators [Sat88]. This showed that the verification that an example of a non-classical Schottky group was what it claimed to be would require two steps: show (1) that it was non-classical on the given set of generators and (2) that it remained non-classical under any change of generators.

In 1990 Yamamoto gave an example of a non-classical Schottky group on two loxodromic generators [Yam91]. Unfortunately this has never been well enough understood to lead to any further examples. In [GW02] an explicit construction is given for a one complex parameter family of non-classical two parabolic generated Schottky groups. The construction is not related to Yamamoto's in any obvious manner. Recently Hidalgo and Maskit have given a theoretical construction of other non-classical Schottky groups [HM04].

1.3. Group Theoretic statement

The easiest way to state the main result of [GW02] is group theoretically:

Theorem 1.1. *Let λ be a complex number with $\lambda = |\lambda| e^{i\theta}$ where $0 < \theta < \pi$. The group G_λ is a classical Schottky group*

$$\Leftrightarrow |\lambda|(1 + \sin\theta) \geq 2.$$

Here $|\ |$ denotes absolute value.

A two-by-two matrix with complex entries acts as a fractional linear transformation on $\hat{\mathbb{C}}$ and if it has determinant one, we let tr denote its trace. While there are many variations on trace inequalities that imply discreteness, necessary and sufficient trace inequalities for discreteness are not common. Equivalent to Theorem 1.1 is

Theorem 1.2. *Let A and B be two-by-two complex matrices each with determinant one with $A \neq I$ and $B \neq I$. The group $G = \langle A, B \rangle$ generated by A and B with $\mathrm{tr}(A) = \mathrm{tr}(B) = 2$ is a classical Schottky group*

$$\Leftrightarrow |\mathrm{tr}(AB) - 2| + |\mathrm{Im}[\mathrm{tr}(AB)]| \geq 4.$$

Here Im denotes the imaginary part of a complex number and I denotes the identity matrix.

Theorem 1.3. *Let $\lambda = x + iy$. The group G_λ is a classical Schottky group and lies on the boundary of classical Schottky parameter space*

$$\Leftrightarrow |y| = 1 - \frac{x^2}{4}.$$

In 1996 Hidalgo [Hid96] proved that for arbitrary genus, the space of Schottky groups with no allowed tangencies between Schottky circles is dense in the space of noded Schottky groups, that is, where tangencies at parabolic fixed points are allowed. Related to the density of Schottky groups with no allowed tangencies between Schottky circles is the work of B. Maskit [Mas81] where it is shown that geometrically finite Kleinian groups which are free groups belong to the closure of Schottky space.

The proof of Theorem 1.1 in [GW02] shows that using the most general definition of Schottky group (section 2) does not change the interiors or the boundaries of the classical Schottky space for groups generated by a pair of parabolics.

Theorem 1.4. *All variations on the definition of classical Schottky group yield spaces with the same interior for a group generated by two parabolics. Further, the closures are the same so that the boundaries are the same. For a group generated by two parabolics, the classical Schottky parameter space is closed when the most general definition of Schottky group is used.*

Following the notation of Lyndon–Ullman [LU69], we let K be the convex hull of the set in the λ-plane consisting of the circle $|z| = 1$ and the points $z = \pm 2$. It is shown in [GW02] that if λ is not in the interior of K, then G_λ is Schottky. More specifically, it is proved that

Theorem 1.5. The Non-classical Schottky family. *If λ lies in the upper-half plane below the Schottky parabola $y = 1 - \frac{x^2}{4}$ but exterior to K, then G_λ is a non-classical Schottky group.*

2. Definitions for Schottky groups

A Schottky group is defined as a group where the generators have a certain geometric action on the complex sphere, $\hat{\mathbb{C}}$, that is easily stated. However, the definition of Schottky group has changed over time. Initially a Schottky group of rank n, $n > 0$ and $n \in \mathbb{Z}$, was defined to be a group generated by elements $g_1, ..., g_n$ for which there where $2n$ disjoint oriented circles $C_1, C_1', ..., C_n, C_n'$ such that $g_i(C_i) = C_i'$ with g_i mapping the exterior of C_i to the interior of C_i', where the common exteriors of the circles bounded a connected region in $\hat{\mathbb{C}}$.

For purely loxodromic groups this definition is a natural extension of the idea of isometric circle. However, for a parabolic transformation, its isometric circle and that of its inverse will be tangent at the fixed point of the transformation. Thus the first generalizations of Schottky group allowed tangencies of paired circles at parabolic fixed points, sometimes called *groups of Schottky type*. Later *marked noded Schottky groups* were studied by Hidalgo. Most recently Hidalgo and Maskit(see [Mas88, Hid96]) have obtained results about neo-classical Schottky groups, which allowed tangencies of Schottky circles, paired or not, as long as the points of tangency are parabolic fixed points. Groups where tangencies of paired Schottky circles at non-parabolic fixed points are allowed were sometimes dubbed [MSW02] *kissing Schottky groups*. Eventually [Rat94] topologists dropped even this restriction and we now allow any tangencies between any Schottky circles, paired or not, at parabolic fixed points or not. It is this definition that allows one to more easily identify the boundary of Schottky space or Schottky parameter space for groups generated by two parabolics. The interiors of these spaces and their boundaries turn out to be the same once the restriction of two parabolic generators is made. Further, some of the technical problems about fundamental domains that motivated the original more restrictive definition can be overcome in this case ([GW02]).

Definition 2.1. A *marked classical Schottky group* is a group G together with a set of n generators $g_1, ..., g_n$ for which there are $2n$ circles $C_1, C_1', ..., C_n, C_n'$ in $\hat{\mathbb{C}}$ whose interiors are all pairwise disjoint and such that $g_i(C_i) = C_i'$ with g_i mapping the exterior of C_i to the interior of C_i'. A finitely generated group of Möbius transformations G is a *classical Schottky group* if it is a marked classical Schottky group on *some* set of generators.

If the requirement that the C_i, C_i' be circles is dropped and the C_i are only required to be Jordan curves, with non-empty common exterior, then G is called a *marked Schottky group* and a *Schottky group* respectively. A *marked non-classical Schottky group* is a marked Schottky group that is not marked classical Schottky with the given marking. A Schottky group G is a *non-classical Schottky group* if it is not a marked classical Schottky on any set of generators.

Further, the region exterior to the $2n$ circles is called a *(classical) Schottky domain* or a *(classical) Schottky configuration* and the transformations g_i are called the *Schottky pairings*.

Remark 2.2. We observe that classical Schottky domains are not necessarily fundamental domains, but we can push out at tangencies that are not at parabolic fixed points to get a non-classical domain that is a fundamental domain. In [GW02] it is shown that the so called *extreme* domains for two parabolic generator classical Schottky groups are fundamental domains.

Remark 2.3. All Schottky groups are geometrically finite, free, discrete groups by the Klein-Maskit combination theorems or the Poincaré Polyhedron theorem [Mas88].

3. The Geometry and Topology: pinching and nodes

We remind the reader that the hyperbolic three-space is given by,

$$\mathbb{H}^3 = \{ (x,y,t) \mid x,y,t \in \mathbb{R} \text{ with } t > 0 \}$$

together with the metric

$$ds = \frac{\sqrt{dx^2 + dy^2 + dt^2}}{t}.$$

The boundary of $\mathbb{H}^3$, $\hat{\mathbb{C}} = \{(x,y,0) \in \mathbb{R}^3\} \cup \{\infty\}$ is the complex sphere and is also called the *sphere at infinity*.

We recall that a discrete group of Möbius transformations G acts on $\hat{\mathbb{C}}$ and divides the complex plane into two disjoint regions, the region consisting of points where G acts discontinuously, called *the region of discontinuity of G* and denoted $\Omega(G)$ and its complement $\Lambda(G)$, *the limit set of G*. A group G acts *discontinuously* at a point $z \in \hat{\mathbb{C}}$ if z has a neighborhood U such that $g(U) \cap U = \emptyset$ for all but finitely many $g \in G$ and *freely discontinuously* at the point if $g(U) \cap U \neq \emptyset \Rightarrow g =$ the identity. The group G is *Kleinian* if it acts freely discontinuously at some point z. If the group is Kleinian and the stabilizer of a point $z \in \Omega(G)$ is finite, then U can be replaced by a *precisely invariant* neighborhood of z (i.e. a neighborhood U where $g(U) \cap U \neq \emptyset \Rightarrow g(U) = U$).

Since the discrete group of Möbius transformations G acts discontinuously on $\mathbb{H}^3$, one can form the quotient, $\overline{M(G)} = (\mathbb{H}^3 \cup \Omega(G))/G$. The groups we consider are finitely generated so that by the Ahlfors finiteness theorem [Ahl64] $S = \Omega(G)/G$, called the ideal boundary of $M(G)$, is a Riemann surface of finite type or a finite union of such surfaces where every boundary component has a neighborhood conformally equivalent to a punctured disc. It may be that $\Omega(G)$ is empty. If G is torsion free, $M(G) = \mathbb{H}^3/G$ is a hyperbolic three manifold. If G has finite torsion, $M(G)$ is a hyperbolic three-orbifold.

In this paper the groups we consider will be finitely generated and either torsion free or have torsion elements of order two. In the later case appropriate modifications of the following can be made as needed.

If G has a parabolic element with fixed point $p \in \hat{\mathbb{C}}$ and G is torsion free Kleinian, then G_p, the stabilizer of p, is an abelian group of rank 1 or 2.

Let $\pi : \Omega(G) \to S$ be the projection. If p_0 is a puncture on the quotient, it has a neighborhood that lifts to a closed circular disc D_p containing p, a parabolic fixed point. D_p is fixed by every element of G_p, moved to a disjoint disc by every element of G not in G_p, and $D_p \cap \Lambda(G) = \{p\}$. The point p belongs to the boundary of D_p. Such a point p is said to *support a horocycle* and D_p is called a horocyclic neighborhood of p. For every puncture on the quotient there is a lift of the point that supports a horocycle. If p supports two horocyclic neighborhoods that are disjoint except for p, then p is said to support a *double horocycle* or be a *double cusp* and the quotient manifold is said to be doubly cusped at p_0.

A Kleinian group is termed *geometrically finite* if there is a finite sided fundamental polyhedron for its action on hyperbolic three space, $\mathbb{H}^3$. By the Poincaré polyhedron theorem, a geometrically finite group is finitely generated.

In terms of the limit set, G is geometrically finite if and only if every point in the limit set is of one of three types: (i) a rank two parabolic fixed point, (ii) a doubly cusped parabolic fixed points or (iii) a point of approximation. It is shown in [Mas88, p. 123] that a parabolic fixed point is not a point of approximation and it is proved

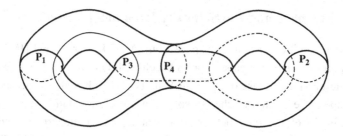

Figure 5: The surface is the quotient of a rank two Schottky group. A noded surface is obtained if the generators are parabolic. In that case the curves P_1 and P_2 are pinched to form a noded surface with two double cusps.

in [JMM79] that a finitely generated free Kleinian group is geometrically finite if and only if each of its parabolic fixed points supports a double horocycle. Thus when we consider free, geometrically finite Kleinian groups, we have that all parabolic fixed points are rank one, doubly cusped.

A *noded surface* is one where every point has a neighborhood which is either isomorphic to a disc in $\mathbb{C}$ or to the set $|z| < 1, |w| < 1, zw = 0$ where (z, w) are coordinates in $\mathbb{C}^2$ [Ber74]. The later points are called *nodes*. Geometrically, a parabolic fixed point corresponds to a node if the parabolic is not *accidental* [Mar77].

If G is a rank two Schottky group with no parabolics, topologically $\overline{M(G)}$ is a solid handlebody of genus two and S is a genus two Riemann surface (see figure 5). If G is *generated* by two parabolics, it will have at least two double cusps and topologically (if it has only those two cusps) S will be a sphere with four punctures. The four punctures are identified in pairs. The two double nodes are obtained from the surface in the figure by *pinching* the curves P_1 and P_2 to points. This is the case we study.

For any given group the surface may or may not be further pinched along some curve(s), but not all curves are *pinchable* (see [HM04]). While every noded surface can be obtained by taking algebraic limits of groups corresponding topologically to pinching [Yam79], the results of [GW02] show that

Theorem 3.1. *The boundary of classical two-parabolic Schottky (parameter) space (see section 4) does not generically come from additional pinching. In particular, there are only four points on the boundary, the points $\lambda = \pm i$ and $\lambda = \pm 2$, where the group G_λ corresponds to additional pinching to a parabolic.*

Related to this theorem is the observation of [HM04] that there are only finitely many different topological types of neo-classical Schottky groups of a fixed genus.

4. Schottky Space and the Schottky Parameter Space

We recall the definition of Schottky space from [Mar74, Mar77] and [JMM79] and refer the reader to Marden's papers for an excellent fuller background. We alert the reader to the fact that the use of the term *Schottky space* has also varied over time. Original papers use the term Schottky space for the representation space, which is how we define it below. Some, but not all, recent papers use the term *Schottky space* to refer to the quotient of the space under the conjugation action of $PSL(2,\mathbb{C})$. We term the latter the call the Schottky *parameter* space to avoid confusion. We define these space with some care.

4.1. The Representation Variety

Following Marden [Mar77] and the Jørgensen-Marden-Maskit paper [JMM79] we discuss the representation space of a finitely generated group and complex projective coordinates for these spaces. To shorten the exposition, we assume the group in question is of rank two, but the statements hold for any group (see [GW02]).

• *The space V of representations for a two generator group.* Let G be the group generated by A_1, A_2 a pair of 2×2 of non-singular matrices. If A_i is the matrix $\begin{pmatrix} a_i & b_i \\ c_i & d_i \end{pmatrix}$ then it determines a point in complex projective space of dimension 3 under the map $A_i \mapsto [(a_i, b_i, c_i, d_i)]$ where $[\ ]$ denotes the projective equivalence class of the four-tuple. If $\mathbb{CP}_3$ denotes complex projective space of dimension 3, the ordered pair (A_1, A_2) determines a point in $\mathbb{CP}_3{}^2$ as does any ordered pair of matrices in $PSL(2,\mathbb{C})$. The image of $(PSL(2,\mathbb{C}))^2$ is the open set $V = \mathbb{CP}_3{}^2 - \mathbf{P}$ where $\mathbf{P}$ is the subvariety given by $\Pi_{i=1}^2 (a_i d_i - b_i c_i) = 0$.

• *The space V(G) for a finitely presented group, G.* More generally, let G be a finitely presented group $G = \langle A_1, A_2 | R_m(A_1, A_2) = 1, \mathrm{tr}^2(P_j) = 4 \rangle$ where $P_j = P_j(A_1, A_2), j = 1, ..., r$ is a complete set of unique representatives for the conjugacy classes of maximal parabolic elements and $R_m, m = 1, ..., k$, is a complete set of relations for the group. Let $V(G)$ be the set of images of A_1, A_2 in $\mathbb{CP}_3{}^2$ as above and let $V(G)^*$ be the set of points in $\mathbb{CP}_3{}^2$ satisfying the polynomial equations induced by the relations R_i and P_j. Then $V(G)^*$ is an algebraic sub-variety, and $V(G)$ is Zariski open with $V(G) = V(G)^* - \mathbf{P}$. $V(G)$ can be represented as an affine algebraic variety, in fact a domain in complex number space of appropriate dimensions.

Each point of $V(G)$ corresponds to a pair of Möbius transformations, (B_1, B_2) satisfying the relations and thus is the image of a homomorphism of G onto the group H generated by the B_i. A homomorphism $\Theta : A_i \mapsto B_i$ that sends each parabolic element either to a parabolic or the identity is called an *allowable* homomorphism. From now on we assume all homomorphisms Θ are allowable and note that $V(G)$ can be written as $V(G) = \{(H, \Theta) |\ \Theta$ is a homomorphism of G onto $H\}$.

The natural topology of $V(G)$ is the topology of point-wise convergence, also called the topology of algebraic convergence. $V(G)$ with this topology is called *the representation space of G*.

• *The space $V_0(G)$ of PSL$(2, \mathbb{C})$ conjugacy classes of representations.* An element $h \in PSL(2, \mathbb{C})$ induces an action on $V(G)$ given by conjugation, (that is, $(H, \Theta) \mapsto (hHh^{-1}, h \circ \Theta \circ h^{-1})$). We let $V_0(G)$ be the quotient of $V(G)$ under this action and call it the *representation parameter space*.

4.2. Spaces of two-generator groups

We can define additional spaces related to $V(G)$ by putting conditions on the group G or the group H or the map Θ and we use the notation $V(G)_X$ to denote that H satisfies condition X. We want to attach names to the spaces defined using this notation. (See section 5 for the definition of nsdc group.)

Consider a fixed group $\widehat{G} = \langle A_1, A_2 \rangle$ so that $\widehat{G}$ has no R_i or P_j relations, that is, a free group of rank two. We define

NSDC-space: $V(\widehat{G})_{nsdc} = \{(H, \Theta) \in V(\widehat{G})|\ H$ is of nsdc-type$\}$

  and its quotient *NSDC-parameter space*: $V_0(\widehat{G})_{nsdc} = V(\widehat{G})_{nsdc}/PSL(2, \mathbb{C})$.

Schottky space: $V(\widehat{G})_{Schottky} = \{(H, \Theta) \in V(\widehat{G}) \mid H$ is a Schottky Group$\}$

  and its quotient *Schottky parameter space*: $V_0(\widehat{G})_{Schottky} = V(\widehat{G})_{Schottky}/PSL(2, \mathbb{C})$

Classical Schottky Space: $V(\widehat{G})_{classical} = \{(H, \Theta) \in V(\widehat{G}) \mid H$ is a Classical Schottky Group.$\}$

  and its quotient *Classical Schottky parameter space*, $V_0(\widehat{G})_{classical} = V(\widehat{G})_{classical}/PSL(2, \mathbb{C})$.

Next, consider a non-elementary free Kleinian group $G^{\mathcal{P}}$ generated by two parabolic transformations A_1 and A_2 without any additional parabolic transformations (i.e. any parabolic transformation is conjugate to a power of A_1 or A_2). $G^{\mathcal{P}} = \langle A_1, A_2 \mid \mathrm{tr}^2 A_1 = \mathrm{tr}^2 A_2 = 4 \rangle$. Such a group is often called a *Riley group* [MSW02]. We call $V(G^{\mathcal{P}})$ the *two-parabolic representation space*. It is a slice of the representation space. We have

Two-parabolic Schottky space:

  $V(G^{\mathcal{P}})_{Schottky} = \{(H, \Theta) \in V(G^{\mathcal{P}}) \mid H$ is a Schottky group$\}$

Two-parabolic classical Schottky space:

$V(G^P)_{classical} = \{(H, \Theta) \in V(G^P) \mid H \text{ is a classical Schottky group}\}.$

Marked two parabolic classical Schottky space:

$V(G^P)_{marked\ classical} = \{(H, \Theta) \in V(G^P) \mid H \text{ is marked classical on } (\Theta(A_1), \Theta(A_2))\}$

Non-classical two-parabolic Schottky space:

$V(G^P)_{non-classical} = \{(H, \Theta) \in V(G^P) \mid H \text{ is a non-classical Schottky group}\}$

It is clear that the two parabolic representation space, $V(G^P)$ is a slice of the rank two representation space $V(\widehat{G})$ and that two-parabolic Schottky space $V(G^P)_{Schottky}$ is a slice of Schottky space $V(\widehat{G})_{Schottky}$. Since trace is a conjugacy invariant, we have $V_0(G^P)_{Schottky} \subset V_0(\widehat{G})_{Schottky}$. It is easy to see that any non-elementary Möbius group generated by two parabolics is conjugate to a group of the form G_λ for some λ where $\text{tr}[A_1, A_2] - 2 = 4\lambda^2$. One can take λ to be the complex parameter for the Schottky parameter space $V_0(G^P)_{Schottky}$.

5. Non-separating disjoint circle groups

The precise definition of non-separating disjoint circle groups appears below in section 5.2. We end this paper with a summary of some related results for non-separating disjoint circle groups for two reasons: (1) the theory here motivated the theory for two-parabolic Schottky groups by indicating how allowing tangencies between any Schottky circles, at non-parabolic points or at parabolic fixed points, made it possible to identify the boundary of the space and (2) it is easy to outline the method here so as to illustrate when non-parabolic tangencies can be *pulled apart*. Troels Jørgensen pointed out the fact that allowing such arbitrary tangencies would yield the tear drop boundary of figure 7.

5.1. $\mathbb{H}^3$ geometry

A hyperbolic line is determined by two points and has two ends on the sphere at infinity. We follow the notation of [Fen89] and denote a hyperbolic line by $[v, v']$ where v and v' may lie on $\widehat{\mathbb{C}}$ or in $\mathbb{H}^3$. An elliptic or loxodromic transformation fixes a unique hyperbolic line, known as its *axis*. A parabolic transformation fixes a unique point point on $\widehat{\mathbb{C}}$. We may see that parabolic fixed point as degeneration of the axis of either a loxodromic or elliptic transformation.

Given any hyperbolic line, L, there is a unique hyperbolic isometry that fixes the line and rotates points in $\mathbb{H}^3$ by an angle of π about the line and which also acts on the boundary. We call this transformation the *half-turn* about L and denote it by H_L.

5.2. Definitions and basic facts

Non-separating disjoint circle groups, known as *nsdc groups* for short, were first defined in [Gil97]. Like Schottky groups, nsdc groups are geometrically defined groups. To define them, we begin with the definition of the *ortho-end* of a group.

For a two generator group $G = \langle A, B \rangle$, we define N, *the perpendicular to A and B* to be the hyperbolic line in $\mathbb{H}^3$ that is the common perpendicular to the axis of A and B if A and B are either loxodromic or elliptic. If either A or B or both are parabolic, we define N to be the perpendicular from the parabolic fix point to the other axis or to the other parabolic fixed point. We assume that G is non-elementary so that the axes of A and B always have a common perpendicular.

We associate to (A, B) its *ortho-end*, the six-tuple of complex numbers (a, a', n, n', b, b') where A factors as $H_{[a,a']}H_{[n,n']}$ and B factors as $H_{[n,n']}H_{[b,b']}$ where N has ends n and n', the axis of A has ends a and a' and the axis of B has ends b and b' (see [Fen89] for more details about this notation). Conversely, an ordered six-tuple of numbers $(a, a', n, n', b, b') \in \hat{\mathbb{C}}^6$ determines an ordered pair of matrices (A, B) where $A = H_{[a,a']}H_{[n,n']}$ and $B = H_{[n,n']}H_{[b,b']}$.

Definition 5.1. We say that a point in $\hat{\mathbb{C}}^6$ has the *non-separating disjoint circle property* if there exist circles C_A, C_D and $C_B \in \hat{\mathbb{C}}$ with disjoint interiors, where C_A passes through a and a', C_B passes through b and b' and C_D passes through n and n' and where no one of the three circles separates the other two. We allow the possibility that some of the three circles are tangent to others.

We say that G is a *marked nsdc group* if its ortho-end has the nsdc property and G is an *nsdc group* if some ortho-end has the nsdc property.

For an non-elementary group G the ortho-end of the group can be defined whether or not the group is discrete. It is shown in [Gil97] that an nsdc group is always discrete. Further it is shown that an nsdc group is free and is always a classical Schottky group. It is clear that one can easily pass back and forth between the matrix entries for A and B to the ortho-ends.

Geometrically when G is nsdc, the three generator group $3G = \langle H_{[a,a']}, H_{[n,n']}, H_{[b,b']} \rangle$ has as its quotient an orbifold whose singular set is the image of three hyperbolic lines. There is a natural projection from $\mathbb{H}^3/G \to \mathbb{H}^3/(3G)$ that comes from factoring out by the action of a hyperbolic isometry of order two.

Remark 5.2. If we begin with a group $G = \langle A, B \rangle$ for which the three circles C_A, C_B and C_D are pairwise disjoint and no one separates the other two, then the group $3G$ is a geometrically finite function group uniformizing a sphere with exactly six points of order two. Such a group is called a *Whittaker group of genus two*. These have been studied by [Kee80] and more recently by [GGD04]. Each Whittaker group of genus two contains a Schottky group of genus two as an index two subgroup. The hyperellptic involution of the uniformized genus two surface is induced by any of the elements of order two in the Whittaker group. Conversely each genus two surface can be obtained in this way. As any simple closed geodesic on a genus two surface is invariant under the hyperelliptic involution, we have this phenomena is still valid after degeneration to the two parabolic Schottky group: the Whittaker groups then degenerate to $3G$ groups with tangencies.

5.3. Methods

The nsdc boundary tear drop (figure 7) is found by analyzing configurations of circles, attaching appropriate parameters to the configurations, and then relating their geometry to the entries in the matrices of the generators for the group. This method is also used to find the boundary for Riley groups that are Schottky groups, but the latter proof involves many more technicalities. The main tools for finding the boundary in the nsdc case are analytic geometry and the inverse function theorem. We give some of the details of this case to illustrate the method and the idea of *pulling tangencies apart*.

An nsdc group generated by two parabolics involves one free parameter $d = x + iy$. The six-tuple of points are $(-2, 0, 2, d, 0, 2)$. The circles C_A and C_B are tangent at the point 0 and, therefore, determine an angle τ. This is the angle that the line connecting their centers makes with the positive x-axis moving in a counter-clockwise direction (see figure 6). Conversely, any angle τ between $\pi/4$ and $-\pi/4$ determines such a pair of tangent circles. Any circle C_D passing through 2 and d will have a center with coordinates (M, N) and radius r. If C_D is tangent to C_A at a point T, then the coordinates of the point (M, N) can be computed as explicit functions of τ as can the radius r. Points on the circle C_D, known as the τ-circle are given by $(x, y) = (M(\tau) + r_\tau \cdot \cos t, N(\tau) + r_\tau \cdot \sin t)$, for some real parameter t, $0 \le t \le 2\pi$.

If C_D is not required to be tangent to C_A but only to pass through d and be tangent to C_B so that we have an nsdc triple, then it is clear that for $d' = (x', y')$ in a small circular neighborhood of $d = (x, y)$ there are circles through (x', y') so that $(-2, 0, 2, d', 0, 2)$ are still nsdc. This shows that non-tangent C_D's correspond to interior points of *NSDC*-space.

If C_D is required to be tangent to C_A, we have that for some d' near the point on $d = (x, y)$ on the τ-circle, C_D, this may or may not be the case.

We consider all points on the τ-circle. One has $x = x(\tau, t) = M(\tau) + r_\tau \cdot \cos t$ and $y = y(\tau, t) = N(\tau) + r_\tau \cdot \sin t$.

Thus one has a map from $\mathbb{R}^2$ to itself: $(\tau, t) \mapsto (x, y)$. One can calculate the Ja-

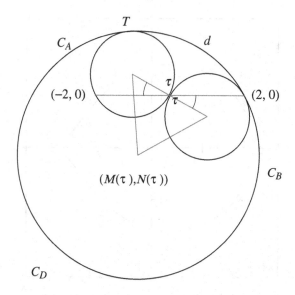

Figure 6: τ - circle configuration with C_D and C_A tangent

cobian of this map to show that (x_0, y_0) is a boundary point of nsdc space precisely when $x_0 = 2 - \frac{\sin^2 \tau}{1 + \sin^2 \tau}$ and $y_0 = 8 \frac{\sin^3 \tau \cos \tau}{1 + \sin^2 \tau}$. The plot of this boundary is the tear drop shown in figure 7. Points $d = x + iy$ in the interior of the tear drop corresponds to (marked) groups G_d that are not nsdc and points in the exterior to G_d that are nsdc groups. A change of parameters maps the tear drop into a parabola and replaces the parameter d by $\lambda = \frac{4d}{d-2}$. Taking marked and non-marked nsdc groups into account yields the region bounded by the two parabolas pictured in figure 4 as nsdc-space, that is $\{\lambda \mid G_\lambda \text{ or } G_{-\lambda} \text{ is an nsdc group}\}$.

References

[Ahl64]   L. V. Ahlfors (1964). Finitely generated Kleinian groups. *Amer. J. Math.* **86**, 413–429.

[Ber74]   L. Bers (1974). On spaces of Riemann surfaces with nodes. *Bull. Amer. Math. Soc.* **80**, 1219–1222.

[Fen89]   W. Fenchel (1989). *Elementary Geometry in Hyperbolic Space, de Gruyter Studies in Mathematics*, volume 11. Walter de Gruyter & Co., Berlin.

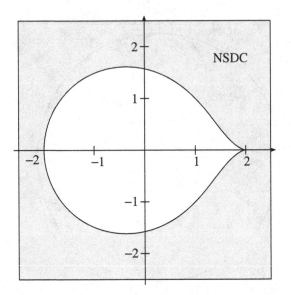

Figure 7: The NSDC tear drop. Points d in the complex plane that lie exterior to the tear drop parameterize groups that are discrete and of non-separating disjoint circle type.

[GGD04]   E. Girondo & G. González-Diez (2004). On a conjecture of Whittaker concerning uniformization of hyperelliptic curves. *Trans. Amer. Math. Soc.* **356** (2), 691–702 (electronic).

[Gil97]   J. Gilman (1997). A discreteness condition for subgroups of $PSL(2, \mathbf{C})$. In *Lipa's Legacy (New York, 1995), Contemp. Math.*, volume 211, pp. 261–267. Amer. Math. Soc., Providence, RI.

[GW02]   J. Gilman & P. Waterman (2002). Parameters for two-parabolic Schottky groups. Preprint, submitted.

[Hid96]   R. A. Hidalgo (1996). The noded Schottky space. *Proc. London Math. Soc. (3)* **73** (2), 385–403.

[HM04]   R. A. Hidalgo & B. Maskit (2004). On neo-classical Schottky groups. Preprint.

[JMM79]   T. Jørgensen, A. Marden & B. Maskit (1979). The boundary of classical Schottky space. *Duke Math. J.* **46** (2), 441–446.

[Kee80]   L. Keen (1980). On hyperelliptic Schottky groups. *Ann. Acad. Sci. Fenn. Ser. A I Math.* **5** (1), 165–174.

[KS94]   L. Keen & C. Series (1994). The Riley slice of Schottky space. *Proc. London Math. Soc. (3)* **69** (1), 72–90.

[LU69]   R. C. Lyndon & J. L. Ullman (1969). Groups generated by two parabolic
          linear fractional transformations. *Canad. J. Math.* **21**, 1388–1403.

[Mar74]   A. Marden (1974). The geometry of finitely generated Kleinian groups.
          Ann. of Math. (2) **99**, 383–462.

[Mar77]   A. Marden (1977). Geometrically finite Kleinian groups and their defor-
          mation spaces. In *Discrete Groups and Automorphic Functions (Proc.
          Conf., Cambridge, 1975)*, pp. 259–293. Academic Press, London.

[Mas81]   B. Maskit (1981). On free Kleinian groups. *Duke Math. J.* **48** (4), 755–765.

[Mas88]   B. Maskit (1988). *Kleinian Groups, Grundlehren der Mathematischen
          Wissenschaften [Fundamental Principles of Mathematical Sciences]*, vol-
          ume 287. Springer-Verlag, Berlin.

[MSW02]   D. Mumford, C. Series & D. Wright (2002). *Indra's Pearls*. Cambridge
          University Press, New York.

[Rat94]   J. G. Ratcliffe (1994). *Foundations of Hyperbolic Manifolds, Graduate
          Texts in Mathematics*, volume 149. Springer-Verlag, New York.

[Sat88]   H. Sato (1988). On a paper of Zarrow. *Duke Math. J.* **57** (1), 205–209.

[Yam79]   H.-o. Yamamoto (1979). Squeezing deformations in Schottky spaces. *J.
          Math. Soc. Japan* **31** (2), 227–243.

[Yam91]   H.-o. Yamamoto (1991). An example of a nonclassical Schottky group.
          Duke Math. J. **63** (1), 193–197.

[Zar75]   R. Zarrow (1975). Classical and non-classical Schottky groups. *Duke
          Math. J.* **42** (4), 717–724.

·

Jane Gilman

Rutgers University
Newark, NJ 07102
U.S.A.

gilman@andromeda.rutgers.edu

AMS Classification: 30F40; 14H30, 20H10

Keywords: Schottky group, classical Schottky group, non-separating disjoint circle groups, Kleinian groups

Spaces of Kleinian Groups   Cambridge University Press

Lond. Math. Soc. Lec. Notes **329**, 301–336   Y. Minsky, M. Sakuma & C. Series (Eds.)

Searching for the cusp

David J. Wright

Abstract

We discuss the process of algebraically finding cusps on the boundaries of deformation spaces of kleinian groups. The geometric starting point is an arrangement of circles with prescribed tangencies and relationships under a set of Möbius transformations. These lead to polynomial equations in several complex variables, which may then be numerically solved for the values which describe the cusp. We will go through this process for several deformation spaces corresponding to "plumbing" constructions of Maskit and Kra, and we will present some of the numerical output. The same techniques can also be used to calculate the coherent spiral hexagonal circle packings discovered by Peter Doyle, and we will compare the similarity factors of those packings to cusps on boundaries of deformation spaces.

1. Introduction

In this paper, we study the theoretical and numerical calculation of *maximal cusp groups* on the boundary of deformation spaces of kleinian groups. Specifically, we are interested in *maximally parabolic groups* which allow no deformations with a greater number of classes of parabolic elements.

The foundational study of cusp groups occurred in Bers' paper [Ber70] on boundaries of Teichmüller spaces. There he proved that the cusp groups form a set of measure zero in the boundary, and that there exist boundary groups he termed *totally degenerate,* for which the ordinary set is a single invariant domain. Then came a windstorm of ideas from Thurston, first in his celebrated *Geometry and Topology of 3-Manifolds* and then in many subsequent talks and notes [Thu80, Thu82, Thu89, KT90]. Thurston introduced the idea of traintracks to describe classes of simple closed curves on surfaces and used that idea to combinatorially describe the boundary of Teichmüller space. He also introduced the basic notions of circle packings and demonstrated their relevance to hyperbolic geometry. The limit sets of maximal cusp groups are all infinite circle packings of the Riemann sphere. These are just a few castoffs of Thurston's vast program in geometry and topology.

There have been very many results on the boundary since then, including McMullen's theorem [McM91] that the maximal cusp groups are dense in the boundary of Bers' embedding of the Teichmüller space of a Riemann surface of finite type. As a result one can try to explore the boundary of Teichmüller space by calculating very many maximal cusps. Similarly, Canary, Culler, Hersonsky and Shalen prove in [CCHS03] that the maximal cusp groups are dense in the boundary of Schottky space.

In the manuscript [Wri87], the author explained numerical calculations of the cusps on the boundary of Maskit's version of the Teichmüller space of once-punctured tori, as well as some theoretical aspects of this calculation. The basics of that calculation now appear in the book *Indra's Pearls* [MSW02], and related material may be found in papers of Keen and Series, for example [KS92, KS93, KS94, KMS93].

In this paper, we have reworked the search for cusp groups in very elementary terms involving the existence of an arrangement of tangent circles (which we call the *circle web* of the group) and a set of Möbius transformations that maps some of the circles to others in a prescribed way. We will explain how maximal cusps on the boundary of the Schottky space of genus g may be described in this manner, and we will present several numerical examples of the geometry of these cusp groups. While leading up to higher genus theory and examples in Section 6, we review the theory of these circle webs for maximal cusps occurring on the boundary of Maskit's Teichmüller space of once-punctured tori in Section 2, on the space of pairs of punctured tori in Section 4, and Riley's slice of two-parabolic-generator groups in Section 5. We also present precise images of these maximal cusp groups for the first time.

In the course of reviewing the cusps on Maskit's $T_{1,1}$, we observed a striking parallel between the geometry of these cusp groups and that of the Doyle spiral circle hexagonal packings. In Section 3, we study spiral circle packings following some of the same techniques used for maximal cusp groups. We show that all the similarity factors of Doyle packings and their square grid analogues are algebraic numbers, and we present detailed calculations revealing a conjectural asymptotic pattern for these similarity factors, which is analogous to the cardioid shape of cusps on Maskit's $T_{1,1}$, proved in [Miy03].

Finally, we would like to extend deep thanks to the Isaac Newton Institute and the organizers Caroline Series, Makoto Sakuma and Yair Minsky for their support during the program on *Spaces of Kleinian Groups*, during which the bulk of this manuscript was prepared.

2. The first example: cusps on the boundary of Maskit's $T_{1,1}$

In [Mas74], Maskit gave a geometric method based on his combination theorems leading to a complex embedding of the Teichmüller space of a Riemann surface of finite type. The specialization of this method to the case of a once-punctured torus yields a family of groups G_μ generated by two Möbius transformations of the form

$$a_\mu(z) = \mu + \frac{1}{z}, \qquad\qquad b(z) = z + 2.$$

We will sometimes denote a_μ by simply a. In addition, when we need to, we shall take the matrix realizations of these transformations to be

$$a_\mu = \begin{pmatrix} -i\mu & -i \\ -i & 0 \end{pmatrix}, \qquad\qquad b = \begin{pmatrix} 1 & 2 \\ 0 & 1 \end{pmatrix}.$$

Maskit proved that the set of values of μ with $\Im\mu > 0$ for which the group G_μ satisfies certain geometric conditions forms a model of the Teichmüller space of marked punctured tori. As a consequence, this set is a simply-connected domain in the upper half-plane, which we will simply refer to as $T_{1,1}$.

Our numerical calculation of this boundary was based on searching for cusps. For μ in $T_{1,1}$, all the elements of the group G_μ are hyperbolic (by which we mean not elliptic or parabolic; hence, this includes "loxodromic"), except for conjugates of powers of the words b and $abAB$. (We shall use the upper case convention to denote inverses of elements in our group. Thus, A and B are the inverses of a and b, respectively.) Cusps correspond to values of μ on the boundary where an additional word becomes "accidentally" parabolic.

Only one additional class of words in G_μ may become parabolic, and that class corresponds to a simple closed curve on the once-punctured torus. These words may be parametrized by rational numbers p/q. Additional details about this theory may be found in *Indra's Pearls* and the papers we have referenced above. There are several possibilities for this parametrization; for this paper, we will use the following.

Definition 2.1. For each fraction p/q in $\widehat{\mathbb{Q}} = \mathbb{Q} \cup \{1/0\}$, we define a word $w_{p/q}$ in the free group generated by a and b (with inverses A and B) by the following recursive rules:

(i) $w_{0/1} = A$ and $w_{1/0} = b$.

(ii) For any pair of fractions $p/q < h/k$ with $hq - kp = 1$, we have

$$w_{\frac{p+h}{q+k}} = w_{p/q} w_{h/k}.$$

The justification of this definition lies in the elementary theory of continued fractions, and we shall leave this to other sources. As a few examples, let us just mention

$$w_{5/8} = AAbAAbAbAAbAb, \qquad\qquad w_{1/15} = A^{15}b,$$

$$w_{-1/1} = AB \quad \text{(due to } w_{-1/1}w_{1/0} = w_{0/1}), \qquad w_{-7/9} = (AB)^4 A(AB)^3 A.$$

In the group G_μ, the matrix entries of each word are polynomials in the complex variable μ. Thus, the condition that the word $w_{p/q}$ be parabolic is equivalent to a polynomial equation $\operatorname{tr} w_{p/q} = \pm 2$. The story of these trace polynomials is told partly in *Indra's Pearls* and in more detail in [KS92]. As a consequence of the analytic theory of Teichmüller space as constructed by Bers, there is precisely one solution of this trace equation lying on the boundary of $T_{1,1}$ and we shall denote this distinguished solution by $\mu(p/q)$. With the assumption that $\Im\mu > 0$, it turns out that in this case that solution satisfies the trace equation equal to 2. This is proved in [KS93] by an analysis of the 'p/q pleating rays' in $T_{1,1}$ which are shown to be curves terminating at the p/q cusp and along which the trace of the p/q word increases from 2 to ∞. The mirror image $\overline{T_{1,1}}$ of $T_{1,1}$ under complex conjugation has cusps on its boundary which are solutions of the other equation $\operatorname{tr} w_{p/q} = -2$; this happens precisely when q is odd. Minsky's work [Min99] on the space of pairs of punctured tori has established the fact that the mapping $p/q \mapsto \mu(p/q)$ continuously extends to an injective continuous map of $\mathbb{R}$ onto the boundary of $T_{1,1}$.

As explained in Chapter 9 of *Indra's Pearls*, the mapping $p/q \mapsto \mu(p/q)$ was used to numerically compute this boundary. It is easy to calculate by hand that $\mu(0/1) = 2i$, as well as various other values, for instance, $\mu(1/1) = 2 + 2i$ and $\mu(1/2) = 1 + \sqrt{3}i$. To calculate other values of $\mu(p/q)$, we enumerate the Farey series of fractions of denominator at most D, for some large integer $D > q$, and we solve the trace equation $\operatorname{tr} w_{p/q} = 2$ for $\mu(p/q)$ by Newton's method using as the seed value $\mu(h/k)$ where h/k is the fraction preceding p/q in the Farey series. This is what we refer to as a *boundary tracing algorithm*. It turns out to be crucial to compute the trace polynomials in the most efficient recursive manner, and then to use a numerical approximation to the derivative that appears in the formula for Newton's method. Nonetheless, the convergence of this algorithm is quite fast. There are now other calculations of this and slices of other Teichmüller spaces by Komori-Sugawa-Wada-Yamashita (see [KS04, KSWY]); their program is based on a criterion for discreteness, rather than a search for cusps.

Now we turn to the most important part of this section, the geometry of the limit sets of the cusps. As an example, we take the $1/15$ cusp approximately given by

$$\mu(1/15) = 0.011278560612 + 1.958591030112\, i$$

corresponding to the word $A^{15}b$. As a maximal cusp, every simple closed curve on the corresponding Riemann surface may be shrunk to a puncture, and a consequence of this is that every component of the ordinary set is a circular disk stabilized by a Fuchsian subgroup of G_μ. There are several distinguished subgroups in G_μ. First, in all these groups we have that $\operatorname{tr} abAB = -2$, regardless of the value of μ. Therefore, the subgroup H_1 generated by b and aBA has the property that $\operatorname{tr} b = \operatorname{tr} aBA = 2$ and $\operatorname{tr} b(aBA) = \operatorname{tr} abAB = -2$. The conditions $\operatorname{tr} u = \operatorname{tr} v = 2$, $\operatorname{tr} uv = -2$ always imply that the group generated by u and v is conjugate to the triply-punctured sphere group generated by

$$\begin{pmatrix} 1 & 2 \\ 0 & 1 \end{pmatrix}, \qquad \begin{pmatrix} 1 & 0 \\ -2 & 1 \end{pmatrix}.$$

In fact, in this case b and aBA are easily seen to be exactly those matrices. The limit set of H_1 is just the extended real line $\hat{\mathbb{R}}$ bounding the lower half-plane D_1, which is a component of the ordinary set. For any element $g \in G_\mu$, we also have all the circles $g(\mathbb{R})$ belonging to the limit set of G_μ, bounding disks stabilized by gH_1g^{-1}.

In a cusp group, another family of circular disks appears. We consider the subgroup H_2 generated by $A^{15}b$ and $A(Ba^{15})a$. This also is seen to be a triply-punctured sphere group from the trace equations

$$\operatorname{tr} A^{15}b = \operatorname{tr} A(Ba^{15})a = 2,$$
$$\operatorname{tr} A(Ba^{15})a A^{15}b = \operatorname{tr} ABab = -2.$$

The limit set of H_2 is another circle bounding a disk D_2 in the ordinary set. The ordinary set of G_μ consists of the two families of disks $g(D_1)$ and $g(D_2)$ for all elements g of G_μ.

We emphasize a particular selection of these disks in Figure 1. We have relabelled D_1 as ε_0 and D_2 as δ_0. In addition, we have plotted certain images of ε_0 (all dark gray) and of δ_0 (all light gray) under elements of the group G_μ.

The numbering of the circles follows these rules:

$$a(\varepsilon_j) = \varepsilon_{j+1} \quad \text{for } 0 \le j \le 0; \qquad b(\varepsilon_j) = \varepsilon_{j+0} \quad \text{for } 0 \le j \le 1;$$
$$a(\delta_j) = \delta_{j+1} \quad \text{for } 0 \le j \le 15; \qquad b(\delta_j) = \delta_{j+15} \quad \text{for } 0 \le j \le 1.$$

We have stated these in a manner to show how they generalize to other cusp groups. This web of circles with mapping relations is enough to imply that the words b, $ABab$ and $A^{15}b$ are all parabolic. The reason is contained in a simple and obvious lemma.

Lemma 2.2. *A non-identity Möbius transformation that fixes each of a pair of tangent circles in the Riemann sphere is parabolic with unique fixed point equal to the tangent point.*

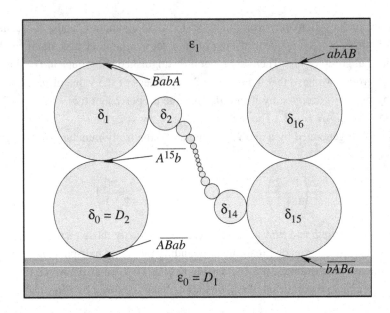

Figure 1: A chart of tangent circles in the ordinary set of the 1/15 cusp group.

To start, the map b fixes both ε_0 and ε_1, which are tangent at ∞; this confirms b is parabolic. Secondly, we can follow the mapping relations to establish

$$ABab(\varepsilon_0) = ABa(\varepsilon_0) = AB(\varepsilon_1) = A(\varepsilon_1) = \varepsilon_0,$$
$$ABab(\delta_0) = ABa(\delta_{15}) = AB(\delta_{16}) = A(\delta_1) = \delta_0.$$

This establishes that $ABab$ is parabolic with fixed point at the intersection point $\varepsilon_0 \cap \delta_0$. Finally, a similar argument proves that $A^{15}b$ fixes both δ_0 and δ_1, and hence its fixed point is $\delta_0 \cap \delta_1$. We have marked the fixed points of various words using the convention that $\overline{w}$ means the attractive fixed point of the Möbius map w.

The external tangency of the disks in the diagram is crucial to identifying this group as the 1/15 cusp. The disk D_1 can be defined as that bounded by the unique circle passing through the points $\overline{ABab}$, $\overline{bABa}$ and $\overline{b}$ ($= \infty$). Similarly, the disk D_2 can be defined as that bounded by the circle tangent to D_1 at $\overline{ABab}$ and passing through $\overline{A^{15}b}$. The other disks are defined by the mapping rules given above. Provided b, $ABab$ and $A^{15}b$ are all parabolic, these disks will all be uniquely and well defined by these stipulations. Thus, there is such a circle web for any solution μ of the 1/15 trace equation $\mathrm{tr}\,A^{15}b = \pm 2$. The particular solution $\mu(1/15)$ is the only solution with $\Im\mu > 0$ for which these disks are distinct and can be chosen to have disjoint interiors. (There is also one solution with $\Im\mu < 0$ which yields a mirror image circle web in the lower half-plane.)

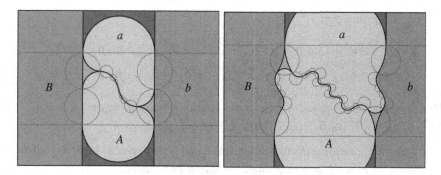

Figure 2: Drawing the orthogonal arcs in the circle web, we arrive at four Schottky *blobs*. The left frame is the 1/15 cusp group; the right is the 5/26 cusp group.

To use the external tangency to establish that this cusp group is discrete and free, draw the orthogonal arcs in each disk between tangent points. The resulting arcs piece together to form four curves bounding "blobs" which are shown in Figure 2. The original web of circles is also shown in outline. Due to the mapping relations given above, the chain of 17 circles surrounding the hatched blob marked *A* is mapped by the transformation *a* onto the chain of 17 circles surrounding the hatched blob labelled *a*. Since the arcs were chosen to be orthogonal to these circles, it follows that *a* maps the boundary of the *A* blob precisely onto the boundary of the *a* blob in such a way that the interior of the *A* blob is mapped onto the exterior of the *a* blob. The analogous argument shows that the transformation *b* maps the interior of the *B* blob (right hatched) onto the exterior of the *b* blob (also right hatched). We should point out that although the picture may not reveal this, the boundaries of both the *B* and *b* blobs consists of four separate circular arcs, two of which are half-lines passing through ∞.

The four blobs therefore satisfy exactly the mapping relations that define a Schottky group of genus two, except for the condition that their closures be disjoint. Much of Klein's combination theorem still applies, with the conclusions that the group generated by *a* and *b* is free and discontinuous, and that the common exterior of the four blobs, which consists of two dark gray circular (or "ideal") triangles (with one vertex at ∞) and two light gray circular triangles, is at least a subset of some fundamental region for the group. In fact, the four triangles do indeed form a fundamental region for the group, and the associated Riemann surface is a pair of triply-punctured spheres.

This example establishes the theme of this paper: defining cusp groups in terms of patterns of tangent circles with given mapping relations under given Möbius transformations. For the Maskit groups, the general mapping relations are a theorem of Keen and Series [KS92].

Theorem 2.3. *Given a fraction $0/1 \leq p/q < 1/0$, for the p/q cusp group $G_\mu = \langle a, b \rangle$ on the boundary of Maskit's $T_{1,1}$ there are disjoint open disks ε_j, $0 \leq j \leq 1$, and δ_j,*

$0 \le j \le p+q$, *such that*

$$a(\varepsilon_j) = \varepsilon_{j+1} \quad for\ 0 \le j \le 0; \qquad\qquad b(\varepsilon_j) = \varepsilon_{j+0} \quad for\ 0 \le j \le 1;$$
$$a(\delta_j) = \delta_{j+p} \quad for\ 0 \le j \le q; \qquad\qquad b(\delta_j) = \delta_{j+q} \quad for\ 0 \le j \le p.$$

Moreover, the following pairs of disks are externally tangent: $(\varepsilon_0, \varepsilon_1)$, (δ_j, δ_{j+1}) *for* $0 \le j \le p+q-1$, $(\varepsilon_0, \delta_0)$, $(\varepsilon_0, \delta_q)$, $(\varepsilon_1, \delta_p)$, $(\varepsilon_1, \delta_{p+q})$.

Conversely, given transformations a_μ and b and disks ε_j and δ_j satisfying all these conditions, $\langle a, b \rangle$ is the p/q cusp group on Maskit's boundary.

There is a similar statement for negative fractions $-p/q$, which can be deduced from the fact that $w_{-p/q}(a,b)$ is conjugate to $w_{p/q}(A,b)$. The more convoluted circle web for the $5/26$ cusp group is shown on the right in Figure 2. In this case, the chain of circles seems to consist of five chains of five circles each, which seems to correlate to the continued fraction expansion $\frac{1}{5+\frac{1}{5}}$ of $5/26$. Similar patterns may be perceived in the circle chains for more complicated fractions. We do not yet know a precise statement of how the circle web reveals the continued fraction expansion. For a very rough statement, we might hazard the following. For a fraction $p/q = \frac{1}{a_1+\frac{r}{s}}$ with both p/q and r/s between 0 and 1, the circle chain of p/q appears to consist of a gentle spiral of a_1 mildly distorted copies of the r/s circle chain. Further analysis of this phenomenon may be found in work of Scorza [Sco].

3. Cusp groups and spiral circle packings

If we apply all powers of the transformation a to the original circles in the circle web shown in Figure 1, we obtain the two double spirals of disks depicted in Figure 3. Both spirals emanate from the repelling fixed point of a and flow into the attractive fixed point. There is a symmetry to this pattern in that each disk is tangent to exactly two disks of each shade. Locally, this packing of circles has the combinatorics of a square grid. Such a packing was called a *square grid circle packing* by Schramm in [Sch97], provided the circle which passes through the tangent points of a ring of four circles corresponding to one square in the grid is also orthogonal to those four circles. This is ever so slightly not true for the $1/n$ Maskit cusp groups. Nonetheless, the pattern is rigidly determined by a small subset of the mapping relations given in the previous section. In this section, we wish to compare these spiral circle packings against the bona fide square grid packings and the more common spiral hexagonal circle packings discovered by Peter Doyle and treated in [BDS94].

In general, the pattern for the $1/n$ cusp group is determined by the arrangement of the six disjoint disks ε_0, ε_1, δ_0, δ_1, η_0 and η_1, as shown in Figure 4. Here is the desired abbreviation of the mapping relations in Theorem 2.3.

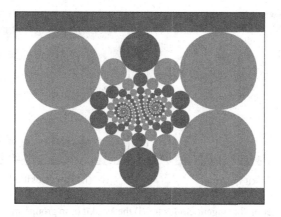

Figure 3: Spirals of circles in the 1/15 cusp group.

Proposition 3.1. *For any complex number μ with imaginary part $\tau = \Im \mu > 0$, let $a(z) = \mu + \frac{1}{z}$, $b(z) = z + 2$, $\varepsilon_0 = \widehat{\mathbb{R}}$, $\varepsilon_1 = \widehat{\mathbb{R}} + i\tau$. There is a unique value of μ such that we have circular disks δ_0, δ_1, η_0, η_1 satisfying the following properties:*

M0: *The six disks ε_j, δ_j, η_j have disjoint interiors, and the following circle pairs are externally tangent:*

$$(\varepsilon_0, \varepsilon_1), \quad (\delta_0, \delta_1), \quad (\eta_0, \eta_1), \quad (\delta_0, \varepsilon_0), \quad (\eta_0, \varepsilon_0), \quad (\delta_1, \varepsilon_1), \quad (\eta_1, \varepsilon_1).$$

M1: $a(\delta_0) = \delta_1$, $a(\varepsilon_0) = \varepsilon_1$ *and* $a(\eta_0) = \eta_1$;

M2: $b(\varepsilon_j) = \varepsilon_j$ *and* $b(\delta_j) = \eta_j$ *for* $j = 0, 1$;

M3: $a^n(\delta_0) = \eta_0$;

M4: *the disks $a^j(\delta_0)$ all have disjoint interiors.*

The group $\langle a, b \rangle$ is the $1/n$ Maskit cusp group.

Proof. Once again these conditions are sufficient to imply that b, $abAB$ and $A^n b$ fix pairs of tangent circles and hence are all parabolic. The last condition completes proof on the basis of Theorem 2.3, since all the circles defined by $\delta_j = a^j(\delta_0)$ for $0 \leq j \leq n+1$ have disjoint interiors. ◻

One feature of these conditions is that, if we start at any circle in the pattern and follow the spiral arm that it lies on from one circle to the next, then after n steps (15 in the example shown in Figure 3) we arrive at a circle that is separated from the original circle by a single tangent circle on the other spiral. This feature can also be found in

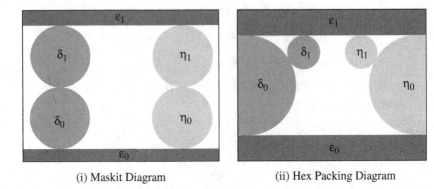

(i) Maskit Diagram   (ii) Hex Packing Diagram

Figure 4: The cluster of six tangent circles for (i) the Maskit cusp groups and (ii) the hexagonal circle packing.

square grid and hexagonal (or Doyle) circle packings. In both of these cases, we will also assume that conditions **M0** and **M1** hold. We will replace **M2** by one of

H2: The pairs of circles $(\delta_0, \varepsilon_1)$ and (η_0, ε_1) are both externally tangent.

SG2: The circles passing through the tangent points of the two chains of four circles $(\varepsilon_0, \delta_0, \delta_1, \varepsilon_1)$ and $(\varepsilon_0, \eta_0, \eta_1, \varepsilon_1)$ are orthogonal to their respective chains of circles.

These are conditions on the geometry of the six circles, and yet both conditions lead to the existence of a second Möbius transformation $b(z)$. Before we come to that fact, we first examine the dependence of the circles δ_j, η_j on the original choice of circles ε_j. In the case of Doyle packings, this analysis overlaps with that found in [BDS94], although in this paper we start with the Möbius symmetry a of the Doyle flower rather than derive it.

Proposition 3.2. *Suppose ε_0 and ε_1 are circular disks which are externally tangent at the point z_0, and $a(z)$ is a Möbius transformation that maps ε_0 onto ε_1 and does not fix z_0. Then, for either choice of **H2** or **SG2** as an additional condition, there is at most one collection of circles δ_0, δ_1, η_0 and η_1 satisfying **M0**, **M1**, and the chosen third condition, as well as the topological condition that $a^{-1}(z_0)$ lies on the one component of the complement of the six disks that touches all six disks.*

The existence of such collections of circles depends on certain algebraic inequalities on μ.

Proof. We can conjugate by a Möbius map so that the tangent point becomes infinity, the inverse image of the tangent point under a becomes 0, and the bounding circles are now the horizontal straight lines $\widehat{\mathbb{R}}$ and $\widehat{\mathbb{R}} + i\tau$ for some $\tau > 0$. That is, ε_0 and ε_1 are now the shaded half-planes shown in either frame of Figure 4. After conjugation, we would have $a(0) = \infty$, which implies that the matrix of a is of the form $\begin{pmatrix} \alpha & \beta \\ -\beta^{-1} & 0 \end{pmatrix}$

or $a(z) = -\alpha\beta - \frac{\beta^2}{z}$. The interiors are mapped in the correct manner only if $-\beta^2$ is positive. Then we can uniquely conjugate by a positive diagonal matrix $\begin{pmatrix} \gamma & 0 \\ 0 & \gamma^{-1} \end{pmatrix}$

so that a becomes the transformation $a(z) = \mu + \frac{1}{z}$ for some complex number μ with $\Im \mu > 0$. (Choose γ so that $-\gamma^4\beta^2 = 1$.) This shows that our starting pair $\varepsilon_0, \varepsilon_1$ in Figure 4 is perfectly general up to Möbius transformation.

The topological condition on $a^{-1}(\infty) = 0$ is designed to ensure that the disks δ_1 and η_1 lie "between" δ_0 and η_0 in the hexagonal packing case shown in Figure 4 (ii). In addition, the tangent points of δ_0 and η_0 to ε_0 must have opposite signs.

Let $\mu = \sigma + i\tau$ with $\tau > 0$. Let δ_0 be the circle tangent to $\varepsilon_0 = \widehat{\mathbb{R}}$ at x with radius $r \leq \tau/2$. Thus, the center of δ_0 is $x + ir$. The circle $\delta_1 = a(\delta_0)$ is then tangent to $a(\varepsilon_0) = \varepsilon_1$ at $a(x) = \mu + \frac{1}{x}$. The center c of δ_1 is symmetric in δ_1 to ∞, and therefore $a^{-1}(c)$ is symmetric in δ_0 to $a^{-1}(\infty) = 0$. Thus,

$$a^{-1}(c) = x + ir + \frac{r^2}{0 - x - ir} = \frac{x^2}{x - ir}$$

which implies the center of δ_1 is

$$c = \mu + \frac{x - ir}{x^2} = \sigma + \frac{1}{x} + i\left(\tau - \frac{r}{x^2}\right).$$

Since δ_1 passes through $a(x) = \mu + \frac{1}{x}$, the radius of δ_1 is

$$|a(x) - c| = \frac{r}{x^2}.$$

Note that, if $x = 0$, then this radius is infinite, and δ_1 would be a half-plane which would not be disjoint from one of ε_0 or ε_1, a violation of **M0**. We conclude that $x \neq 0$. Furthermore, since δ_1 is required to be in the horizontal strip between 0 and μ, we deduce that $\frac{r}{x^2} \leq \frac{\tau}{2}$, or $x^2 \geq \frac{2r}{\tau}$.

The requirement that δ_0 and δ_1 be externally tangent is equivalent to

$$|c - (x + ir)| = r + \frac{r}{x^2},$$

or

$$\left|\left(\sigma + \frac{1}{x} - x\right) + i\left(\tau - r - \frac{r}{x^2}\right)\right| = r + \frac{r}{x^2}. \tag{3.1}$$

Now let us consider the third condition **SG2**. The common orthogonals to ε_0 and ε_1 are the vertical lines. Since δ_0 and δ_1 are bisected by the same vertical line, the real parts of their centers must be the same. That implies $\sigma + \frac{1}{x} = x$, or equivalently the quadratic equation

$$x^2 - \sigma x - 1 = 0, \tag{3.2}$$

which has one positive and one negative solution. From equation (3.1), the radius is

$$r = \frac{\tau}{2} \frac{x^2}{1 + x^2}. \tag{3.3}$$

Thus, the choice of circles δ_0 and η_0 is uniquely determined. The inequalities that guarantee the existence of the collection of six circles as described come from the requirements that the δ_j's be disjoint from the η_j's. For each such pair of circles with centers c_1, c_2 and radii r_1, r_2 we derive an inequality from $|c1 - c2| \geq r_1 + r_2$. The inequalities are rather generous, even if they are tedious to write down.

On the other hand, consider the other choice of third condition **H2**, that δ_0 is also tangent to ε_1. This implies that the radius r of δ_0 is $\tau/2$ (half the width of the horizontal strip between ε_0 and ε_1). Then equation (3.1) becomes

$$\left| \left(\sigma + \frac{1}{x} - x \right) + i \frac{\tau}{2} \left(1 - \frac{1}{x^2} \right) \right| = \frac{\tau}{2} \left(1 + \frac{1}{x^2} \right).$$

Square both sides and simplify and we are left with

$$x^2 - \sigma x - 1 = \pm \tau.$$

Let's suppose the tangent points of δ_0 and η_0 to ε_0 are $x < 0$ and $y > 0$, respectively. The tangent points of δ_1 and η_1 to ε_1 must have real parts between x and y. This implies $\sigma + \frac{1}{x} > x$ and $\sigma + \frac{1}{y} < y$. Thus, by our equation and the condition $\tau > 0$, we have $\sigma x + 1 - x^2 = (\sigma + \frac{1}{x} - x)x = -\tau$, or $x^2 - \sigma x - \tau - 1 = 0$. Analogous reasoning implies that $y^2 - \sigma y - \tau - 1 = 0$. Thus, the tangent points of the circles δ_0 and η_0 to ε_0 are precisely the two solutions to

$$x^2 - \sigma x - \tau - 1 = 0, \tag{3.4}$$

one of which is positive and one of which is negative. This again establishes the uniqueness of the collection of six circles. Inequalities guaranteeing existence may again be derived from the disjointness of the δ_j's and η_j's. □

The next proposition reveals the second symmetry of these patterns of six circles. In fact, the existence of this symmetry is more general.

Proposition 3.3. *Let ε_j, δ_j, $j = 0, 1$, be circular disks with disjoint interiors such that the following pairs are externally tangent:*

$$(\varepsilon_0, \varepsilon_1), \quad (\delta_0, \delta_1), \quad (\delta_0, \varepsilon_0), \quad (\delta_1, \varepsilon_1).$$

Suppose there is a Möbius transformation $a(z)$ such that $a(\delta_0) = \delta_1$ and $a(\varepsilon_0) = \varepsilon_1$. Then there is a unique Möbius transformation $b(z)$ such that $b(\varepsilon_j) = \delta_j$ for $j = 0, 1$ and b commutes with a.

Proof. To begin with, we may conjugate so that $a(z) = z + 1$ (the parabolic case) or so that $a(z) = \lambda z$ for some λ of absolute value at least 1. In the parabolic case, we may then conjugate by a translation so that ε_0, ε_1 are the circles of radius $\frac{1}{2}$ centered at 0 and 1, respectively. Since $a(\delta_0) = \delta_1$, it follows similarly that δ_0 and δ_1 are circles of radius $\frac{1}{2}$ centered at points c and $c + 1$, respectively. Finally, since ε_0 and δ_0 are externally tangent, it follows that $|c| = 1$. Any commuting transformation is a translation, and the unique one that satisfies the statement is then $b(z) = z + c$.

In the non-parabolic case, we may conjugate by a similarity $z \mapsto \gamma z$ so that the disk ε_0 is then centered at 1. Let r, s be the radii of ε_0, δ_0, respectively. Since $\varepsilon_1 = a(\varepsilon_0)$, the center of ε_1 is λ and the radius is $|\lambda| r$. Since ε_0 is tangent to ε_1, we conclude that

$$r = \frac{|\lambda - 1|}{|\lambda| + 1}.$$

Let κ be the center of the disk δ_0; since $\delta_1 = a(\delta_0)$ the center and radius of δ_1 are $\kappa\lambda$ and $|\lambda| s$, respectively. A similar argument to that for the ε's based on the tangency of δ_0 and δ_1 proves that

$$s = \frac{|\kappa\lambda - \kappa|}{|\lambda| + 1} = |\kappa| r.$$

This establishes that the similarity $b(z) = \kappa z$ maps ε_j to δ_j for $j = 0, 1$. The uniqueness follows from the fact that any transformation that commutes with a nontrivial similarity $a(z) = \lambda z$ is another similarity of the same form. □

Remark 3.4. In the non-parabolic case of Proposition 3.3 where the transformations have the form $a(z) = \lambda z$ and $b(z) = \kappa z$ after conjugation, the similarity factors are related by the equation

$$\frac{|\lambda - 1|}{|\lambda| + 1} = \frac{|\kappa - 1|}{|\kappa| + 1}$$

which comes from the calculation of the radius of ε_0 by using the tangency of ε_0 to $\varepsilon_1 = a(\varepsilon_0)$ and $\delta_0 = b(\varepsilon_0)$.

With the two transformations a and b, we can extend our original six circles by considering all image circles $a^n b^m(\varepsilon_0)$ for integers n, m and thereby obtain the cor-

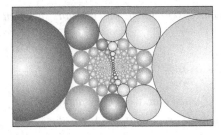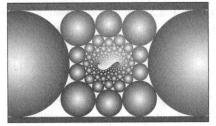

Figure 5: A spiral hexagonal circle packing. All the circles belong to one infinite double spiral generated by a Möbius transformation a. After 16 steps along this spiral, we arrive at a circle tangent to the original circle. The second picture shows the circles shaded with a different choice of exactly 16 shades, showing how the circles align themselves into spirals invariant under a second Möbius transformation b.

responding circle packing. The extension of the cluster in Figure 4 (ii) is shown in Figure 5. The surprise is that all the circles belong to the same double spiral. That is, all the circles are $a^n(\varepsilon_0)$ for all integers n. We have shaded the disks $a^j(\delta_0)$ for $0 \leq j \leq 15$ and $A^j(\eta_1)$ for $0 \leq j \leq 16$ differently from the rest to show how the spiral arm nestles against itself. In particular, after 16 steps from any given circle on the spiral arm, we come to a circle tangent to the original circle.

To highlight the combinatorics of this circle packing, we give each disk one of 16 different shades in succession along the spiral arm. The shades align themselves in 16 new spiral arms shown in Figure 5. The second Möbius transformation b moves the circles along these new spirals. In fact, we have $a^{16} = b$.

A diagram of the basic Doyle "flower" for the spiral packing of Figure 5 after conjugation is shown in Figure 6. We have marked the points $1, \lambda, \lambda^2, \kappa$ and κ/λ to show the arrangement of the similarity factors we have discussed.

We now introduce the analogue of conditions **M3** and **M4** for these circle packings.

Definition 3.5. We say the family of disks $\{a^n b^m(\varepsilon_0)\}$ is *coherent* if any pair of disks are either the same or have disjoint interiors.

In the parabolic case, all the disks have disjoint interiors. In the non-parabolic case, the coherent Doyle packings were analyzed in [BDS94], and the coherent square grid packings are treated in [Sch97]. In these cases, coincidence of the circles amounts to an equation of the form

$$a^q = b^p \tag{3.5}$$

for some pair of integers (p, q), not both zero. Generally, we will restrict ourselves to the case where the multipliers of a and b are of absolute value at least 1, and so the integers (p, q) are both nonnegative.

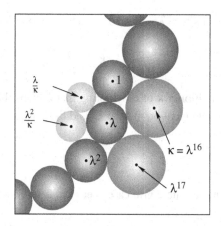

Figure 6: A Doyle flower with centers marked and labelled in terms of the similarity factors λ and κ. By repeatedly applying a to the circle centered at 1 until after 16 steps we arrive back at the circle centered at $\kappa = \lambda^{16}$.

In [BDS94], it is proved there is a unique coherent Doyle packing satisfying equation (3.5) for any such integer pair (p, q) with $p \neq 0$, excluding a few small ones such as $(1, 1), (1, 2), (2, 1), (2, 2)$. The uniqueness is based on a circle packing rigidity theorem of Schramm. The rigidity of the square grid packings is treated in [Sch97]. We would like to make one small observation that might not have been noticed before.

Proposition 3.6. *The similarity factors of coherent spiral Doyle packings and coherent spiral square grid packings are algebraic numbers.*

Proof. First, let us conjugate the transformations so that $a(z) = \mu + 1/z$ again, and so that $\varepsilon_0 = \widehat{\mathbb{R}}$ and $\varepsilon_1 = \mu + \widehat{\mathbb{R}}$. The multiplier λ of a satisfies $\sqrt{\lambda} + 1/\sqrt{\lambda} = -i\mu$, by considering the trace of a. In Proposition 3.2, we have determined the two choices for δ_0. The tangent point x of δ_0 to ε_0 satisfies the quadratic equation (3.4) in the Doyle case, and (3.2) in the square grid case. Picking one of these solutions x, the radius of δ_0 is $\tau/2$ in the Doyle case and given by (3.3) in the square grid case.

To determine the second transformation b, we note that it shares fixed points with a, which are given by

$$z_1, z_2 = \frac{1}{2}(\mu \pm \sqrt{\mu^2 + 4}).$$

Suppose that the signs are chosen so that

$$\frac{a(z) - z_1}{a(z) - z_2} = \lambda \frac{z - z_1}{z - z_2}.$$

Since $a(0) = \infty$, this means $z_2 = \lambda z_1$. Then the transformation b is given by

$$\frac{b(z) - z_1}{b(z) - z_2} = \kappa \frac{z - z_1}{z - z_2}.$$

We can determine the multiplier κ of b by noting that b maps the tangent point of ε_0 and ε_1, which is ∞, to the tangent point z_0 of δ_0 and δ_1. This means we have

$$\kappa = \frac{z_0 - z_1}{z_0 - z_2}.$$

The tangent point in the square grid case is easy to determine:

$$z_0 = x + 2ir = x + i\tau \frac{x^2}{1 + x^2}.$$

The centers c_0, c_1 and radii r_0, r_1 of δ_0 and δ_1 in either case were determined in the proof of Proposition 3.2. Thus, the tangent point can be determined on the line segment between c_0 and c_1 as $z_0 = \frac{r_1 c_0 + r_0 c_1}{r_0 + r_1}$. After a small amount of algebra, we find the tangent point in the Doyle case to be

$$z_0 = \frac{x(2 + \mu x)}{1 + x^2}.$$

Since σ, τ lie in $\mathbb{Q}(\mu, \bar{\mu}, i) \subset \mathbb{Q}(\sqrt{\lambda}, \sqrt{\bar{\lambda}}, i)$, and x lies in a quadratic extension of $\mathbb{Q}(\sigma, \tau)$, we see that κ lies in a solvable algebraic extension of at most degree 16 of $\mathbb{Q}(\lambda)$. The coherence equation $\lambda^q = \kappa^p$ then may be simplified to a polynomial equation for λ over $\mathbb{Q}$. $\qquad\square$

We shall give some examples of the algebraic determination of these similarity factors. The first is the case of a Doyle packing with $p = 1$, $q = 4$, which is known as the Brooks spiral (see [Ste05, Bro85]). In the configuration of Figure 4 (ii), we should have $\varepsilon_4 = a^4(\varepsilon_0) = \delta_0$, which is tangent to ε_0 and ε_1. This implies $\varepsilon_2 = a(\varepsilon_1)$ is tangent to $\varepsilon_{-1} = A(\varepsilon_0)$ and ε_{-2}. That inspires the following calculation.

Lemma 3.7.   (i) *The image $\varepsilon_{-1} = A(\varepsilon_0)$ is the circle of radius $\frac{1}{2\tau}$ centered at $\frac{i}{2\tau}$.*

(ii) *The image $\varepsilon_2 = a(\varepsilon_1)$ is the circle of radius $\frac{1}{2\tau}$ centered at $\mu - \frac{i}{2\tau}$.*

Proof. Let ε_{-1} have center c and radius r. Since c is symmetric with ∞ in ε_{-1}, we will have $a(c)$ symmetric with $a(\infty) = \mu$ in ε_0. Thus, $a(c) = \bar{\mu}$, and hence $c = A(\bar{\mu}) = \frac{1}{\bar{\mu} - \mu} = \frac{i}{2\tau}$. Since ε_{-1} is tangent to ε_0 at $A(\infty) = 0$, the radius of ε_{-1} is $\frac{1}{2\tau}$.

For the second part, note that the order two rotation $r(z) = \mu - z$ conjugates a into A, that is, $rar = A$. Thus, $\varepsilon_2 = a^2(\varepsilon_0) = rA^2r(\varepsilon_0) = rA^2(\varepsilon_1) = r(\varepsilon_{-1})$. Hence, ε_2 has center $\mu - \frac{i}{2\tau}$ and radius $\frac{1}{2\tau}$. $\qquad\square$

From Lemma 3.7, the disks ε_{-1} and ε_2 are tangent if and only if $|\mu - \frac{i}{2\tau} - \frac{i}{2\tau}| = 2\frac{1}{2\tau}$, which simplifies to $|\mu| = \sqrt{2}$.

Suppose now the center of ε_{-2} is c. Then $a(c)$ is symmetric with $a(\infty) = \mu$ in ε_{-1}. Thus,

$$a(c) = \frac{i}{2\tau} + \frac{(1/(2\tau))^2}{\mu - \frac{i}{2\tau}} = \frac{\bar{\mu}}{1 - 2i\tau\bar{\mu}}.$$

Since $a(c) = \mu + \frac{1}{c}$ and $\mu\bar{\mu} = |\mu|^2 = 2$, we can simplify the above to find $c = -\bar{\mu} - \frac{i}{2\tau}$. Since ε_{-2} is tangent to ε_1 and $\Im(-\bar{\mu}) = \Im\mu$, we see the radius is $\frac{1}{2\tau}$. The tangency between ε_{-2} and ε_2 now yields

$$\left| \mu - \frac{i}{2\tau} - \left(-\bar{\mu} - \frac{i}{2\tau} \right) \right| = \frac{1}{\tau}$$

This reduces to $2|\sigma| = |\mu + \bar{\mu}| = \frac{1}{\tau}$. Inserting this into $\sigma^2 + \tau^2 = 2$, it is a simple matter to solve for σ and τ, and we find

$$\mu(1,4) = \frac{\sqrt{3} - 1}{2} + i\frac{1 + \sqrt{3}}{2}.$$

The similarity factor λ may then be found by solving $-i\mu = \sqrt{\lambda} + \frac{1}{\sqrt{\lambda}}$, although the exact expression is rather messy. Numerically, we have

$$\mu(1,4) = 0.366025403784440 + 1.36602540378444\, i,$$
$$\lambda = 0.194235974222535 + 1.61161245101347\, i,$$
$$|\lambda| = 1.62327517874938.$$

The Brooks spiral is shown in Figure 7, colored with 4 shades in succession. One can detect the progression of the spiral arm by observing that ε_j and ε_{j+1} always have distinct shades, while ε_j and ε_{j+4} are always tangent disks of the same shade.

Remark 3.8. The requirements that ε_2 and ε_{-1} lie in the strip between ε_0 and ε_1 are therefore equivalent with the condition $\tau = \Im\mu \geq 1$. If we require that ε_2 and ε_{-1} also have disjoint interiors, from the above lemma we may derive the inequality

$$|\mu| \geq \sqrt{2}.$$

This gives us some preliminary estimates for values of μ corresponding to coherent circle packings of either type. The same estimates apply to Maskit's $T_{1,1}$ as well.

A more symmetric version of the equations determining λ and κ may be extracted from Proposition 3.6 (see [BDS94]). This version would be the coherence equation

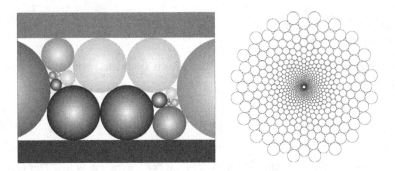

Figure 7: Two Doyle spirals: The left frame is the Brooks spiral, corresponding to $p = 1$, $q = 4$. The right frame is a portion of a pattern explored by R. Weedon. It corresponds to $p = 20, q = 40$.

together with the tangency equations.

$$\frac{|\lambda - 1|}{|\lambda| + 1} = \frac{|\kappa - 1|}{|\kappa| + 1} = \frac{|\lambda - \kappa|}{|\lambda| + |\kappa|}. \tag{3.6}$$

R. Weedon of the United Kingdom has explored the families of Doyle packings where $q = p$ or $q = 2p$. A general observation in [BDS94] is that when $r = \mathrm{GCD}(p, q) > 1$ there is a rotational symmetry of order r in the corresponding Doyle packing. That is, $\lambda^{p/r}\kappa^{-q/r}$ is an r-th root of unity. For $q = p$ the the equations (3.6) and (3.5) may be solved to yield

$$w = \tan^2 \frac{\pi}{p} + \sec \frac{\pi}{p}, \qquad\qquad \ell = w + \sqrt{w^2 - 1}.$$

For $q = 2p$, the solution may be obtained by the following recipe.

$$c = \cos \frac{\pi}{p}, \qquad w = \sqrt{\frac{\sqrt{5 + 4c} - 2c - 1}{2}}, \qquad \ell = \frac{1 + w}{c + w^2},$$

In both cases, the multiplier is given by $\lambda = \ell \exp \frac{i\pi}{p}$. The pattern for $p = 20, q = 40$ is shown in Figure 7.

The complexity of the equations in general seems to make exact formulas difficult to obtain. However, using Newton's method again, it is possible to numerically solve the algebraic equations derived in Proposition 3.6. We made use of several Maple procedures to accomplish this. First, two procedures `multiplier` and `similar` calculate the corresponding λ and κ to a given μ; these are simply a matter of extracting square roots carefully. The iterative part comes in solving the coherence equation

$$L_{p,q}(\mu) = \mathtt{multiplier}(\mu)^q - \mathtt{similar}(\mu)^p = 0.$$

Unlike the case of the Maskit trace polynomials, this equation is only a real analytic function of μ, not complex analytic. We use a two-dimensional real Newton's method to solve for the solution μ. We define a two-dimensional vector valued function:

$$F\begin{pmatrix}\sigma\\\tau\end{pmatrix} = \begin{pmatrix}\Re L_{p,q}(\sigma+i\tau)\\\Im L_{p,q}(\sigma+i\tau)\end{pmatrix}$$

One pass of Newton's method consists of starting with a seed value $\begin{pmatrix}\sigma_0\\\tau_0\end{pmatrix}$, numerically computing the jacobian matrix

$$J_F = \begin{pmatrix}\frac{\partial F_1}{\partial\sigma} & \frac{\partial F_1}{\partial\tau}\\\frac{\partial F_2}{\partial\sigma} & \frac{\partial F_2}{\partial\tau}\end{pmatrix}$$

by means of some very small increments in σ and τ, and then solving the linear system $J_F\begin{pmatrix}\Delta\sigma\\\Delta\tau\end{pmatrix} = -F\begin{pmatrix}\sigma_0\\\tau_0\end{pmatrix}$ for the change in σ, τ. The refined approximation of the solution is then $\sigma_0 + \Delta\sigma$, $\tau_0 + \Delta\tau$. With a good initial guess, the process produces a solution very quickly.

By moderate diligence, we compiled the constants $\mu(p,q)$ for $p \le 6$ and $q \le 20$ for both the Doyle and the spiral square grid packings. We have displayed these values for the Doyle packings in Figure 8. Typically, ten digits of accuracy requires less than ten Newton passes, once a good seed value has been found. The chart shows that the values of $\mu(2,q)$ follow closely the shape of the Maskit $T_{1,1}$ boundary (which is plausible since both circle packings are composed of two double spirals), while the values of $\mu(1,q)$ follow the shape of the boundary of another deformation space of Koebe groups (see [Par95]; Wada communicated that this is essentially the same as the Riley slice).

The chart makes clear that the constants $\mu(p,q)$ have a strong asymptotic pattern. We may elucidate this pattern by first computing $\sqrt{\lambda}$ from $\sqrt{\lambda} + \frac{1}{\sqrt{\lambda}} = -i\mu$ (so that λ is the multiplier), and then plotting the expression $\frac{i}{1-\sqrt{\lambda}}$ for all the data points shown above. Figure 9 shows that the points for Doyle packings lie on an approximate hexagonal lattice, while those for the square grid packings lie on an approximate

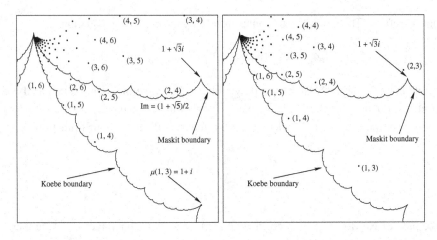

Figure 8: Chart of the constants μ for the (p,q) coherent Doyle packings (left) and square grid packings (right). Superimposed on the plot are traces of Maskit's boundary of $T_{1,1}$ and of the boundary of the Koebe slice of groups $\langle a,b \rangle$ with $abAB$ parabolic and b elliptic of order 2.

square lattice. In fact, the asymptotic formulas appear to be

$$\frac{i}{1-\sqrt{\lambda(p,q)}} = \frac{i}{2} + \frac{1}{\pi}(q - p\,e^{\pi i/3}) \qquad (Doyle)$$

$$\frac{i}{1-\sqrt{\lambda(p,q)}} = \frac{3i}{2\pi} + \frac{1}{\pi}(q - pi) \qquad (SquareGrid)$$

For large p and q, only a few passes of Newton's method are required to obtain the value $\mu(p,q)$ to high precision from this seed value. In [Wri87], a similar asymptotic formula for the Maskit cusps $\mu(1/n)$ was conjectured based on algebraic manipulation of the corresponding trace polynomials. In the above form, this asymptotic formula would be expressed as

$$\frac{i}{1-\sqrt{\lambda(1/n)}} = \frac{(\pi-4)i}{2\pi} + \frac{n}{\pi} \qquad (Maskit)$$

The main term (and probably more) was established by Miyachi in [Miy02, Miy03]. It's possible similar geometric methods could be used to establish the asymptotics of the coherent circle packings.

4. Double cusp groups on the space of pairs of punctured tori

For the remainder of the paper, we concentrate on the problem of maximal cusps on the boundary of deformation spaces. In this section, we consider the space of qua-

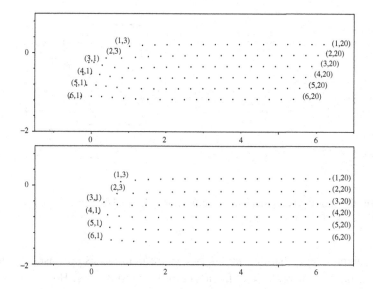

Figure 9: Plot of $\frac{i}{1-\sqrt{\lambda}}$ for $\sqrt{\lambda}+\frac{1}{\sqrt{\lambda}}=-i\mu(p,q)$ for the Doyle packings (top) and square grid packings (bottom). The points follow asymptotic hexagonal and square lattices, respectively, edge length $\frac{1}{\pi}$. Some points are labelled with the corresponding (p,q). The value of q increases from left to right, and the value of p increases downward.

sifuchsian groups $\langle a,b\rangle$ with $\mathrm{tr}\,abAB=-2$ which represent a pair of once-punctured tori. We'll refer to this as PPT space for brevity. One choice of complex parameters for this space is the pair of traces $(\mathrm{tr}\,a,\mathrm{tr}\,b)$. This is a two complex-dimensional space, and there is consequently a much more complicated boundary. In the interior of this space, all the words in G are hyperbolic except for conjugates of powers of $abAB$. A maximal cusp group on the boundary has two additional classes of words becoming accidentally parabolic, corresponding to a simple closed curve on each torus being pinched to a point. The two accidentally parabolic words are again conjugate to powers of the words of the form we defined in Section 2, $w_{p/q}$ and $w_{r/s}$ for some pair of different fractions p/q, r/s. Given such fractions, we may seek a maximal cusp by solving the two algebraic equations

$$\mathrm{tr}\,w_{p/q}=\pm 2, \qquad\qquad \mathrm{tr}\,w_{r/s}=\pm 2.$$

for solutions $(\mathrm{tr}\,a,\mathrm{tr}\,b)$ lying on the boundary of quasifuchsian space. There will be one solution corresponding to pinching the p/q curve on one torus and the r/s curve on the other torus, and a second solution corresponding to pinching the r/s curve on the first torus and the p/q curve on the second. By Theorem III of [KMS93], these two points are represented by anticonformally conjugate groups in $\mathrm{PSL}(2,\mathbb{C})$.

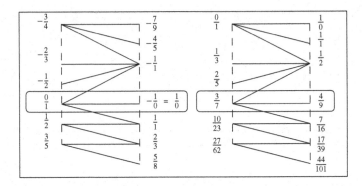

Figure 10: Chart of the Farey steps from the fraction $-7/9$ to $5/8$, and the computation of the analogous Maskit fraction $44/101$.

One might suspect that the extra degree of freedom makes this task significantly more difficult than the problem of finding cusps for Maskit's $T_{1,1}$. However, there is a very useful, labor-saving fact: every maximal cusp on the boundary of quasifuchsian space is conjugate to a maximal cusp on the boundary of Maskit's $T_{1,1}$ or its mirror image $\overline{T_{1,1}}$ under complex conjugation. Essentially, no new groups need to be unearthed to calculate the maximal cusps for quasifuchsian space. We will give only a few examples to show how to calculate cusps on quasifuchsian space, and we shall again emphasize the circle web that reveals the limiting Schottky nature of these groups.

The key to finding the cusp corresponding to the pair fractions $(p/q, r/s)$ lies in the Farey tessellation diagram shown in Figure 9.7 of *Indra's Pearls*, which gives the structure of the rational numbers relative to the process of the Farey addition $\frac{p+r}{q+s}$ of fractions p/q, r/s which are Farey neighbors, meaning that $rq - sp = \pm 1$. Rather than reproduce all the Farey theory explained in *Indra's Pearls*, we shall just show a chart indicating the use of Farey addition to move from one fraction to the next.

As a first example, we take the fractions $(5/8, -7/9)$. The fraction $-q/p$ is something like the "antipodal" fraction to p/q, so that this pair, while not antipodal, is in some sense spread far apart. In *Indra's Pearls*, we show some antipodal Fibonacci fraction pairs approximating a doubly degenerate boundary group. We begin by charting the edges in the Farey tessellation separating the two fractions as shown in Figure 10.

Each roughly horizontal edge in the chart connects a pair of fractions $(p/q, r/s)$ with $rq - sp = \pm 1$, i.e. Farey neighbors. The change from one pair of fractions to the next as we follow the chart from top to bottom consists of a single Farey addition (upward or downward). This chart is always uniquely determined by the original pair of fractions. To construct the righthand chart, we consider the same sequence of "pivots" or edges in the Farey tessellation starting from the edge between $0/1$ and $1/0$, with $1/0$ positioned at one of the original fractions (in this case $-7/9$). At the end of the sequence of moves, in the position of the other fraction in the original pair there will be an equivalent fraction h/k, defining a Maskit cusp $\mu = \mu(h/k)$ for which $G_\mu = \langle a_0, b_0 \rangle$ has the properties that $w_{1/0} = B$ and $w_{h/k}$ are both parabolic. Now we turn to the fractions in the righthand chart that are in the same positions as $0/1$ and $1/0$ in the lefthand chart (both pairs are circled). In this example, we construct a group $\langle a, b \rangle$ such that

$$\operatorname{tr} a = \operatorname{tr} w_{3/7}(a_0, b_0), \qquad \operatorname{tr} B = \operatorname{tr} w_{4/9}(a_0, b_0), \qquad \operatorname{tr} aB = \operatorname{tr} w_{7/16}(a_0, b_0).$$

Since $w_{3/7}$ and $w_{4/9}$ are generators of G_μ, it follows that the new group is conjugate to G_μ. This new group will have the property that $\operatorname{tr} w_{-7/9}(a, b) = \operatorname{tr} B_0 = 2$ and $\operatorname{tr} w_{5/8}(a, b) = \operatorname{tr} w_{44/101}(a_0, b_0) = \pm 2$, and thus corresponds to a maximal cusp, associated to $(\frac{5}{8}, -\frac{7}{9})$ or $(-\frac{7}{9}, \frac{5}{8})$. The same argument succeeds for any pair of distinct fractions.

An easier way around the arithmetic is to carry out these operations inside the group $\mathrm{SL}_2(\mathbb{Z})$. For each pair of fractions on an edge $(\frac{p}{q}, \frac{r}{s})$ from left to right, associate the matrix $\begin{pmatrix} r & s \\ p & q \end{pmatrix}$, which must have determinant ± 1. Multiply the top row by -1 if the determinant is -1. Suppose A is the matrix so obtained for the top edge of the chart and B is the matrix for the bottom edge of the chart. Then there is a matrix U in $\mathrm{SL}_2(\mathbb{Z})$ such that $UA = B$. The equivalent fraction we calculated above is just the top row of this matrix $U = BA^{-1} = \begin{pmatrix} h & k \\ * & * \end{pmatrix}$. The fractions $3/7$, $4/9$ that were used at the end to find the traces of the desired maximal cusp group come from the rows of the matrix A^{-1}.

To continue with our example, our boundary tracing program computes the Maskit cusp

$$\mu(44/101) = 0.88884237349 + 1.632028510111i.$$

From that value, we can calculate

$$\operatorname{tr} w_{3/7} = 1.5720620876 - 0.6328874943i,$$
$$\operatorname{tr} w_{4/9} = 1.4493809001 + 1.4142149587i,$$
$$\operatorname{tr} w_{7/16} = 1.5223946854 + 0.9431138430i.$$

Formulas such as those given in *Indra's Pearls* may now be used to generate the matrices a, b of the maximal cusp group.

We are interested in the chains of circular disks that occur in the ordinary set of the maximal cusp group associated to $(p/q, r/s)$. The same reasoning as in Section 2 shows that the group generated by $ABab$ and $w_{p/q}(a, b)$ is a fuchsian triply-punctured sphere group. This group stabilizes a disk passing through the fixed points of *ABab* and $w_{p/q}(a,b)$ and stabilized by the commutator *ABab* (a convenient third condition since the commutator is usually easy to compute). That gives us one disk $D_1 = \delta_0$ in the ordinary set, and we obtain the circle chain by applying *a* and *b* according to the dynamical rules stated in Theorem 2.3. However, the same process applied to the disk $D_2 = \varepsilon_0$ stabilized by *ABab* and passing through the fixed point of $w_{r/s}(a,b)$ gives a second chain for the fraction r/s. A corollary of the Keen-Series Theorem 2.2 is that the two chains link up in a meaningful way. First, we will state the pattern only for positive fractions.

Theorem 4.1. *Given a pair of distinct fractions $0/1 \leq p/q, r/s \leq 1/0$, for the p/q, r/s cusp group $G = \langle a, b \rangle$ on the boundary of PPT space there are disjoint open disks ε_j, $0 \leq j \leq r+s$, and δ_j, $0 \leq j \leq p+q$, such that*

$$a(\varepsilon_j) = \varepsilon_{j+r} \quad for \ 0 \leq j \leq s; \qquad b(\varepsilon_j) = \varepsilon_{j+s} \quad for \ 0 \leq j \leq r;$$
$$a(\delta_j) = \delta_{j+p} \quad for \ 0 \leq j \leq q; \qquad b(\delta_j) = \delta_{j+q} \quad for \ 0 \leq j \leq p.$$

Moreover, the following pairs of disks are externally tangent:

- (δ_j, δ_{j+1}) *for* $0 \leq j \leq p+q-1$;

- $(\varepsilon_j, \varepsilon_{j+1})$ *for* $0 \leq j \leq r+s-1$;

- $(\varepsilon_0, \delta_0), (\varepsilon_r, \delta_p), (\varepsilon_s, \delta_q), (\varepsilon_{r+s}, \delta_{p+q})$.

Conversely, given transformations a, b, disks ε_j, δ_j satisfying all these conditions, $\langle a, b \rangle$ is conjugate to the $(\frac{p}{q}, \frac{r}{s})$ cusp group.

Unlike Theorem 2.3 where the question of existence is a little more subtle, this theorem is simply an exercise in the arithmetic of $SL_2(\mathbb{Z})$. The third set of tangencies listed are at the commutator fixed points $\overline{ABab}$, $\overline{BabA}$, $\overline{bABa}$ and $\overline{abAB}$ (respectively).

Since our examples will involve some negative fractions, we now state the modification necessary for one negative fraction.

Theorem 4.2. *Given a pair of fractions $0/1 \leq p/q \leq 1/0$, $-1/0 < -r/s < 0/1$, for the p/q, r/s cusp group $G = \langle a, b \rangle$ on the boundary of PPT space there are disjoint open disks ε_j, $0 \leq j \leq r+s$, and δ_j, $0 \leq j \leq p+q$, such that*

$$a(\varepsilon_j) = \varepsilon_{j+r} \quad for \ 0 \leq j \leq s; \qquad B(\varepsilon_j) = \varepsilon_{j+s} \quad for \ 0 \leq j \leq r;$$
$$a(\delta_j) = \delta_{j+p} \quad for \ 0 \leq j \leq q; \qquad b(\delta_j) = \delta_{j+q} \quad for \ 0 \leq j \leq p.$$

Moreover, the following pairs of disks are externally tangent:

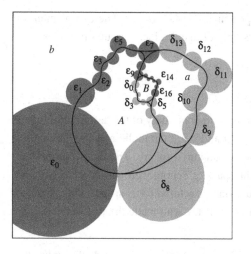
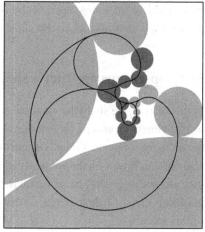

Figure 11: Chains of circles in maximal cusp groups on the boundary of quasifuchsian space. Left frame: the $(\frac{5}{8}, -\frac{7}{9})$ group; right frame: the $(\frac{3}{4}, \frac{1}{5})$ group.

- (δ_j, δ_{j+1}) *for* $0 \leq j \leq p+q-1$;

- $(\varepsilon_j, \varepsilon_{j+1})$ *for* $0 \leq j \leq r+s-1$;

- $(\varepsilon_s, \delta_0)$, $(\varepsilon_{r+s}, \delta_p)$, $(\varepsilon_0, \delta_q)$, $(\varepsilon_r, \delta_{p+q})$.

The converse holds as in Theorem 4.1.

Figure 11 shows the circle chains for the $5/8$, $-7/9$ maximal cusp group. The δ circle chain for $5/8$ (lightly shaded) and the ε chain for $-7/9$ (darker) have been numbered according to the above theorem.

Again we connect all pairs of tangent points on a given disk by orthogonal arcs to the bounding circle. These arcs piece together into four simple closed curves bounding the Schottky blobs. The combinatorial mapping rules given in Theorem 4.2 imply the Schottky relations, that a maps the interior of the A blob onto the exterior of the a blob and b maps the B blob onto the exterior of the b blob. Any pair of transformations that map a web of circles with this combinatorics must generate the $(5/8, -7/9)$ double cusp group, up to conjugacy.

There may be external tangencies that do not play a role in the definition of the Schottky curves. As an example, Figure 11 shows the $(\frac{3}{4}, \frac{1}{5})$ double cusp with the circle chains shaded darkly for $3/4$ and lightly for $1/5$. The orthogonal arcs are also drawn, and once again delineate four simple curves that obey the Schottky mapping conditions. One may see that there are two extra tangencies between dark and light disks not accounted for (nor precluded) by our theorem. The situation is more extreme when the pair of fractions $(\frac{p}{q}, \frac{r}{s})$ consists of Farey neighbors, because every such double cusp is conjugate to the Apollonian Gasket group. The complement of the corresponding web of circles consists only of disjoint ideal circular triangles.

5. Cusp groups on Riley's slice

In this section, we study the maximal cusps on the boundary of the deformation space of two-generator groups $\langle a,b \rangle$ where both a and b are parabolic and the quotient of the ordinary set by the group is a four-times punctured sphere. These cusp groups are also maximal cusps on the boundary of the space of Schottky groups of genus two, but they are not in general cusps on quasifuchsian punctured torus space. This is called the *Riley slice* of Schottky space after Robert Riley who made the first detailed map of this space.

This is a one-complex-dimensional deformation space, and we shall employ the parametrization

$$a = a_\rho = \begin{pmatrix} 1 & 0 \\ \rho & 1 \end{pmatrix} \qquad\qquad b = \begin{pmatrix} 1 & 2 \\ 0 & 1 \end{pmatrix} \qquad (5.1)$$

of the associated groups $G_\rho = \langle a,b \rangle$ (we will recycle some notation from Section 2, since no use of that material will be made). When $|\rho|$ is sufficiently large, the ordinary set of this group is connected, and the quotient surface is a four-times punctured sphere. The domain of ρ for which this is true is topologically a punctured disk (with the puncture at ∞), with a cuspy roughly diamond-shaped inner boundary symmetrically arranged around the origin. An analysis of this boundary and the cusp groups is found in [KS98] (a revision of [KS94]).

The cusps on the boundary of Riley's slice are again associated to rational numbers p/q, $0/1 \le p/q \le 2/1$, with $\rho(2/1) = \rho(0/1)$. The rational number again describes the simple closed curve on the four-times punctured sphere that has been pinched to a point, as well as the word in the group G_ρ that has become accidentally parabolic. What we shall focus on here is how the rational number manifests itself in a web of circles in the limit set.

The following theorem is justified in [KS94].

Theorem 5.1. *Given a fraction $0/1 \le p/q < 2/1$, let $G_\rho = \langle a,b \rangle$ be the p/q cusp group on the boundary of Riley's slice. There are disjoint open disks δ_j, $0 \le j < 2q$, such that, we have*

$$a(\delta_j) = \delta_{2q-j} \quad \text{for } 0 \le j \le q; \qquad B(\delta_j) = \delta_{2p+2q-j} \quad \text{for } p \le j \le p+q,$$

where we define δ_j for all integers j by the periodicity condition $\delta_{j+2q} = \delta_j$. Moreover, *the following pairs of disks are externally tangent:*

- (δ_j, δ_{j+1}) *for* $0 \le j < 2q$;

- (δ_0, δ_q), (δ_p, δ_{p+q}).

With the normalization (5.1) of G_ρ, the boundaries of δ_p and δ_{p+q} are horizontal lines (tangent at ∞), and δ_0 and δ_q are tangent at 0, the fixed point of a. The disks δ_j and δ_k are equivalent under G_ρ if and only if j, k have the same parity.

Again, given such a circle web for a_ρ, b, we can conclude ρ is the p/q cusp on Riley's slice.

We illustrate this first in Figure 12 with the $1/3$ cusp group which, in our setup, turns out to have

$$\rho = -0.75 - 0.6614378278i.$$

The circles are labelled δ_0 through δ_5 in accordance with the theorem, with δ_6 being the same as δ_0. The shades of the disks indicate their equivalence class under the group; the picture shows clearly how the equivalence class alternates in order around the chain of circles.

As before, we have also connected all the tangent points in the chain with orthogonal arcs to the circles. These piece together into four Schottky blobs which are paired by a and b according to the Schottky mapping rules. For example, the tangent chain $\delta_0 \to \delta_1 \to \delta_2 \to \delta_3$ is mapped by a onto the chain $\delta_6 \to \delta_5 \to \delta_4 \to \delta_3$, and that implies that the interior of the A blob is mapped onto the exterior of the a blob. Again, the geometry of these external tangencies and Schottky curves immediately implies by Klein's combination theorem that the group is discontinuous and freely generated by a and b. With a little more work, this circle web identifies the group as the $1/3$ cusp group. The figure also shows all the visible part of the limit set of the group.

We have not revealed the recursive definition of the Riley p/q words comparable to the one given for Maskit p/q words in Section 2, but the dynamical definition can be read off the web of circles. The key is to keep track of a pair of tangent circles under the mapping rules for a and b. As a shorthand notation, we use $i \wedge j$ to denote the pair of disks δ_i, δ_j, which are presumed to be one of the tangent pairs given in Theorem 5.1. Each pair of tangent disks lies on the boundary of precisely two of the four Schottky blobs. That means we can apply either of the transformations taking those blobs to the paired blobs and we will obtain another tangent pair of disks in the web of circles. That new pair will also lie on the boundary of precisely two blobs, and again there will be two possible transformations to apply. However, one will just lead back to the original pair of circles. Thus, if we continue to choose the new direction,

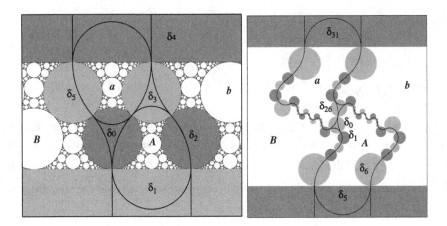

Figure 12: Chains of circles in cusp groups on Riley's slice. Left frame: the $\frac{1}{3}$ group; right frame: the $\frac{5}{26}$ group. The left includes the complete limit set in this frame.

we will gradually build up a word in the group that maps our original pair of circles to itself. By Lemma 2.2, that word is parabolic and stabilizes the tangent point of the pair of circles. That will be our Riley p/q word, up to conjugacy.

Let's begin this journey for the $1/3$ group starting with the pair $0 \wedge 1$ at the lower left in Figure 12. That lies on the boundaries of the A and B blobs, so that we can apply either a or b. We choose to begin with a; the other choice would simply lead to the inverse of the word we will eventually come up with. Our mapping rules are $a(j) = 6 - j$, for $0 \le j \le 3$, and $B(j) = 8 - j$, for $1 \le j \le 4$, with that for A and b being the inverse of these. Then going forward we obtain:

$$0 \wedge 1 \xrightarrow{a} 0 \wedge 5 \xrightarrow{b} 2 \wedge 3 \xrightarrow{a} 4 \wedge 3 \xrightarrow{B} 4 \wedge 5 \xrightarrow{A} 2 \wedge 1 \xrightarrow{B} 0 \wedge 1$$

Reading from right to left, the word that stabilizes $0 \wedge 1$ is *BABaba*, which can be taken to be the $1/3$ Riley word. The stabilizers of all the tangent pairs other than $0 \wedge 3$ (fixed by a) and $1 \wedge 4$ are cyclic permutations of this word. In particular, the stabilizer of $1 \wedge 2$ is $b(BABaba)B = ABabaB$. The fuchsian subgroup of $\langle a, b \rangle$ that stabilizes the lower half-plane δ_1 is the group generated by b and $BABaba$, or alternatively by B and $ABaba$. Since $\operatorname{tr} ABaba = \operatorname{tr} a = 2$, as they are conjugate, and we have set $\operatorname{tr} b = 2$, we conclude the cusp group occurs at a solution ρ to $\operatorname{tr} BABaba = -2$. The reason is that a triply-punctured sphere group $\langle u, v \rangle$ with u, v and uv parabolic can never have a representation in which $\operatorname{tr} u = \operatorname{tr} v = \operatorname{tr} uv = 2$. Note that the fractions $0/1$ and $1/1$ are somewhat special cases, as the groups are fuchsian (*the* triply-punctured sphere group), and the above procedure has to be suitably interpreted to produce the correct words. The second picture in Figure 12 shows the more complicated $5/26$ cusp group.

6. Maximal cusps in the Schottky space of genus g

In this final section, we wish to cull the common themes running through our previous examples into a model of the geometry of maximal cusps on the boundary of Schottky space of genus g. In particular, we will define the *circle web* of a maximal cusp and describe some of its basic properties. Let G be a discrete group freely generated by the Möbius transformations $a_1, \ldots, a_g$. We shall assume that G is a maximally parabolic group. By Theorem II of [KMS93], G is geometrically finite. By this and Theorem 2 in [Mas81], we know that G is an accessible point on the boundary of the Schottky space of genus g, meaning that G is an algebraic limit of Schottky groups. By Theorem I of [KMS93], every component of the ordinary set of G is a circular disk, on which G represents a triply-punctured sphere. The circle web is a finite selection of these disks determined by the nature of the limit points on their bounding circles.

The infinite words in the generators $a_1^{\pm 1}, \ldots, a_g^{\pm 1}$ are the usual infinite sequences of choices from the generators or their inverses, satisfying the rule that no two consecutive terms are inverse to each other. The space of infinite words is endowed with the dictionary topology. This space is also known as the *completion* (or *boundary*) $\overline{G}$ of the group G relative to the given generator set. This notion was introduced in [Flo80], where it was proved that there is a continuous mapping from $\overline{G}$ to the limit set $\Lambda(G)$ of G. In this case, the mapping is defined by sending the infinite word $w = u_1 u_2 u_3 \cdots$ to the limit point $\phi(w) = \lim_{n \to \infty} u_1 u_2 \cdots u_n(z)$ for almost any seed point z. Floyd also proved that the mapping ϕ is injective everywhere except at the inverse images of fixed points of parabolic elements of G, where it is two-to-one. If z is the fixed point of a parabolic word p with inverse P, then $\phi^{-1}(z)$ consists of the two infinite words $pppp \cdots$ and $PPPP \cdots$, allowing for some cancellation.

With this notation, we may now divide the limit set $\Lambda(G)$ into subsets defined by

$$\Lambda(a) = \{z \in \Lambda \mid z = \phi(aw) \text{ for some infinite word } aw \text{ beginning with } a\}$$

for any choice of a from $\{a_1^{\pm 1}, \ldots, a_g^{\pm 1}\}$. We now define the *circle web* as the circles that separate these subsets of $\Lambda(G)$.

Definition 6.1. For a maximal cusp group G on the boundary of Schottky space, we define the *circle web of* G, relative to the designated generators $a_j^{\pm 1}$, to consist of each circular disk in the ordinary set of G whose boundary circle contains points from at least two different subsets $\Lambda(a)$, $a \in \{a_1^{\pm 1}, \ldots, a_g^{\pm 1}\}$.

In the topology of the space of infinite words, the subset of those words beginning with a fixed prefix is closed. Since the mapping ϕ is continuous by Floyd's theorem, each of the subsets $\Lambda(a)$ is closed as well.

In [Mas83], Maskit gave a general description of the collections of words that may become parabolic in a deformation of a function group. In the case of a maximal cusp group on the boundary of Schottky space of genus g, all parabolic words are conjugate to a power of one of a list of $3g - 3$ words corresponding to a maximal system of isotopy classes of disjoint simple closed curves on a surface of genus g, or alternatively a pair-of-pants decomposition of this surface. Each parabolic word is doubly cusped (see [Mas81]), meaning that it fixes two disks in the ordinary set, which are externally tangent at the fixed point of the word.

In the free group, a word u of minimum length in any conjugacy class must be such that u and its inverse U begin with different choices from the generators $a_j^{\pm 1}$, $1 \le j \le g$. Otherwise, u would be of the form avA, and thus conjugate to a shorter word v. If u has length n which is minimal in the conjugacy class of u, there are at most n elements of length n in the class of u, namely, all the cyclic permutations of u. Furthermore, if u is not a power of another element, then all the cyclic permutations of u are distinct words. The fixed points of a cyclic permutation w of u correspond to two distinct infinite words $www\cdots$ and $WWW\cdots$, which begin with different generator letters, and thus belong to two different subsets $\Lambda(a)$. Since the mapping from infinite words to limit points is at most 2-to-1, this also implies the fixed points of distinct cyclic permutations of u are different, given that u is not a power.

Each of these fixed points of minimal length parabolic words is a tangent point between two circles in the circle web. If w_j, $1 \le j \le 3g - 3$, is a maximal list of non-conjugate and non-inverse minimal length parabolic words in the maximal cusp group G, the number of these tangent points is precisely $\sum_j |w_j|$, where $|w|$ denotes the length of the word w. Each circle in the circle web must contain at least two of these fixed points, dividing the circle into arcs each belonging entirely to one of the subsets $\Lambda(a)$ of the limit set. These observations prove the following.

Proposition 6.2. *There are only finitely many disks in the circle web of a maximal cusp group. To be precise, let w_j, $1 \le j \le 3g - 3$, be a maximal list of minimal length parabolic words which are non-conjugate and non-inverse. The fixed points of all the cyclic permutations of these words are tangent points between pairs of disks in the circle web. The number of tangent points is thus the sum of the lengths of this list of parabolic words, and the number of circles is at most this same sum.*

Each circle in the web contains at least two of these tangent points. Connect the consecutive tangent points going around the circle by arcs which are interior to the circle and orthogonal to it. We will call these the *orthogonal arcs* to the circle web. Label an orthogonal arc with generator a if there is an arc of the boundary circle connecting the endpoints of the orthogonal arc and belonging entirely to the subset $\Lambda(a)$ of the limit set. Each orthogonal arc will then have one or two generator labels. If it has two labels, the corresponding circle in the web contains only that one orthogonal arc. The orthogonal arcs, without the labels, are all drawn in Figures 2, 11, 12.

Figure 13: Possible meetings of the orthogonal arcs at a tangent point between two circles in the web.

Consider the orthogonal arcs meeting a given tangent point between two circles in the web. The tangent point is the fixed point of a parabolic word beginning with generator a and ending with generator B, and a and B are non-inverse. Thus, the orthogonal arcs ending at that point are labelled a or b or both. Then, in each circle, there are one or two orthogonal arcs ending at that tangent point. The three possibilities for these arcs and their labels are shown in Figure 13.

For a particular generator a, now consider only the collection of orthogonal arcs labelled with a. By the previous paragraph, these arcs form a combinatorial graph in which each vertex has degree 2. Thus, this graph is a disjoint union of loops. We would like to prove there is exactly one loop. There must be at least one since some of the list of parabolic words must include a as a term, for otherwise we could further deform the group to one in which a is also parabolic, contrary to our assumption of maximality. (Since the list of parabolic words does not include a as a term, we are free to define a to be a parabolic transformation that identifies two very small tangent circles inside a fundamental region for the group generated by the other generators.)

To prove there is only one a loop, we need a more precise description of the list of parabolic words $w_1, \ldots w_{3g-3}$. We consider the corresponding simple closed curves $\omega_1, \ldots, \omega_{3g-3}$ on the associated surface of genus g. At the same time, there are disjoint simple closed curves γ_j on the surface corresponding to the generators a_j, $1 \leq j \leq g$. These are the curves coming from the pair of loops in the plane that are paired by the generators of the Schottky group. We shall assume that all the curves are arranged to have minimal intersections. An example of a maximal system of six disjoint curves is sketched in Figure 14.

The word associated to each simple closed curve ω_i may be discovered by following the curve through its intersection points with each of the generator curves γ_j. Label one side of γ_j by the generator a_j and the other side by the inverse a_j^{-1} according to the loops in the plane that are paired by a_j. Starting at one intersection point of the curve ω_i, we pick one direction and follow the curve to the next intersection point in that direction. If the curve passes from the a_j^{-1} to the a_j side at that intersection point, we write down the letter a_j; otherwise, we write down a_j^{-1}. Continuing to the next intersection point in that direction, we make the analogous observation and write the next generator to the left of the previous way. In this way, when we return to our

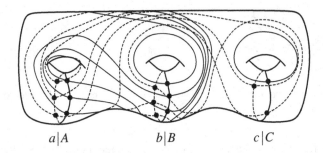

$$a|A \qquad\qquad b|B \qquad\qquad c|C$$

Figure 14: A system of six disjoint loops on a surface of genus 3. The generator curves are labelled by the respective generator a, b and c. The curves correspond to the words b, c, bC, $acAB$, baa and $abAB$. Dotted lines signify parts of the curve on the back side of the surface.

starting point we will have written down the word corresponding to this curve with the given starting point and given direction.

This associates to each intersection point two words, one for each direction. The collection of words obtained in this way are all the cyclic permutations of the original list $w_1, \ldots, w_{3g-3}$ and their inverses. The words associated to the intersection points lying on the generator curve γ_j are all those words beginning with a_j and all those beginning with a_j^{-1} (in the reverse direction). As we follow the curve γ_j around its handle, adjacent intersection points either belong to the same curve ω_i or they belong to two distinct curves that bound the same pairs of pants.

If the two words u_1 and u_2 (both beginning with a_j) for those adjacent intersection points bound the same pair of pants, then there is a third word u_3 corresponding to the third bounding curve of the pair of pants such that u_3 is conjugate to $u_1^{-1}u_2$. It follows then that in the maximal cusp group G, when all three words are parabolic, the words u_1 and u_2 generate a triply punctured sphere group (since $u_1^{-1}u_2$ is also parabolic) corresponding to one of the circular disks in the ordinary set of G. Since u_1 and u_2 both begin with a_j and do not end with a_j^{-1}, it follows that they are connected by an orthogonal arc labelled a_j.

If the adjacent intersection points actually lie on the same curve ω_i and the curve passes through γ_j in the same direction at the two points, then this curve forms two boundary components of a pair of pants. The third boundary component again corresponds to a conjugate of $u_1^{-1}u_2$. The same argument as before again proves that u_1 and u_2 generate a triply-punctured sphere group in the maximal cusp, and that the fixed points of u_1 and u_2 are connected by an orthogonal arc.

Finally, if the same curve passes through adjacent points on γ_j in opposite directions, then u_1 and u_2^{-1} are conjugate words, and the conjugating word u_3 corresponds to one of the other two boundary components of the pair of pants bounded by our original curve. That is, $u_2^{-1} = u_3 u_1 u_3^{-1}$. In the maximal cusp group u_1, u_2 and u_3 are all parabolic, implying the again that the fixed points of u_1 and u_2 are connected by an orthogonal arc in the circle web.

Thus, the orthogonal arcs labelled with a_j pass through in order the fixed points of the words corresponding to the intersection points lying on γ_j, which are all the words beginning with a_j. This proves there is exactly one loop of orthogonal arcs labelled a_j. We summarize our work as follows.

Theorem 6.3. *Let $G = \langle a_1, \ldots, a_g \rangle$ be a maximal cusp group. Let $w_1, \ldots, w_{3g-3}$ be the complete list of minimal length parabolic words which are non-conjugate and non-inverse. Label each disk δ in the circle web of G with each generator 'a' among the $a_j^{\pm 1}$'s for which δ has nonempty intersection with $\Lambda(a)$. The following are true.*

(i) *The disks labelled 'a' form a loop of disjoint disks $\delta_1, \ldots \delta_n$ in which each δ_i is tangent to δ_{i+1}, and δ_n is tangent to δ_1. The number n of these disks equals the number of a's and a^{-1}'s occurring in the expressions of the minimal length parabolic words w_k in terms of the generators of G. Call this loop the 'a' loop of circles. The 'a' and 'a^{-1}' loops contain the same number of circles.*

(ii) *The orthogonal arcs labelled 'a' form a simple closed loop passing through the tangent points of the 'a' loop of circles. The subset $\Lambda(a)$ of the limit set consists precisely of all the limit points contained in one component of the complement of this simple closed curve.*

(iii) *The generator 'a' maps the 'a^{-1}' loop of circles onto the 'a' loop of circles in reverse order relative to the subsets $\Lambda(a^{\pm 1})$ they enclose.*

The circle web can be described by a combinatorial graph where the vertices are the disks and the edges correspond to pairs of disks which are tangent at fixed points of minimal length parabolic words. (The right side of Figure 11 shows there may be other tangent points, not corresponding to edges.) The previous theorem shows that this is a planar graph with $2g$ faces labelled by each of the generators and their inverses. The degrees of paired faces, meaning the number of vertices around each face, are equal.

On the other hand, by taking a planar graph with $2g$ paired faces, with paired faces of equal degree, we may choose mappings of the vertices around the a_j^{-1} face onto those around the a_j face in reverse order. Then by following these mappings around the graph we may calculate the word stabilizing each vertex. This process was described at the end of Section 5 for a Riley group. If we end up with a collection of $3g - 3$ words corresponding to a maximal system of simple closed curves on a surface of genus g, we may solve the polynomial equations derived from setting the traces of these words equal to ± 2 to find matrices that generate a corresponding maximal cusp.

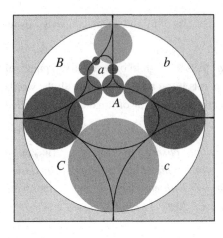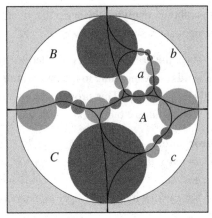

Figure 15: Circle webs for maximal cusps on the Schottky space of genus 3. The orthogonal arcs connecting the tangent points again piece together to form Schottky "blobs." The paired blobs are marked $a \leftrightarrow A$, $b \leftrightarrow B$, $c \leftrightarrow C$, in such a way that a takes the interior of the A blob onto the exterior of the a blob, etc. The shade of the disk indicates its equivalence class. The collection of minimal length parabolic words for the righthand picture is $\{b, c, bC, acAB, BacBAbCAba, BacBaaa\}$.

In this way, in a future work we hope to calculate a catalog of many cusps and some idea of the geometry of the boundary of Schottky space.

Two precise examples are shown in Figure 15. The left corresponds to the system of curves sketched in Figure 14. The second comes from a more complicated system of curves.

References

[BDS94]   A. F. Beardon, T. Dubejko & K. Stephenson (1994). Spiral hexagonal circle packings in the plane. *Geom. Dedicata* **49** (1), 39–70.

[Ber70]   L. Bers (1970). On boundaries of Teichmüller spaces and on Kleinian groups. I. *Ann. of Math. (2)* **91**, 570–600.

[Bro85]   R. Brooks (1985). On the deformation theory of classical Schottky groups. *Duke Math. J.* **52** (4), 1009–1024.

[CCHS03] R. D. Canary *et al.* (2003). Approximation by maximal cusps in boundaries of deformation spaces of Kleinian groups. *J. Differential Geom.* **64** (1), 57–109.

[Flo80]   W. J. Floyd (1980). Group completions and limit sets of Kleinian groups. *Invent. Math.* **57** (3), 205–218.

[KMS93]   L. Keen, B. Maskit & C. Series (1993). Geometric finiteness and uniqueness for Kleinian groups with circle packing limit sets. *J. Reine Angew.
Math.* **436**, 209–219.

[KS92]   L. Keen & C. Series (1992). Pleating coordinates for the Teichmüller
space of a punctured torus. *Bull. Amer. Math. Soc. (N.S.)* **26** (1), 141–146.

[KS93]   L. Keen & C. Series (1993). Pleating coordinates for the Maskit embedding of the Teichmüller space of punctured tori. *Topology* **32** (4), 719–749.

[KS94]   L. Keen & C. Series (1994). The Riley slice of Schottky space. *Proc.
London Math. Soc. (3)* **69** (1), 72–90.

[KS98]   Y. Komori & C. Series (1998). The Riley slice revisited. In *The Epstein
Birthday Schrift, Geom. Topol. Monogr.*, volume 1, pp. 303–316 (electronic). Geom. Topol. Publ., Coventry.

[KS04]   Y. Komori & T. Sugawa (2004). Bers embedding of the Teichmüller space
of a once-punctured torus. *Conform. Geom. Dyn.* **8**, 115–142.

[KSWY]   Y. Komori *et al.* Drawing Bers embeddings of the Teichmüller space of
once-punctured tori. To appear.

[KT90]   S. P. Kerckhoff & W. P. Thurston (1990). Noncontinuity of the action of
the modular group at Bers' boundary of Teichmüller space. *Invent. Math.*
100 (1), 25–47.

[Mas74]   B. Maskit (1974). Moduli of marked Riemann surfaces. *Bull. Amer. Math.
Soc.* **80**, 773–777.

[Mas81]   B. Maskit (1981). On free Kleinian groups. *Duke Math. J.* **48** (4), 755–
765.

[Mas83]   B. Maskit (1983). Parabolic elements in Kleinian groups. *Ann. of Math.
(2)* **117** (3), 659–668.

[McM91]   C. McMullen (1991). Cusps are dense. *Ann. of Math. (2)* **133** (1), 217–
247.

[Min99]   Y. N. Minsky (1999). The classification of punctured-torus groups. *Ann.
of Math. (2)* **149** (2), 559–626.

[Miy02]   H. Miyachi (2002). On the horocyclic coordinate for the Teichmüller
space of once punctured tori. *Proc. Amer. Math. Soc.* **130** (4), 1019–1029
(electronic).

[Miy03]   H. Miyachi (2003). Cusps in complex boundaries of one-dimensional Teichmüller space. *Conform. Geom. Dyn.* **7**, 103–151 (electronic).

[MSW02]   D. Mumford, C. Series & D. Wright (2002). *Indra's Pearls*. Cambridge University Press, New York.

[Par95]   J. Parkkonen (1995). Geometric complex analytic coordinates for deformation spaces of Koebe groups. *Ann. Acad. Sci. Fenn. Math. Diss.* **102**, iv+50.

[Sch97]   O. Schramm (1997). Circle patterns with the combinatorics of the square grid. *Duke Math. J.* **86** (2), 347–389.

[Sco]   I. Scorza. The core chain of circles in Maskit's embedding for once-punctured torus groups. Preprint.

[Ste05]   K. Stephenson (2005). Introduction to circle packing and the theory of discrete analytic functions. To appear.

[Thu80]   W. Thurston (1980). *The Geometry and Topology of 3-Manifolds*. Princeton University Press, Princeton.

[Thu82]   W. P. Thurston (1982). Three-dimensional manifolds, Kleinian groups and hyperbolic geometry. *Bull. Amer. Math. Soc. (N.S.)* **6** (3), 357–381.

[Thu89]   W. Thurston (1989). Groups, tilings, and finite state automata. Technical Report GCG1, Geometry Supercomputer Project Research Report. Summer 1989 AMS Colloquium Lectures.

[Wri87]   D. J. Wright (1987). The shape of the boundary of the Teichmüller space of once-punctured tori in Maskit's embedding. Preprint.

David J. Wright

Department of Mathematics
Oklahoma State University
Stillwater, OK 74078
U.S.A.

wrightd@math.okstate.edu

AMS Classification: 30F40, 22E40, 57S30

Keywords: Kleinian groups

Spaces of Kleinian Groups
Lond. Math. Soc. Lec. Notes **329**, 337–353

Cambridge University Press
Y. Minsky, M. Sakuma & C. Series (Eds.)

Circle packings on surfaces with projective structures: a survey

Sadayoshi Kojima, Shigeru Mizushima and Ser Peow Tan

Abstract

This paper surveys our on-going study of the moduli space of pairs of a surface with a complex projective structure, on which the circle makes sense, and a circle packing on it whose combinatorics is fixed. A conjectural picture, the results obtained so far and a list of problems for further study are discussed.

1. Introduction

The study of circle packings dates back to antiquity, but has seen a surge of activity in the last thirty years, especially after Thurston's [Thu80] re-interpretation and generalization of the circle packing theorem on the sphere by Koebe [Koe36] and Andreev [And70] to surfaces of higher genus. Classically, this involves the study of circle patterns as defined on a surface with a Riemannian metric. Circle packings have found applications to various fields including complex analysis, hyperbolic geometry and even probability theory. On the other hand, the study of surfaces with complex projective structures, called *projective Riemann surfaces* in this paper, is relatively modern (we avoid the terminology projective surface as it has a different meaning in algebraic geometry, the terminology $\mathbb{CP}^1$-surface is also used in the literature). It has also seen much activity in the last couple of decades due to its connections with hyperbolic geometry, Kleinian groups and Teichmüller theory.

We are interested in the interplay between these two fields, specifically, circle packings on projective Riemann surfaces. The first observation is that circles/disks are fundamental geometric objects in 1-dimensional complex projective geometry. This is despite the fact that $\mathrm{PGL}_2(\mathbb{C})\,(= \mathrm{PSL}_2(\mathbb{C}))$ does not preserve a spherical metric (in fact, any metric) on the Riemann sphere $\hat{\mathbb{C}}$. Thus "circles" on a projective Riemann surface are not metric circles in the usual sense, rather, they are homotopically trivial simple closed curves which develop onto (round) circles in $\hat{\mathbb{C}}$ via the developing map. This is an important distinction as methods which rely on parameters such as the radii of the circles do not come into play here. Circles on a projective Riemann surface are well-defined, since $\mathrm{PGL}_2(\mathbb{C})$ acts as Möbius transformations of $\hat{\mathbb{C}}$ and Möbius transformations preserve circles.

Our main focus will be on the moduli space of all pairs (S,P) consisting of a projective Riemann surface S, and a circle packing P on S with a fixed combinatorial pattern. Using the Koebe-Andreev-Thurston theorem as a starting point and prototype, and setting up a conjectural picture which we will discuss in §5 based on the second author's result [Miz00], we studied the intrinsic and extrinsic properties of such moduli spaces in [KMT03a, KMT03b]. This paper surveys these on-going studies by describing a foundational basis, the results obtained so far, and a list of problems for further study.

2. Projective Structures

Throughout this paper, we use Σ_g to denote a closed, orientable topological surface of genus g without any auxiliary structure. A projective structure on Σ_g is a geometric structure modeled on the pair of the Riemann sphere $\widehat{\mathbb{C}}$ and the projective linear group $\mathrm{PGL}_2(\mathbb{C})$ acting on $\widehat{\mathbb{C}}$ by orientation preserving projective transformations, that is, the Möbius transformations. Since Möbius transformations are in particular holomorphic and one-to-one, a projective structure automatically induces an underlying complex structure. However, requiring the transition maps to be Möbius transformations is far more rigid than merely requiring that they be holomorphic one-to-one maps, so different projective structures can have the same underlying complex structure.

Let S be a surface with a projective structure, called a projective Riemann surface, homeomorphic to Σ_g. The notation S is thus to denote not just the topological surface Σ_g, but one equipped with a projective structure. We always attach to S a reference homeomorphism $h : \Sigma_g \to S$ for marking. Then two projective Riemann surfaces, say (S_1, h_1) and (S_2, h_2), are considered to be marked projectively equivalent if there exists a projective isomorphism $\varphi : S_1 \to S_2$ such that $\varphi \circ h_1$ is homotopic to h_2.

Associated to S are the developing map,

$$\mathrm{dev} : \widetilde{S} \longrightarrow \widehat{\mathbb{C}},$$

defined up to composition with projective transformations, where $\widetilde{S}$ is the universal cover of S, and the holonomy representation,

$$\rho : \pi_1(S) \longrightarrow \mathrm{PGL}_2(\mathbb{C}),$$

defined up to conjugation by projective transformations. The developing map is equivariant with respect to the holonomy representation, that is,

$$\mathrm{dev}(x.g) = (\mathrm{dev}(x)).\rho(g)$$

for all $g \in \pi_1(S)$, $x \in \tilde{S}$, where $\pi_1(S)$ acts as deck transformations on $\tilde{S}$. We note that both the developing map and the holonomy representation can be extremely complicated. In general, the developing map is not injective to its image and the holonomy representation may not be discrete or faithful. One of the main questions arising in this theory is the question of which representations can occur as the holonomy of a projective structure. This was settled by Gallo, Kapovich and Marden in [GKM00], where they showed that any representation whose image is non-elementary and which lifts to $SL_2(\mathbb{C})$ occurs as the holonomy representation of some projective structure.

There are many natural projective structures, for example, the canonical complex structure on the Riemann sphere. This projective structure on the sphere is unique, since the conformal automorphisms of the Riemann sphere with the underlying complex structure are precisely $PGL_2(\mathbb{C})$. Similarly, a Euclidean structure on the torus also naturally defines a projective structure, since Euclidean isometries form a subgroup of $PGL_2(\mathbb{C})$, and the Euclidean plane is a subset of the Riemann sphere. Likewise, a hyperbolic structure on Σ_g defines a projective structure, for the analogous reason that the isometries of $\mathbb{H}^2$ can be identified with a subgroup of $PGL_2(\mathbb{C})$. More interesting examples arise from, say, taking the quotient of one of the components of the domain of discontinuity of a quasi-Fuchsian group by the group, or by performing grafting on a hyperbolic structure, see for example [Gol87].

3. Circle Packings

A projective structure on a surface is not a metric structure, but the circle still makes sense since a projective (Möbius) transformation maps a circle on $\widehat{\mathbb{C}}$ to a circle on $\widehat{\mathbb{C}}$. Hence we define a circle on a projective Riemann surface S to be a homotopically trivial curve on S such that its lifts on $\tilde{S}$ are mapped to circles in $\widehat{\mathbb{C}}$ by the developing map.

Let S be a projective Riemann surface. A circle packing P on S is a collection of circles in S such that each circle bounds a disk and the interior of the disks are all disjoint. The circle packings we will consider hereafter are also assumed to have the additional property:

$$\text{Complementary regions are all triangular.} \qquad (3.1)$$

Note that the circles are allowed to have self-tangency points.

To each circle packing, we associate a graph τ on S called the nerve of P. It is obtained by assigning a vertex to each circle and an edge between two (not necessarily distinct) vertices for each tangency point. It is easy to see that τ is covered by an honest triangulation $\tilde{\tau}$ in the universal cover $\tilde{S}$.

Conversely, suppose we are given a topological graph τ on Σ_g such that τ is covered by an honest triangulation in the universal cover. We then would like to find a complete description of the set of all pairs (S, P), where S is a projective Riemann surface equipped with a reference homeomorphism $h : \Sigma_g \to S$, and P is a circle packing on S such that its nerve is isotopic to $h(\tau)$. Here two packings P_1 and P_2 on the same surface S are equivalent if there is a projective automorphism of S isotopic to the identity which takes P_1 to P_2, and we are interested in the equivalence classes of packings. Henceforth, for clarity of exposition, we shall simply say that P has nerve τ, or nerve $h(\tau)$, as the case may be, and suppress mentioning isotopy.

We can ask the same question for Riemannian surfaces of constant curvature, that is, to give a complete description of pairs (S', P') where S' is a surface with a constant curvature Riemannian metric and P' is a circle packing on S', consisting of circles with respect to the Riemannian metric, and whose nerve is isotopic to τ. Indeed, this is the context addressed in most papers on circle packings. Here two packings P_1' and P_2' on the same Riemannian surface S' are equivalent if there exists a conformal automorphism of the surface (with respect to the Riemannian metric) isotopic to the identity which takes P_1' to P_2', and we are interested in the equivalence classes of circle packings. Then the Koebe-Andreev-Thurston theorem [Koe36, And70, Thu80] says that given τ, there is a unique such pair (S', P') up to scalar multiple and isotopy of the metric. In the case where $g = 0$, we assume the metric is scaled to have constant curvature 1 and where $g \geq 2$ we assume that the metric has been scaled to have constant curvature -1. More precisely, when $g = 0$, there is a circle packing P on the unit sphere with nerve τ, unique up to conformal automorphisms of the sphere (that is, up to the action of $\mathrm{PGL}_2(\mathbb{C})$ on the Riemann sphere). In the case where $g = 1$, there is a unique Euclidean torus up to scaling with a circle packing with nerve τ and the packing is rigid up to translation. And when $g \geq 2$, there is a unique hyperbolic surface with a circle packing with nerve τ and the packing is rigid. Observe that the solution (S', P') also provides a solution (S, P) to our original question in the category of projective Riemann surfaces. We shall call this solution the KAT pair associated to τ. The main interest of our study is to see how much the KAT pair for τ can be deformed with a deformation of the projective structure.

Our first clue that there was a rich deformation theory arose from results of computer experiments. Figures 1 and 2 show the developing images of two projective circle packings on a genus 2 surface which have the same nerve τ (up to isotopy). This is an example where τ has exactly one vertex. Figure 1 represents the KAT solution for τ, figure 2 a small deformation. In both cases the developing maps are injective, but there are also many examples where the developing map is not injective.

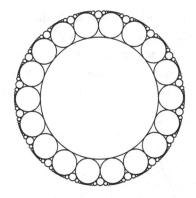

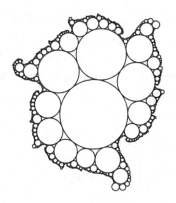

Figure 1: Hyperbolic example   **Figure 2:** Deformed example

4. Cross Ratios

Projective geometry is not a metric geometry and metric concepts such as radii do not make sense in general. Hence we need to use some other invariants which would help to quantify and parameterize the space of pairs (S, P) and allow us to analyze the space more systematically. We introduce a projective invariant of a circle packing based on the cross ratio. It can also be found in the works by He and Schramm [HS98] and separately by Schramm [Sch97] as well in a different context.

Suppose that (S, P) is a pair of a projective Riemann surface S and a circle packing P on S with the property (3.1). The invariant of (S, P) which we will define is a map

$$\mathbf{x} : E_\tau \longrightarrow \mathbb{R},$$

where E_τ is the set of edges of τ. To each edge e of τ, we choose a lift $\widetilde{e}$ in $\widetilde{\tau}$ and associate a configuration of four circles on $\widehat{\mathbb{C}}$ in the developed image about $\mathrm{dev}(\widetilde{e})$, see Figure 3. Recall that the cross ratio of four distinct ordered points in $\widehat{\mathbb{C}}$ is given in [Ahl53] by

$$(z_1, z_2, z_3, z_4) = \frac{(z_1 - z_3)(z_2 - z_4)}{(z_1 - z_4)(z_2 - z_3)}.$$

It is the value of the image of z_1 under the projective transformation which takes z_2, z_3 and z_4 to $1, 0$ and ∞ respectively. The value assigned to the edge e will be the imaginary part of the cross ratio of the four contact points $(p_{14}, p_{23}, p_{12}, p_{13})$ of the configuration chosen as in Figure 3 with orientation convention. Note that the cross ratio of these four points is always purely imaginary with positive imaginary part since the projective transformation taking the ordered triple (p_{23}, p_{12}, p_{13}) to $(1, 0, \infty)$ maps C_1 to the imaginary axis, and hence takes p_{14} to a point on the positive imaginary axis due to the nature of the configuration. Since the cross ratio is a projective invariant, the value does not depend on the choice of lift $\tilde{e}$ and the developing map. Collecting the values for each edge, we obtain the map $\mathbf{x}$ of E_τ, which we call a cross ratio parameter. The cross ratio of the edge e determines the position of the circle C_4 in Figure 3 once the positions of C_1, C_2 and C_3 are fixed, and if the cross ratio of e approaches ∞, then C_4 approaches p_{13}.

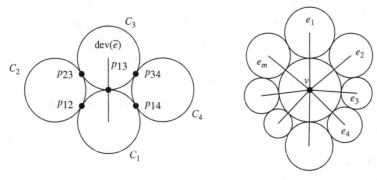

Figure 3: Four circle configuration

Figure 4: A surrounded circle

Of course, not all real valued maps of E_τ can be the cross ratio parameter for some circle packing. There must be some conditions which ensure that for each circle in $\tilde{S}$, the developing image of the surrounding circles closes up neatly. The key to obtaining these conditions in a concise and useful form is to consider a normalized picture of a circle with its surrounding circles. The normalization we chose maps the central circle to the real line and one of the adjoining interstices to the standard interstice with vertices at ∞, 0 and $\sqrt{-1}$. This led us to introduce an associated matrix $A \in \mathrm{SL}_2(\mathbb{R})$ to each edge $e \in E_\tau$. If the value of a cross ratio parameter at e is x, A is defined to be $\begin{pmatrix} 0 & 1 \\ -1 & x \end{pmatrix} \in \mathrm{SL}_2(\mathbb{R})$. The relationship between the associated matrix and the configuration of four circles corresponding to an edge with cross ratio x can then be seen by first normalizing the configuration in Fig 3 by sending the triple (p_{12}, p_{13}, p_{23}) to $(\infty, 0, \sqrt{-1})$, so that we are considering C_1 as the central circle and taking the normalized picture. In the normalized configuration, C_1 and C_2 are horizontal lines through 0 and $\sqrt{-1}$ respectively, C_3 is a circle of radius $1/2$ with center at $\sqrt{-1}/2$ and C_4 is a circle tangent to C_1 at the point $1/x$, and is also tangent to C_3. Then a simple computation shows that the associated matrix A represents a transformation which sends the left triangular interstice of this configuration to the right triangular interstice.

Let v be a vertex of τ with valence m. We read off the edges $e_1, \cdots, e_m$ incident to v in a clockwise direction to obtain a sequence of assigned values $x_1, \ldots, x_m$ of cross ratio parameters. Let

$$W_j = A_1 A_2 \cdots A_j = \begin{pmatrix} a_j & b_j \\ c_j & d_j \end{pmatrix}, \quad j = 1, \ldots, m,$$

where A_i is the matrix $\begin{pmatrix} 0 & 1 \\ -1 & x_i \end{pmatrix}$ associated to e_i. Then, by a careful study of the normalized picture, and the composition of moves that shifts the standard interstice to the interstices on its right step by step, until it finally returns to itself, it was verified in [KMT03a] that for each vertex v of τ, we have

$$W_v = A_1 A_2 \cdots A_m = \begin{pmatrix} -1 & 0 \\ 0 & -1 \end{pmatrix}, \tag{4.1}$$

and

$$\begin{cases} a_j, c_j < 0, \ b_j, d_j > 0 \ \text{for} \ 1 \le j \le m-1 \\ \text{except for} \ a_1 = d_{m-1} = 0. \end{cases} \tag{4.2}$$

The first condition comes from the fact that the chain of circles surrounding the circle corresponding to v closes up. The second condition is a no overwinding condition, and it eliminates the case where the chain surrounds the central circle more than once. Notice here that the associated matrices are in $SL_2(\mathbb{R})$ and not in $PSL_2(\mathbb{R})$, so that the inequalities of (4.2) do make sense.

On the other hand, given a real valued map $\mathbf{x}$ of E_τ satisfying (4.1) and (4.2) for each vertex of τ, it is relatively routine to construct a pair (S, P) of a projective Riemann surface S and a circle packing P on S so that its cross ratio parameter is $\mathbf{x}$ (see [KMT03a] for details). We thus set

$$\mathcal{C}_\tau = \{\mathbf{x} : E_\tau \to \mathbb{R} \,|\, \mathbf{x} \text{ satisfies (4.1) and (4.2) for each vertex}\},$$

and call it the *cross ratio parameter space*.

Since the condition (4.1) gives a set of polynomial equations for the x_i's and (4.2) are polynomial inequalities in the x_i's, $\mathcal{C}_\tau$ is a semi-algebraic set by definition, and we define the topology on $\mathcal{C}_\tau$ to be the one induced by the tautological inclusion $\iota : \mathcal{C}_\tau \to \mathbb{R}^{E_\tau}$. It turns out that this naive construction gives us a correct parameterization of the moduli space of pairs (S, P) where S is a projective Riemann surface and P is a circle packing on S with nerve τ.

Lemma 4.1. *We have the following :*

(i) **(Lemma 2.17 in [KMT03a])** $\mathcal{C}_\tau$ *corresponds bijectively to the moduli space of all pairs (S, P) where S is a projective Riemann surface and P is a circle packing* *on S combinatorially with nerve* τ, *up to marked projective equivalence.*

(ii) (**Lemma 3.2 in [KMT03b]**) *The tautological inclusion* $\iota : C_\tau \to \mathbb{R}^{E_\tau}$ *is proper.*

The above result states that we can identify C_τ with the moduli space of all pairs (S, P) with nerve τ, which we do from now on. The study of the moduli space then reduces to the study of the semi-algebraic set C_τ.

5. Conjectures

Since the KAT pair represents a point in C_τ, the moduli space is certainly nonempty. However, this is the only fact we know up to this stage, and we are far from knowing what C_τ looks like. To understand C_τ better, we relate it with some other spaces and formulate a conjectural picture.

Let $\mathcal{P}_g$ be the space of all projective Riemann surfaces homeomorphic to Σ_g up to marked projective equivalence. In other words, it is the space of all marked projective structures on Σ_g. To each pair (S, P) in C_τ, assign only its first component and we obtain the forgetting map

$$f : C_\tau \longrightarrow \mathcal{P}_g.$$

Thus the image $f(C_\tau)$ consists of all projective Riemann surfaces which admit a circle packing with nerve τ and it is not difficult to see that the injectivity of f is equivalent to the rigidity of the circle packings with nerve τ on such projective Riemann surfaces.

Let $\mathcal{T}_g$ be the space of all complex structures on Σ_g up to marked conformal equivalence, which is commonly called the Teichmüller space. To each projective Riemann surface, assign its underlying complex structure and we obtain the projection map

$$p : \mathcal{P}_g \longrightarrow \mathcal{T}_g.$$

Teichmüller space is known to be homeomorphic to the Euclidean space of dimension 2 or $6g - 6$ according to whether $g = 1$ or $g \geq 2$. The projection map p is a vector bundle projection where the fiber over each conformal class consists of its holomorphic quadratic differentials, and each fiber is a vector space of the same dimension as $\mathcal{T}_g$. In particular, $\mathcal{P}_g$ is homeomorphic to Euclidean space of dimension 4 or $12g - 12$ according to whether $g = 1$ or $g \geq 2$.

To each conformal class of a Riemann surface homeomorphic to Σ_g, the Koebe-Riemann Uniformization Theorem states that there is a unique metric of constant curvature $1, 0$ or -1 on the surface, up to scaling if $g = 1$, in the conformal class. Hence, we obtain a natural section

$$s : \mathcal{T}_g \longrightarrow \mathcal{P}_g$$

to the projection map $p : \mathcal{P}_g \to \mathcal{T}_g$. This section is none other than the space of marked Euclidean structures on the torus up to scaling in the case where $g = 1$, and the space of marked hyperbolic structures on Σ_g in the case $g \geq 2$, which, we recall from §2, are projective structures by definition.

The Koebe-Andreev-Thurston theorem implies that $f(C_\tau)$ intersects $s(\mathcal{T}_g)$ only at $f(\{KAT\})$ and furthermore, the rigidity of the circle packing on $f(\{KAT\})$ means that the inverse image of this point under f consists of exactly one point. It is natural to conjecture that the rigidity of the circle packings holds for all projective Riemann surfaces in $f(C_\tau)$, that is, each projective Riemann surface S admits at most one circle packing with nerve τ up to projective automorphisms of S isotopic to the identity. This can be stated as follows:

Conjecture 1. *The forgetting map* $f : C_\tau \to \mathcal{P}_g$ *is injective.*

This does not however give a possible description of the space C_τ. To get a more detailed understanding of C_τ and its image under f, we formulate a stronger conjecture, which certainly implies the first one, as follows:

Conjecture 2. *The composition* $p \circ f : C_\tau \to \mathcal{T}_g$ *is a homeomorphism.*

The motivation for this rather strong conjecture goes back to the second author's result in [Miz00], where it was shown to be true when $g = 1$ and τ has only one vertex. Here are some implications of the conjecture, which we believe to be true, and have verified in certain special cases. They serve as further motivation for the conjecture:

  (i) The moduli space C_τ is homeomorphic to Euclidean space of dimension 2 or $6g - 6$ according to whether $g = 1$ or $g \geq 2$.

  (ii) $f(C_\tau)$ and $s(\mathcal{T}_g)$ are middle dimensional proper submanifolds in $\mathcal{P}_g$ which intersect only at $f(\{KAT\})$. Probably the intersection can be shown to be transverse with a little more argument.

  (iii) The image $f(C_\tau)$ of the forgetting map defines a new natural section or a slice to $p : \mathcal{P}_g \to \mathcal{T}_g$. In words, this would mean that for each conformal class of a Riemann surface, there exists a unique projective structure within the conformal class which admits a circle packing with nerve τ.

6. Local Results

We have shown that part of the conjecture is true, at least topologically in a neighborhood of the KAT pair.

Theorem 6.1 (Theorem 1 in [KMT03a]). *There is a neighborhood U of the KAT pair in C_τ such that*

(i) *U is homeomorphic to Euclidean space of dimension* 2 *or* $6g - 6$ *according to whether* $g = 1$ *or* $g \geq 2$,

(ii) *the restriction of f to U is injective.*

This result was proved by comparing the deformation of a hyperbolic 3-manifold constructed from the KAT pair and the deformation of a projective Riemann surface admitting a circle packing with nerve τ. Hyperbolic Dehn filling theory and quasi-Fuchsian deformation theory were used for the cases $g = 1$ and $g \geq 2$ respectively.

Since the neighborhood of U is defined constructively, we roughly know how large it is. When $g = 1$, U is chosen so that the image of U under f is identified with hyperbolic Dehn surgery space of a corresponding cusped 3-manifold, which can be embedded as a 2-dimensional subspace in $\mathcal{P}_1$. Since hyperbolic Dehn surgery space omits only a finite number of classical Dehn surgery coefficients, U in this case would be fairly large. When $g \geq 2$, U is chosen to be the preimage of the space of all quasi-Fuchsian deformations of a corresponding hyperbolic 3-manifold by f. Since this space can be identified with the connected component of discrete representations containing the one coming from the KAT pair, U in this case is also fairly large.

7. Global Results

The work to prove the global results stated in Conjectures 1 and 2 is still in progress. We discuss in this section some of the partial results we have obtained thus far in [KMT03a] and [KMT03b]. To begin, it is useful to consider Thurston's parameterization of $\mathcal{P}_g$ which we will describe shortly, since it is more geometric than the projection map $p : \mathcal{P}_g \to \mathcal{T}_g$. We first describe two more spaces closely related with $\mathcal{P}_g$.

The first is the space of non-elementary representations of $\pi_1(\Sigma_g)$ in $\mathrm{PGL}_2(\mathbb{C})(= \mathrm{PSL}_2(\mathbb{C}))$ up to conjugation, which has a natural structure of a $(6g - 6)$-dimensional complex analytic manifold [Gun67]. Since the holonomy representation of a projective structure is not only non-elementary but lifts to $\mathrm{SL}_2(\mathbb{C})$, we will focus on the open subset $\mathcal{X}_g$ consisting of representations which lift to $\mathrm{SL}_2(\mathbb{C})$ up to conjugation. Then, assigning its holonomy representation to each projective Riemann surface, we obtain the map

$$\mathrm{hol} : \mathcal{P}_g \longrightarrow \mathcal{X}_g.$$

hol is known to be a local homeomorphism by Hejhal [Hej75].

The second is the space of isotopy classes of measured laminations on Σ_g $(g \geq 2)$, which we denote by $\mathcal{ML}_g$. A measured lamination is defined to be a closed subset on Σ_g locally homeomorphic to a product of a totally disconnected subset of the interval with an interval, together with a transverse measure. A simple closed curve on Σ_g

with the counting measure for transverse arcs is an elementary, but important and fundamental example of a measured lamination. In fact, the set of weighted simple closed curves is dense in $\mathcal{ML}_g$. See [Thu80, Thu88] for details.

Although a measured lamination is a topological concept, once we put a hyperbolic metric on Σ_g, its support is canonically realized as a disjoint union of simple geodesics which forms a closed subset on the surface. Such a lamination is called a geodesic lamination with transverse measure.

We now describe Thurston's parameterization of $\mathcal{P}_g$. Thurston has shown that any projective Riemann surface corresponds uniquely to a hyperbolic surface pleated along a geodesic lamination with a fixed bending measure. Following [KT92], we briefly review his parameterization. Start with a projective structure on S which is not a hyperbolic structure. Consider the set of maximal disks in the universal cover $\widetilde{S}$. Each maximal disk is naturally endowed with the hyperbolic metric, the boundary of each disk intersects the ideal boundary of $\widetilde{S}$ in two or more points and we can take the convex hull of these ideal boundary points. It can be shown that this gives a stratification of $\widetilde{S}$ by ideal polygons, and ideal bigons foliated by "parallel lines" joining the two ideal vertices of the bigons. The polygonal parts support a canonical hyperbolic metric. Collapsing each bigon foliated by parallel lines in $\widetilde{S}$ to a line and taking the quotient of the result by the action of the fundamental group, we obtain a hyperbolic surface H. This defines a hyperbolization map

$$\pi : \mathcal{P}_g \to s(\mathcal{T}_g).$$

Also the stratification defines a geodesic lamination λ on H by taking the union of collapsed lines. Moreover, using the convex hull of the ideal points of the maximal disk not in the disk but in the 3-dimensional hyperbolic space, we can assign a transverse bending measure supported on λ. This defines a pleating map

$$\beta : \mathcal{P}_g \to \mathcal{ML}_g.$$

The pair of these maps (π, β) becomes a homeomorphism of $\mathcal{P}_g$ onto $s(\mathcal{T}_g) \times \mathcal{ML}_g$.

Figure 5 shows the related spaces and the maps between them as discussed above.

Now by using Thurston's parameterization of $\mathcal{P}_g$, we can analyze $f : \mathcal{C}_\tau \to \mathcal{P}_g$ by looking at $\pi \circ f$ and $\beta \circ f$ separately. We have the following result:

Lemma 7.1 (Lemma 4.1 in [KMT03b]). *If $g \geq 2$, then the composition $\beta \circ f : \mathcal{C}_\tau \to \mathcal{ML}_g$ has bounded image.*

This is proved by observing how the developed image of a projective Riemann surface admitting a circle packing with nerve τ is controlled by the combinatorial data

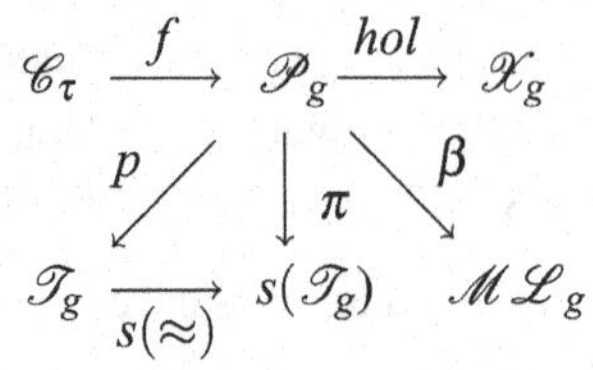

$$\mathscr{C}_\tau \xrightarrow{\ f\ } \mathscr{P}_g \xrightarrow{\ hol\ } \mathscr{X}_g$$

$$p \diagup \quad \Big\downarrow \pi \quad \diagdown \beta$$

$$\mathscr{T}_g \xrightarrow[s(\approx)]{} s(\mathscr{T}_g) \quad \mathscr{ML}_g$$

Figure 5: Related Spaces

of τ. The key is that there is a relation between the circle packing on $dev(\tilde{S})$ and the maximal disks in $dev(\tilde{S})$ which can be exploited to control the image $\beta \circ f(\mathscr{C}_\tau)$.

Apart from this general result, the best global results towards the conjecture we have obtained so far are for the case when τ has only one vertex. This arose from our attempt to understand the cross ratio parameter space $\mathscr{C}_\tau$ concretely in the simplest settings. Note that in this case, $\mathscr{C}_\tau$ is defined by just one matrix equation and set of inequalities corresponding to (4.1) and (4.2) respectively.

Theorem 7.2. *If τ has one vertex and $g \geq 2$, then*

  (i) **(Theorem 2 and Lemma 5.1 in [KMT03a])** *$\mathscr{C}_\tau$ is homeomorphic to $\mathbb{R}^{6g-6}$ and $\mathrm{hol} \circ f : \mathscr{C}_\tau \to \mathscr{X}_g$ is injective. In particular, $f : \mathscr{C}_\tau \to \mathscr{P}_g$ is injective.*

  (ii) **(Theorem 1.1 in [KMT03b])** *$p \circ f : \mathscr{C}_\tau \to \mathscr{T}_g$ is proper.*

Theorem 7.2 comes fairly close to proving Conjecture 2 for the one circle packing case. What is missing is a proof that p restricted to $f(\mathscr{C}_\tau)$ is locally injective. An argument similar to the one worked out by Scannell and Wolf in [SW02] is expected to complete this case.

The arguments used to prove Theorem 7.2 depend technically on the simplifying assumption that τ has only one vertex. At the moment, we do not know how to generalize these arguments to prove the conjecture in general. In fact, the proof of Theorem 7.2 (i) given in [KMT03a] involves a careful study of the cross ratio parameter space, showing that one can always choose a set of $6g - 6$ free parameters lying in a convex subset of $\mathbb{R}^{6g-6}$ which completely parameterizes $\mathscr{C}_\tau$. This requires a very good understanding of the equations (4.1) and inequalities (4.2). Further analysis then shows that in fact $\mathrm{hol} \circ f$ is injective, from which we conclude that $f : \mathscr{C}_\tau \to \mathscr{P}_g$ is injective.

The proof of Theorem 7.2 (ii) uses Theorem 7.2 (i), but otherwise does not rely on the assumption that τ has only one vertex. The main ingredient is the result of Tanigawa in [Tan97] which relates quantities associated to $p \circ f$, $\pi \circ f$, and $\beta \circ f$. We give a brief sketch of the proof. First note that the fact that $hol \circ f$ is injective, stated in

Theorem 7.2 (i), immediately implies that $\pi \circ f$ is proper. Now choose a sequence x_n in C_τ which escapes every compact subset. Since $\pi \circ f$ is proper, the hyperbolic surfaces $H_n = \pi \circ f(x_n)$ escape every compact subset of $\mathcal{T}_g$. Let $h_n : X_n \to H_n$ be the harmonic map, where X_n is the Riemann surface $p \circ f(x_n)$. Then Tanigawa's inequalities [Tan97] imply that either the extremal lengths $E_{\lambda_n}(X_n)$ of $\lambda_n = \beta \circ f(x_n)$ in X_n diverge, or that the energies of the maps h_n remain bounded. In the first case, since by Lemma 7.1 λ_n is bounded in $\mathcal{ML}_g$, it must be that X_n escapes every compact subset. In the second case, the result of Wolf in [Wol89] implies that X_n is again unbounded. This concludes the proof that $\pi \circ f$, and hence $p \circ f$ are proper.

8. Cone Projective Structures

This section is just to give a small remark not mentioned in [KMT03a, KMT03b]. It is natural to study the extension of circle packings on projective Riemann surfaces to cone projective Riemann surfaces, where the cone points are in the vertex set of τ, as was done in the metric structure case. This would include circle packings on orbifold surfaces and branched surfaces, and can also be used to connect a circle packing on a compact surface and a horocycle packing on a cusped surface continuously.

The analysis using the cross ratio parameter can be applied to the study of circle packings on projective Riemann surfaces with cone singularities, with little modification. In the case of cone structures, the equation (4.1) is replaced by an equation involving the trace, arising from the cone angle condition assuming the cone points are "centers" of circles. However, it can be much more complicated if we allow cone angles $> 2\pi$. To demonstrate the complications, we describe here the moduli space of circle packings by one circle on the torus with a cone point where the cone point is the center of the circle.

Let θ (≥ 0) be a cone angle and τ the nerve of a circle packing. We here regard a cone point with $\theta = 0$ as a cusp. Let us denote the cross ratios corresponding to the three edges of τ by x, y, z and their associated matrices by X, Y, Z. Then the condition corresponding to (4.1) in this case is replaced by the condition that $(XYZ)^2$ is conjugate to a θ-rotation when $\theta > 0$ and a translation when $\theta = 0$. This implies the equation,

$$|\mathrm{tr}(XYZ)| = |2\cos(\theta/4)|,$$

by which we can compute the cross ratio parameter space $C_{\tau,\theta}$ concretely as follows.

$$C_{\tau,\theta} = \begin{cases} \{xyz - x - y - z = 2, \, xy > 1, \, x > 0\} & \text{if } \theta = 0. \\ \{xyz - x - y - z = 2\cos\frac{\theta}{4}, \, xy > 1, \, x > 0\} & \text{if } \theta \in (0, 4\pi). \\ \{(1,1,1)\} & \text{if } \theta = 4\pi. \\ \{xyz - x - y - z = 2\cos\frac{\theta}{4}, \, xy < 1\} & \text{if } \theta \in (4\pi, 8\pi). \\ \{(-1,-1,-1)\} & \text{if } \theta = 8\pi. \\ \{xyz - x - y - z = 2\cos\frac{\theta}{4}, \, xy > 1, \, x < 0\} & \text{if } \theta \in (8\pi, 12\pi). \end{cases}$$

When $\theta = 0$, 4π and 8π, the trace condition is not sufficient to determine the conjugacy class of XYZ, and we need a little argument to obtain the above representations.

The equation $xyz - x - y - z = t$ has the unique solution z for given x, y, t unless $xy = 1$, and $C_{\tau,\theta}$ is homeomorphic to $\mathbb{R}^2$ except for $\theta = 4\pi, 8\pi$. Note that it degenerates to a point in these two exceptional cases.

9. Problems

Here we list up some open problems and directions for further study in the subject.

(i) Can we use a variational method to study the conjecture, similar to the methods used by Colin de Verdiere in [CdV91] ?

(ii) If the conjecture were true, the resulting new section in $\mathcal{P}_g$ should find some applications in the study of Kleinian group theory. What are they ?

(iii) In the one circle packing case, the moduli space C_τ can be identified with a very nice convex subset of $\mathbb{R}^{6g-6}$. This is as opposed to many other embeddings of $\mathcal{T}_g$ which have fractal type boundary. Do C_τ or $f(C_\tau)$ have nice compactifications with interesting geometric interpretations?

(iv) Brooks [Bro86] has shown that the set of hyperbolic surfaces admitting a circle packing with the property (3.1) is dense in Teichmüller space $\mathcal{T}_g$. A natural question is whether projective Riemann surfaces admitting a circle packing with the property (3.1) are dense in $\mathcal{P}_g$?

(v) Study the intersection of $f(C_\tau)$ with the space of quasi-Fuchsian structures for fixed g and τ. For example, we may ask whether this intersection is connected, and what is the boundary of this space like ?

(vi) Study the geometry of $\mathcal{P}_g$ from $f(C_\tau)$ for various τ.

(vii) The cross ratios are naturally positive real numbers, but it is possible to make sense of negative cross ratios if we allow overlapping (overwinding). Is it possible to make geometric sense of the cross ratio as a complex number ?

(viii) Study the space of "triangulations" of Σ_g, which lift to honest triangulations of $\tilde{\Sigma}_g$. In particular, it would be interesting to generate a complex from the triangulations, similar to the complex of curves studied by Harvey in [Har92].

References

[Ahl53]   L. V. Ahlfors (1953). *Complex Analysis. An introduction to the theory of analytic functions of one complex variable.* McGraw-Hill Book Company, Inc., New York-Toronto-London.

[And70]   E. M. Andreev (1970). Convex polyhedra of finite volume in Lobačevskiĭ space. *Mat. Sb. (N.S.)* **83 (125)**, 256–260. *Math. USSR-Sb.* (English), **12**, 255–259.

[Bro86]   R. Brooks (1986). Circle packings and co-compact extensions of Kleinian groups. *Invent. Math.* **86** (3), 461–469.

[CdV91]   Y. Colin de Verdière (1991). Un principe variationnel pour les empilements de cercles. *Invent. Math.* **104** (3), 655–669.

[GKM00]   D. Gallo, M. Kapovich & A. Marden (2000). The monodromy groups of Schwarzian equations on closed Riemann surfaces. *Ann. of Math. (2)* **151** (2), 625–704.

[Gol87]   W. M. Goldman (1987). Projective structures with Fuchsian holonomy. *J. Differential Geom.* **25** (3), 297–326.

[Gun67]   R. C. Gunning (1967). *Lectures on Vector Bundles over Riemann Surfaces.* University of Tokyo Press, Tokyo.

[Har92]   W. J. Harvey (1992). Modular groups – geometry and physics. In *Discrete Groups and Geometry (Birmingham, 1991), London Math. Soc. Lecture Note Ser.*, volume 173, pp. 94–103. Cambridge Univ. Press, Cambridge.

[Hej75]   D. A. Hejhal (1975). Monodromy groups and linearly polymorphic functions. *Acta Math.* **135** (1), 1–55.

[HS98]   Z.-X. He & O. Schramm (1998). The C^∞-convergence of hexagonal disk packings to the Riemann map. *Acta Math.* **180** (2), 219–245.

[KMT03a]   S. Kojima, S. Mizushima & S. P. Tan (2003). Circle packings on surfaces with projective structures. *J. Differential Geom.* **63** (3), 349–397.

[KMT03b]   S. Kojima, S. Mizushima & S. P. Tan (2003). Circle packings on surfaces with projective structures and uniformization. Preprint, `math.GT/ 0308147`.

[Koe36]   P. Koebe (1936). Kontaktprobleme der konformen abbildung. *Ber. Sachs. Akad. Wiss. Leipzig, Math-Phys. Klasse* **88**, 141–164.

[KT92]   Y. Kamishima & S. P. Tan (1992). Deformation spaces on geometric structures. In *Aspects of Low-dimensional Manifolds, Adv. Stud. Pure Math.*, volume 20, pp. 263–299. Kinokuniya, Tokyo.

[Miz00]   S. Mizushima (2000). Circle packings on complex affine tori. *Osaka J. Math.* **37** (4), 873–881.

[Sch97]   O. Schramm (1997). Circle patterns with the combinatorics of the square grid. *Duke Math. J.* **86** (2), 347–389.

[SW02]   K. P. Scannell & M. Wolf (2002). The grafting map of Teichmüller space. *J. Amer. Math. Soc.* **15** (4), 893–927 (electronic).

[Tan97]   H. Tanigawa (1997). Grafting, harmonic maps and projective structures on surfaces. *J. Differential Geom.* **47** (3), 399–419.

[Thu80]   W. P. Thurston (1980). The Geometry and Topology of 3-Manifolds. Lecture notes, Princeton University. Available at `www.msri.org/publications/books/gt3m/`.

[Thu88]   W. P. Thurston (1988). On the geometry and dynamics of diffeomorphisms of surfaces. *Bull. Amer. Math. Soc. (N.S.)* **19** (2), 417–431.

[Wol89]   M. Wolf (1989). The Teichmüller theory of harmonic maps. *J. Differential Geom.* **29** (2), 449–479.

Sadayoshi Kojima

Department of Mathematical and Computing Sciences
Tokyo Institute of Technology
Ohokayama Meguro Tokyo 152-8552
Japan

sadayosi@is.titech.ac.jp

Shigeru Mizushima

Department of Mathematical and Computing Sciences
Tokyo Institute of Technology
Ohokayama Meguro Tokyo 152-8552
Japan

mizusima@is.titech.ac.jp

Ser Peow Tan

Department of Mathematics
National University of Singapore
Singapore 117543
Singapore

mattansp@nus.edu.sg

AMS Classification: primary 52C15; secondary 30F99, 57M50

Keywords: circle packing, projective structure

Spaces of Kleinian Groups   Cambridge University Press

Lond. Math. Soc. Lec. Notes **329**, 355–373   Y. Minsky, M. Sakuma & C. Series (Eds.)

Grafting and components of quasi-fuchsian projective structures

Kentaro Ito

Abstract

We give an expository account of our results in [Ito00] and [Itoa] on the bumping and self-bumping of components of quasi-fuchsian projective structures from the view point of [Itob] on continuity of grafting maps at boundary groups.

1. Introduction

We consider the space of projective structures $P(S)$ on a closed surface S of hyperbolic type and its open subset $Q(S)$ consists of projective structures with quasi-fuchsian holonomy. It is known that $Q(S)$ have infinitely many connected components. The aim of this note is to outline and explain how components of $Q(S)$ lies in $P(S)$, especially how these components bump or self-bump. Here we say that components Q, Q' of $Q(S)$ *bump* if they have intersecting closures and that a component Q *self-bumps* if there is a point $\Sigma \in \partial Q$ such that $U \cap Q$ is disconnected for any sufficiently small neighborhood U of Σ. Studying how $Q(S)$ lies in $P(S)$ is closely related to studying how the quasi-fuchsian space $Q\mathcal{F} = Q\mathcal{F}(S)$ lies in the representation space $R(S)$, where $R(S)$ is the set of conjugacy classes of representations $\rho : \pi_1(S) \to \mathrm{PSL}_2(\mathbb{C})$ with the algebraic topology and $Q\mathcal{F} \subset R(S)$ is the subspace of faithful representations with quasi-fuchsian images.

Now let Γ be a geometrically finite Kleinian group with non-trivial space $AH(\Gamma)$ of conjugacy classes of discrete faithful representations $\Gamma \to \mathrm{PSL}_2(\mathbb{C})$. Then the bumping of components of the interior of $AH(\Gamma)$ are characterized by the topological data of the quotient manifold $\mathbb{H}^3/\Gamma$ by Anderson, Canary and McCullough [AC96], [ACM00]. In our setting, it is known that the quasi-fuchsian space $Q\mathcal{F}$ is the interior of the space $AH(S)$ of discrete faithful representations and that $Q\mathcal{F}$ consists of exactly one connected component. Nevertheless, McMullen [McM98] showed that $Q\mathcal{F}$ self-bumps by using projective structures and ideas of Anderson and Canary [AC96]. In his argument, he used the fact that the local structure of the boundary of $Q\mathcal{F}$ is equal to that of $Q(S) = \mathrm{hol}^{-1}(Q\mathcal{F})$ because the holonomy map $\mathrm{hol} : P(S) \to R(S)$, assigning a projective structure to its holonomy representation, is a local homeomorphism (see [Hej75]). After that, Bromberg and Holt [BH01] showed that each component of the interior of $AH(\Gamma)$ self-bumps for more general Kleinian groups Γ without using projective structures. We refer the reader to a survey written by Canary [Can04] for further information on the bumping and self-bumping of deformation spaces of Kleinian groups.

In this note, we push ahead with the observation in [McM98] and studied the bumping and self-bumping of components of $Q(S) = \text{hol}^{-1}(Q\mathcal{F})$. By Goldman's grafting theorem (Theorem C in [Gol87]), the set of components of $Q(S)$ is in one-to-one correspondence with the set $\mathcal{ML}_{\mathbb{N}}$ of integral measured laminations on S. Thus we obtain a decomposition $\bigsqcup_{\lambda \in \mathcal{ML}_{\mathbb{N}}} Q_\lambda$ of $Q(S)$, where Q_λ is the connected component of $Q(S)$ associated to $\lambda \in \mathcal{ML}_{\mathbb{N}}$. Especially, the component Q_0 for zero-lamination $0 \in \mathcal{ML}_{\mathbb{N}}$ consists of all quasi-fuchsian projective structures with injective developing map. We know that the map $\text{hol}|_{Q_\lambda} : Q_\lambda \to Q\mathcal{F}$ is biholomorphic for each $\lambda \in \mathcal{ML}_{\mathbb{N}}$ and let $\Psi_\lambda : Q\mathcal{F} \to Q_\lambda$ denote the univalent local branch of hol^{-1}, which is called the *grafting map* for λ. In §3, we discuss conditions under which the map Ψ_λ is extended continuously to a boundary point of $Q\mathcal{F}$. Recall that Bers' simultaneous uniformization gives a bijection $B : T(S) \times T(S) \to Q\mathcal{F}$, where $T(S)$ denote the Teichmüller space of S. Suppose that a sequence $\rho_n = B(X_n, Y_n) \in Q\mathcal{F}$ converges to $\rho_\infty \in \partial Q\mathcal{F}$. Then we say that the convergence $\rho_n \to \rho_\infty$ is *standard* if there exists a compact subset K of $T(S)$ which contains all X_n or all Y_n; otherwise it is *exotic*. Then we have the following:

Theorem 1.1 ([Itob]). *For every $\lambda \in \mathcal{ML}_{\mathbb{N}}$, the grafting map $\Psi_\lambda : Q\mathcal{F} \to P(S)$ takes every standardly convergent sequence to a convergent sequence in $\widehat{P}(S)$, where $\widehat{P}(S) = P(S) \cup \{\infty\}$ denote the one-point compactification of $P(S)$.*

On the other hand, in §4, we outline the following result, which is obtained by making use of exotically convergent sequences constructed by Anderson and Canary [AC96] and McMullen [McM98]:

Theorem 1.2 ([Ito00, Itoa]).

 (i) *Any two components of $Q(S)$ bump,*

 (ii) *Every component of $Q(S)$ except for Q_0 self-bumps, and*

 (iii) *For any $n \in \mathbb{N}$, there exist n-components of $Q(S)$ which bump simultaneously.*

The same argument as in Theorem 1.2 reveals that the grafting map $\Psi_\lambda : Q\mathcal{F} \to P(S)$ does not extend continuously to $\partial Q\mathcal{F}$; see Theorem 5.1. Then Theorem 1.1 implies that only exotically convergent sequences cause this non-continuity and the bumping of distinct components of $Q(S)$.

2. Preliminaries

2.1. Quasi-fuchsian space

We let $R(S)$ denote the space of conjugacy classes $[\rho]$ of representations $\rho : \pi_1(S) \to$ $\mathrm{PSL}_2(\mathbb{C})$ with non-abelian image $\rho(\pi_1(S))$. (For simplicity, we denote $[\rho]$ by ρ if there is no confusion.) The space $R(S)$ is endowed with the algebraic topology and is known to be a complex manifold (see for example [MT98]). *Quasi-fuchsian space* $Q\mathcal{F}$ is the subset of $R(S)$ of conjugacy classes of faithful representations whose images are quasi-fuchsian groups. Then $Q\mathcal{F}$ is open, connected and contractible in $R(S)$. Let $\rho \in Q\mathcal{F}$ with quasi-fuchsian image $\Gamma = \rho(\pi_1(S))$. Then the region of discontinuity Ω_Γ of Γ decomposes into two invariant components Ω_Γ^+ and Ω_Γ^-, and the representation ρ determines a pair $(\Omega_\Gamma^+/\Gamma, \Omega_\Gamma^-/\Gamma)$ in $T(S) \times T(\bar{S})$. Here $T(S)$ is the Teichmüller space of S, and $\bar{S}$ denotes S with orientation reversed. On the contrary, it was shown by Bers [Ber60] that each pair $(X, \bar{Y}) \in T(S) \times T(\bar{S})$ has its unique simultaneous uniformization $\rho = B(X, \bar{Y}) \in Q\mathcal{F}$. Thus we have a parameterization

$$B : T(S) \times T(\bar{S}) \to Q\mathcal{F}$$

of $Q\mathcal{F}$. We define *vertical* and *horizontal Bers slices* in $Q\mathcal{F}$ by $B_X = \{B(X, \bar{Y}) : \bar{Y} \in T(\bar{S})\}$ and $B_{\bar{Y}} = \{B(X, \bar{Y}) : X \in T(S)\}$. Bers showed that both B_X and $B_{\bar{Y}}$ are precompact in $R(S)$, whose frontiers are denoted by ∂B_X and $\partial B_{\bar{Y}}$.

2.2. Space of projective structures

A projective structure on S is a (G, X)-structure, where X is a Riemann sphere $\widehat{\mathbb{C}}$ and $G = \mathrm{PSL}_2(\mathbb{C})$ is the group of projective automorphisms of $\widehat{\mathbb{C}}$. We let $P(S)$ denote the space of marked projective structures on S. A projective structure $\Sigma \in P(S)$ determines its underlying conformal structure $\pi(\Sigma) \in T(S)$. It is known that $P(S)$ is a holomorphic affine bundle over $T(S)$ with the projection $\pi : P(S) \to T(S)$ and that each fiber $\pi^{-1}(X)$ for $X \in T(S)$ can be identified with the space of holomorphic quadratic differentials on X. As an usual (G, X)-structure, a projective structure $\Sigma \in P(S)$ determines a pair (f_Σ, ρ_Σ) of a developing map $f_\Sigma : \tilde{S} \to \widehat{\mathbb{C}}$ and a holonomy representation $\rho_\Sigma : \pi_1(S) \to \mathrm{PSL}_2(\mathbb{C})$, which is uniquely determined up to $\mathrm{PSL}_2(\mathbb{C})$. We now define the *holonomy map*

$$\mathrm{hol} : P(S) \to R(S)$$

by $\Sigma \mapsto [\rho_\Sigma]$. Hejhal [Hej75] showed that the map hol is a local homeomorphism and Earle [Ear81] and Hubbard [Hub81] independently showed that the map is holomorphic.

In this note, we are mainly concerned with the subset $Q(S) = \text{hol}^{-1}(Q\mathcal{F})$ of $P(S)$. An element of $Q(S)$ is said to be *standard* if its developing map is injective; otherwise it is *exotic*. We denote by $Q_0 \subset Q(S)$ the subset of standard projective structures. For a quasi-fuchsian representation $\rho = B(X, \bar{Y})$ with $\Gamma = \rho(\pi_1(S))$, the quotient surface $\Sigma = \Omega_\Gamma^+/\Gamma$ is regarded as a standard projective structure on S with bijective developing map $f_\Sigma : \tilde{\Sigma} \to \Omega_\Gamma^+$, with holonomy representation $\rho_\Sigma = \rho$, and with underlying complex structure $X \in T(S)$. Let

$$\Psi_0 : Q\mathcal{F} \to Q_0$$

be the map defined by the correspondence $\rho \mapsto \Omega_\Gamma^+/\Gamma$ as above. Then the map Ψ_0 turns out to be a univalent local branch of hol^{-1} onto the connected component Q_0 of $Q(S)$, which is called the *standard component*. It is known by Bers that every Bers slice $B_X \subset Q\mathcal{F}$ is embedded by the map Ψ_0 into a bounded domain $\Psi_0(B_X)$ of the fiber $\pi^{-1}(X) \subset P(S)$.

2.3. Grafting

We let $\mathcal{ML}_\mathbb{N} = \mathcal{ML}_\mathbb{N}(S)$ denote the set of integral measured laminations, or the set of formal summation $\sum_{i=1}^l k_i c_i$ of homotopically distinct simple closed curves c_i on S with positive integer k_i weights. A *realization* $\hat{\lambda}$ of $\lambda = \sum_{i=1}^l k_i c_i \in \mathcal{ML}_\mathbb{N}$ is a disjoint union of simple closed curves which realize each weighted simple closed curve $k_i c_i$ by k_i parallel disjoint simple closed curves homotopic to c_i. For two element $\lambda, \mu \in \mathcal{ML}_\mathbb{N}$, the geometric intersection number is denoted by $i(\lambda, \mu)$.

Let λ be non-zero element of $\mathcal{ML}_\mathbb{N}$. We now explain haw to obtain the grafting map

$$\text{Gr}_\lambda : Q_0 \to P(S),$$

which satisfies $\text{hol} \circ \text{Gr}_\lambda \equiv \text{hol}$ on Q_0. In this note, we shall give two equivalent definitions of grafting operation, the one given here is as usual, and the one given in §3.1 is due to Bromberg [Bro02]. We assume that λ is a simple closed curve c of weight one for simplicity and fix our notation as follows:

Notation 2.1. Let $\rho \in Q\mathcal{F}$ with $\Gamma = \rho(\pi_1(S))$. Then we have projective structures $\Sigma = \Omega_\Gamma^+/\Gamma = \Psi_0(\rho)$ on S and $\Sigma^- = \Omega_\Gamma^-/\Gamma$ on $\bar{S}$. Suppose that $c^+ \subset \Sigma$ and $c^- \subset \Sigma^-$ are simple closed curves associated to $c \subset S$ and that $\gamma \in \Gamma \cong \pi_1(S)$ is a representative of the homotopy class of c. Let $\tilde{c}^+ \subset \Omega_\Gamma^+$ and $\tilde{c}^- \subset \Omega_\Gamma^-$ be the $\langle\gamma\rangle$-invariant lifts of $c^+ \subset \Sigma$ and $c^- \subset \Sigma^-$, respectively.

Definition 2.2 (Grafting I). We adopt Notation 2.1. Let A_c be a cylinder $(\hat{\mathbb{C}} - \tilde{c}^+)/\langle\gamma\rangle$ equipped with a projective structure induced from that of $\hat{\mathbb{C}}$. Then the *grafting* $\text{Gr}_c(\Sigma)$

is obtained by cutting Σ along c and inserting A_c at the cut locus without twisting.

For general $\lambda \in \mathcal{ML}_{\mathbb{N}}$, the grafting $\Sigma' = \mathrm{Gr}_\lambda(\Sigma)$ of Σ along λ is also defined by linearity. Then it is important to note that $\rho_\Sigma = \rho_{\Sigma'}$ is always satisfied and that the pull-back $\Lambda_{\Sigma'} := f_{\Sigma'}^{-1}(\Lambda_\Gamma)/\pi_1(\Sigma') \subset \Sigma'$ of the limit set Λ_Γ of the holonomy image $\Gamma = \rho_\Sigma(\pi_1(S)) = \rho_{\Sigma'}(\pi_1(S))$ is a realization of 2λ (see [Gol87]). Since $\mathrm{hol} \circ \mathrm{Gr}_\lambda \equiv \mathrm{hol}$ is satisfied on $\mathcal{Q}_0$, the grafting map Gr_λ takes $\mathcal{Q}_0$ biholomorphically onto the connected component $\mathcal{Q}_\lambda := \mathrm{Gr}_\lambda(\mathcal{Q}_0)$ of $Q(S)$. Thus we have a univalent local branch

$$\Psi_\lambda := \mathrm{Gr}_\lambda \circ \Psi_0 : Q\mathcal{F} \to \mathcal{Q}_\lambda$$

of hol^{-1}. By abuse of terminology, we also call $\Psi_\lambda(\rho)$ the *grafting* of ρ along λ and Ψ_λ the *grafting map* for λ. By Goldman's grafting theorem [Gol87] below, we obtain the decomposition $\bigsqcup_{\lambda \in \mathcal{ML}_{\mathbb{N}}} \mathcal{Q}_\lambda$ of $Q(S)$ into its connected components.

Theorem 2.3 (Goldman [Gol87]). *For every* $\rho \in Q\mathcal{F}$, *we have*

$$\mathrm{hol}^{-1}(\rho) = \{\Psi_\lambda(\rho) : \lambda \in \mathcal{ML}_{\mathbb{N}}\}.$$

2.4. Sequences of quasi-fuchsian representations

Now we introduce the notion of standard and exotic convergence for a sequence $\rho_n \in Q\mathcal{F}$ tending to a limit $\rho_\infty \in \partial Q\mathcal{F}$.

Definition 2.4 (Standard and exotic convergence). Suppose that a sequence $\rho_n = B(X_n, \bar{Y}_n) \in Q\mathcal{F}$ converges to $\rho_\infty \in \partial Q\mathcal{F}$. Then the convergence $\rho_n \to \rho_\infty$ is said to be *standard* if (i) there exist compact set $K \subset T(S)$ such that $\{X_n\} \subset K$, or (ii) there exist compact set $\bar{K} \subset T(\bar{S})$ such that $\{\bar{Y}_n\} \subset \bar{K}$. Otherwise, we say that the convergence is *exotic*.

We let $\partial^+ Q\mathcal{F}$ and $\partial^- Q\mathcal{F}$ denote the subsets of $\partial Q\mathcal{F}$ of standard convergent limits of type (i) and (ii) respectively, and set $\partial^\pm Q\mathcal{F} = \partial^+ Q\mathcal{F} \sqcup \partial^- Q\mathcal{F}$. An element $\rho \in \partial Q\mathcal{F}$ is called a *b-group* if the image $\Gamma = \rho(\pi_1(S))$ is a *b*-group, i.e., there exists exactly one simply connected invariant component of Ω_Γ. Then we remark that the set $\partial^\pm Q\mathcal{F}$ is equals to the set of all *b*-groups in $\partial Q\mathcal{F}$ and that the following hold (see [Itob]):

$$\partial^+ Q\mathcal{F} = \bigsqcup_{X \in T(S)} \partial B_X, \quad \partial^- Q\mathcal{F} = \bigsqcup_{\bar{Y} \in T(\bar{S})} \partial B_{\bar{Y}}.$$

As we will explain in §2.5, there exists a sequence in $Q\mathcal{F}$ which converges exotically into $\partial^\pm Q\mathcal{F}$. On the other hand, the set $\partial Q\mathcal{F} - \partial^\pm Q\mathcal{F}$ is not empty, for instance, it contains a limit of a sequence which appears in Thurston's double limit theorem.

As a consequence of the following lemma, we see that the map $\mathrm{hol}|_{\overline{Q_0}} : \overline{Q_0} \to QF \sqcup \partial^+ QF$ is bijective, where $\overline{Q_0}$ is the closure of Q_0 in $P(S)$.

Lemma 2.5. *The map $\Psi_0 : QF \to Q_0$ takes every standardly convergent sequence $\rho_n \in QF$ with $\lim \rho_n \in \partial^+ QF$ to a convergent sequence $\Sigma_n \in Q_0$ with $\lim \Sigma_n \in \partial Q_0$.*

Proof. Suppose that a sequence $\rho_n = B(X_n, \bar{Y}_n) \in QF$ converges standardly to some $\rho_\infty \in \partial^+ QF$. Then we first show that $X_n \to X$ and that $\rho_\infty \in \partial B_X$ for some $X \in T(S)$. In fact, there exists a subsequence of $\{X_n\}$, denoted by the same symbol, which converges to some $X \in T(S)$. Now we take a new sequence $\rho'_n = B(X, \bar{Y}_n)$ in B_X. Then the sequence $\{\rho'_n\}$ also converges to ρ_∞, since maximal dilatations of quasiconformal automorphisms of $\widehat{\mathbb{C}}$ conjugating ρ_n to ρ'_n tend to 1 as $n \to \infty$. This implies that $\rho_\infty \in \partial B_X$. Since $\partial B_{X_1} \cap \partial B_{X_2} = \emptyset$ if $X_1 \neq X_2$, $X_n \to X$ without passing to a subsequence. Therefore, any accumulation point $\Sigma_\infty \in \partial Q_0$ of the precompact set $\{\Sigma_n \in \pi^{-1}(X_n) : n \in \mathbb{N}\}$ is contained in $\pi^{-1}(X)$. From the injectivity of the map $\mathrm{hol}|_{\pi^{-1}(X)} : \pi^{-1}(X) \to R(S)$ (see [Kra71]), we see that Σ_∞ is uniquely determined by the condition $\mathrm{hol}(\Sigma_\infty) = \rho_\infty$, and thus $\Sigma_n \to \Sigma_\infty$ without passing to a subsequence. $\square$

We collect in Table 1 below the equivalent conditions with standard/exotic convergence of quasi-fuchsian representations, as a consequence of Lemma 2.5 and [Ito00, Proposition 3.4] (see also [McM98, Appendix A]). The situation in which we consider is as follows: suppose that a sequence $\rho_n \in QF$ converges to $\rho_\infty \in \partial^+ QF$ and that the sequence $\Gamma_n = \rho_n(\pi_1(S))$ converges geometrically to a Kleinian group $\widehat{\Gamma}$, which contains the algebraic limit $\Gamma_\infty = \rho_\infty(\pi_1(S))$. Let Σ_∞ be the unique projective structure in ∂Q_0 with holonomy ρ_∞ and let $\Phi : U \to P(S)$, $\rho_\infty \mapsto \Sigma_\infty$ be a univalent local branch of hol^{-1} which is defined on a neighborhood U of ρ_∞. Then the sequence $\Sigma_n = \Phi(\rho_n)$ converges to $\Sigma_\infty = \Phi(\rho_\infty)$. We denote by $\Omega_{\widehat{\Gamma}}^+$ the unique invariant component of the region of discontinuity Ω_{Γ_∞} of Γ_∞, which is equals to the image of the injective developing map $f_{\Sigma_\infty} : \widetilde{\Sigma}_\infty \to \widehat{\mathbb{C}}$. In this situation, all conditions in the same line in Table 1 are equivalent.

Table 9: Equivalent conditions with standard/exotic convergence.

$\rho_n \to \rho_\infty$: standard	$\rho_n \to \rho_\infty$: exotic
$\Omega_{\widehat{\Gamma}}^+ \cap \Lambda(\widehat{\Gamma}) = \emptyset$	$\Omega_{\widehat{\Gamma}}^+ \cap \Lambda(\widehat{\Gamma}) \neq \emptyset$
Σ_n are standard $(n \gg 0)$	Σ_n are exotic $(n \gg 0)$

2.5. ACM-sequences

We will explain a typical example of a sequence $\rho_n \in Q\mathcal{F}$ converging exotically to $\rho_\infty \in \partial^+ Q\mathcal{F}$, which we call an ACM-sequence named after Anderson-Canary [AC96] and McMullen [McM98]. We remark that all the known such a sequence is basically obtained by their technique. We give here a brief survey and refer to [McM98] or [Ito00] for more details. Let c be a simple closed curve on S and let $\tau = \tau_c$ be the Dehn twist along c. Then an *ACM-sequence* in $Q\mathcal{F}$ for c with a starting point $(X, \bar{Y}) \in T(S) \times T(\bar{S})$ is defined by

$$\rho_n = B(\tau^n X, \tau^{2n} \bar{Y}) \quad (n \in \mathbb{Z}),$$

which is known to converge to some $\rho_\infty \in \partial^+ Q\mathcal{F}$. For example, the convergence can be observed as follows: let us consider sequences $\eta_n = B(X, \tau^n \bar{Y})$ in B_X and $\eta'_n = B(\tau^{-n} X, \bar{Y})$ in $B_{\bar{Y}}$. Then $\rho_n = \eta_n \circ \tau_*^{-n}$ and $\eta'_n = \eta_n \circ \tau_*^n$ hold for all n, where τ_* is the group automorphism of $\pi_1(S)$ induced by τ. Since both sequences η_n and η'_n converge up to subsequence, the same argument in [KT90] (see also [Bro97]) reveals that the sequence ρ_n also converges up to subsequence. Moreover, we know that the sequences η_n, η'_n and ρ_n converge without passing to a subsequence from the Dehn filling construction; see [AC96] and [McM98]. Similarly, we obtain a convergent sequence

$$\rho_n = B(\tau^{kn} X, \tau^{(k+1)n} \bar{Y}) = \eta_n \circ \tau_*^{-kn} \quad (n \in \mathbb{Z})$$

for each $k \in \mathbb{Z}$, which converges standardly to its limit if and only if $k = 0, -1$ and whose limit is in $\partial^+ Q\mathcal{F}$ if $k \geq 0$ and in $\partial^- Q\mathcal{F}$ if $k \leq -1$.

We now define an ACM-sequence for general element $\lambda = \sum_{i=1}^l k_i c_i \in \mathcal{ML}_{\mathbb{N}}$, whose *support* is denoted by $\underline{\lambda} = \sqcup_i c_i$. The Dehn twist for λ is defined by $\tau_\lambda = \tau_{c_1}^{k_1} \circ \cdots \circ \tau_{c_l}^{k_l}$, and thus $\tau_{\underline{\lambda}} \circ \tau_\lambda = \tau_{c_1}^{k_1+1} \circ \cdots \circ \tau_{c_l}^{k_l+1}$. Then an ACM-sequence for λ is defined by

$$\rho_n = B(\tau_\lambda^n X, (\tau_{\underline{\lambda}} \circ \tau_\lambda)^n \bar{Y}) \quad (n \in \mathbb{Z}), \tag{2.1}$$

which converges exotically to some $\rho_\infty \in \partial^+ Q\mathcal{F}$. (An ACM-sequence converging to some $\rho_\infty \in \partial^- Q\mathcal{F}$ is also obtained by the same way, but we do not discuss such a sequence in this note.) We now recall some basic fact of the ACM-sequence $\{\rho_n\}$ as in (2,1): by passing to a subsequence if necessary, we may assume that the sequence $\Gamma_n = \rho_n(\pi_1(S))$ of quasi-fuchsian groups converges geometrically to a Kleinian group $\widehat{\Gamma}$, whose Kleinian manifold $N_{\widehat{\Gamma}}$ is homeomorphic to $S \times [-1, 1] - \cup_i (c_i \times \{0\})$ and have conformal boundary $X \sqcup \bar{Y}$ up to marking. Then the algebraic limit $\Gamma_\infty = \rho_\infty(\pi_1(S))$ is a proper subgroup of $\widehat{\Gamma}$ which carried by an immersed surface $\varphi(S) \subset N_{\widehat{\Gamma}}$. Here the immersion $\varphi : S \to N_{\widehat{\Gamma}}$ up to homotopy is obtained from the identity map $S \to S \times \{-1\}$ by adding annulus which wraps around $c_i \times \{0\}$ for k_i-times for every i (see Figure 1). Then one can see that Γ_∞ is a b-group and that $\rho_\infty \in \partial^+ Q\mathcal{F}$. On the contrary, an ACM-sequence as in (2.1) is obtained from a Kleinian manifold $N_{\widehat{\Gamma}}$ and an immersion $\varphi : S \to N_{\widehat{\Gamma}}$ as above by simultaneous $(1, n)$-Dehn filling at every rank-two cusps of $N_{\widehat{\Gamma}}$.

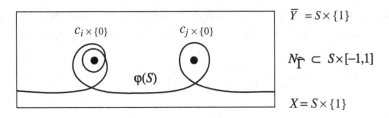

Figure 1: An immersion $\varphi : S \to N_{\widehat{\Gamma}}$.

Note that the bottom and top conformal structures of the ACM-sequence $\rho_n = B(\tau_c^n X, \tau_c^{2n} \bar{Y})$ converge to the same projective lamination $[c] \in \mathbb{P}\mathcal{M}L(S)$ in the Thurston compactifications of $T(S)$ and $T(\bar{S})$ but in different speeds. On the other hand, we remark that a sequence $\rho_n = B(\tau_c^n X, \tau_c^n \bar{Y})$ diverges and its top and bottom structures converge to $[c] \in \mathbb{P}\mathcal{M}L$ in the same speed. Moreover, Ohshika [Ohs98] showed that any sequence $\rho_n = B(X_n, \bar{Y}_n)$ diverge if the sequences X_n and $\bar{Y}_n$ converge to maximal and connected projective laminations $[\mu]$, $[\nu] \in \mathbb{P}\mathcal{M}L(S)$ with the same support.

2.6. Pull-backs of limit sets

Suppose that a sequence $\Sigma_n \in Q(S)$ of quasi-fuchsian projective structures converges to $\Sigma_\infty \in \overline{Q(S)}$. Here we explain our fundamental idea on how to know what component of $Q(S)$ contains Σ_n. Note that the sequence $\rho_{\Sigma_n} \in Q\mathcal{F}$ of their holonomy converges to $\rho_{\Sigma_\infty} \in \overline{Q\mathcal{F}}$ and set $\Gamma_n = \rho_{\Sigma_n}(\pi_1(S))$ and $\Gamma_\infty = \rho_{\Sigma_\infty}(\pi_1(S))$. In addition, we assume that Γ_n converges geometrically to a Kleinian group $\widehat{\Gamma}$, which contains the algebraic limit Γ_∞. Since the sequence Λ_{Γ_n} converges to $\Lambda_{\widehat{\Gamma}}$ in the sense of Hausdorff ([KT90]), one see that the sequence $\Lambda_{\Sigma_n} \subset \Sigma_n$ of pull-backs also converges to $\widehat{\Lambda}_{\Sigma_\infty} \subset \Sigma_\infty$ in the sense of Hausdorff, where $\Lambda_{\Sigma_n} = f_{\Sigma_n}^{-1}(\Lambda_{\Gamma_n})/\pi_1(\Sigma_n)$ and $\widehat{\Lambda}_{\Sigma_\infty} = f_{\Sigma_\infty}^{-1}(\Lambda_{\widehat{\Gamma}})/\pi_1(\Sigma_\infty)$ (see Lemma 3.3 in [Ito00]). Here the sets $\Lambda_{\Sigma_n} \subset \Sigma_n$ and $\widehat{\Lambda}_{\Sigma_\infty} \subset \Sigma_\infty$ are compared via K_n-quasi-isometry maps $q_n : \Sigma_\infty \to \Sigma_n$ between hyperbolic surfaces Σ_∞ and Σ_n such that $K_n \to 1$ as $n \to \infty$. Now recall that Σ_n is in Q_{λ_n} if and only if Λ_{Σ_n} is a realization of $2\lambda_n$ on Σ_n. Therefore, the shape of $\widehat{\Lambda}_{\Sigma_\infty}$ in Σ_∞ give us information on the shape of $\Lambda_{\Sigma_n} \subset \Sigma_n$ and hence on the lamination $\lambda_n \in \mathcal{M}L_{\mathbb{N}}$ such that $\Sigma_n \in Q_{\lambda_n}$.

3. Standardly convergent sequence in $Q\mathcal{F}$

In this section, we survey our results in [Itob], one of which states that the grafting map $\Psi_\lambda : Q\mathcal{F} \to P(S)$ takes every standardly convergent sequence to a convergent sequence.

3.1. Grafting for boundary groups

Let $\widehat{P}(S)$ denotes the one-point compactification $P(S) \cup \{\infty\}$ of $P(S)$. We will extend the grafting map $\Psi_\lambda : Q\mathcal{F} \to Q$ to $\Psi_\lambda : Q\mathcal{F} \sqcup \partial^\pm Q\mathcal{F} \to \widehat{P}(S)$. To this end, we first recall another (but equivalent) definition of the grafting operation which was introduced by Bromberg [Bro02] so that it also makes sense for elements of $\partial^- Q\mathcal{F}$.

Definition 3.1 (Grafting II). We adopt Notation 2.1. Here we further assume that c separates S into two surfaces S_1 and S_2 with boundaries. (The non-separating case is described precisely in [Bro02].) Accordingly, Σ and Σ^- decompose into $\Sigma - c^+ = \Sigma_1 \sqcup \Sigma_2$ and $\Sigma^- - c^- = \Sigma_1^- \sqcup \Sigma_2^-$, respectively. Let i denotes either 1 or 2 and let $\Delta_i \subset \Omega_\Gamma^-$ be the connected component of the inverse image of $\Sigma_i^- \subset \Sigma^-$ whose closure $\overline{\Delta}_i$ contains $\tilde{c}^-$. Then the stabilizer subgroup $\Gamma_i = \mathrm{Stab}_\Gamma(\Delta_i)$ of $\Gamma \cong \pi_1(S)$ is identified with $\pi_1(S_i)$. Since Γ_i is a purely loxodromic free group with non-empty region of discontinuity, Maskit's result [Mas67] implies that Γ_i is a Schottky group. Note that the conformal boundary $\Omega_{\Gamma_i}/\Gamma_i$ of $M_{\Gamma_i} = \mathbb{H}^3/\Gamma_i$ with natural projective structure is containing both projective surfaces Σ_i and Σ_i^-. Then the *grafting* $\mathrm{Gr}_c(\Sigma)$ is obtained from projective surfaces $\Omega_{\Gamma_1}/\Gamma_1 - \Sigma_1^-$ and $\Omega_{\Gamma_2}/\Gamma_2 - \Sigma_2^-$ by gluing their boundaries without twisting (see Figure 2).

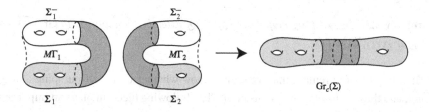

Figure 2: The grafting $\mathrm{Gr}_c(\Sigma)$ of Σ along c.

Observe that Definition 2.2 works well even for $\rho \in \partial^+ Q\mathcal{F}$ whenever γ is loxodromic, because there still exists a $\langle\gamma\rangle$-invariant simple arc $\tilde{c}^+$ in non-degenerate component Ω_Γ^+, for which $(\widehat{\mathbb{C}} - \tilde{c}^+)/\langle\gamma\rangle$ is still an annulus. On the other hand, Definition 3.1 works well for $\rho \in \partial^- Q\mathcal{F}$ whenever every connected component of the parabolic locus $\mathrm{para}(\rho)$ of ρ intersects c essentially. In fact, in this case, Γ_1 and Γ_2

in Definition 3.1 are still Schottky groups, γ is still loxodromic, and there still exists a $\langle\gamma\rangle$-invariant simple arc $\tilde{c}^-$ in the non-degenerate component Ω_Γ^-. For general $\lambda \in \mathcal{ML}_\mathbb{N}$, we also obtain the grafting $\Psi_\lambda(\rho) \in P(S)$ of ρ along λ if the pair (λ, ρ) is *admissible*, or satisfies the following condition:

- $\rho \in \partial^+ Q\mathcal{F}$ and $\text{para}(\rho)$ and λ have no parallel component in common, or

- $\rho \in \partial^- Q\mathcal{F}$ and every component of $\text{para}(\rho)$ intersects λ essentially.

Otherwise, we set $\Psi_\lambda(\rho) = \infty \in \widehat{P}(S)$.

3.2. Continuity of grafting maps

One may expect that the extended grafting map $\Psi_\lambda : Q\mathcal{F} \sqcup \partial^\pm Q\mathcal{F} \to \widehat{P}(S)$ is also continuous at $\partial^\pm Q\mathcal{F}$, but this is not the case for every $\lambda \in \mathcal{ML}_\mathbb{N}$; see Theorem 5.1. On the contrary, we have the following theorem, which implies that only exotically convergent sequences cause the non-continuity of the extended grafting maps.

Theorem 3.2 ([Itob]). *Let $\rho_n \in Q\mathcal{F}$ be a sequence converging standardly to $\rho_\infty \in \partial^\pm Q\mathcal{F}$. Then the sequence $\Psi_\lambda(\rho_n)$ also converges to $\Psi_\lambda(\rho_\infty)$ in $\widehat{P}(S)$ for every $\lambda \in \mathcal{ML}_\mathbb{N}$.*

Here and throughout, we let $\overline{Q_\lambda}$ denote the closure of the component Q_λ of $Q(S)$ in $P(S)$, *not* in $\widehat{P}(S)$, and set $\partial Q_\lambda = \overline{Q_\lambda} - Q_\lambda$. Then the above theorem tells us that $\Psi_\lambda(\rho)$ is surely contained in ∂Q_λ if the pair of $\lambda \in \mathcal{ML}_\mathbb{N}$ and $\rho \in \partial^\pm Q\mathcal{F}$ is admissible. Recall that, as we observed in §2.4, a sequence $\Sigma_n \in Q_0$ converges to $\Sigma_\infty \in \partial Q_0$ if and only if $\rho_{\Sigma_n} \in Q\mathcal{F}$ converges standardly to $\rho_{\Sigma_\infty} \in \partial^+ Q\mathcal{F}$. Thus we obtain the following:

Corollary 3.3. *The grafting map $\text{Gr}_\lambda : Q_0 \to Q_\lambda$ extends continuously to $\text{Gr}_\lambda : \overline{Q_0} \to \widehat{P}(S)$ for every $\lambda \in \mathcal{ML}_\mathbb{N}$.*

It is important to remark that we do not know whether Q_0 is self-bumping at ∂Q_0 or not, and thus we have to avoid this point. The following theorem plays an important roll in the proof of Theorem 3.2:

Theorem 3.4. *For a given $B(X, \bar{Y}) \in Q\mathcal{F}$, set $\mathcal{B} = B_X \cup B_{\bar{Y}}$. Suppose that $\{\lambda_n\}$ is a sequence of distinct elements of $\mathcal{ML}_\mathbb{N}$. Then the sequence $\{\pi \circ \Psi_{\lambda_n}(\mathcal{B})\}$ eventually escapes any compact subset K of $T(S)$; that is, $\pi \circ \Psi_{\lambda_n}(\mathcal{B}) \cap K = \emptyset$ for all large enough n.*

We now outline the proof of Theorem 3.2. We only consider the case where the pair (λ, ρ_∞) is admissible and set $\Sigma_\infty = \Psi_\lambda(\rho_\infty) \in P(S)$. Let $\Phi : U \to P(S)$, $\rho_\infty \mapsto \Sigma_\infty$.

be a univalent local branch of hol^{-1} defined on a neighborhood U of ρ_∞. Then the sequence $\Sigma_n = \Phi(\rho_n)$ converges to Σ_∞. If $\Sigma_n \in Q_\lambda$ for all large enough n, then $\Sigma_n = \Psi_\lambda(\rho_n)$, and then we obtain the desired convergence $\Sigma_n = \Psi_\lambda(\rho_n) \to \Sigma_\infty = \Psi_\lambda(\rho_\infty)$. We show that $\Sigma_n \in Q_\lambda$ by using the idea in §2.6; that is, we show that the pull backs $\Lambda_{\Sigma_n} \subset \Sigma_n$ of the limit sets Λ_{Γ_n} are realizations of 2λ. We assume that the sequence $\Gamma_n = \rho_n(\pi_1(S))$ converges geometrically to a Kleinian group $\widehat{\Gamma}$, which contains the algebraic limit $\Gamma_\infty = \rho_\infty(\pi_1(S))$. Then $\Lambda_{\Sigma_n} \subset \Sigma_n$ converges to the pull-back $\widehat{\Lambda}_{\Sigma_\infty} \subset \Sigma_\infty$ of $\Lambda_{\widehat{\Gamma}}$. Although it is difficult to understand the shape of $\widehat{\Lambda}_{\Sigma_\infty}$, we know a rough sketch of the subset $\Lambda_{\Sigma_\infty} \subset \widehat{\Lambda}_{\Sigma_\infty}$ in relation to λ from the definition of the grafting $\Sigma_\infty = \Psi_\lambda(\rho_\infty)$. Moreover, we can see that each connected component of $\widehat{\Lambda}_{\Sigma_\infty}$ contains that of Λ_{Σ_∞} by using [ACCS96, Lemma 2.4]. By combining the above observations, we see that Λ_{Σ_n} are realizations of 2λ in Σ_n for all large enough n. At this stage, we make use of Theorem 3.4 essentially, which asserts that the sequence Σ_n is contained in a finite union of components of $Q(S)$.

As a consequence of Theorems 3.2 and 3.4, Goldman's grafting theorem for quasi-fuchsian groups (Theorem 2.3) extends to all boundary b-groups, which is conjectured by Bromberg in [Bro02].

Theorem 3.5 ([Itob]). *For $\rho \in \partial^\pm QF$, we have*

$$\text{hol}^{-1}(\rho) = \{\Psi_\lambda(\rho) \,|\, \lambda \in \mathcal{ML}_\mathbb{N}, \Psi_\lambda(\rho) \neq \infty\}.$$

4. Exotically convergent sequence in QF

In this section, we shall show that ACM-sequences cause the bumping and self-bumping of components of $Q(S)$. Throughout this section, Figures 3 and 4 should be helpful for the reader to understand the arguments.

4.1. Exotic components bump to the standard one

We first show the following:

Theorem 4.1 ([Ito00]). *For any non-zero $\lambda \in \mathcal{ML}_\mathbb{N}$, we have $\overline{Q_0} \cap \overline{Q_\lambda} \neq \emptyset$.*

Let $\{\rho_n\} \subset QF$ be the ACM-sequence for λ as in (2.1), which converges exotically to some $\rho_\infty \in \partial^+ QF = \text{hol}(\partial Q_0)$. Let Σ_∞ be the unique point in ∂Q_0 with $\text{hol}(\Sigma_\infty) = \rho_\infty$ and let $\Phi : U \to P(S), \rho_\infty \mapsto \Sigma_\infty$ be a univalent local branch of hol^{-1} defined on a neighborhood U of ρ_∞. Then the sequence $\Sigma_n = \Phi(\rho_n)$ converges to $\Sigma_\infty = \Phi(\rho_\infty)$. Since the convergence $\rho_n \to \rho_\infty$ is exotic, we see from Table 1 that Σ_n are exotic for all $|n| \gg 0$ (see also Theorem A.2 in [McM98]). Moreover, we see that $\Sigma_n \in Q_\lambda$ for all $|n| \gg 0$ by using the idea in §2.6. In fact, one can observe that

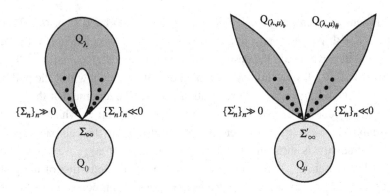

Figure 3: Sequences $\{\Sigma_n\}_{n\in\mathbb{Z}}$ and $\{\Sigma'_n\}_{n\in\mathbb{Z}}$.

$\widehat{\Lambda}_{\Sigma_\infty} \subset \Sigma_\infty$ is a "decorated realization" of 2λ, that is, $\widehat{\Lambda}_{\Sigma_\infty}$ contains a realization of 2λ and is contained in a regular neighborhood of a realization of 2λ (see the left side of Figure 4). Since $\Lambda_{\Sigma_n} \subset \Sigma_n$ converge to $\widehat{\Lambda}_{\Sigma_\infty} \subset \Sigma_\infty$, the sets Λ_{Σ_n} turn out to be realizations of 2λ and thus $\Sigma_n \in Q_\lambda$ for all $|n| \gg 0$. This implies that $\overline{Q_0} \cap \overline{Q_\lambda} \neq \emptyset$. We also remark that $\Sigma_n = \Psi_\lambda(\rho_n)$ hold for all $|n| \gg 0$.

4.2. Simultaneous bumping

We extend Theorem 4.1 to the following:

Theorem 4.2 ([Ito00]). *Let $\{\lambda_i\}_{i=1}^m$ be a finite subset of $\mathcal{ML}_\mathbb{N} - \{0\}$ such that $i(\lambda_i, \lambda_j) = 0$ for every $1 \leq i < j \leq m$. Then we have $\overline{Q_0} \cap \overline{Q_{\lambda_1}} \cap \cdots \cap \overline{Q_{\lambda_m}} \neq \emptyset$.*

In fact, we can construct ACM-sequences

$$\rho_n^{(i)} = B(\tau_{\lambda_i}^n X_i, (\tau_{\lambda_i} \circ \tau_{\lambda_i})^n \bar{Y}_i)$$

for λ_i for each $1 \leq i \leq m$ so that all of these sequences $\rho_n^{(i)}$ $(i = 1,\ldots,m)$ have the same algebraic limit $\rho_\infty \in \partial^+ Q\mathcal{F}$; see §5 in [Ito00]. Let $\Phi : U \to P(S)$ be a univalent local branch of hol^{-1} such that $\Phi(\rho_\infty) \in \partial Q_0$ as in §4.1. Then for each i, we have a convergent sequence $\Phi(\rho_n^{(i)}) \to \Phi(\rho_\infty)$, which turns out to be $\Phi(\rho_n^{(i)}) \in Q_{\lambda_i}$ for all $|n| \gg 0$ by the same argument as in §4.1. Thus we have $\overline{Q_0} \cap \left(\bigcap_{i=1}^m \overline{Q_{\lambda_i}}\right) \neq \emptyset$.

4.3. Self-bumping of exotic components

Here we outline the proof of the following:

Theorem 4.3 ([Itoa]). *For any non-zero $\lambda \in \mathcal{ML}_\mathbb{N}$, Q_λ self-bumps.*

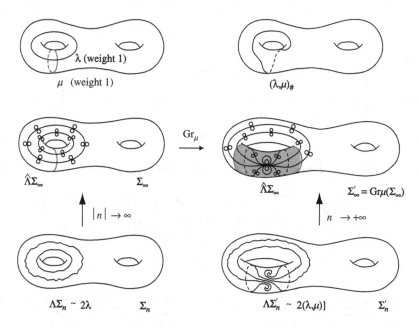

Figure 4: Schematic figure explaining the proof of Theorems 4.1 and 4.2.

Let $\{\rho_n\}$ be the ACM-sequence for λ as in (2.1). Then we actually show that the subsequences $\{\rho_n\}_{n\gg0}$, $\{\rho_n\}_{n\ll0}$ of $\{\rho_n\}$ are contained in distinct components of $U \cap \mathcal{QF}$ for any sufficiently small neighborhood U of ρ_∞. This is a consequence of the following fact: in the same notation as in §2.6, both sequences $\{\Lambda_{\Gamma_n}\}_{n\gg0}$, $\{\Lambda_{\Gamma_n}\}_{n\ll0}$ converges to $\Lambda_{\widehat{\Gamma}}$ in the sense of Hausdorff but Λ_{Γ_n} $(n \gg 0)$ and Λ_{Γ_n} $(n \ll 0)$ are spiraling in opposite directions at each fixed point of rank-two parabolic subgroups of $\widehat{\Gamma}$.

Before outlining the proof, we recall the definition of operations $(\cdot,\cdot)_\sharp$, $(\cdot,\cdot)_\flat$: $\mathcal{ML}_{\mathbb{N}} \times \mathcal{ML}_{\mathbb{N}} \to \mathcal{ML}_{\mathbb{N}}$, which is closely observed in [Luo01]. For any two elements $\lambda, \mu \in \mathcal{ML}_{\mathbb{N}}$, new elements $(\lambda,\mu)_\sharp$ and $(\lambda,\mu)_\flat$ in $\mathcal{ML}_{\mathbb{N}}$ are obtained by taking realizations $\widehat{\lambda}, \widehat{\mu}$ of λ, μ so that the geometric intersection number of $\widehat{\lambda}$ and $\widehat{\mu}$ is minimal, and drawing "zigzag" paths on $\widehat{\lambda} \cup \widehat{\mu}$ under the rules in Figure 5 (see also Figure 6). Now let $\lambda, \mu \in \mathcal{ML}_{\mathbb{N}}$. We collect here some of basic properties of these operations:

(i) $(\lambda,\mu)_\sharp = (\mu,\lambda)_\flat$.

(ii) $(\lambda,\mu)_\sharp \neq (\lambda,\mu)_\flat$ if and only if $i(\lambda,\mu) \neq 0$. If $i(\lambda,\mu) = 0$ then $(\lambda,\mu)_\sharp = (\lambda,\mu)_\flat = \lambda + \mu$.

(iii) Assume that every components of μ intersects λ. Then $((\lambda,\mu)_\sharp,\mu)_\flat = ((\lambda,\mu)_\flat,\mu)_\sharp = \lambda$, $((\lambda,\mu)_\sharp,\mu)_\sharp = (\lambda,2\mu)_\sharp$ and $((\lambda,\mu)_\flat,\mu)_\flat = (\lambda,2\mu)_\flat$.

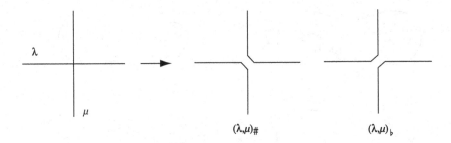

Figure 5: Rules to construct $(\lambda,\mu)_\sharp$ and $(\lambda,\mu)_\flat$.

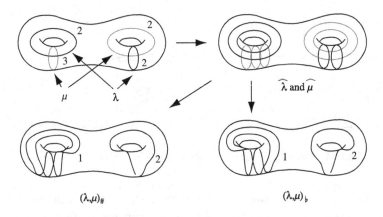

Figure 6: Examples of $(\lambda,\mu)_\sharp$ and $(\lambda,\mu)_\flat$.

We now go back to the proof of Theorem 4.3. Let ρ_n be the ACM-sequence for λ as in (2.1). We now take another non-zero element $\mu \in \mathcal{ML}_\mathbb{N}$ such that λ and μ have no parallel component in common. Then the pair (μ,ρ_∞) is admissible, and thus we have the grafting $\Psi_\mu(\rho_\infty) \in P(S)$ of ρ_∞ along μ. Let $\Phi' : U \to P(S)$, $\rho_\infty \mapsto \Psi_\mu(\rho_\infty)$ be a univalent local branch of hol^{-1} defined on a neighborhood U of ρ_∞. Then the sequence $\Sigma'_n = \Phi'(\rho_n)$ converges to $\Sigma'_\infty = \Psi_\mu(\rho_\infty)$ as $|n| \to \infty$. One of the crucial observations in [Itoa] is the following:

Proposition 4.4. $\Sigma'_n \in Q_{(\lambda,\mu)_\sharp}$ *for all* $n \gg 0$ *and* $\Sigma'_n \in Q_{(\lambda,\mu)_\flat}$ *for all* $n \ll 0$.

Skech of proof. If $i(\lambda,\mu) = 0$, it is easy to see that the set $\widehat{\Lambda}_{\Sigma'_\infty} \subset \Sigma'_\infty$ is a decorated realization of $\lambda + \mu = (\lambda,\mu)_\sharp = (\lambda,\mu)_\flat$ and that $\Sigma'_n \in Q_{\lambda+\mu}$ for all $|n| \gg 0$. Suppose that $i(\lambda,\mu) \neq 0$. Then $\widehat{\Lambda}_{\Sigma'_\infty} \subset \Sigma'_\infty$ is a "decorated train-track" whose switches in $\widehat{\Lambda}_{\Sigma'_\infty}$ are pull-backs of rank-two parabolic fixed points in $\Lambda_{\widehat{\Gamma}}$ (see the right side of Figure 4). Since $\Lambda_{\Sigma'_n} \to \widehat{\Lambda}_{\Sigma'_\infty}$ as $|n| \to \infty$ and since $\{\Lambda_{\Sigma'_n}\}_{n\gg 0}$ and $\{\Lambda_{\Sigma'_n}\}_{n\ll 0}$ are spiraling in opposite directions at each switch of $\widehat{\Lambda}_{\Sigma'_\infty} \subset \Sigma'_\infty$, we see that the sets $\Lambda_{\Sigma'_n} \subset \Sigma'_n$ are realizations of $2(\lambda,\mu)_\sharp$ if $n \gg 0$ and of $2(\lambda,\mu)_\flat$ if $n \ll 0$. $\qquad\square$

Therefore, if we further assume that $i(\lambda,\mu) \neq 0$, the sequences $\{\Sigma'_n\}_{n\gg0}$, $\{\Sigma'_n\}_{n\ll0}$ are contained in distinct components of $Q(S)$. This implies that $\{\rho_n\}_{n\gg0}$, $\{\rho_n\}_{n\ll0}$ are contained in distinct components of $U \cap Q\mathcal{F}$, and hence that the sequences $\{\Sigma_n = \Phi(\rho_n)\}_{n\gg0}$, $\{\Sigma_n = \Phi(\rho_n)\}_{n\ll0}$, obtained in §4.1, are contained in distinct components of $\Phi(U) \cap Q_\lambda$. Therefore Q_λ self-bumps at $\rho_\infty \in \overline{Q_0} \cap \overline{Q_\lambda}$. We also remark that we have $\Sigma'_n = \Psi_{(\lambda,\mu)_\sharp}(\rho_n)$ for all $n \gg 0$ and $\Sigma'_n = \Psi_{(\lambda,\mu)_\flat}(\rho_n)$ for all $n \ll 0$.

4.4. Bumping of any two components

As a consequence of the above arguments in this section, we obtain the following:

Theorem 4.5. *For any* $\lambda, \mu \in \mathcal{ML}_\mathbb{N}$, *we have* $\overline{Q_\lambda} \cap \overline{Q_\mu} \neq \emptyset$.

For the convenience of the reader, we give here the same proof as in [Itoa].

Proof of Theorem 4.5. If $i(\lambda,\mu) = 0$, we obtain the result from Theorem 4.2. Hence, we assume that $i(\lambda,\mu) \neq 0$. We decompose μ into $\mu = \mu' + \mu''$ so that $\mu', \mu'' \in \mathcal{ML}_\mathbb{N}$ and that $i(\lambda,\mu) = i(\lambda,\mu')$. We first consider the case where $\mu'' = 0$. As observed in §4.1, there exists an element $\rho_\infty \in \overline{Q_0} \cap \overline{Q_{(\lambda,\mu)_\flat}}$ which is a limit of an ACM-sequence for $(\lambda,\mu)_\flat$. Then the same argument as in §4.3 reveals that $\Psi_\mu(\rho_\infty) \in \overline{Q_\mu} \cap \overline{Q_\lambda}$ since $\lambda = ((\lambda,\mu)_\flat,\mu)_\sharp$. We next consider the case where $\mu'' \neq 0$. Since $i(\lambda,\mu') = 0$ and $i(\mu',\mu'') = 0$, we have $i((\lambda,\mu')_\flat,\mu'') = 0$. Then as observed in §4.2, there exists a limit $\rho_\infty \in \overline{Q_0} \cap \overline{Q_{(\lambda,\mu')_\flat}} \cap \overline{Q_{\mu''}}$ of ACM-sequences. Then we have $\Psi_{\mu'}(\rho_\infty) \in \overline{Q_\lambda} \cap \overline{Q_\mu}$ since $\lambda = ((\lambda,\mu')_\flat,\mu')_\sharp$ and since $\mu = \mu' + \mu''$. $\qquad\square$

5. Additional observations

Throughout this section, we suppose that $\{\rho_n\}$ is the ACM-sequence for λ as in (2.1). We have observed that $\lim_{n\to\pm\infty} \Psi_\lambda(\rho_n) = \Psi_0(\rho_\infty)$ in §4.1 and that

$$\lim_{n\to+\infty} \Psi_{(\lambda,\mu)_\sharp}(\rho_n) = \lim_{n\to-\infty} \Psi_{(\lambda,\mu)_\flat}(\rho_n) = \Psi_\mu(\rho_\infty) \tag{5.1}$$

in §4.3. Now let ν be an element of $\mathcal{ML}_\mathbb{N}$ such that $i(\nu,\lambda) \neq 0$ and that ν and λ have no parallel component in common. Then we have

$$\lim_{n\to+\infty} \Psi_\nu(\rho_n) = \Psi_{(\nu,\lambda)_\sharp}(\rho_\infty), \quad \lim_{n\to-\infty} \Psi_\nu(\rho_n) = \Psi_{(\nu,\lambda)_\flat}(\rho_\infty) \tag{5.2}$$

by substituting $\mu = (\nu,\lambda)_\sharp$ or $\mu = (\nu,\lambda)_\flat$ in (5.1). Since $i(\nu,\lambda) \neq 0$, we have $\Psi_{(\nu,\lambda)_\sharp}(\rho_\infty) \neq \Psi_{(\nu,\lambda)_\flat}(\rho_\infty)$. By choosing λ suitably for every ν, we obtain the following:

Theorem 5.1. *The grafting map* $\Psi_\nu : Q\mathcal{F} \to P(S)$ *does not extend continuously to* $\partial Q\mathcal{F}$ *for every* $\nu \in \mathcal{ML}_\mathbb{N}$.

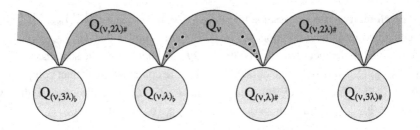

Figure 7: Analytic continuation of Ψ_ν along α.

We now observe some properties of analytic continuations of local branches of hol^{-1}. Suppose that U is a sufficiently small neighborhood of $\rho_\infty \in \partial^+ Q\mathcal{F}$. Let $\alpha : S^1 = \mathbb{R} \cup \{\infty\} \to Q\mathcal{F} \cup U$ be a continuous map such that $\alpha(n) = \rho_n$ for all $n \in \mathbb{Z}$ and that $\alpha(\infty) = \rho_\infty$. (We do not know whether we can choose α so that $\alpha(S^1) \subset \overline{Q\mathcal{F}}$.) The closed curve $\alpha(S^1)$ in $R(S)$ is also denote by α. Since $\Psi_\lambda(\alpha(n)) \to \Psi_0(\rho_\infty)$ as $|n| \to \infty$, the branch $\Psi_\lambda : Q_\lambda \to P(S)$ of hol^{-1} is continued analytically to a univalent local branch $\Phi : U \to P(S)$, $\rho_\infty \mapsto \Psi_0(\rho_\infty)$ along both the paths $\alpha(\mathbb{R}_{\geq 0})$ and $\alpha(\mathbb{R}_{\leq 0})$, and hence there exists a lift $\tilde{\alpha} \subset Q_\lambda \cup \Phi(U)$ of α for which $\text{hol}|_{\tilde{\alpha}} : \tilde{\alpha} \to \alpha$ is one-to-one. Note that the above argument does not imply that the map Ψ_λ extends to a univalent local branch $Q\mathcal{F} \cup U \to P(S)$ of hol^{-1}. In fact, if $\eta_n \in Q\mathcal{F} \cap U$ is a sequence converging standardly to ρ_∞, then $\Psi_\lambda(\eta_n)$ converges to ∞ in $\widehat{P}(S)$, not to $\Psi_0(\rho_\infty) \in P(S)$. Since $P(S)$ is contractible, $\tilde{\alpha}$ is contractible in $P(S)$, and hence α is contractible in $R(S)$. This implies that the bumping at $\rho_\infty \in \partial Q\mathcal{F}$ of the two arms of $Q\mathcal{F}$ containing $\{\rho_n\}_{n \gg 0}$ and $\{\rho_n\}_{n \ll 0}$ yields no non-trivial element of $\pi_1(R(S))$. On the other hand, let take $\nu \in \mathcal{ML}_\mathbb{N}$ as above and let us consider the analytic continuation of Ψ_ν. In this case, since $\lim_{n \to +\infty} \Psi_\nu(\alpha(n)) \neq \lim_{n \to -\infty} \Psi_\nu(\alpha(n))$ from (5.2), the analytic continuation of the local branch Ψ_ν along $\alpha \subset R(S)$ yields succeeding sequence of local branches

$$\ldots, \Psi_{(\nu,4\lambda)_\flat}, \Psi_{(\nu,2\lambda)_\flat}, \Psi_\nu, \Psi_{(\nu,2\lambda)_\sharp}, \Psi_{(\nu,4\lambda)_\sharp}, \ldots$$

(see Figure 7). Thus we obtain a lift $\tilde{\alpha}'$ of α in $P(S)$ for which $\text{hol}|_{\tilde{\alpha}'} : \tilde{\alpha}' \to \alpha$ is infinite-to-one. We sum up the arguments in this section:

Theorem 5.2. *There exists a contractible closed curve α in $R(S)$ whose pre-image $\text{hol}^{-1}(\alpha) \subset P(S)$ has connected components $\tilde{\alpha}, \tilde{\alpha}'$ such that the map $\text{hol}|_{\tilde{\alpha}} : \tilde{\alpha} \to \alpha$ is one-to-one and the map $\text{hol}|_{\tilde{\alpha}'} : \tilde{\alpha}' \to \alpha$ is infinite-to-one. Especially, the lift $\widetilde{\text{hol}} : P(S) \to \widetilde{R(S)}$ of the map $\text{hol} : P(S) \to R(S)$ to the universal cover is not an embedding.*

References

[AC96]   J. W. Anderson & R. D. Canary (1996). Algebraic limits of Kleinian groups which rearrange the pages of a book. *Invent. Math.* **126** (2), 205–214.

[ACCS96] J. W. Anderson *et al.* (1996). Free Kleinian groups and volumes of hyperbolic 3-manifolds. *J. Differential Geom.* **43** (4), 738–782.

[ACM00] J. W. Anderson, R. D. Canary & D. McCullough (2000). The topology of deformation spaces of Kleinian groups. *Ann. of Math. (2)* **152** (3), 693–741.

[Ber60]   L. Bers (1960). Simultaneous uniformization. *Bull. Amer. Math. Soc.* **66**, 94–97.

[BH01]   K. Bromberg & J. Holt (2001). Self-bumping of deformation spaces of hyperbolic 3-manifolds. *J. Differential Geom.* **57** (1), 47–65.

[Bro97]   J. F. Brock (1997). Iteration of mapping classes on a Bers slice: examples of algebraic and geometric limits of hyperbolic 3-manifolds. In *Lipa's Legacy (New York, 1995), Contemp. Math.*, volume 211, pp. 81–106. Amer. Math. Soc., Providence, RI.

[Bro02]   K. Bromberg (2002). Projective structures with degenerate holonomy and the Bers density conjecture. Preprint, math.GT/0211402.

[Can04]   R. D. Canary (2004). Pushing the boundary. In W. Abikoff & A. Haas (eds.), *In the Tradition of Ahlfors and Bers, III, Contemp. Math.*, volume 355, pp. 109–121. Amer. Math. Soc.

[Ear81]   C. J. Earle (1981). On variation of projective structures. In *Riemann Surfaces and Related Topics: Proceedings of the 1978 Stony Brook Conference (State Univ. New York, Stony Brook, N.Y., 1978), Ann. of Math. Stud.*, volume 97, pp. 87–99. Princeton Univ. Press, Princeton, N.J.

[Gol87]   W. M. Goldman (1987). Projective structures with Fuchsian holonomy. *J. Differential Geom.* **25** (3), 297–326.

[Hej75]   D. A. Hejhal (1975). Monodromy groups and linearly polymorphic functions. *Acta Math.* **135** (1), 1–55.

[Hub81]   J. H. Hubbard (1981). The monodromy of projective structures. In *Riemann Surfaces and Related Topics: Proceedings of the 1978 Stony Brook Conference (State Univ. New York, Stony Brook, N.Y., 1978), Ann. of Math. Stud.*, volume 97, pp. 257–275. Princeton Univ. Press, Princeton, N.J.

[Itoa]   K. Ito. Exotic projective structures and quasi-fuchsian spaces II. Preprint. www.math.nagoya-u.ac.jp/~itoken/index.html.

[Itob]   K. Ito. On continuous extension of grafting maps. Preprint, math.GT/ 0411133.

[Ito00]   K. Ito (2000). Exotic projective structures and quasi-Fuchsian space. *Duke Math. J.* **105** (2), 185–209.

[Kra71]   I. Kra (1971). A generalization of a theorem of Poincaré. *Proc. Amer. Math. Soc.* **27**, 299–302.

[KT90]   S. P. Kerckhoff & W. P. Thurston (1990). Noncontinuity of the action of the modular group at Bers' boundary of Teichmüller space. *Invent. Math.* **100** (1), 25–47.

[Luo01]   F. Luo (2001). Some applications of a multiplicative structure on simple loops in surfaces. In *Knots, Braids, and Mapping Class Groups – papers dedicated to Joan S. Birman (New York, 1998), AMS/IP Stud. Adv. Math.*, volume 24, pp. 123–129. Amer. Math. Soc., Providence, RI.

[Mas67]   B. Maskit (1967). A characterization of Schottky groups. *J. Analyse Math.* **19**, 227–230.

[McM98]   C. T. McMullen (1998). Complex earthquakes and Teichmüller theory. *J. Amer. Math. Soc.* **11** (2), 283–320.

[MT98]   K. Matsuzaki & M. Taniguchi (1998). *Hyperbolic Manifolds and Kleinian Groups.* Oxford Mathematical Monographs. The Clarendon Press Oxford University Press, New York.

[Ohs98]   K. Ohshika (1998). Divergent sequences of Kleinian groups. In *The Epstein Birthday Schrift, Geom. Topol. Monogr.*, volume 1, pp. 419–450 (electronic). Geom. Topol. Publ., Coventry.

Kentaro Ito

Graduate School of Mathematics
Nagoya University
Nagoya 464-8602
Japan

itoken@math.nagoya-u.ac.jp

AMS Classification: 30F40, 57M50

Keywords: Kleinian groups, quasi-fuchsian space, projective structures, grafting

Spaces of Kleinian Groups
Lond. Math. Soc. Lec. Notes **329**, 375–390

Cambridge University Press
Y. Minsky, M. Sakuma & C. Series (Eds.)

Computer experiments on the discreteness locus in projective structures

Yasushi Yamashita

Abstract

This article has two purposes. First, we give a brief exposition of a method for producing computer pictures of Bers embeddings of the Teichmüller space of once punctured tori. (See [KSWY].) Then we describe how the work of Bowditch [Bow98] can be used to improve the algorithm mentioned above. Our second purpose is to present several pictures produced using the algorithm and discuss the related topics.

1. Introduction

This article has two purposes. First, we give a brief exposition of a method for producing computer pictures of Bers embeddings of the Teichmüller space of once punctured tori. For the full details of the algorithm, we refer to [KSWY]. Then we describe how the work of Bowditch [Bow98] can be used to improve the algorithm mentioned above. Our second purpose is to present several pictures produced using the algorithm and discuss the related topics.

Let us begin by outlining the algorithm for drawing a Bers slice. Let Γ be a Fuchsian group acting on the unit disk $\mathbb{D}$ uniformizing a once-punctured torus T, and $B_2(\mathbb{D}, \Gamma)$ the complex Banach space of holomorphic quadratic differentials for Γ on $\mathbb{D}$ with finite norm. It is well known that the complex dimension of $B_2(\mathbb{D}, \Gamma)$ is one and the Teichmüller space $\mathcal{T}(\Gamma)$ of Γ can be realized as a bounded contractible open subset in $B_2(\mathbb{D}, \Gamma)$ through the Bers embedding. In [KSWY], we presented a method for producing computer pictures of $\mathcal{T}(\Gamma)$. The algorithm consists of two steps. First, for each point in $B_2(\mathbb{D}, \Gamma) \simeq \mathbb{C}$, we compute the corresponding holonomy representation of $\pi_1(T)$ by numerical integration of the Schwarzian equation. Second, we decide whether the image of the representation is discrete or not and plot the discreteness loci. One component of this locus is $\mathcal{T}(\Gamma)$.

The second step of the calculation is based on Jørgensen's theory of punctured torus groups [Jør03]. It describes the Ford domain for quasifuchsian groups which the algorithm attempts to find. In our discreteness algorithm, we have adopted a heuristic method in one step. But the heuristic is not completely reliable and sometimes our software is unable to determine whether or not the holonomy group is discrete even though the corresponding point seems to be in the discreteness locus. ("Small red island" in the postcard of Bers slice sold in the workshop on "Spaces of Kleinian Groups" is an example.) In this paper, we modify the algorithm so as to use aspects of Bowditch's theory of punctured torus groups [Bow98]. With the modification, if a given holonomy representation is discrete, we can detect its discreteness by the modified algorithm. See Proposition 4.4.

The second purpose of this article is to present some pictures that were shown in the Workshop on Spaces of Kleinian Groups, which took place in August 2003, at the Isaac Newton Institute in Cambridge. These include some images of three-dimensional objects. We hope that these pictures give us some new insight for the shape of the Bers embedding of Teichmüller space.

The software made by the author, which was used to produce the images, is available from the author upon request.

This paper is organized as follows. Section 2 is dedicated to the background material, especially the Markoff maps and type preserving representations. In section 3, we give a brief exposition of our old method for producing pictures of Bers embeddings and point out the problem in the original algorithm. In section 4, we review Bowditch's theory and describe our modified algorithm. In section 5, we present the results of our computer experiments.

I thank the referee for his or her careful reading of the manuscript and a number of very helpful comments.

2. Markoff maps of type preserving representations

In this section, we introduce the definitions and basic facts about Markoff maps and type preserving representations for the punctured torus. We begin with Markoff map.

Let T be a once punctured torus. We fix standard generators α and β of $\pi_1(T)$. The commutator $[\alpha, \beta] = \alpha\beta\alpha^{-1}\beta^{-1}$ represents a loop around the puncture.

Recall that a *slope* in T is the isotopy class of an essential and nonperipheral (i.e. it bounds neither a disk nor a once punctured disk.) simple closed curve on T. We identify T with the quotient space $(\mathbb{R}^2 - \mathbb{Z}^2)/\mathbb{Z}^2$. Then the slopes in T are in one-to-one correspondence with $\mathbb{Q} \cup \{1/0(= \infty)\}$. To fix our notation, we choose α and β so that the slope of α and β are $1/0$ and $0/1$ respectively. For a slope $q \in \mathbb{Q} \cup \{1/0\}$, set $S_q = \{g \in \pi_1(T) \mid \text{slope of } g = q\}$. Note that $\alpha \in S_{1/0}, \beta \in S_{0/1}$ and $\alpha\beta \in S_{1/1}$. We identify the set of slopes as a subset of $\partial\mathbb{H}^2$. Two rational numbers p/q and r/s are *Farey neighbors* if $|ps - qr| = 1$. By joining all pairs of Farey neighbors by geodesics in $\mathbb{H}^2$, we get the *Farey tessellation* of $\mathbb{H}^2$ by ideal triangles which will be called *Farey triangles.* Note that the slopes of α, β and $\alpha\beta$ form an Farey triangle. By taking the dual graph of this triangulation, we have a trivalent graph Σ properly embedded in $\mathbb{H}^2$.

By a *complementary region* of Σ, we mean the closure of a connected component of the complement. (These notations are taken directly from [Bow98].) These look like polygonal horodisks, and the boundary of each one consists of a bi-infinite sequence of Dehn twists around a fixed curve. Note that each complementary region corresponds to a slope in T, so we will use rational number to denote a complementary region. Each edge e of Σ meets four complementary regions X, Y, Z, W in such a way that $e = X \cap Y$, and $e \cap Z$ and $e \cap W$ are vertices of Σ. We denote this edge e by $(X, Y; Z, W)$. We denote by $V(\Sigma)$, $E(\Sigma)$ and Ω the set of vertices, edges and complementary regions of Σ respectively.

Definition 2.1. A triple of complex numbers (x, y, z) is called a *Markoff triple* if they satisfy the following equation:

$$x^2 + y^2 + z^2 = xyz. \tag{2.1}$$

A map $\phi : \Omega \to \mathbb{C}$ is called a *Markoff map* if

(i) for all vertices $v \in V(\Sigma)$, the triple $(\phi(X), \phi(Y), \phi(Z))$ is a Markoff triple, where $X, Y, Z \in \Omega$ are the three complementary regions meeting v.

(ii) If $e \in E(\Sigma)$, we have

$$xy = w + z \tag{2.2}$$

where $e = (X, Y; Z, W)$ and $x = \phi(X), y = \phi(Y), z = \phi(Z), w = \phi(W)$.

We denote by Φ the set of all Markoff maps. There is a bijective correspondence between the set of all Markoff triples and the set of all Markoff maps which is given as follows. For $\phi \in \Phi$, we get Markoff triple $(\phi(1/0), \phi(0/1), \phi(1/1))$. If we have a Markoff triple (x, y, z), set $\phi(1/0) = x, \phi(0/1) = y, \phi(1/1) = z$. We can extend this map to other complementary regions by using relation (2.2). The conditions for vertices are automatically satisfied.

On Φ, observe that there is an involution A of changing signs of two entries in a Markoff triple.

Fix a Markoff map $\phi \in \Phi$. For each edge $e \in E(\Sigma)$, we assign a direction on e as follows: if $e = (X, Y; Z, W)$ and $|\phi(Z)| > |\phi(W)|$, then we have the arrow from $e \cap Z$ to $e \cap W$. If the norms are equal, the edge can be oriented arbitrarily.

Next, we consider the $\mathrm{PSL}_2(\mathbb{C})$ representations of $\pi_1(T)$.

Definition 2.2. A representation $\rho : \pi_1(T) \to \mathrm{PSL}_2(\mathbb{C})$ is called *type preserving* if $\mathrm{tr}\,\rho([\alpha, \beta]) = -2$.

Note that the sign of the trace of commutator in $\mathrm{PSL}_2(\mathbb{C})$ is well defined. Also note that -2 (and not 2) is the right choice if one wants to include the Fuchsian uniformization of a punctured torus among the representations considered.

Set $x = \mathrm{tr}\,\rho(\alpha)$, $y = \mathrm{tr}\,\rho(\beta)$ and $z = \mathrm{tr}\,\rho(\alpha\beta)$. The triple (x,y,z) is well defined up to changing the signs of any two entries which corresponds to the action A mentioned above. Recall the well known trace identities for $\mathrm{SL}_2(\mathbb{C})$:

$$\mathrm{tr}\,X\,\mathrm{tr}\,Y = \mathrm{tr}\,XY + \mathrm{tr}\,XY^{-1},$$

$$2 + \mathrm{tr}[X,Y] = (\mathrm{tr}\,X)^2 + (\mathrm{tr}\,Y)^2 + (\mathrm{tr}\,XY)^2 - \mathrm{tr}\,X\,\mathrm{tr}\,Y\,\mathrm{tr}\,XY.$$

Since our representation is type preserving, these equation implies that (x,y,z) is a Markoff triple and the map defined by $\phi(X) = \mathrm{tr}\,\rho(g)$, where $g \in \pi_1(T)$ represents the slope which corresponds to $X \in \Omega$, is a Markoff map.

Conversely, given any Markoff triple (x,y,z), we can reconstruct the type preserving representation up to conjugacy. This representation can be realized by using Jørgensen's normalization and denoted by $\rho_{x,y,z}$.

$$\rho_{x,y,z}(\alpha) = \frac{1}{x}\begin{pmatrix} xy - z & y/x \\ xy & z \end{pmatrix}, \quad \rho_{x,y,z}(\beta) = \frac{1}{x}\begin{pmatrix} xz - y & -z/x \\ -xz & y \end{pmatrix}. \tag{2.3}$$

A type preserving representation ρ is quasifuchsian if it is discrete, faithful and geometrically finite without accidental parabolics. The set of quasifuchsian representations is open and dense in the discreteness locus. This is also true in a Bers slice. (See [ST99].) We denote by Φ_{QF} the set of Markoff maps corresponding to quasifuchsian representations.

3. Drawing the Bers slice

In this section, we give a brief exposition of a method for producing pictures of Bers embeddings which consists of two steps. In subsection 3.1, we describe the first step which is about calculating holonomy. In subsection 3.2, we describe the second step. This is about discreteness of the holonomy and includes a difficulty which will be addressed in subsection 3.3.

3.1. Holonomy of projective structures

Let Γ be a Fuchsian group acting on the unit disk $\mathbb{D}$ uniformizing a once punctured torus T. A projective structure ϕ on T is, by definition, a geometric structure modeled on $(\widehat{\mathbb{C}}, \mathrm{PSL}_2(\mathbb{C}))$. Such a projective structure ϕ can be expressed by the corresponding developing map $f_\phi : \widetilde{T} \simeq \mathbb{D} \to \widehat{\mathbb{C}}$ and the holonomy representation $\rho_\phi : \pi_1(T) \simeq \Gamma \to$ $\mathrm{PSL}_2(\mathbb{C})$ defined by $f_\phi \circ \gamma = \rho_\phi(\gamma) \circ f$ ($\gamma \in \Gamma$) up to Möbius conjugacy.

Let $\mathcal{P}$ be the space of (marked) projective structures on T and $\mathcal{T}$ the Teichmüller space of T. Then we have a natural projection $\pi : \mathcal{P} \to \mathcal{T}$. For an element $X \in \mathcal{T}$, define $K(X) := \{\phi \subset \pi^{-1}(X) \,|\, \rho_\phi(\pi_1(T))$ is discrete in $\mathrm{PSL}_2(\mathbb{C})\}$. Recall that a projective structure ϕ with quasifuchsian holonomy is called standard if the developing map f_ϕ is injective and exotic otherwise. There exists an unique component K_0 of $\mathrm{int}\,K(X)$ which consists of the standard structures. This is the image of Teichmüller space $\mathcal{T}$ of once punctured torus T in the Bers embedding.

Let $X := \{\rho \in \mathrm{Hom}(\pi_1(T), \mathrm{PSL}_2(\mathbb{C})) | \text{irreducible and type preserving}\} / \sim$ where $\sim$ is the equivalence relation of $\mathrm{PSL}_2(\mathbb{C})$-conjugacy. The first step in imaging the Bers slice is to calculate the map from $\pi^{-1}(X)$ to X. By taking the Schwarzian derivative of the developing map, $\pi^{-1}(X)$ is identified with $B_2(\mathbb{D}, \Gamma)$ the complex Banach space of holomorphic quadratic differentials for Γ on $\mathbb{D}$ with finite norm, where Γ is a Fuchsian group such that X is uniformized by Γ. The space $B_2(\mathbb{D}, \Gamma)$ is identified with $B_2(S_\lambda)$, an affine subspace of the space of meromorphic differentials on S_λ where S_λ is the 4-times punctured sphere $\widehat{\mathbb{C}} - \{0, 1, \infty, \lambda\}$ which is commensurable to X. The pole structure of the differential is prescribed by the commensurate coverings, and the accessory parameter is the point on this affine space that corresponds to the zero differential on the torus. The relation between the above two is explained in [KSWY]. Then we have

$$B_2(S_\lambda) = \left\{ t \cdot \frac{1}{z(z-1)(z-\lambda)} \,\middle|\, t \in \mathbb{C} \right\}. \tag{3.1}$$

Our idea is to compute the developing map on S_λ instead of $\mathbb{D}$. We take a branch P of p^{-1} around $p(0)$, where $p : \mathbb{D} \to S_\lambda$ is the covering projection, so that $P(p(0)) = 0$. Put $g_\phi(z) := f_\phi(P(z))$. (Recall that f_ϕ is the developing map of $\phi \in \pi^{-1}$.) Then we have $\{g_\phi, z\} = t \cdot \frac{1}{z(z-1)(z-\lambda)} + \{P, z\}$ for some $t \in \mathbb{C}$, where $\{,\}$ denotes the Schwarzian derivative. We have $\{P, z\} = \frac{1}{2z^2} + \frac{(1-\lambda)^2}{2(z-1)^2(z-\lambda)^2} + \frac{c(\lambda)}{z(z-1)(z-\lambda)}$, where $c(\lambda)$ is the accessory parameter. (See Lemma 2.1 and the related arguments in [KSWY].) For each element $t \cdot \frac{1}{z(z-1)(z-\lambda)} \in B_2(S_\lambda)$, we solve the following differential equation, which corresponds to the Schwarzian derivative of g_ϕ, numerically.

$$2y'' + \left(\frac{1}{2z^2} + \frac{(1-\lambda)^2}{2(z-1)^2(z-\lambda)^2} + \frac{t'}{z(z-1)(z-\lambda)} \right) y = 0, \tag{3.2}$$

In the above equation, we set $t' = t + c(\lambda)$. Using a pair of fundamental solutions of (3.2) along a certain loops, we get the generators for the image of the holonomy representation $\rho_\phi \in X$.

3.2. Jørgensen's theory of once punctured tori

The second step of the algorithm is to decide the discreteness of the holonomy representation. Some sufficient conditions for a given holonomy representation to be discrete and conditions to be indiscrete are known. Well known Jørgensen's inequality is a sufficient condition for indiscreteness [Jør76]. In our algorithm we have used a more elementary version due to Shimizu and Leutbecher [Shi63]. On the other hand, if we can construct a fundamental region for $\rho_\phi(\pi_1(T))$ and can apply Poincaré's polygon theorem (See for instance [Rat94].), the image is discrete and an element of $K(X)$. Difficulties arise because, a priori, we have to consider infinitely many group elements in Γ to apply Shimizu-Leutbecher's Lemma or to construct a fundamental polyhedron. Here, Jørgensen's theory of punctured torus group [Jør03] plays a crucial rule. Let us review this theory very briefly.

We use the notations introduced in section 2. For each vertex v in Σ we can associate a subset S_v of $\pi_1(T)$ by

$$S_v = S_{q_1} \cup S_{q_2} \cup S_{q_3},$$

where slopes $q_1, q_2, q_3 \in \mathbb{Q} \cup \{\infty\}$ are the ideal vertices of the triangle in Farey tessellation which is dual to v. (The notion S_q, where q is a slope, was introduced in section 2.) Set $I_v = \{$isometric hemisphere of $g \mid g \in S_v\}$. I_v is an infinite set of hemispheres bounded by equally horizontally spaced circles in the complex plane.

The principal result of Jørgensen's theory is that if the image of the holonomy representation $\rho_{x,y,z}$ is discrete, then there is a path P in Σ which depends on (x,y,z) such that the boundary of the Ford region is given by $\bigcup_{v \in P} I_v$. After Jørgensen's normalization, which was introduced in the previous subsection, we can define a direction of "upward" / "downward" in P. We will say that some vertex $v' \in P$ is the upper/lower neighbor of $v \in P$ if v' is adjacent to v and the direction from v to v' is upward/downward. We will also use terms like "upper end point" / "lower end point" of P for end points of P. See Figure 1. The left figure is the Farey tessellation and the dual graph Σ. This figure depicts the case where the value of (x,y,z) is approximately $(2.536 - 1.115i, 2.616 - 0.645i, 2.203 + 0.660i)$. (Note that, under small deformation of the trace, the combinatorial structure of the Ford domain is stable.) Jørgensen's path is illustrated by the thick path in Σ. The vertex which is dual to $\triangle(1/1, 0/1, 1/2)$ is the upper end point and the vertex which is dual to $\triangle(1/0, 0/1, -1/1)$ is the lower end point. Isometric hemispheres are depicted in the right figure.

Jørgensen's theory allows one to identify the Ford domain given a vertex in the Farey tree that belongs to corresponding Jørgensen's path. A key point here is that for a quasifuchsian group, Jørgensen's path is finite, and that there is a way to identify this path starting from any of its vertices with finitely many operations. This presents

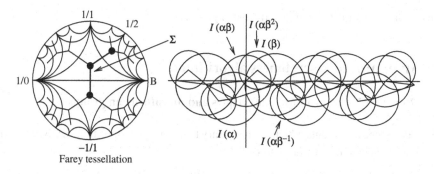

Farey tessellation

Figure 1: A path in the Farey tree (at left, with the dual tessellation) corresponds to a set of isometric hemispheres bounding the Ford domain (right) in Jørgensen's theory

a problem if one does not know any of these vertices.

3.3. A problem in the original algorithm

As explained in the previous subsection, Jørgensen's theory tells us how to choose group elements to construct the fundamental region if $\rho_\phi(\pi_1(T))$ is discrete and if isometric hemispheres of α, β and $\alpha\beta$ are part of the boundary of the Ford region.

Of course, we can not expect the latter condition in general and we used an ad hoc method to deal with these situations. But the algorithm sometimes fails to determine whether or not the groups is discrete. See Figure 2.

Figure 2: The original algorithm sometimes fails to recognize holonomy groups that are discrete, as seen in the gray points surrounded by two ovals.

In Figure 2, white region is where the output was discrete and black region, indiscrete. But we can see "gray pixels" surrounded by two ovals. These points seem to correspond to discrete holonomy because of the stability of the combinatorial structure of the Ford domain under small deformation of the trace. But our algorithm could not determine that these are discrete.

In this note, we suggest a method to deal with this situation based on Bowditch's theory of Markoff triple and quasifuchsian groups.

4. An improved discreteness algorithm

4.1. Bowditch's theory of Markoff triples and quasifuchsian groups

We recall some arguments which are necessary to modify our algorithm. Both notation and arguments are taken from [Bow98].

Given Markoff map ϕ and some $k \geq 0$, define $\Omega(k) = \Omega_\phi(k) \subset \Omega$ by

$$\Omega_\phi(k) = \left\{ X \in \Omega \mid |\phi(X)| \leq k \right\}.$$

Set

$$\Phi_Q := \left\{ \phi \in \Phi \mid \phi^{-1}([-2,2]) = \emptyset, \ \Omega_\phi(2) \text{is finite set} \right\}.$$

From Lemma 4.4 (or see the paragraph before Lemma 4.6.) of [Bow98], we have that $\Phi_{QF} \subset \Phi_Q$. Bowditch proposed an interesting conjecture.

Conjecture 4.1 ([Bow98] Conjecture A). A Markoff map lies in Φ_Q if and only if it is a quasifuchsian representation.

Bowditch described a finite criterion for recognizing that a given Markoff map lies in Φ_Q. To state his criterion, let us introduce some definitions.

Fix a Markoff map ϕ.

Suppose $X \in \Omega$. Then ∂X is a bi-infinite path consisting of edges of the form $X \cap Y_n$, where $(Y_n)_{n \in \mathbb{Z}}$ is a bi-infinite sequence of complementary regions. Set $x = \phi(X), y_n = \phi(Y_n) \ (n \in \mathbb{Z}), y = y_0$. Suppose that $x \in \mathbb{C} \backslash \{-2, 2\}$ and we have $x = \lambda + \lambda^{-1}$ with $|\lambda| > 1$. Then there are constants $A, B \in \mathbb{C}$ with $AB = x^2/(x^2 - 4)$ such that $y_n = A\lambda^n + B\lambda^{-n}$. (See [Bow98] paragraph before Lemma 3.3.) Note that $|y_n|$ grows exponentially as $n \to \infty$ and as $n \to -\infty$. (Lemma 3.3 (1) [Bow98])

Then we can find a continuous function $H : \mathbb{C} \backslash \{-2, 2\} \to (0, \infty)$ such that there are numbers $n_0 \leq n_1 \in \mathbb{Z}$ so that $|y_n| \leq H(x)$ if and only if $n_0 \leq n \leq n_1$ and so that $|y_n|$ is monotonically decreasing on $(-\infty, n_0 - 1]$ and monotonically increasing on $[n_1 + 1, \infty)$. We also assume that $H(x) \geq 2$ for all x. (See [Bow98] paragraph before Lemma 3.14.) Bowditch did not provide a formula for $H(x)$ in his paper. In our implementation of the algorithm, we defined $H(x)$ as follows:

$$H(x) := \sqrt{|x^2/(x^2 - 4.0)| \, (|\lambda| + 1.0) \, |\lambda|/(|\lambda| - 1.0)},$$

where λ is defined from x as mentioned above.

Define a subtree $T(t)$ $(t \geq 0)$ of Σ as follows. Let e be an edge of Σ with $e = X \cap Y \in E(\Sigma)$ where X and Y are complementary regions. Then e is an edge of $T(t)$ if and only if ether $|x| \leq 2 + t$ and $|y| \leq H(x) + t$ or $|y| \leq 2 + t$ and $|x| \leq H(y) + t$.

Lemma 4.2 ([Bow98] Lemma 3.15). *For any fixed $t \geq 0$, we have $\phi \in \Phi_Q$ if and only if $T(t)$ is finite.*

$T(t)$ has the following nice properties.

Lemma 4.3.

(i) *If $t \geq 1$, $T(t) \neq \emptyset$. ([Bow98] Theorem 1(1))*

(ii) *$T(t)$ is connected. ([Bow98] Lemma 3.11)*

(iii) *If $T(t) \neq \emptyset$, the arrow on any edge not in $T(t)$ points toward $T(t)$. ([Bow98] Lemma 3.11)*

Here, we give a sketch of the proof that $T(t)$ can be identified in finitely many steps. Set $t = 1$. Using Lemma 4.3, we can check that $T(1)$ is finite ($\Leftrightarrow \phi \in \Phi_Q$ by Lemma 4.2) if it is by the following algorithm. (If $T(1)$ is infinite, the algorithm will not stop.)

Let v_0 be the vertex of Σ which is dual to $\triangle(1/0, 0/0, 1/1)$. Starting from v_0, if we follow the arrows on edges, we can reach $T(1)$ because it is not empty and the arrow on any edge not in $T(1)$ points toward $T(1)$. Then we can check that $T(1)$ is finite if it is since $T(1)$ is connected.

4.2. Modified algorithm

In this section, we discuss an algorithm that starts with a Markoff map and produces one of three outputs: "discrete", "not discrete", or "undecided". One possible algorithm for discreteness is, believing that Conjecture 4.1 true, just follow the algorithm described in the previous subsection—one would find the Bowditch tree $T(t)$ and call a representation discrete if it is finite.

But our approach here is a little bit more careful which is defined as follows.

(i) Starting from v_0, we locate $T(1)$. This is done by (substep a) following the arrows on edges in Σ to reach one vertex in $T(1)$ and (substep b) using depth first search to find all the vertices in $T(1)$. The above (substep b) may not finish in a finite time because $T(1)$ can be infinite. If the number of vertices in $T(1)$ found by this search becomes bigger than a fixed constant (say 10000), we stop our calculation and go to step (iv). We also have to consider about indiscreteness. Recall that for each vertex in Σ there are three complementary regions (slopes) adjacent to v. When we come to a new vertex in (substep a) and (substep b), we test Shimizu-Leutbecher's Lemma for each complementary region associated to the vertex and output "indiscrete" if necessary.

(ii) Suppose that we could find that $T(1)$ is finite. Consider the complementary regions which are adjacent to $T(1)$. Choose one complementary region X (i.e., slope) whose value of modulus of Markoff map $|\phi(X)|$ is the smallest. This slope X corresponds to the isometric hemispheres with the largest radius, so X must be a part of the walls of the Ford region if it is discrete. Observe that there is a finite sequence of complementary regions $\{Y_n\}$ such that each Y_i is adjacent to X and the edge adjacent to both X and Y_i is in $T(1)$. From this set, choose one complementary region Y_i whose modulus $|\phi(Y_i)|$ is the smallest. Then two vertices v_1, v_2 in $T(1)$ which are adjacent to both X and Y_i must be in Jørgensen's path if it is discrete.

(iii) Then start Jørgensen's algorithm from v_i ($i = 1$ or 2) to find Jørgensen's path P. Again, when we come to a new vertex in this process, we test Shimizu-Leutbecher's Lemma for each complementary region associated to the vertex and output "indiscrete" if necessary. If this process will finish successfully and we can construct the Ford region, the output is "discrete". But, this process may not finish in a finite time because P can be infinite for geometrically infinite group, even though we can almost ignore this possibility. If the number of vertices in P found by this search becomes bigger than a fixed constant (say 10000), we stop our calculation and go to step (iv). Jørgensen's algorithm does not cover all the possible cases of the configuration of isometric hemispheres, because it only works when the group is discrete. Thus, in some cases, we can not continue Jørgensen's algorithm and go to (iv).

(iv) Try to show that the group is indiscrete. Using depth first search, we test Shimizu-Leutbecher's Lemma for each complementary region of each vertex in Σ. The output is "indiscrete" if we can find a complementary region which satisfies the condition of the lemma. This process may not finish in a finite time. We stop after a fixed number of search and output "unknown".

Suppose that $\phi \in \Phi_{QF}$. Since $\Phi_{QF} \subset \Phi_Q$, $\phi \in \Phi_Q$ and our algorithm must to be able to find that $T(1)$ is finite. If we start from v_1 or v_2 mentioned above, we can find Jørgensen's path P. Therefore, we have:

Proposition 4.4. *If $\phi \in \Phi_{QF}$, the algorithm described above will determine this in finite time that $\phi \in \Phi_{QF}$.*

We remark that in the course of computer experiments with this algorithm, we have not yet found any counterexamples to Conjecture 4.1.

5. Computer experiments on the discreteness locus

Throughout the development of Kleinian group theory, exploratory computer graphics have consistently played an important role. The pioneering work was done by Mumford, McMullen, and Wright [MMW] that later lead to [MSW02].

In this section we present some pictures of our computer experiments of the discreteness locus.

5.1. Three-dimensional slice of quasifuchsian space

Before the workshop in Newton Institute, we produced pictures of many kinds of slices of punctured torus groups. At the suggestion of Wada, we used the software package DeltaViewer [Wad] to produce three-dimensional pictures of a family of slices. In our talk in the workshop, using DeltaViewer, we presented a three-dimensional picture made from one parameter family of trace-constant slices with $\mathrm{tr}(\alpha)$ from 2 to 10 (See Figure 3.), and a three-dimensional picture made from one parameter family of Bers slices (See Figure 4.).

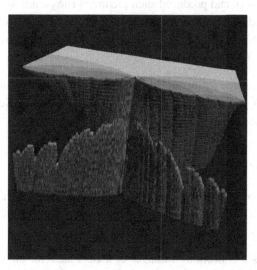

Figure 3: One parameter family of trace constant slices. $(2 < \mathrm{tr}(\alpha) < 10)$ For top face, $\mathrm{tr}(\alpha) = 2$ (Maskit slice) and bottom $\mathrm{tr}(\alpha) = 10$. This is a screen shot produced by DeltaViewer.

Series also suggested that we produce a three-dimensional image from a family of slices each of which has a trace fixed and changing the value in the imaginary

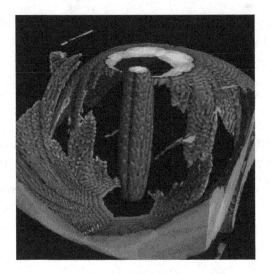

Figure 4: One parameter family of Bers slices: This is a family of rectangular punctured tori with the square torus Bers slice corresponding to a horizontal plane through the center of the picture. The standard component K_0 (See 3.1.) is located at the center and two other (exotic) components of discreteness loci are nearly touching the standard component at the top face and the bottom face. This is a screen shot produced by DeltaViewer.

direction. We thought that an interesting example may be from $\mathrm{tr}(\alpha) = 2$ (Maskit slice) to $\mathrm{tr}(\alpha) = 2 + 6i$ and produced such picture. (The picture for the case $\mathrm{tr}(\alpha) = 2 + 6i$ was presented by McMullen in [McM98] to illustrate the topological complexity of the closure of the space of quasifuchsian groups in $\mathcal{X}$. See also Figure 10.20 of [MSW02].)

DeltaViewer is a free software which runs only on Apple Macintosh computers. The data for the figures in this section and the one requested by Series are available from the author upon request.

5.2. Overhang in the boundary of Bers slice

For the boundary of the Maskit slice, the phenomenon of "overhang" is observed. (See Figure 10.5 "Overhang in the Maskit boundary" of [MSW02].) Motivated by this figure, we have carried out a computer experiment.

Figure 5 shows a collection of images of a Bers slice for the square torus. The center of the picture is $0.6166004915862 - 0.0011501283077i$ and width and height is 0.2. (Recall that the coordinate of the plane is given by (3.1).) Then we zoom in the picture with factor $= 10$ having the same center to get the next picture. We continue the process of this zooming in with center and factor fixed. For the last picture, the width and height of the region is 0.000000000002. The order of the picture is from

Figure 5: Overhang in the Bers slice boundary for square torus. The center for all the picture is $0.6166004915862 - 0.0011501283077i$. For upper left, the width and the height is 0.2. Then we zoom in with factor 10. The order of the picture is from left to right and from up to bottom.

left to right and from top to bottom. Therefore, for Bers slice, we can also see the overhang in the boundary.

We remark that the idea for this particular choice of center mentioned above is very simple. We can see the triangulation (except the center square) of the Bers slice for square torus in our picture. (The color is determined according to the length of Jørgensen's path *P*.) First, suppose that we are sitting at the center of the Bers slice and facing upward (north). We move to the adjacent "right" triangle. Again we move to the adjacent "right" triangle. We continue this process 16 times more. Then we move to "left" adjacent triangle. Then we move to "right" adjacent triangle 18 times and next move to "left". We repeated this "right turn 18 times and left turn one times" period several times and get the value of the above center. The number 18 was chosen because the period of the color of our picture happened to be 18.

In [KSWY], we have observed numerically the asymptotic self-similarity of the Bers slice for square torus, too. This phenomena was first discovered by McMullen for Maskit slice [McM96].

5.3. Final remark

The aim of producing these pictures is to find the presence of new phenomena.

Let us present an example. During the poster session of the workshop, Ito presented his theorem (see [Ito] for necessary notations and details.)

Theorem 5.1 ([Ito] Theorem A). *For any* $\lambda \in \mathcal{ML}_{\mathbb{Z}}(S) - \{0\}$, *there exists a point* $Y \in \overline{Q_0} \cap \overline{Q_\lambda}$ *such that* $U \cap \overline{Q_\lambda}$ *consists of more than 2 components for any sufficiently small neighborhood U of Y.*

See also the related work on bumping and self-bumping of deformation spaces by Anderson-Canary [AC96], McMullen [McM98] and Bromberg-Holt [BH01].

In Figure 6, the Bers embedding is visible and seems to be touched by thin "arms". The "arms" belong to other components because other standard projective structure can not appear in this plane π^{-1}(square torus). (See subsection 3.1.) These "arms" are studied in Ito's paper,

He used our picture, which is like the one in Figure 6, in his paper and said, "This graphic seems to guarantee the claim of the theorem".

We hope that computer graphics can stimulate research in this field.

Figure 6: Bers slice for $\widehat{\mathbb{C}} - \{0, 1, \infty, 0.5 + 10^8 i\}$. The center is $0.5 + 4.0 \times 10^7 i$ and width and height is 2.0×10^7.

References

[AC96]   J. W. Anderson & R. D. Canary (1996). Algebraic limits of Kleinian groups which rearrange the pages of a book. *Invent. Math.* **126** (2), 205–214.

[BH01]   K. Bromberg & J. Holt (2001). Self-bumping of deformation spaces of hyperbolic 3-manifolds. *J. Differential Geom.* **57** (1), 47–65.

[Bow98] B. H. Bowditch (1998). Markoff triples and quasi-Fuchsian groups. *Proc. London Math. Soc. (3)* **77** (3), 697–736.

[Ito]   K. Ito. Exotic projective structures and quasi-fuchsian spaces II. Preprint. www.math.nagoya-u.ac.jp/~itoken/index.html.

[Jør76] T. Jørgensen (1976). On discrete groups of Möbius transformations. *Amer. J. Math.* **98** (3), 739–749.

[Jør03] T. Jørgensen (2003). On pairs of once-punctured tori. In *Kleinian Groups and Hyperbolic 3-Manifolds (Warwick, 2001), London Math. Soc. Lecture Note Ser.*, volume 299, pp. 183–207. Cambridge Univ. Press, Cambridge.

[KSWY]   Y. Komori *et al.* Drawing Bers embeddings of the Teichmüller space of once-punctured tori. Preprint.

[McM96] C. T. McMullen (1996). *Renormalization and 3-manifolds which Fiber over the Circle, Annals of Mathematics Studies*, volume 142. Princeton University Press, Princeton, NJ.

[McM98]   C. T. McMullen (1998). Complex earthquakes and Teichmüller theory. *J. Amer. Math. Soc.* **11** (2), 283–320.

[MMW]   D. Mumford, C. T. McMullen & D. J. Wright. Limit sets of free two-generator kleinian groups. Preprint, Version 0.95.

[MSW02]   D. Mumford, C. Series & D. Wright (2002). *Indra's Pearls*. Cambridge University Press, New York.

[Rat94]   J. G. Ratcliffe (1994). *Foundations of Hyperbolic Manifolds, Graduate Texts in Mathematics*, volume 149. Springer-Verlag, New York.

[Shi63]   H. Shimizu (1963). On discontinuous groups operating on the product of the upper half planes. *Ann. of Math. (2)* **77**, 33–71.

[ST99]   H. Shiga & H. Tanigawa (1999). Projective structures with discrete holonomy representations. *Trans. Amer. Math. Soc.* **351** (2), 813–823.

[Wad]   M. Wada. Delta Viewer software. `vivaldi.ics.nara-wu.ac.jp/~wada/DeltaViewer/`.

Yasushi Yamashita

Department of Information and Computer Sciences
Nara Women's University
Kita-uoya-nishi-machi, Nara 630-8506
Japan

`yamasita@ics.nara-wu.ac.jp`

AMS Classification: 57M50, 30F40, 65D18

Keywords: Kleinian groups, projective structures, Bers slice

Printed in the United States
by Baker & Taylor Publisher Services